A HISTORY OF GREEK MATHEMATICS

SIR THOMAS HEATH

VOLUME I

FROM THALES TO EUCLID

An independent world,
Created out of pure intelligence.
—Wordsworth

Dover Publications, Inc.
New York

Published in Canada by General Publishing Company, Ltd., 30 Lesmill Road, Don Mills, Toronto, Ontario.
Published in the United Kingdom by Constable and Company, Ltd., 10 Orange Street, London WC2H 7EG.

This Dover edition, first published in 1981, is an unabridged republication of the work first published in 1921 by the Clarendon Press, Oxford. For this edition the errata of the first edition have been corrected.

International Standard Book Number: 0-486-24073-8
Library of Congress Catalog Card Number: 80-70126

Manufactured in the United States of America
Dover Publications, Inc.
180 Varick Street
New York, N.Y. 10014

PREFACE

THE idea may seem quixotic, but it is nevertheless the author's confident hope that this book will give a fresh interest to the story of Greek mathematics in the eyes both of mathematicians and of classical scholars.

For the mathematician the important consideration is that the foundations of mathematics and a great portion of its content are Greek. The Greeks laid down the first principles, invented the methods *ab initio*, and fixed the terminology. Mathematics in short is a Greek science, whatever new developments modern analysis has brought or may bring.

The interest of the subject for the classical scholar is no doubt of a different kind. Greek mathematics reveals an important aspect of the Greek genius of which the student of Greek culture is apt to lose sight. Most people, when they think of the Greek genius, naturally call to mind its masterpieces in literature and art with their notes of beauty, truth, freedom and humanism. But the Greek, with his insatiable desire to know the true meaning of everything in the universe and to be able to give a rational explanation of it, was just as irresistibly driven to natural science, mathematics, and exact reasoning in general or logic. This austere side of the Greek genius found perhaps its most complete expression in Aristotle. Aristotle would, however, by no means admit that mathematics was divorced from aesthetic; he could conceive, he said, of nothing more beautiful than the objects of mathematics. Plato delighted in geometry and in the wonders of numbers; *ἀγεωμέτρητος μηδεὶς εἰσίτω*, said the inscription over the door of the Academy. Euclid was a no less typical Greek. Indeed, seeing that so much of Greek is mathematics, it is arguable that, if one would understand the Greek genius fully, it would be a good plan to begin with their geometry.

The story of Greek mathematics has been written before. Dr. James Gow did a great service by the publication in 1884 of his *Short History of Greek Mathematics*, a scholarly and useful work which has held its own and has been quoted with respect and appreciation by authorities on the history of mathematics in all parts of the world. At the date when he wrote, however, Dr. Gow had necessarily to rely upon the works of the pioneers Bretschneider, Hankel, Allman, and Moritz Cantor (first edition). Since then the subject has been very greatly advanced; new texts have been published, important new documents have been discovered, and researches by scholars and mathematicians in different countries have thrown light on many obscure points. It is, therefore, high time for the complete story to be rewritten.

It is true that in recent years a number of attractive histories of mathematics have been published in England and America, but these have only dealt with Greek mathematics as part of the larger subject, and in consequence the writers have been precluded, by considerations of space alone, from presenting the work of the Greeks in sufficient detail.

The same remark applies to the German histories of mathematics, even to the great work of Moritz Cantor, who treats of the history of Greek mathematics in about 400 pages of vol. i. While no one would wish to disparage so great a monument of indefatigable research, it was inevitable that a book on such a scale would in time prove to be inadequate, and to need correction in details; and the later editions have unfortunately failed to take sufficient account of the new materials which have become available since the first edition saw the light.

The best history of Greek mathematics which exists at present is undoubtedly that of Gino Loria under the title *Le scienze esatte nell' antica Grecia* (second edition 1914,

Ulrico Hoepli, Milano). Professor Loria arranges his material in five Books, (1) on pre-Euclidean geometry, (2) on the Golden Age of Greek geometry (Euclid to Apollonius), (3) on applied mathematics, including astronomy, sphaeric, optics, &c., (4) on the Silver Age of Greek geometry, (5) on the arithmetic of the Greeks. Within the separate Books the arrangement is chronological, under the names of persons or schools. I mention these details because they raise the question whether, in a history of this kind, it is best to follow chronological order or to arrange the material according to subjects, and, if the latter, in what sense of the word 'subject' and within what limits. As Professor Loria says, his arrangement is 'a compromise between arrangement according to subjects and a strict adherence to chronological order, each of which plans has advantages and disadvantages of its own'.

In this book I have adopted a new arrangement, mainly according to subjects, the nature of which and the reasons for which will be made clear by an illustration. Take the case of a famous problem which plays a great part in the history of Greek geometry, the doubling of the cube, or its equivalent, the finding of two mean proportionals in continued proportion between two given straight lines. Under a chronological arrangement this problem comes up afresh on the occasion of each new solution. Now it is obvious that, if all the recorded solutions are collected together, it is much easier to see the relations, amounting in some cases to substantial identity, between them, and to get a comprehensive view of the history of the problem. I have therefore dealt with this problem in a separate section of the chapter devoted to 'Special Problems', and I have followed the same course with the other famous problems of squaring the circle and trisecting any angle.

Similar considerations arise with regard to certain well-defined subjects such as conic sections. It would be inconvenient to interrupt the account of Menaechmus's solution of the problem of the two mean proportionals in order to consider the way in which he may have discovered the conic sections and their fundamental properties. It seems to me much better to give the complete story of the origin and development of the geometry of the conic sections in one place, and this has been done in the chapter on conic sections associated with the name of Apollonius of Perga. Similarly a chapter has been devoted to algebra (in connexion with Diophantus) and another to trigonometry (under Hipparchus, Menelaus and Ptolemy).

At the same time the outstanding personalities of Euclid and Archimedes demand chapters to themselves. Euclid, the author of the incomparable *Elements*, wrote on almost all the other branches of mathematics known in his day. Archimedes's work, all original and set forth in treatises which are models of scientific exposition, perfect in form and style, was even wider in its range of subjects. The imperishable and unique monuments of the genius of these two men must be detached from their surroundings and seen as a whole if we would appreciate to the full the pre-eminent place which they occupy, and will hold for all time, in the history of science.

The arrangement which I have adopted necessitates (as does any other order of exposition) a certain amount of repetition and cross-references; but only in this way can the necessary unity be given to the whole narrative.

One other point should be mentioned. It is a defect in the existing histories that, while they state generally the contents of, and the main propositions proved in, the great treatises of Archimedes and Apollonius, they make little attempt to describe the procedure by which the results are obtained. I have therefore taken pains, in the most significant cases, to show the course of the argument in sufficient detail to enable a competent mathematician to grasp the method used and to apply it, if he will, to other similar investigations.

The work was begun in 1913, but the bulk of it was written, as a distraction, during the first three years of the war, the hideous course of which seemed day by day to enforce the profound truth conveyed in the answer of Plato to the Delians. When they consulted him on the problem set them by the Oracle, namely that of duplicating the cube, he replied, 'It must be supposed, not that the god specially wished this problem solved, but that he would have the Greeks desist from war and wickedness and cultivate the Muses, so that, their passions being assuaged by philosophy and mathematics, they might live in innocent and mutually helpful intercourse with one another'.

Truly

> Greece and her foundations are
> Built below the tide of war,
> Based on the crystàlline sea
> Of thought and its eternity.

T. L. H.

CONTENTS OF VOL. I

I

INTRODUCTORY

The Greeks and mathematics.

It is an encouraging sign of the times that more and more effort is being directed to promoting a due appreciation and a clear understanding of the gifts of the Greeks to mankind. What we owe to Greece, what the Greeks have done for civilization, aspects of the Greek genius : such are the themes of many careful studies which have made a wide appeal and will surely produce their effect. In truth all nations, in the West at all events, have been to school to the Greeks, in art, literature, philosophy, and science, the things which are essential to the rational use and enjoyment of human powers and activities, the things which make life worth living to a rational human being. 'Of all peoples the Greeks have dreamed the dream of life the best.' And the Greeks were not merely the pioneers in the branches of knowledge which they invented and to which they gave names. What they began they carried to a height of perfection which has not since been surpassed; if there are exceptions, it is only where a few crowded centuries were not enough to provide the accumulation of experience required, whether for the purpose of correcting hypotheses which at first could only be of the nature of guesswork, or of suggesting new methods and machinery.

Of all the manifestations of the Greek genius none is more impressive and even awe-inspiring than that which is revealed by the history of Greek mathematics. Not only are the range and the sum of what the Greek mathematicians actually accomplished wonderful in themselves; it is necessary to bear in mind that this mass of original work was done in an almost incredibly short space of time, and in spite of the comparative inadequacy (as it would seem to us) of the only methods at their disposal, namely those of pure geometry, supplemented, where necessary, by the ordinary arithmetical operations.

Let us, confining ourselves to the main subject of pure geometry by way of example, anticipate so far as to mark certain definite stages in its development, with the intervals separating them. In Thales's time (about 600 B.C.) we find the first glimmerings of a theory of geometry, in the theorems that a circle is bisected by any diameter, that an isosceles triangle has the angles opposite to the equal sides equal, and (if Thales really discovered this) that the angle in a semicircle is a right angle. Rather more than half a century later Pythagoras was taking the first steps towards the theory of numbers and continuing the work of making geometry a theoretical science; he it was who first made geometry one of the subjects of a liberal education. The Pythagoreans, before the next century was out (i.e. before, say, 450 B.C.), had practically completed the subject-matter of Books I–II, IV, VI (and perhaps III) of Euclid's *Elements*, including all the essentials of the 'geometrical algebra' which remained fundamental in Greek geometry; the only drawback was that their theory of proportion was not applicable to incommensurable but only to commensurable magnitudes, so that it proved inadequate as soon as the incommensurable came to be discovered. In the same fifth century the difficult problems of doubling the cube and trisecting any angle, which are beyond the geometry of the straight line and circle, were not only mooted but solved theoretically, the former problem having been first reduced to that of finding two mean proportionals in continued proportion (Hippocrates of Chios) and then solved by a remarkable construction in three dimensions (Archytas), while the latter was solved by means of the curve of Hippias of Elis known as the *quadratrix*; the problem of squaring the circle was also attempted, and Hippocrates, as a contribution to it, discovered and squared three out of the five lunes which can be squared by means of the straight line and circle. In the fourth century Eudoxus discovered the great theory of proportion expounded in Euclid, Book V, and laid down the principles of the *method of exhaustion* for measuring areas and volumes; the conic sections and their fundamental properties were discovered by Menaechmus; the theory of irrationals (probably discovered, so far as $\sqrt{2}$ is concerned, by the early Pythagoreans) was generalized by Theaetetus; and the geometry of the sphere was worked out in systematic treatises. About the end of the century Euclid wrote his *Elements* in thirteen Books. The next century, the third, is that of Archimedes, who may be said to have anticipated the integral calculus, since, by performing what are practically *integrations*, he found the area of a parabolic segment and of a spiral, the surface and volume of a sphere and a segment of a sphere, the volume of any segment of the solids of revolution of the second degree, the centres of gravity of a semicircle, a parabolic segment, any segment of a paraboloid of revolution, and any segment of a sphere or spheroid. Apollonius of Perga, the 'great geometer', about 200 B.C., completed the theory of geometrical conics, with specialized investigations of normals as maxima and minima leading quite easily to the determination of the circle of curvature at any point of a conic and of the equation of the evolute of the conic, which with us is part of analytical conics. With Apollonius the main body of Greek geometry is complete, and we may therefore fairly say that four centuries sufficed to complete it.

But some one will say, how did all this come about? What special aptitude had the Greeks for mathematics? The answer to this question is that their genius for mathematics was simply one aspect of their genius for philosophy. Their mathematics indeed constituted a large part of their philosophy down to Plato. Both had the same origin.

Conditions favouring the development of philosophy among the Greeks.

All men by nature desire to know, says Aristotle.[1] The Greeks, beyond any other people of antiquity, possessed the love of knowledge for its own sake; with them it amounted to an instinct and a passion.[2] We see this first of all in their love of adventure. It is characteristic that in the *Odyssey* Odysseus is extolled as the hero who had 'seen the cities of many men and learned their mind',[3] often even taking his life in his hand, out of a pure passion for extending his horizon,

[1] Arist. *Metaph.* A. 1, 980 a 21.
[2] Cf. Butcher, *Some Aspects of the Greek Genius*, 1892, p. 1.
[3] *Od.* i. 3.

as when he went to see the Cyclopes in order to ascertain 'what sort of people they were, whether violent and savage, with no sense of justice, or hospitable and godfearing'.[1] Coming nearer to historical times, we find philosophers and statesmen travelling in order to benefit by all the wisdom that other nations with a longer history had gathered during the centuries. Thales travelled in Egypt and spent his time with the priests. Solon, according to Herodotus,[2] travelled 'to see the world' (θεωρίης εἵνεκεν), going to Egypt to the court of Amasis, and visiting Croesus at Sardis. At Sardis it was not till 'after he had seen and examined everything' that he had the famous conversation with Croesus; and Croesus addressed him as the Athenian of whose wisdom and peregrinations he had heard great accounts, proving that he had covered much ground in seeing the world and pursuing philosophy. (Herodotus, also a great traveller, is himself an instance of the capacity of the Greeks for assimilating anything that could be learnt from any other nations whatever; and, although in Herodotus's case the object in view was less the pursuit of philosophy than the collection of interesting information, yet he exhibits in no less degree the Greek passion for seeing things as they are and discerning their meaning and mutual relations; 'he compares his reports, he weighs the evidence, he is conscious of his own office as an inquirer after truth'.) But the same avidity for learning is best of all illustrated by the similar tradition with regard to Pythagoras's travels. Iamblichus, in his account of the life of Pythagoras,[3] says that Thales, admiring his remarkable ability, communicated to him all that he knew, but, pleading his own age and failing strength, advised him for his better instruction to go and study with the Egyptian priests. Pythagoras, visiting Sidon on the way, both because it was his birthplace and because he properly thought that the passage to Egypt would be easier by that route, consorted there with the descendants of Mochus, the natural philosopher and prophet, and with the other Phoenician hierophants, and was initiated into all the rites practised in Biblus, Tyre, and in many parts of Syria, a regimen to which he submitted, not out of religious

[1] *Od.* ix. 174–6. [2] Herodotus, i. 30.
[3] Iamblichus, *De vita Pythagorica*, cc. 2–4.

enthusiasm, '*as you might think*' (ὡς ἄν τις ἁπλῶς ὑπολάβοι), but much more through love and desire for philosophic inquiry, and in order to secure that he should not overlook any fragment of knowledge worth acquiring that might lie hidden in the mysteries or ceremonies of divine worship; then, understanding that what he found in Phoenicia was in some sort an offshoot or descendant of the wisdom of the priests of Egypt, he concluded that he should acquire learning more pure and more sublime by going to the fountain-head in Egypt itself.

'There', continues the story, 'he studied with the priests and prophets and instructed himself on every possible topic, neglecting no item of the instruction favoured by the best judges, no individual man among those who were famous for their knowledge, no rite practised in the country wherever it was, and leaving no place unexplored where he thought he could discover something more. . . . And so he spent 22 years in the shrines throughout Egypt, pursuing astronomy and geometry and, of set purpose and not by fits and starts or casually, entering into all the rites of divine worship, until he was taken captive by Cambyses's force and carried off to Babylon, where again he consorted with the Magi, a willing pupil of willing masters. By them he was fully instructed in their solemn rites and religious worship, and in their midst he attained to the highest eminence in arithmetic, music, and the other branches of learning. After twelve years more thus spent he returned to Samos, being then about 56 years old.'

Whether these stories are true in their details or not is a matter of no consequence. They represent the traditional and universal view of the Greeks themselves regarding the beginnings of their philosophy, and they reflect throughout the Greek spirit and outlook.

From a scientific point of view a very important advantage possessed by the Greeks was their remarkable capacity for accurate observation. This is attested throughout all periods, by the similes in Homer, by vase-paintings, by the ethnographic data in Herodotus, by the 'Hippocratean' medical books, by the biological treatises of Aristotle, and by the history of Greek astronomy in all its stages. To take two commonplace examples. Any person who examines the under-side of a horse's hoof, which we call a 'frog' and the

Greeks called a 'swallow', will agree that the latter is the more accurate description. Or again, what exactness of perception must have been possessed by the architects and workmen to whom we owe the pillars which, seen from below, appear perfectly straight, but, when measured, are found to bulge out (ἔντασις).

A still more essential fact is that the Greeks were a race of *thinkers*. It was not enough for them to know the fact (the ὅτι); they wanted to know the why and wherefore (the διὰ τί), and they never rested until they were able to give a rational explanation, or what appeared to them to be such, of every fact or phenomenon. The history of Greek astronomy furnishes a good example of this, as well as of the fact that no visible phenomenon escaped their observation. We read in Cleomedes[1] that there were stories of extraordinary lunar eclipses having been observed which 'the more ancient of the mathematicians' had vainly tried to explain; the supposed 'paradoxical' case was that in which, while the sun appears to be still above the western horizon, the *eclipsed* moon is seen to rise in the east. The phenomenon was seemingly inconsistent with the recognized explanation of lunar eclipses as caused by the entrance of the moon into the earth's shadow; how could this be if both bodies were above the horizon at the same time? The 'more ancient' mathematicians tried to argue that it was possible that a spectator standing on an *eminence* of the spherical earth might see along the generators of a *cone*, i.e. a little downwards on all sides instead of merely in the plane of the horizon, and so might see both the sun and the moon although the latter was in the earth's shadow. Cleomedes denies this, and prefers to regard the whole story of such cases as a fiction designed merely for the purpose of plaguing astronomers and philosophers; but it is evident that the cases had actually been observed, and that astronomers did not cease to work at the problem until they had found the real explanation, namely that the phenomenon is due to atmospheric refraction, which makes the sun visible to us though it is actually beneath the horizon. Cleomedes himself gives this explanation, observing that such cases of atmospheric refraction were especially

[1] Cleomedes, *De motu circulari*, ii. 6, pp. 218 sq.

noticeable in the neighbourhood of the Black Sea, and comparing the well-known experiment of the ring at the bottom of a jug, where the ring, just out of sight when the jug is empty, is brought into view when water is poured in. We do not know who the 'more ancient' mathematicians were who were first exercised by the 'paradoxical' case; but it seems not impossible that it was the observation of this phenomenon, and the difficulty of explaining it otherwise, which made Anaxagoras and others adhere to the theory that there are other bodies besides the earth which sometimes, by their interposition, cause lunar eclipses. The story is also a good illustration of the fact that, with the Greeks, pure theory went hand in hand with observation. Observation gave data upon which it was possible to found a theory; but the theory had to be modified from time to time to suit observed new facts; they had continually in mind the necessity of 'saving the phenomena' (to use the stereotyped phrase of Greek astronomy). Experiment played the same part in Greek medicine and biology.

Among the different Greek stocks the Ionians who settled on the coast of Asia Minor were the most favourably situated in respect both of natural gifts and of environment for initiating philosophy and theoretical science. When the colonizing spirit first arises in a nation and fresh fields for activity and development are sought, it is naturally the younger, more enterprising and more courageous spirits who volunteer to leave their homes and try their fortune in new countries; similarly, on the intellectual side, the colonists will be at least the equals of those who stay at home, and, being the least wedded to traditional and antiquated ideas, they will be the most capable of striking out new lines. So it was with the Greeks who founded settlements in Asia Minor. The geographical position of these settlements, connected with the mother country by intervening islands, forming stepping-stones as it were from the one to the other, kept them in continual touch with the mother country; and at the same time their geographical horizon was enormously extended by the development of commerce over the whole of the Mediterranean. The most adventurous seafarers among the Greeks of Asia Minor, the Phocaeans, plied their trade successfully as far as the Pillars of Hercules, after they had explored the Adriatic sea, the west coast of Italy, and the coasts of the Ligurians and Iberians. They are said to have founded Massalia, the most important Greek colony in the western countries, as early as 600 B.C. Cyrene, on the Libyan coast, was founded in the last third of the seventh century. The Milesians had, soon after 800 B.C., made settlements on the east coast of the Black Sea (Sinope was founded in 785); the first Greek settlements in Sicily were made from Euboea and Corinth soon after the middle of the eighth century (Syracuse 734). The ancient acquaintance of the Greeks with the south coast of Asia Minor and with Cyprus, and the establishment of close relations with Egypt, in which the Milesians had a large share, belongs to the time of the reign of Psammetichus I (664–610 B.C.), and many Greeks had settled in that country.

The free communications thus existing with the whole of the known world enabled complete information to be collected with regard to the different conditions, customs and beliefs prevailing in the various countries and races; and, in particular, the Ionian Greeks had the inestimable advantage of being in contact, directly and indirectly, with two ancient civilizations, the Babylonian and the Egyptian.

Dealing, at the beginning of the *Metaphysics*, with the evolution of science, Aristotle observes that science was preceded by the arts. The arts were invented as the result of general notions gathered from experience (which again was derived from the exercise of memory); those arts naturally came first which are directed to supplying the necessities of life, and next came those which look to its amenities. It was only when all such arts had been established that the sciences, which do not aim at supplying the necessities or amenities of life, were in turn discovered, and this happened first in the places where men began to have leisure. This is why the mathematical arts were founded in Egypt; for there the priestly caste was allowed to be at leisure. Aristotle does not here mention Babylon; but, such as it was, Babylonian science also was the monopoly of the priesthood.

It is in fact true, as Gomperz says,[1] that the first steps on the road of scientific inquiry were, so far as we know from

[1] *Griechische Denker*, i, pp. 36, 37.

history, never accomplished except where the existence of an organized caste of priests and scholars secured the necessary industry, with the equally indispensable continuity of tradition. But in those very places the first steps were generally the last also, because the scientific doctrines so attained tend, through their identification with religious prescriptions, to become only too easily, like the latter, mere lifeless dogmas. It was a fortunate chance for the unhindered spiritual development of the Greek people that, while their predecessors in civilization had an organized priesthood, the Greeks never had. To begin with, they could exercise with perfect freedom their power of unerring eclecticism in the assimilation of every kind of lore. 'It remains their everlasting glory that they discovered and made use of the serious scientific elements in the confused and complex mass of exact observations and superstitious ideas which constitutes the priestly wisdom of the East, and threw all the fantastic rubbish on one side.'[1] For the same reason, while using the earlier work of Egyptians and Babylonians as a basis, the Greek genius could take an independent upward course free from every kind of restraint and venture on a flight which was destined to carry it to the highest achievements.

The Greeks then, with their 'unclouded clearness of mind' and their freedom of thought, untrammelled by any 'Bible' or its equivalent, were alone capable of creating the sciences as they did create them, i.e. as living things based on sound first principles and capable of indefinite development. It was a great boast, but a true one, which the author of the *Epinomis* made when he said, 'Let us take it as an axiom that, whatever the Greeks take from the barbarians, they bring it to fuller perfection'.[2] He has been speaking of the extent to which the Greeks had been able to explain the relative motions and speeds of the sun, moon and planets, while admitting that there was still much progress to be made before absolute certainty could be achieved. He adds a characteristic sentence, which is very relevant to the above remarks about the Greek's free outlook:

'Let no Greek ever be afraid that we ought not at any time to study things divine because we are mortal. We ought to

[1] Cumont, *Neue Jahrbücher*, xxiv, 1911, p. 4. [2] *Epinomis*, 987 D.

maintain the very contrary view, namely, that God cannot possibly be without intelligence or be ignorant of human nature: rather he knows that, when he teaches them, men will follow him and learn what they are taught. And he is of course perfectly aware that he does teach us, and that we learn, the very subject we are now discussing, number and counting; if he failed to know this, he would show the greatest want of intelligence; the God we speak of would in fact not know himself, if he took it amiss that a man capable of learning should learn, and if he did not rejoice unreservedly with one who became good by divine influence.'[1]

Nothing could well show more clearly the Greek conviction that there could be no opposition between religion and scientific truth, and therefore that there could be no impiety in the pursuit of truth. The passage is a good parallel to the statement attributed to Plato that θεὸς ἀεὶ γεωμετρεῖ.

Meaning and classification of mathematics.

The words μαθήματα and μαθηματικός do not appear to have been definitely appropriated to the special meaning of mathematics and mathematicians or things mathematical until Aristotle's time. With Plato μάθημα is quite general, meaning any subject of instruction or study; he speaks of καλὰ μαθήματα, good subjects of instruction, as of καλὰ ἐπιτηδεύματα, good pursuits, of women's subjects as opposed to men's, of the Sophists hawking sound μαθήματα; what, he asks in the *Republic*, are the greatest μαθήματα? and he answers that the greatest μάθημα is the Idea of the Good.[2] But in the *Laws* he speaks of τρία μαθήματα, three subjects, as fit for freeborn men, the subjects being arithmetic, the science of measurement (geometry), and astronomy[3]; and no doubt the pre-eminent place given to mathematical subjects in his scheme of education would have its effect in encouraging the habit of speaking of these subjects exclusively as μαθήματα. The Peripatetics, we are told, explained the special use of the word in this way; they pointed out that, whereas such things as rhetoric and poetry and the whole of popular μουσική can be understood even by one who has not learnt them, the subjects called by the special name of μαθήματα cannot be known

[1] *Epinomis*, 988 A. [2] *Republic*, vi. 505 A. [3] *Laws*, vii. 817 E.

by any one who has not first gone through a course of instruction in them; they concluded that it was for this reason that these studies were called μαθηματική.[1] The special use of the word μαθηματική seems actually to have originated in the school of Pythagoras. It is said that the esoteric members of the school, those who had learnt the theory of knowledge in its most complete form and with all its elaboration of detail, were known as μαθηματικοί, mathematicians (as opposed to the ἀκουσματικοί, the exoteric learners who were entrusted, not with the inner theory, but only with the practical rules of conduct); and, seeing that the Pythagorean philosophy was mostly mathematics, the term might easily come to be identified with the mathematical subjects as distinct from others. According to Anatolius, the followers of Pythagoras are said to have applied the term μαθηματική more particularly to the two subjects of geometry and arithmetic, which had previously been known by their own separate names only and not by any common designation covering both.[2] There is also an apparently genuine fragment of Archytas, a Pythagorean and a contemporary and friend of Plato, in which the word μαθήματα appears as definitely appropriated to mathematical subjects:

'The mathematicians (τοὶ περὶ τὰ μαθήματα) seem to me to have arrived at correct conclusions, and it is not therefore surprising that they have a true conception of the nature of each individual thing: for, having reached such correct conclusions regarding the nature of the universe, they were bound to see in its true light the nature of particular things as well. Thus they have handed down to us clear knowledge about the speed of the stars, their risings and settings, and about geometry, arithmetic, and sphaeric, and last, not least, about music; for these μαθήματα seem to be sisters.'[3]

This brings us to the Greek classification of the different branches of mathematics. Archytas, in the passage quoted, specifies the four subjects of the Pythagorean *quadrivium*, geometry, arithmetic, astronomy, and music (for 'sphaeric' means astronomy, being the geometry of the sphere con-

[1] Anatolius in Hultsch's Heron, pp. 276–7 (Heron, vol. iv, Heiberg, p. 160. 18–24).

[2] Heron, ed. Hultsch, p. 277; vol. iv, p. 160. 24–162. 2, Heiberg.

[3] Diels, *Vorsokratiker*, i³, pp. 330–1.

sidered solely with reference to the problem of accounting for the motions of the heavenly bodies); the same list of subjects is attributed to the Pythagoreans by Nicomachus, Theon of Smyrna, and Proclus, only in a different order, arithmetic, music, geometry, and sphaeric; the idea in this order was that arithmetic and music were both concerned with number (ποσόν), arithmetic with number in itself, music with number in relation to something else, while geometry and sphaeric were both concerned with magnitude (πηλίκον), geometry with magnitude at rest, sphaeric with magnitude in motion. In Plato's curriculum for the education of statesmen the same subjects, with the addition of stereometry or solid geometry, appear, arithmetic first, then geometry, followed by solid geometry, astronomy, and lastly harmonics. The mention of stereometry as an independent subject is Plato's own idea; it was, however, merely a formal addition to the curriculum, for of course solid problems had been investigated earlier, as a part of geometry, by the Pythagoreans, Democritus and others. Plato's reason for the interpolation was partly logical. Astronomy treats of the motion of solid bodies. There is therefore a gap between plane geometry and astronomy, for, after considering plane figures, we ought next to add the third dimension and consider solid figures in themselves, before passing to the science which deals with such figures in motion. But Plato emphasized stereometry for another reason, namely that in his opinion it had not been sufficiently studied. 'The properties of solids do not yet seem to have been discovered.' He adds:

'The reasons for this are two. First, it is because no State holds them in honour that these problems, which are difficult, are feebly investigated; and, secondly, those who do investigate them are in need of a superintendent, without whose guidance they are not likely to make discoveries. But, to begin with, it is difficult to find such a superintendent, and then, even supposing him found, as matters now stand, those who are inclined to these researches would be prevented by their self-conceit from paying any heed to him.'[1]

I have translated ὡς νῦν ἔχει ('as matters now stand') in this passage as meaning 'in present circumstances', i.e. so

[1] Plato, *Republic*, vii. 528 A–C.

long as the director has not the authority of the State behind him: this seems to be the best interpretation in view of the whole context; but it is possible, as a matter of construction, to connect the phrase with the preceding words, in which case the meaning would be 'and, even when such a superintendent has been found, as is the case at present', and Plato would be pointing to some distinguished geometer among his contemporaries as being actually available for the post. If Plato intended this, it would presumably be either Archytas or Eudoxus whom he had in mind.

It is again on a logical ground that Plato made harmonics or music follow astronomy in his classification. As astronomy is the motion of bodies (φορὰ βάθους) and appeals to the eye, so there is a harmonious motion (ἐναρμόνιος φορά), a motion according to the laws of harmony, which appeals to the ear. In maintaining the sisterhood of music and astronomy Plato followed the Pythagorean view (cf. the passage of Archytas above quoted and the doctrine of the 'harmony of the spheres').

(α) *Arithmetic and logistic.*

By arithmetic Plato meant, not arithmetic in our sense, but the science which considers numbers in themselves, in other words, what we mean by the Theory of Numbers. He does not, however, ignore the art of calculation (arithmetic in our sense); he speaks of number and calculation (ἀριθμὸν καὶ λογισμόν) and observes that 'the art of calculation (λογιστική) and arithmetic (ἀριθμητική) are both concerned with number'; those who have a natural gift for calculation (οἱ φύσει λογιστικοί) have, generally speaking, a talent for learning of all kinds, and even those who are slow are, by practice in it, made smarter.[1] But the art of calculation (λογιστική) is only preparatory to the true science; those who are to govern the city are to get a grasp of λογιστική, not in the popular sense with a view to use in trade, but only for the purpose of knowledge, until they are able to contemplate the nature of number in itself by thought alone.[2] This distinction between ἀριθμητική (the theory of numbers) and λογιστική (the art of

[1] *Republic*, vii. 522 C, 525 A, 526 B.
[2] *Ib.* vii. 525 B, C.

calculation) was a fundamental one in Greek mathematics. It is found elsewhere in Plato,[1] and it is clear that it was well established in Plato's time. Archytas too has λογιστική in the same sense; the art of calculation, he says, seems to be far ahead of other arts in relation to wisdom or philosophy, nay it seems to make the things of which it chooses to treat even clearer than geometry does; moreover, it often succeeds even where geometry fails.[2] But it is later writers on the classification of mathematics who alone go into any detail of what λογιστική included. Geminus in Proclus, Anatolius in the *Variae Collectiones* included in Hultsch's Heron, and the scholiast to Plato's *Charmides* are our authorities. Arithmetic, says Geminus,[3] is divided into the theory of linear numbers, the theory of plane numbers, and the theory of solid numbers. It investigates, in and by themselves, the species of number as they are successively evolved from the unit, the formation of plane numbers, similar and dissimilar, and the further progression to the third dimension. As for the λογιστικός, it is not in and by themselves that he considers the properties of numbers but with reference to sensible objects; and for this reason he applies to them names adapted from the objects measured, calling some (numbers) μηλίτης (from μῆλον, a sheep, or μῆλον, an apple, more probably the latter) and others φιαλίτης (from φιάλη, a bowl).[4] The scholiast to the *Charmides* is fuller still:[5]

'Logistic is the science which deals with numbered things, not numbers; it does not take number in its essence, but it presupposes 1 as unit, and the numbered object as number, e.g. it regards 3 as a triad, 10 as a decad, and applies the theorems of arithmetic to such (particular) cases. Thus it is logistic which investigates on the one hand what Archimedes called the cattle-problem, and on the other hand *melites* and *phialites* numbers, the latter relating to bowls, the former to flocks (he should probably have said "apples"); in other kinds too it investigates the numbers of sensible bodies, treating them as absolute (ὡς περὶ τελείων). Its subject-matter is everything that is numbered. Its branches include the so-called Greek and Egyptian methods in multiplications and divisions,[6] the additions and decompositions

[1] Cf. *Gorgias*, 451 B, C; *Theaetetus*, 145 A with 198 A, &c.
[2] Diels, *Vorsokratiker*, i³, p. 337. 7–11.
[3] Proclus on Eucl. I, p. 39. 14–20.
[4] *Ib.*, p. 40. 2–5.
[5] On *Charmides*, 165 E.
[6] See Chapter II, pp. 52–60.

of fractions; which methods it uses to explore the secrets of the theory of triangular and polygonal numbers with reference to the subject-matter of particular problems.'

The content of *logistic* is for the most part made fairly clear by the scholia just quoted. First, it comprised the ordinary arithmetical operations, addition, subtraction, multiplication, division, and the handling of fractions; that is, it included the elementary parts of what we now call *arithmetic.* Next, it dealt with problems about such things as sheep (or apples), bowls, &c.; and here we have no difficulty in recognizing such problems as we find in the arithmetical epigrams included in the Greek anthology. Several of them are problems of dividing a number of apples or nuts among a certain number of persons; others deal with the weights of bowls, or of statues and their pedestals, and the like; as a rule, they involve the solution of simple equations with one unknown, or easy simultaneous equations with two unknowns; two are indeterminate equations of the first degree to be solved in positive integers. From Plato's allusions to such problems it is clear that their origin dates back, at least, to the fifth century B.C. The cattle-problem attributed to Archimedes is of course a much more difficult problem, involving the solution of a 'Pellian' equation in numbers of altogether impracticable size. In this problem the sums of two pairs of unknowns have to be respectively a square and a triangular number; the problem would therefore seem to correspond to the description of those involving 'the theory of triangular and polygonal numbers'. Tannery takes the allusion in the last words to be to problems in indeterminate analysis like those of Diophantus's *Arithmetica.* The difficulty is that most of Diophantus's problems refer to numbers such that their sums, differences, &c., are *squares,* whereas the scholiast mentions only triangular and polygonal numbers. Tannery takes squares to be included among polygons, or to have been accidentally omitted by a copyist. But there is only one use in Diophantus's *Arithmetica* of a triangular number (in IV. 38), and none of a polygonal number; nor can the *τριγώνους* of the scholiast refer, as Tannery supposes, to right-angled triangles with sides in rational numbers (the main subject of Diophantus's Book VI), the use of the masculine showing that only τριγώνους ἀριθμούς, triangular *numbers*, can be meant. Nevertheless there can, I think, be no doubt that Diophantus's *Arithmetica* belongs to *Logistic*. Why then did Diophantus call his thirteen books *Arithmetica*? The explanation is probably this. Problems of the Diophantine type, like those of the arithmetical epigrams, had previously been enunciated of concrete numbers (numbers of apples, bowls, &c.), and one of Diophantus's problems (V. 30) is actually in epigram form, and is about measures of wine with prices in drachmas. Diophantus then probably saw that there was no reason why such problems should refer to numbers of any one particular thing rather than another, but that they might more conveniently take the form of finding numbers *in the abstract* with certain properties, alone or in combination, and therefore that they might claim to be part of arithmetic, the abstract science or theory of numbers.

It should be added that to the distinction between *arithmetic* and *logistic* there corresponded (up to the time of Nicomachus) different methods of treatment. With rare exceptions, such as Eratosthenes's κόσκινον, or sieve, a device for separating out the successive prime numbers, the theory of numbers was only treated in connexion with geometry, and for that reason only the geometrical form of proof was used, whether the figures took the form of dots marking out squares, triangles, gnomons, &c. (as with the early Pythagoreans), or of straight lines (as in Euclid VII–IX); even Nicomachus did not entirely banish geometrical considerations from his work, and in Diophantus's treatise on Polygonal Numbers, of which a fragment survives, the geometrical form of proof is used.

(β) *Geometry and geodaesia.*

By the time of Aristotle there was separated out from geometry a distinct subject, γεωδαισία, *geodesy*, or, as we should say, *mensuration*, not confined to land-measuring, but covering generally the practical measurement of surfaces and volumes, as we learn from Aristotle himself,[1] as well as from a passage of Geminus quoted by Proclus.[2]

[1] Arist. *Metaph.* B. 2, 997 b 26, 31.
[2] Proclus on Eucl. I, p. 39. 20–40. 2.

(γ) *Physical subjects, mechanics, optics, harmonics, astronomy, and their branches.*

In applied mathematics Aristotle recognizes optics and mechanics in addition to astronomy and harmonics. He calls optics, harmonics, and astronomy the *more physical* (branches) of mathematics,[1] and observes that these subjects and mechanics depend for the proofs of their propositions upon the pure mathematical subjects, optics on geometry, mechanics on geometry or stereometry, and harmonics on arithmetic; similarly, he says, *Phaenomena* (that is, observational astronomy) depend on (theoretical) astronomy.[2]

The most elaborate classification of mathematics is that given by Geminus.[3] After arithmetic and geometry, which treat of non-sensibles, or objects of pure thought, come the branches which are concerned with sensible objects, and these are six in number, namely mechanics, astronomy, optics, geodesy, *canonic* (κανονική), *logistic*. Anatolius distinguishes the same subjects but gives them in the order *logistic*, geodesy, optics, *canonic*, mechanics, astronomy.[4] *Logistic* has already been discussed. Geodesy too has been described as *mensuration*, the practical measurement of surfaces and volumes; as Geminus says, it is the function of geodesy to measure, not a cylinder or a cone (as such), but heaps as cones, and tanks or pits as cylinders.[5] *Canonic* is the theory of the musical intervals as expounded in works like Euclid's κατατομὴ κανόνος, *Division of the canon.*

Optics is divided by Geminus into three branches.[6] (1) The first is Optics proper, the business of which is to explain why things appear to be of different sizes or different shapes according to the way in which they are placed and the distances at which they are seen. Euclid's *Optics* consists mainly of propositions of this kind; a circle seen edgewise looks like a straight line (Prop. 22), a cylinder seen by one eye appears less than half a cylinder (Prop. 28); if the line joining the eye to the centre of a circle is perpendicular

[1] Arist. *Phys.* ii. 2, 194 a 8.
[2] Arist. *Anal. Post.* i. 9, 76 a 22–5; i. 13, 78 b 35–9.
[3] Proclus on Eucl. I, p. 38. 8–12.
[4] See Heron, ed. Hultsch, p. 278; ed. Heiberg, iv, p. 164.
[5] Proclus on Eucl. I, p. 39. 23–5.
[6] *Ib.*, p. 40. 13–22.

to the plane of the circle, all its diameters will look equal (Prop. 34), but if the joining line is neither perpendicular to the plane of the circle nor equal to its radius, diameters with which it makes unequal angles will appear unequal (Prop. 35); if a visible object remains stationary, there exists a locus such that, if the eye is placed at any point on it, the object appears to be of the same size for every position of the eye (Prop. 38). (2) The second branch is *Catoptric*, or the theory of mirrors, exemplified by the *Catoptrica* of Heron, which contains, e.g., the theorem that the angles of incidence and reflexion are equal, based on the assumption that the broken line connecting the eye and the object reflected is a minimum. (3) The third branch is σκηνογραφική or, as we might say, *scene-painting*, i.e. applied perspective.

Under the general term of mechanics Geminus[1] distinguishes (1) ὀργανοποιϊκή, the art of making engines of war (cf. Archimedes's reputed feats at the siege of Syracuse and Heron's βελοποιϊκά), (2) θαυματοποιϊκή, the art of making *wonderful machines*, such as those described in Heron's *Pneumatica* and *Automatic Theatre*, (3) Mechanics proper, the theory of centres of gravity, equilibrium, the mechanical powers, &c., (4) *Sphere-making*, the imitation of the movements of the heavenly bodies; Archimedes is said to have made such a sphere or orrery. Last of all,[2] astronomy is divided into (1) γνωμονική, the art of the gnomon, or the measurement of time by means of the various forms of sun-dials, such as those enumerated by Vitruvius,[3] (2) μετεωροσκοπική, which seems to have included, among other things, the measurement of the heights at which different stars cross the meridian, (3) διοπτρική, the use of the *dioptra* for the purpose of determining the relative positions of the sun, moon, and stars.

Mathematics in Greek education.[4]

The elementary or primary stage in Greek education lasted till the age of fourteen. The main subjects were letters (reading and writing followed by dictation and the study of

[1] Proclus on Eucl. I, p. 41. 3–18.
[2] *Ib.*, pp. 41. 19–42. 6.
[3] Vitruvius, *De architectura*, ix. 8.
[4] Cf. Freeman, *Schools of Hellas*, especially pp. 100–7, 159.

literature), music and gymnastics; but there is no reasonable doubt that practical arithmetic (in our sense), including weights and measures, was taught along with these subjects. Thus, at the stage of spelling, a common question asked of the pupils was, How many letters are there in such and such a word, e.g. Socrates, and in what order do they come?[1] This would teach the cardinal and ordinal numbers. In the same connexion Xenophon adds, 'Or take the case of numbers. Some one asks, What is twice five?'[2] This indicates that counting was a part of learning letters, and that the multiplication table was a closely connected subject. Then, again, there were certain games, played with cubic dice or knucklebones, to which boys were addicted and which involved some degree of arithmetical skill. In the game of knucklebones in the *Lysis* of Plato each boy has a large basket of them, and the loser in each game pays so many over to the winner.[3] Plato connects the art of playing this game with mathematics[4]; so too he associates πεττεία (games with πεσσοί, somewhat resembling draughts or chess) with arithmetic in general.[5] When in the *Laws* Plato speaks of three subjects fit for freeborn citizens to learn, (1) calculation and the science of numbers, (2) mensuration in one, two and three dimensions, and (3) astronomy in the sense of the knowledge of the revolutions of the heavenly bodies and their respective periods, he admits that profound and accurate knowledge of these subjects is not for people in general but only for a few.[6] But it is evident that practical arithmetic was, after letters and the lyre, to be a subject for all, so much of arithmetic, that is, as is necessary for purposes of war, household management, and the work of government. Similarly, enough astronomy should be learnt to enable the pupil to understand the calendar.[7] Amusement should be combined with instruction so as to make the subjects attractive to boys. Plato was much attracted by the Egyptian practice in this matter:[8]

'Freeborn boys should learn so much of these things as vast multitudes of boys in Egypt learn along with their

[1] Xenophon, *Econ.* viii. 14.
[2] Xenophon, *Mem.* iv. 4. 7.
[3] Plato, *Lysis*, 206 E; cf. Apollonius Rhodius, iii. 117.
[4] *Phaedrus*, 274 C–D.
[5] *Politicus*, 299 E; *Laws*, 820 C.
[6] *Laws*, 817 E–818 A.
[7] *Ib.* 809 C, D.
[8] *Ib.* 819 A–C.

letters. First there should be calculations specially devised as suitable for boys, which they should learn with amusement and pleasure, for example, distributions of apples or garlands where the same number is divided among more or fewer boys, or (distributions) of the competitors in boxing or wrestling matches on the plan of drawing pairs with byes, or by taking them in consecutive order, or in any of the usual ways[1]; and again there should be games with bowls containing gold, bronze, and silver (coins?) and the like mixed together,[2] or the bowls may be distributed as undivided units; for, as I said, by connecting with games the essential operations of practical arithmetic, you supply the boy with what will be useful to him later in the ordering of armies, marches and campaigns, as well as in household management; and in any case you make him more useful to himself and more wide awake. Then again, by calculating measurements of things which have length, breadth, and depth, questions on all of which the natural condition of all men is one of ridiculous and disgraceful ignorance, they are enabled to emerge from this state.'

It is true that these are Plato's ideas of what elementary education *should* include; but it can hardly be doubted that such methods were actually in use in Attica.

Geometry and astronomy belonged to secondary education, which occupied the years between the ages of fourteen and eighteen. The pseudo-Platonic *Axiochus* attributes to Prodicus a statement that, when a boy gets older, i.e. after he has

[1] The Greek of this clause is (διανομαὶ) πυκτῶν καὶ παλαιστῶν ἐφεδρείας τε καὶ συλλήξεως ἐν μέρει καὶ ἐφεξῆς καὶ ὡς πεφύκασι γίγνεσθαι. So far as I can ascertain, ἐν μέρει (by itself) and ἐφεξῆς have always been taken as indicating alternative methods, 'in turn and in consecutive order'. But it is impossible to get any satisfactory contrast of meaning between 'in turn' and 'in consecutive order'. It is clear to me that we have here merely an instance of Plato's habit of changing the order of words for effect, and that ἐν μέρει must be taken with the genitives ἐφεδρείας καὶ συλλήξεως; i.e. we must translate as if we had ἐν ἐφεδρείας τε καὶ συλλήξεως μέρει, '*by way of* byes and drawings'. This gives a proper distinction between (1) drawings with byes and (2) taking competitors in consecutive order.

[2] It is difficult to decide between the two possible interpretations of the phrase φιάλας ἅμα χρυσοῦ καὶ χαλκοῦ καὶ ἀργύρου καὶ τοιούτων τινῶν ἄλλων κεραννύντες. It may mean 'taking bowls made of gold, bronze, silver and other metals mixed together (in certain proportions)' or 'filling bowls with gold, bronze, silver, &c. (*sc.* objects such as coins) mixed together'. The latter version seems to agree best with παίζοντες (making a game out of the process) and to give the better contrast to 'distributing the bowls *as wholes*' (ὅλας πως διαδιδόντες).

passed the primary stage under the *paidagogos*, *grammatistes*, and *paidotribes*, he comes under the tyranny of the 'critics', the *geometers*, the tacticians, and a host of other masters.[1] Teles, the philosopher, similarly, mentions arithmetic and geometry among the plagues of the lad.[2] It would appear that geometry and astronomy were newly introduced into the curriculum in the time of Isocrates. 'I am so far', he says,[3] 'from despising the instruction which our ancestors got, that I am a supporter of that which has been established in our time, I mean geometry, astronomy, and the so-called eristic dialogues.' Such studies, even if they do no other good, keep the young out of mischief, and in Isocrates's opinion no other subjects could have been invented more useful and more fitting; but they should be abandoned by the time that the pupils have reached man's estate. Most people, he says, think them idle, since (say they) they are of no use in private or public affairs; moreover they are forgotten directly because they do not go with us in our daily life and action, nay, they are altogether outside everyday needs. He himself, however, is far from sharing these views. True, those who specialize in such subjects as astronomy and geometry get no good from them unless they choose to teach them for a livelihood; and if they get too deeply absorbed, they become unpractical and incapable of doing ordinary business; but the study of these subjects up to the proper point trains a boy to keep his attention fixed and not to allow his mind to wander; so, being practised in this way and having his wits sharpened, he will be capable of learning more important matters with greater ease and speed. Isocrates will not give the name of 'philosophy' to studies like geometry and astronomy, which are of no immediate use for producing an orator or man of business; they are rather means of training the mind and a preparation for philosophy. They are a more manly discipline than the subjects taught to boys, such as literary study and music, but in other respects have the same function in making them quicker to learn greater and more important subjects.

[1] *Axiochus*, 366 E.

[2] Stobaeus, *Ecl.* iv. 34, 72 (vol. v, p. 848, 19 sq., Wachsmuth and Hense).

[3] See Isocrates, *Panathenaicus*, §§ 26–8 (238 b–d); Περὶ ἀντιδόσεως, §§ 261–8.

It would appear therefore that, notwithstanding the influence of Plato, the attitude of cultivated people in general towards mathematics was not different in Plato's time from what it is to-day.

We are told that it was one of the early Pythagoreans, unnamed, who first taught geometry for money: 'One of the Pythagoreans lost his property, and when this misfortune befell him he was allowed to make money by teaching geometry.'[1] We may fairly conclude that Hippocrates of Chios, the first writer of *Elements*, who also made himself famous by his quadrature of lunes, his reduction of the duplication of the cube to the problem of finding two mean proportionals, and his proof that the areas of circles are in the ratio of the squares on their diameters, also taught for money and for a like reason. One version of the story is that he was a merchant, but lost all his property through being captured by a pirate vessel. He then came to Athens to prosecute the offenders and, during a long stay, attended lectures, finally attaining such proficiency in geometry that he tried to square the circle.[2] Aristotle has the different version that he allowed himself to be defrauded of a large sum by custom-house officers at Byzantium, thereby proving, in Aristotle's opinion, that, though a good geometer, he was stupid and incompetent in the business of ordinary life.[3]

We find in the Platonic dialogues one or two glimpses of mathematics being taught or discussed in school- or classrooms. In the *Erastae*[4] Socrates is represented as going into the school of Dionysius (Plato's own schoolmaster[5]) and finding two lads earnestly arguing some point of astronomy; whether it was Anaxagoras or Oenopides whose theories they were discussing he could not catch, but they were drawing circles and imitating some inclination or other with their hands. In Plato's *Theaetetus*[6] we have the story of Theodorus lecturing on surds and proving separately, for the square root of every non-square number from 3 to 17, that it is incommensurable with 1, a procedure which set Theaetetus and the

[1] Iamblichus, *Vit. Pyth.* 89.

[2] Philoponus on Arist. *Phys.*, p. 327 b 44–8, Brandis.

[3] *Eudemian Ethics*, H. 14, 1247 a 17.

[4] *Erastae*, 32 A, B.

[5] Diog. L. iii. 5.

[6] *Theaetetus*, 147 D–148 B.

younger Socrates thinking whether it was not possible to comprehend all such surds under one definition. In these two cases we have advanced or selected pupils discussing among themselves the subject of lectures they had heard and, in the second case, trying to develop a theory of a more general character.

But mathematics was not only taught by regular masters in schools; the Sophists, who travelled from place to place giving lectures, included mathematics (arithmetic, geometry, and astronomy) in their very wide list of subjects. Theodorus, who was Plato's teacher in mathematics and is described by Plato as a master of geometry, astronomy, *logistic* and music (among other subjects), was a pupil of Protagoras, the Sophist, of Abdera.[1] Protagoras himself, if we may trust Plato, did not approve of mathematics as part of secondary education; for he is made to say that

'the other Sophists maltreat the young, for, at an age when the young have escaped the arts, they take them against their will and plunge them once more into the arts, teaching them the art of calculation, astronomy, geometry, and music—and here he cast a glance at Hippias—whereas, if any one comes to me, he will not be obliged to learn anything except what he comes for.'[2]

The Hippias referred to is of course Hippias of Elis, a really distinguished mathematician, the inventor of a curve known as the *quadratrix* which, originally intended for the solution of the problem of trisecting any angle, also served (as the name implies) for squaring the circle. In the *Hippias Minor*[3] there is a description of Hippias's varied accomplishments. He claimed, according to this passage, to have gone once to the Olympian festival with everything that he wore made by himself, ring and seal (engraved), oil-bottle, scraper, shoes, clothes, and a Persian girdle of expensive type; he also took poems, epics, tragedies, dithyrambs, and all sorts of prose works. He was a master of the science of calculation (*logistic*), geometry, astronomy, 'rhythms and harmonies and correct writing'. He also had a wonderful system of mnemonics enabling him, if he once heard a string of fifty

[1] *Theaetetus*, 164 E, 168 E. [2] *Protagoras*, 318 D, E.
[3] *Hippias Minor*, pp. 366 C–368 E.

names, to remember them all. As a detail, we are told that he got no fees for his lectures in Sparta, and that the Spartans could not endure lectures on astronomy or geometry or *logistic*; it was only a small minority of them who could even count; what they liked was history and archaeology.

The above is almost all that we know of the part played by mathematics in the Greek system of education. Plato's attitude towards mathematics was, as we have seen, quite exceptional; and it was no doubt largely owing to his influence and his inspiration that mathematics and astronomy were so enormously advanced in his school, and especially by Eudoxus of Cnidos and Heraclides of Pontus. But the popular attitude towards Plato's style of lecturing was not encouraging. There is a story of a lecture of his on 'The Good' which Aristotle was fond of telling.[1] The lecture was attended by a great crowd, and 'every one went there with the idea that he would be put in the way of getting one or other of the things in human life which are usually accounted good, such as Riches, Health, Strength, or, generally, any extraordinary gift of fortune. But when they found that Plato discoursed about mathematics, arithmetic, geometry, and astronomy, and finally declared the One to be the Good, no wonder they were altogether taken by surprise; insomuch that in the end some of the audience were inclined to scoff at the whole thing, while others objected to it altogether.' Plato, however, was able to pick and choose his pupils, and he could therefore insist on compliance with the notice which he is said to have put over his porch, 'Let no one unversed in geometry enter my doors';[2] and similarly Xenocrates, who, after Speusippus, succeeded to the headship of the school, could turn away an applicant for admission who knew no geometry with the words, 'Go thy way, for thou hast not the means of getting a grip of philosophy'.[3]

The usual attitude towards mathematics is illustrated by two stories of Pythagoras and Euclid respectively. Pythagoras, we are told,[4] anxious as he was to transplant to his own country the system of education which he had seen in opera-

[1] Aristoxenus, *Harmonica*, ii *ad init.*
[2] Tzetzes, *Chiliad.* viii. 972.
[3] Diog. L. iv. 10.
[4] Iamblichus, *Vit. Pyth.* c. 5.

tion in Egypt, and the study of mathematics in particular, could get none of the Samians to listen to him. He adopted therefore this plan of communicating his arithmetic and geometry, so that it might not perish with him. Selecting a young man who from his behaviour in gymnastic exercises seemed adaptable and was withal poor, he promised him that, if he would learn arithmetic and geometry systematically, he would give him sixpence for each 'figure' (proposition) that he mastered. This went on until the youth got interested in the subject, when Pythagoras rightly judged that he would gladly go on without the sixpence. He therefore hinted that he himself was poor and must try to earn his daily bread instead of doing mathematics; whereupon the youth, rather than give up the study, volunteered to pay sixpence himself to Pythagoras for each proposition. We must presumably connect with this story the Pythagorean motto, 'a figure and a platform (from which to ascend to the next higher step), not a figure and sixpence'.[1]

The other story is that of a pupil who began to learn geometry with Euclid and asked, when he had learnt one proposition, 'What advantage shall I get by learning these things?' And Euclid called the slave and said, 'Give him sixpence, since he must needs gain by what he learns.'

We gather that the education of kings in the Macedonian period did not include much geometry, whether it was Alexander who asked Menaechmus, or Ptolemy who asked Euclid, for a short-cut to geometry, and got the reply that 'for travelling over the country there are royal roads and roads for common citizens: but in geometry there is one road for all'.[2]

[1] Proclus on Eucl. I, p. 84. 16.
[2] Stobaeus, *Ecl.* ii. 31, 115 (vol. ii, p. 228, 30, Wachsmuth).

II

GREEK NUMERICAL NOTATION AND ARITHMETICAL OPERATIONS

The decimal system.

THE Greeks, from the earliest historical times, followed the decimal system of numeration, which had already been adopted by civilized peoples all the world over. There are, it is true, traces of *quinary* reckoning (reckoning in terms of five) in very early times; thus in Homer πεμπάζειν (to 'five') is used for 'to count'.[1] But the counting by fives was probably little more than auxiliary to counting by tens; five was a natural halting-place between the unit and ten, and the use of five times a particular power of ten as a separate category intermediate between that power and the next was found convenient in the earliest form of numerical symbolism established in Greece, just as it was in the Roman arithmetical notation. The reckoning by five does not amount to such a variation of the decimal system as that which was in use among the Celts and Danes; these peoples had a vigesimal system, traces of which are still left in the French *quatre-vingts*, *quatre-vingt-treize*, &c., and in our *score*, three-score and ten, twenty-one, &c.

The natural explanation of the origin of the decimal system, as well as of the quinary and vigesimal variations, is to suppose that they were suggested by the primitive practice of reckoning with the fingers, first of one hand, then of both together, and after that with the ten toes in addition (making up the 20 of the vigesimal system). The subject was mooted in the Aristotelian *Problems*,[2] where it is asked:

'Why do all men, whether barbarians or Greeks, count up to ten, and not up to any other number, such as 2, 3, 4, or 5, so that, for example, they do not say one-*plus*-five (for 6),

[1] Homer, *Od.* iv. 412.

[2] xv. 3, 910 b 23–911 a 4.

two-*plus*-five (for 7), as they say one-*plus*-ten (ἕνδεκα, for 11), two-*plus*-ten (δώδεκα, for 12), while on the other hand they do not go beyond ten for the first halting-place from which to start again repeating the units? For of course any number is the next before it *plus* 1, or the next before that *plus* 2, and so with those preceding numbers; yet men fixed definitely on ten as the number to count up to. It cannot have been chance; for chance will not account for the same thing being done always: what is always and universally done is not due to chance but to some natural cause.'

Then, after some fanciful suggestions (e.g. that 10 is a 'perfect number'), the author proceeds:

'Or is it because men were born with ten fingers and so, because they possess the equivalent of pebbles to the number of their own fingers, come to use this number for counting everything else as well?'

Evidence for the truth of this latter view is forthcoming in the number of cases where the word for 5 is either the same as, or connected with, the word for 'hand'. Both the Greek χείρ and the Latin *manus* are used to denote 'a number' (of men). The author of the so-called geometry of Boëtius says, moreover, that the ancients called all the numbers below ten by the name *digits* ('fingers').[1]

Before entering on a description of the Greek numeral signs it is proper to refer briefly to the systems of notation used by their forerunners in civilization, the Egyptians and Babylonians.

Egyptian numerical notation.

The Egyptians had a purely decimal system, with the signs 𓏺 for the unit, 𓎆 for 10, 𓍢 for 100, 𓆼 for 1,000, 𓂭 for 10,000, 𓆐 for 100,000. The number of each denomination was expressed by repeating the sign that number of times; when the number was more than 4 or 5, lateral space was saved by arranging them in two or three rows, one above the other. The greater denomination came before the smaller. Numbers could be written from left to right or from right to left; in the latter case the above signs were turned the opposite way. The fractions in use were all submultiples or single aliquot

[1] Boëtius, *De Inst. Ar.*, &c., p. 395. 6–9, Friedlein.

parts, except $\frac{2}{3}$, which had a special sign or ; the submultiples were denoted by writing over the corresponding whole number; thus

$= \frac{1}{23}$, $= \frac{1}{324}$, $= \frac{1}{2190}$.

Babylonian systems.

(α) Decimal. *(β) Sexagesimal.*

The ancient Babylonians had two systems of numeration. The one was purely decimal based on the following signs. The simple wedge represented the unit, which was repeated up to nine times: where there were more than three, they were placed in two or three rows, e.g. $= 4$, $= 7$. 10 was represented by ; 11 would therefore be . 100 had the compound sign , and 1000 was expressed as 10 hundreds, by , the prefixed (10) being here multiplicative. Similarly, the was regarded as one sign, and denoted not 2000 but 10000, the prefixed being again multiplicative. Multiples of 10000 seem to have been expressed as multiples of 1000: at least, 120000 seems to be attested in the form $100.1000 + 20.1000$. The absence of any definite unit above 1000 (if it was really absent) must have rendered the system very inconvenient as a means of expressing large numbers.

Much more interesting is the second Babylonian system, the sexagesimal. This is found in use on the Tables of Senkereh, discovered by W. K. Loftus in 1854, which may go back as far as the time between 2300 and 1600 B.C. In this system numbers above the units (which go from 1 to 59) are arranged according to powers of 60. 60 itself was called *sussu* (= *soss*), 60^2 was called *sar*, and there was a name also (*ner*) for the intermediate number $10.60 = 600$. The multiples of the several powers of 60, 60^2, 60^3, &c., contained in the number to be written down were expressed by means of the same wedge-notation as served for the units, and the multiples were placed in columns side by side, the columns being appropriated to the successive powers of 60. The unit-term was followed by similar columns appropriated, in order, to the successive submultiples $\frac{1}{60}$, $\frac{1}{60^2}$, &c., the number of sixtieths, &c., being again denoted by the ordinary wedge-numbers. Thus [cuneiform] represents $44.60^2 + 26.60 + 40 = 160,000$; [cuneiform] $= 27.60^2 + 21.60 + 36 = 98,496$. Similarly we find [cuneiform] representing $30 + \frac{30}{60}$ and [cuneiform] representing $30 + \frac{27}{60}$; the latter case also shows that the Babylonians, on occasion, used the subtractive plan, for the 27 is here written 30 *minus* 3.

The sexagesimal system only required a definite symbol for 0 (indicating the absence of a particular denomination), and a fixed arrangement of columns, to become a complete position-value system like the Indian. With a sexagesimal system 0 would occur comparatively seldom, and the Tables of Senkereh do not show a case; but from other sources it appears that a gap often indicated a zero, or there was a sign used for the purpose, namely [cuneiform], called the 'divider'. The inconvenience of the system was that it required a multiplication table extending from 1 times 1 to 59 times 59. It had, however, the advantage that it furnished an easy means of expressing very large numbers. The researches of H. V. Hilprecht show that $60^4 = 12,960,000$ played a prominent part in Babylonian arithmetic, and he found a table containing certain quotients of the number [cuneiform] – – – – – – – $= 60^8 + 10.60^7$, or 195,955,200,000,000. Since the number of units of any denomination are expressed in the purely decimal notation, it follows that the latter system preceded the sexagesimal. What circumstances led to the adoption of 60 as the base can only be conjectured, but it may be presumed that the authors of the system were fully alive to the convenience of a base with so many divisors, combining as it does the advantages of 12 and 10.

Greek numerical notation.

To return to the Greeks. We find, in Greek inscriptions of all dates, instances of numbers and values written out in full; but the inconvenience of this longhand, especially in such things as accounts, would soon be felt, and efforts would be made to devise a scheme for representing numbers more concisely by means of conventional signs of some sort. The Greeks conceived the original idea of using the letters of the ordinary Greek alphabet for this purpose.

(α) *The 'Herodianic' signs.*

There were two main systems of numerical notation in use in classical times. The first, known as the Attic system and used for cardinal numbers exclusively, consists of the set of signs somewhat absurdly called 'Herodianic' because they are described in a fragment[1] attributed to Herodian, a grammarian of the latter half of the second century A.D. The authenticity of the fragment is questioned, but the writer says that he has seen the signs used in Solon's laws, where the prescribed pecuniary fines were stated in this notation, and that they are also to be found in various ancient inscriptions, decrees and laws. These signs cannot claim to be numerals in the proper sense; they are mere compendia or abbreviations; for, except in the case of the stroke Ι representing a unit, the signs are the first letters of the full words for the numbers, and all numbers up to 50000 were represented by combinations of these signs. Ι, representing the unit, may be repeated up to four times; Π (the first letter of πέντε) stands for 5, Δ (the first letter of δέκα) for 10, Η (representing ἕκατον) for 100, Χ (χίλιοι) for 1000, and Μ (μύριοι) for 10000. The half-way numbers 50, 500, 5000 were expressed by combining Π (five) with the other signs respectively; 𐅄, 𐅄, 𐅄, made up of Π (5) and Δ (10), = 50; 𐅅, made up of Π and Η, = 500; 𐅆 = 5000; and 𐅇 = 50000. There are thus six simple and four compound symbols, and all other numbers intermediate between those so represented are made up by juxtaposition on an additive basis, so that each of the simple signs may be repeated not more than four times; the higher numbers come before the lower. For example, ΠΙ = 6, ΔΙΙΙΙ = 14, ΗΠ = 105, ΧΧΧΧ𐅅ΗΗΗΗ𐅄ΔΔΔΔΠΙΙΙΙ = 4999. Instances of this system of notation are found in Attic inscriptions from 454 to about 95 B.C. Outside Attica the same system was in use, the precise form of the symbols varying with the form of the letters in the local alphabets. Thus in Boeotian inscriptions 𐅄 or 𐅄 = 50, ⊢E = 100, Π⊢E = 500,

[1] Printed in the Appendix to Stephanus's *Thesaurus*, vol. viii.

Ψ = 1000, 𐅆 = 5000; and 𐅆𐅅ΗΗΗΗΔΔΙΙΙ = 5823. But, in consequence of the political influence of Athens, the Attic system, sometimes with unimportant modifications, spread to other states.[1]

In a similar manner compendia were used to denote units of coinage or of weight. Thus in Attica Τ = τάλαντον (6000 drachmae), Μ = μνᾶ (1000 drachmae), Σ or ≶ = στατήρ (1/3000th of a talent or 2 drachmae), ⊢ = δραχμή, Ι = ὀβολός (1/6th of a drachma), C = ἡμιωβέλιον (1/12th of a drachma), Ɔ or Τ = τεταρτημόριον (1/4th of an obol or 1/24th of a drachma), Χ = χαλκοῦς (1/8th of an obol or 1/48th of a drachma). Where a number of one of these units has to be expressed, the sign for the unit is written on the left of that for the number; thus ⊢𐅄ΔΙ = 61 drachmae. The two compendia for the numeral and the unit are often combined into one; e.g. 𐅈, 𐅈 = 5 talents, 𐅉 = 50 talents, 𐅊 = 100 talents, 𐅋 = 500 talents, 𐅌 = 1000 talents, 𐅍 = 10 minas, 𐅐 = 5 drachmae, 𐅎, 𐅎, 𐅎 = 10 staters, &c.

(β) The ordinary alphabetic numerals.

The second main system, used for all kinds of numerals, is that with which we are familiar, namely the alphabetic system. The Greeks took their alphabet from the Phoenicians. The Phoenician alphabet contained 22 letters, and, in appropriating the different signs, the Greeks had the happy inspiration to use for the vowels, which were not written in Phoenician, the signs for certain spirants for which the Greeks had no use; Aleph became A, He was used for E, Yod for I, and Ayin for O; when, later, the long E was differentiated, Cheth was used, 𐌇 or H. Similarly they utilized superfluous signs for sibilants. Out of Zayin and Samech they made the letters Z and Ξ. The remaining two sibilants were Ssade and Shin. From the latter came the simple Greek Σ (although the name Sigma seems to correspond to the Semitic Samech, if it is not simply the 'hissing' letter, from σίζω). Ssade, a softer sibilant (= σσ), also called San in early times, was taken over by the Greeks in the place it occupied after Π, and written in the form Μ or Ͷ. The form Ͳ (= σσ) appearing in inscriptions of Halicarnassus

[1] Larfeld, *Handbuch der griechischen Epigraphik*, vol. i, p. 417.

(e.g. Ἁλικαρναϡ[έων] = Ἁλικαρνασσέων) and Teos ([θ]αλάϡης; cf. θάλασσαν in another place) seems to be derived from some form of Ssade; this ϡ, after its disappearance from the literary alphabet, remained as a numeral, passing through the forms ʌ, ϻ, ϻ, ϡ, and ϡ to the fifteenth century form ϡ, to which in the second half of the seventeenth century the name Sampi was applied (whether as being the San which followed Pi or from its resemblance to the cursive form of π). The original Greek alphabet also retained the Phoenician Vau (Ϝ) in its proper place between E and Z and the Koppa = Qoph (ϙ) immediately before P. The Phoenician alphabet ended with T; the Greeks first added Υ, derived from Vau apparently (notwithstanding the retention of Ϝ), then the letters Φ, X, Ψ and, still later, Ω. The 27 letters used for numerals are divided into three sets of nine each; the first nine denote the units, 1, 2, 3, &c., up to 9; the second nine the tens, from 10 to 90; and the third nine the hundreds, from 100 to 900. The following is the scheme:

Α = 1	Ι = 10	Ρ = 100
Β = 2	Κ = 20	Σ = 200
Γ = 3	Λ = 30	Τ = 300
Δ = 4	Μ = 40	Υ = 400
Ε = 5	Ν = 50	Φ = 500
Ϛ [ϛ] = 6	Ξ = 60	Χ = 600
Ζ = 7	Ο = 70	Ψ = 700
Η = 8	Π = 80	Ω = 800
Θ = 9	Ϙ = 90	ϡ [ϡ] = 900

The sixth sign in the first column (Ϛ) is a form of the digamma Ϝ Ϝ. It came, in the seventh and eighth centuries A. D., to be written in the form Ϛ and then, from its similarity to the cursive ϛ (= στ), was called Stigma.

This use of the letters of the alphabet as numerals was original with the Greeks; they did not derive it from the Phoenicians, who never used their alphabet for numerical purposes but had separate signs for numbers. The earliest occurrence of numerals written in this way appears to be in a Halicarnassian inscription of date not long after 450 B.C. Two caskets from the ruins of a famous mausoleum built at Halicarnassus in 351 B.C., which are attributed to the time of Mausolus, about 350 B.C., are inscribed with the letters

ΨΝΔ = 754 and ΣϘΓ = 293. A list of priests of Poseidon at Halicarnassus, attributable to a date at least as early as the fourth century, is preserved in a copy of the second or first century, and this copy, in which the numbers were no doubt reproduced from the original list, has the terms of office of the several priests stated on the alphabetical system. Again, a stone inscription found at Athens and perhaps belonging to the middle of the fourth century B.C. has, in five fragments of columns, numbers in tens and units expressed on the same system, the tens on the right and the units on the left.

There is a difference of opinion as to the approximate date of the actual formulation of the alphabetical system of numerals. According to one view, that of Larfeld, it must have been introduced much earlier than the date (450 B.C. or a little later) of the Halicarnassus inscription, in fact as early as the end of the eighth century, the place of its origin being Miletus. The argument is briefly this. At the time of the invention of the system all the letters from Α to Ω, including Ϝ and Ϙ in their proper places, were still in use, while Ssade (Ͳ, the double *ss*) had dropped out; this is why the last-named sign (afterwards ϡ) was put at the end. If Ϲ (= 6) and Ϙ (= 90) had been no longer in use as letters, they too would have been put, like Ssade, at the end. The place of origin of the numeral system must have been one in which the current alphabet corresponded to the content and order of the alphabetic numerals. The order of the signs Φ, Χ, Ψ shows that it was one of the *Eastern* group of alphabets. These conditions are satisfied by one alphabet, and one only, that of Miletus, at a stage which still recognized the Vau (Ϝ) as well as the Koppa (Ϙ). The Ϙ is found along with the so-called complementary letters including Ω, the latest of all, in the oldest inscriptions of the Milesian colony Naucratis (about 650 B.C.); and, although there are no extant Milesian inscriptions containing the Ϝ, there is at all events one very early example of Ϝ in Ionic, namely *Ἀγα-σιλέϜο* (*Ἀγασιλήϝου*) on a vase in the Boston (U.S.) Museum of Fine Arts belonging to the end of the eighth or (at latest) the middle of the seventh century. Now, as Ω is fully established at the date of the earliest inscriptions at Miletus (about 700 B.C.) and Naucratis (about 650 B.C.), the earlier extension of the alphabet by the letters Φ X Ψ must have taken place not later than 750 B.C. Lastly, the presence in the alphabet of the Vau indicates a time which can hardly be put later than 700 B.C. The conclusion is that it was about this time, if not earlier, that the numerical alphabet was invented.

The other view is that of Keil, who holds that it originated in Dorian Caria, perhaps at Halicarnassus itself, about 550–425 B.C., and that it was artificially put together by some one who had the necessary knowledge to enable him to fill up his own alphabet, then consisting of twenty-four letters only, by taking over Ϝ and Ϙ from other alphabets and putting them in their proper places, while he completed the numeral series by adding Ͳ at the end.[1] Keil urges, as against Larfeld, that it is improbable that Ϝ and Ω ever existed together in the Milesian alphabet. Larfeld's answer[2] is that, although Ϝ had disappeared from ordinary language at Miletus towards the end of the eighth century, we cannot say exactly when it disappeared, and even if it was practically gone at the time of the formulation of the numerical alphabet, it would be in the interest of instruction in schools, where Homer was read, to keep the letter as long as possible in the official alphabet. On the other hand, Keil's argument is open to the objection that, if the Carian inventor could put the Ϝ and Ϙ into their proper places in the series, he would hardly have failed to put the Ssade Ͳ in its proper place also, instead of at the end, seeing that Ͳ is found in Caria itself, namely in a Halicarnassus (Lygdamis) inscription of about 453 B.C., and also in Ionic Teos about 476 B.C.[3] (see pp. 31–2 above).

It was a long time before the alphabetic numerals found general acceptance. They were not officially used until the time of the Ptolemies, when it had become the practice to write, in inscriptions and on coins, the year of the reign of the ruler for the time being. The conciseness of the signs made them particularly suitable for use on coins, where space was limited. When coins went about the world, it was desirable that the notation should be uniform, instead of depending on local alphabets, and it only needed the support of some paramount

[1] *Hermes*, 29, 1894, p. 265 sq. [2] Larfeld, *op. cit.*, i, p. 421.
[3] *Ib.*, i, p. 358.

political authority to secure the final triumph of the alphabetic system. The alphabetic numerals are found at Alexandria on coins of Ptolemy II, Philadelphus, assigned to 266 B.C. A coin with the inscription *Ἀλεξάνδρου* ΚΔ (twenty-fourth year after Alexander's death) belongs, according to Keil, to the end of the third century.[1] A very old Graeco-Egyptian papyrus (now at Leyden, No. 397), ascribed to 257 B.C., contains the number $\kappa\theta = 29$. While in Boeotia the Attic system was in use in the middle of the third century, along with the corresponding local system, it had to give way about 200 B.C. to the alphabetic system, as is shown by an inventory from the temple of Amphiaraus at Oropus[2]; we have here the first official use of the alphabetic system in Greece proper. From this time Athens stood alone in retaining the archaic system, and had sooner or later to come into line with other states. The last certainly attested use of the Attic notation in Athens was about 95 B.C.; the alphabetic numerals were introduced there some time before 50 B.C., the first example belonging to the time of Augustus, and by A.D. 50 they were in official use.

The two systems are found side by side in a number of papyrus-rolls found at Herculaneum (including the treatise of Philodemus *De pietate,* so that the rolls cannot be older than 40 or 50 B.C.); these state on the title page, after the name of the author, the number of books in alphabetic numerals, and the number of lines in the Attic notation, e.g. ΕΠΙΚΟΥΡΟΥ | ΠΕΡΙ | ΦΥΣΕΩΣ | ΙΕ *ἀριθ* . . ΧΧΧΗΗ (where ΙΕ = 15 and ΧΧΧΗΗ = 3200), just as we commonly use Roman figures to denote *Books* and Arabic figures for *sections* or *lines*.[3]

[1] *Hermes*, 29, 1894, p. 276 *n.*
[2] Keil in *Hermes*, 25, 1890, pp. 614–15.
[3] Reference should be made, in passing, to another, *quasi*-numerical, use of the letters of the ordinary alphabet, as current at the time, for numbering particular things. As early as the fifth century we find in a Locrian bronze-inscription the letters Α to ⊕ (including Ϝ then and there current) used to distinguish the nine paragraphs of the text. At the same period the Athenians, instead of following the old plan of writing out ordinal numbers in full, adopted the more convenient device of denoting them by the letters of the alphabet. In the oldest known example ὅρος Κ indicated 'boundary stone No. 10'; and in the fourth century the tickets of the ten panels of jurymen were marked with the letters Α to Κ. In like manner the Books in certain works of Aristotle (the *Ethics*, *Metaphysics*, *Politics*, and *Topics*) were at some time

(γ) *Mode of writing numbers in the ordinary alphabetic notation.*

Where, in the alphabetical notation, the number to be written contained more than one denomination, say, units with tens, or with tens and hundreds, the higher numbers were, as a rule, put before the lower. This was generally the case in European Greece; on the other hand, in the inscriptions of Asia Minor, the smaller number comes first, i.e. the letters are arranged in alphabetical order. Thus 111 may be represented either by PIA or by AIP; the arrangement is sometimes mixed, as PAI. The custom of writing the numbers in descending order became more firmly established in later times through the influence of the corresponding Roman practice.[1]

The alphabetic numerals sufficed in themselves to express all numbers from 1 to 999. For thousands (up to 9000) the letters were used again with a distinguishing mark; this was generally a sloping stroke to the left, e.g. ʹA or ͵A = 1000, but other forms are also found, e.g. the stroke might be combined with the letter as ⅄ = 1000 or again ʽA = 1000, ʽϹ = 6000. For tens of thousands the letter M (*μύριοι*) was borrowed from the other system, e.g. 2 myriads would be BM, MB, or $\overset{\mathrm{B}}{\mathrm{M}}$.

To distinguish letters representing numbers from the letters of the surrounding text different devices are used: sometimes the number is put between dots ⁝ or :, or separated by spaces from the text on both sides of it. In Imperial times distinguishing marks, such as a horizontal stroke above the letter, become common, e.g. *ἡ βουλὴ τῶν* X̄, other variations being ·X·, ·X·, X̌ and the like.

In the cursive writing with which we are familiar the

numbered on the same principle; so too the Alexandrine scholars (about 280 B.C.) numbered the twenty-four Books of Homer with the letters A to Ω. When the number of objects exceeded 24, doubled letters served for continuing the series, as AA, BB, &c. For example, a large quantity of building-stones have been found; among these are stones from the theatre at the Piraeus marked AA, BB, &c., and again AA|BB, BB|BB, &c. when necessary. Sometimes the numbering by double letters was on a different plan, the letter A denoting the full number of the first set of letters (24); thus AP would be 24 + 17 = 41.

[1] Larfeld. *op. cit.*, i, p. 426.

orthodox way of distinguishing numerals was by a horizontal stroke above each sign or collection of signs; the following was therefore the scheme (with ϛ substituted for Ϝ representing 6, and with ϡ = 900 at the end):

units (1 to 9) $\bar{\alpha}$, $\bar{\beta}$, $\bar{\gamma}$, $\bar{\delta}$, $\bar{\epsilon}$, $\bar{ϛ}$, $\bar{\zeta}$, $\bar{\eta}$, $\bar{\theta}$;
tens (10 to 90) $\bar{\iota}$, $\bar{\kappa}$, $\bar{\lambda}$, $\bar{\mu}$, $\bar{\nu}$, $\bar{\xi}$, $\bar{o}$, $\bar{\pi}$, $\bar{ϙ}$;
hundreds (100 to 900) $\bar{\rho}$, $\bar{\sigma}$, $\bar{\tau}$, $\bar{\upsilon}$, $\bar{\phi}$, $\bar{\chi}$, $\bar{\psi}$, $\bar{\omega}$, $\bar{ϡ}$;
thousands (1000 to 9000) ,$\bar{\alpha}$, ,$\bar{\beta}$, ,$\bar{\gamma}$, ,$\bar{\delta}$, ,$\bar{\epsilon}$, ,$\bar{ϛ}$, ,$\bar{\zeta}$, ,$\bar{\eta}$, ,$\bar{\theta}$;

(for convenience of printing, the horizontal stroke above the sign will hereafter, as a rule, be omitted).

(δ) *Comparison of the two systems of numerical notation.*

The relative merits of the two systems of numerical notation used by the Greeks have been differently judged. It will be observed that the *initial*-numerals correspond closely to the Roman numerals, except that there is no formation of numbers by subtraction as IX, XL, XC; thus

ΧΧΧΧ𐅆ΗΗΗΗ𐅅ΔΔΔΔ𐅃ΙΙΙΙ = MMMMDCCCCLXXXXVIIII

as compared with MMMMCMXCIX = 4999. The absolute inconvenience of the Roman system will be readily appreciated by any one who has tried to read Boëtius (Boëtius would write the last-mentioned number as $\overline{\text{IV}}$.DCCCCXCVIIII). Yet Cantor[1] draws a comparison between the two systems much to the disadvantage of the alphabetic numerals. 'Instead', he says, 'of an advance we have here to do with a decidedly retrograde step, especially so far as its suitability for the further development of the numeral system is concerned. If we compare the older "Herodianic" numerals with the later signs which we have called alphabetic numerals, we observe in the latter two drawbacks which do not attach to the former. There now had to be more signs, with values to be learnt by heart; and to reckon with them required a much greater effort of memory. The addition

ΔΔΔ + ΔΔΔΔ = 𐅅ΔΔ (30 + 40 = 70)

could be coordinated in one act of memory with that of

ΗΗΗ + ΗΗΗΗ = 𐅆ΗΗ (300 + 400 = 700)

in so far as the sum of 3 and 4 units of the same kind added

[1] Cantor, *Gesch. d. Math.* I[3], p. 129.

up to 5 and 2 units of the same kind. On the other hand $\lambda + \mu = o$ did not at all immediately indicate that $\tau + \upsilon = \psi$. The new notation had only one advantage over the other, namely that it took less space. Consider, for instance, 849, which in the "Herodianic" form is ΠΗΗΗΔΔΔΔΠΙΙΙΙ, but in the alphabetic system is ωμθ. The former is more self-explanatory and, for reckoning with, has most important advantages.' Gow follows Cantor, but goes further and says that 'the alphabetical numerals were a fatal mistake and hopelessly confined such nascent arithmetical faculty as the Greeks may have possessed'![1] On the other hand, Tannery, holding that the merits of the alphabetic numerals could only be tested by using them, practised himself in their use until, applying them to the whole of the calculations in Archimedes's *Measurement of a Circle*, he found that the alphabetic notation had practical advantages which he had hardly suspected before, and that the operations took little longer with Greek than with modern numerals.[2] Opposite as these two views are, they seem to be alike based on a misconception. Surely we do not 'reckon with' the numeral *signs* at all, but with the *words* for the numbers which they represent. For instance, in Cantor's illustration, we do not conclude that the *figure* 3 and the *figure* 4 added together make the *figure* 7; what we do is to say 'three and four are seven'. Similarly the Greek would not say to himself 'γ and $\delta = \zeta$' but τρεῖς καὶ τέσσαρες ἑπτά; and, notwithstanding what Cantor says, this *would* indicate the corresponding addition 'three hundred and four hundred are seven hundred', τριακόσιοι καὶ τετρακόσιοι ἑπτακόσιοι, and similarly with multiples of ten or of 1000 or 10000. Again, in using the multiplication table, we say 'three times four is twelve', or 'three multiplied by four = twelve'; the Greek would say τρὶς τέσσαρες, or τρεῖς ἐπὶ τέσσαρας, δώδεκα, and this would equally indicate that '*thirty* times *forty* is *twelve* hundred or one thousand two hundred', or that '*thirty* times *four* hundred is *twelve* thousand or a myriad and two thousand' (τριακοντάκις τεσσαράκοντα χίλιοι καὶ διακόσιοι, or τριακοντάκις τετρακόσιοι μύριοι καὶ δισχίλιοι).

[1] Gow, *A Short History of Greek Mathematics*, p. 46.

[2] Tannery, *Mémoires scientifiques* (ed. Heiberg and Zeuthen), i, pp. 200–1.

The truth is that in mental calculation (whether the operation be addition, subtraction, multiplication, or division), we reckon with the corresponding *words*, not with the symbols, and it does not matter a jot to the calculation how we choose to write the figures down. While therefore the alphabetical numerals had the advantage over the 'Herodianic' of being so concise, their only disadvantage was that there were more signs (twenty-seven) the meaning of which had to be committed to memory: truly a very slight disadvantage. The one real drawback to the alphabetic system was the absence of a sign for 0 (zero); for the O for οὐδεμία or οὐδέν which we find in Ptolemy was only used in the notation of sexagesimal fractions, and not as part of the numeral system. If there had been a sign or signs to indicate the absence in a number of a particular denomination, e.g. units or tens or hundreds, the Greek symbols could have been made to serve as a position-value system scarcely less effective than ours. For, while the position-values are clear in such a number as 7921 (,ζϡκα), it would only be necessary in the case of such a number as 7021 to show a blank in the proper place by writing, say, ,ζ - κα. Then, following Diophantus's plan of separating any number of myriads by a dot from the thousands, &c., we could write ζϡκα . ,ςτπδ for 79216384 or ,ζ - - - . - τ - δ for 70000304, while we could continually add sets of four figures to the left, separating each set from the next following by means of a dot.

(ε) *Notation for large numbers.*

Here too the orthodox way of writing tens of thousands was by means of the letter M with the number of myriads above it, e.g. $\overset{\beta}{\mathrm{M}}$ = 20000, $\overset{,\zeta\rho o\epsilon}{\mathrm{M}}$,εωοε = 71755875 (Aristarchus of Samos); another method was to write M or $\overset{\Upsilon}{\mathrm{M}}$ for the myriad and to put the number of myriads after it, separated by a dot from the remaining thousands, &c., e.g.

$$\overset{\Upsilon}{\mathrm{M}}\ \rho\nu . ,\zeta\text{ϡ}\pi\delta = 1507984$$

(Diophantus, IV. 28). Yet another way of expressing myriads was to use the symbol representing the number of myriads with two dots over it; thus α̈,ηφϙβ = 18592 (Heron, *Geometrica*, 17. 33). The word μυριάδες could, of course, be written in full, e.g. μυριάδες ͵βσοη καὶ ϡιβ = 22780912 (*ib.* 17. 34). To express still higher numbers, powers of myriads were used; a myriad (10000) was a *first myriad* (πρώτη μυριάς) to distinguish it from a *second myriad* (δευτέρα μυριάς) or 10000^2, and so on; the words πρῶται μυριάδες, δεύτεραι μυριάδες, &c., could either be written in full or expressed by $\overset{\Upsilon}{M}$, $M\overset{\Upsilon}{M}$, &c., respectively; thus δεύτεραι μυριάδες ις πρῶται (μυριάδες) ͵βϡνη $\overset{o}{M}$ ͵ϛφξ = 16 2958 6560 (Diophantus, V. 8), where $\overset{o}{M}$ = μονάδες (units) is inserted to distinguish the ͵βϡνη, the number of the 'first myriads', from the ͵ϛφξ denoting 6560 *units*.

(i) Apollonius's 'tetrads'.

The latter system is the same as that adopted by Apollonius in an arithmetical work, now lost, the character of which is, however, gathered from the elucidations in Pappus, Book II; the only difference is that Apollonius called his *tetrads* (sets of four digits) μυριάδες ἁπλαῖ, διπλαῖ, τριπλαῖ, &c., 'simple myriads', 'double', 'triple', &c., meaning 10000, 10000^2, 10000^3, and so on. The abbreviations for these successive powers in Pappus are μ^{α}, μ^{β}, μ^{Γ}, &c.; thus μ^{Γ} ͵ευξβ καὶ μ^{β} ͵γχ καὶ μ^{α} ͵ϛυ = 5462 3600 6400 0000. Another, but a less convenient, method of denoting the successive powers of 10000 is indicated by Nicolas Rhabdas (fourteenth century A.D.) who says that, while a pair of dots above the ordinary numerals denoted the number of myriads, the 'double myriad' was indicated by two pairs of dots one above the other, the 'triple myriad' by three pairs of dots, and so on. Thus ϡ̈ = 9000000, β̈̈ = 2 $(10000)^2$, μ̈̈̈ = 40 $(10000)^3$, and so on.

(ii) Archimedes's system (by octads).

Yet another special system invented for the purpose of expressing very large numbers is that of Archimedes's *Psammites* or *Sand-reckoner*. This goes by *octads*:

$$10000^2 = 100000000 = 10^8,$$

and all the numbers from 1 to 10^8 form the *first order*; the last number, 10^8, of the *first order* is taken as the unit of the *second order*, which consists of all the numbers from

10^8, or 100000000, to 10^{16}, or 100000000^2; similarly 10^{16} is taken as the unit of the *third order*, which consists of all numbers from 10^{16} to 10^{24}, and so on, the *100000000th order* consisting of all the numbers from $(100000000)^{99999999}$ to $(100000000)^{100000000}$, i.e. from $10^{8.(10^8-1)}$ to $10^{8.10^8}$. The aggregate of all the *orders* up to the 100000000th form the *first period*; that is, if $P \equiv (100000000)^{10^8}$, the numbers of the *first period* go from 1 to P. Next, P is the unit of the *first order* of the *second period*; the *first order* of the *second period* then consists of all numbers from P up to 100000000 P or $P.10^8$; $P.10^8$ is the unit of the *second order* (of the *second period*) which ends with $(100000000)^2$ P or $P.10^{16}$; $P.10^{16}$ begins the *third order* of the *second period*, and so on; the *100000000th order* of the *second period* consists of the numbers from $(100000000)^{99999999}$ P or $P.10^{8.(10^8-1)}$ to $(100000000)^{100000000}$ P or $P.10^{8.10^8}$, i.e. P^2. Again, P^2 is the unit of the *first order* of the *third period*, and so on. The *first order* of the *100000000th period* consists of the numbers from P^{10^8-1} to $P^{10^8-1}.10^8$, the *second order* of the same *period* of the numbers from $P^{10^8-1}.10^8$ to $P^{10^8-1}.10^{16}$, and so on, the (10^8)th *order* of the (10^8)th *period*, or the *period* itself, ending with $P^{10^8-1}.10^{8.10^8}$, i.e. P^{10^8}. The last number is described by Archimedes as a 'myriad-myriad units of the myriad-myriadth order of the myriad-myriadth period (*αἱ μυριακισμυριοστᾶς περιόδου μυριακισμυριοστῶν ἀριθμῶν μύριαι μυριάδες*)'. This system was, however, a *tour de force*, and has nothing to do with the ordinary Greek numerical notation.

Fractions.

(α) *The Egyptian system*

We now come to the methods of expressing fractions. A fraction may be either a submultiple (an 'aliquot part', i.e. a fraction with numerator unity) or an ordinary proper fraction with a number not unity for numerator and a greater number for denominator. The Greeks had a preference for expressing ordinary proper fractions as the sum of two or more submultiples; in this they followed the Egyptians, who always expressed fractions in this way, with the exception that they had a single sign for $\frac{2}{3}$, whereas we should have expected them to split it up into $\frac{1}{2}+\frac{1}{6}$, as $\frac{3}{4}$ was split up into $\frac{1}{2}+\frac{1}{4}$. The orthodox sign for a submultiple was the letter for the corresponding number (the denominator) but with an accent instead of a horizontal stroke above it; thus $\gamma' = \frac{1}{3}$, the full expression being γ' *μέρος* = *τρίτον μέρος*, a third part (γ' is in fact short for *τρίτος*, so that it is also used for the ordinal number 'third' as well as for the fraction $\frac{1}{3}$, and similarly with all other accented numeral signs); $\lambda\beta' = \frac{1}{32}$, $\rho\iota\beta' = \frac{1}{112}$, &c. There were special signs for $\frac{1}{2}$, namely L′ or C′,[1] and for $\frac{2}{3}$, namely ω'. When a number of submultiples are written one after the other, the sum of them is meant, and similarly when they follow a whole number; e.g. $L'\,\delta' = \frac{1}{2}\,\frac{1}{4}$ or $\frac{3}{4}$ (Archimedes); $\kappa\theta\ \omega'\ \iota\gamma'\ \lambda\theta' = 29\frac{2}{3}\,\frac{1}{13}\,\frac{1}{39} = 29\frac{2}{3}+\frac{1}{13}+\frac{1}{39}$ or $29\frac{10}{13}$;

$$\mu\theta\ L'\ \iota\zeta'\ \lambda\delta'\ \nu\alpha' = 49\tfrac{1}{2}\,\tfrac{1}{17}\,\tfrac{1}{34}\,\tfrac{1}{51} = 49\tfrac{31}{51}$$

(Heron, *Geom.* 15. 8, 13). But $\iota\gamma'$ *τὸ* $\iota\gamma'$ means $\frac{1}{13}$th times $\frac{1}{13}$ or $\frac{1}{169}$ (*ibid.* 12. 5), &c. A less orthodox method found in later manuscripts was to use two accents and to write, e.g., ζ'' instead of ζ', for $\frac{1}{7}$. In Diophantus we find a different mark in place of the accent; Tannery considers the genuine form of it to be ×, so that $\gamma^{\times} = \frac{1}{3}$, and so on.

(β) *The ordinary Greek form, variously written.*

An ordinary proper fraction (called by Euclid *μέρη*, *parts*, in the plural; as meaning a certain number of aliquot parts, in contradistinction to *μέρος*, *part*, in the singular, which he restricts to an aliquot part or submultiple) was expressed in various ways. The first was to use the ordinary cardinal number for the numerator followed by the accented number representing the denominator. Thus we find in Archimedes $\bar{\iota}\ o\alpha' = \frac{10}{71}$ and $\overline{{}_{,}\alpha\omega\lambda\eta}\ \bar{\theta}\ \iota\alpha' = 1838\frac{9}{11}$: (it should be noted, however, that the $\bar{\iota}\ o\alpha'$ is a correction from $o\iota\alpha$, and this seems to indicate that the original reading was $\overset{o\alpha}{\iota}$, which would accord with Diophantus's and Heron's method of writing fractions). The method illustrated by these cases is open to objection as likely to lead to confusion, since $\iota\ o\alpha'$

[1] It has been suggested that the forms C and Ↄ for $\frac{1}{2}$ found in inscriptions may perhaps represent half an O, the sign, at all events in Boeotia, for 1 obol.

would naturally mean $10\frac{1}{71}$ and $\theta\ \iota\alpha'$ $9\frac{1}{11}$; the context alone shows the true meaning. Another form akin to that just mentioned was a little less open to misconstruction; the numerator was written in full with the accented numeral (for the denominator) following, e.g. δύο με′ for 2/45ths (Aristarchus of Samos). A better way was to turn the aliquot part into an abbreviation for the ordinal number with a termination superposed to represent the *case*, e.g. $\delta^{\omega\nu}\ \varsigma = \frac{6}{4}$ (Dioph. Lemma to V. 8), $\nu\ \kappa\gamma^{\omega\nu} = \frac{50}{23}$ (*ibid.* I. 23), $\rho\kappa\alpha^{\omega\nu}$ ͵αωλδ𐅵′ $= 1834\frac{1}{2}/121$ (*ibid.* IV. 39), just as $\gamma^{o\varsigma}$ was written for the ordinal τρίτος (cf. τὸ $\varsigma^{o\nu}$, the $\frac{1}{6}$th part, Dioph. IV. 39; αἴρω τὰ $\iota\gamma^{\alpha}$ 'I remove the 13ths', i.e. I multiply up by the denominator 13, *ibid.* IV. 9). But the trouble was avoided by each of two other methods.

(1) The accented letters representing the denominator were written twice, along with the cardinal number for the numerator. This method is mostly found in the *Geometrica* and other works of Heron: cf. ε ιγ′ ιγ′ $= \frac{5}{13}$, τὰ ϛ ζ′ζ′ $= \frac{6}{7}$. The fractional signification is often emphasized by adding the word λεπτά ('fractions' or 'fractional parts'), e.g. in λεπτὰ ιγ′ ιγ′ ιβ $= \frac{12}{13}$ (*Geom.* 12. 5), and, where the expression contains units as well as fractions, the word 'units' (μονάδες) is generally added, for clearness' sake, to indicate the integral number, e.g. μονάδες ιβ καὶ λεπτὰ ιγ′ ιγ′ ιβ $= 12\frac{12}{13}$ (*Geom.* 12. 5), μονάδες ρμδ λεπτὰ ιγ′ ιγ′ σϙθ $= 144\frac{299}{13}$ (*Geom.* 12. 6). Sometimes in Heron fractions are alternatively given in this notation and in that of submultiples, e.g. β γ′ ιε′ ἤτοι β καὶ β ε′ ε′ = '$2\frac{1}{3}\ \frac{1}{15}$ or $2\frac{2}{5}$' (*Geom.* 12. 48); ζ 𐅵′ ι′ ιε′ οε′ ἤτοι μονάδες ζ ε′ ε′ γ καὶ β ε′ ε′ τῶν ε′ ε′ = '$7\frac{1}{2}\ \frac{1}{10}\ \frac{1}{15}\ \frac{1}{75}$ or $7\frac{3}{5} + \frac{2}{5} \times \frac{1}{5}$', i.e. $7\frac{3}{5} + \frac{2}{25}$ (*ibid.*); η 𐅵′ ι′ κε′ ἤτοι μονάδες η ε′ ε′ γ καὶ ε′ τὸ ε′ = '$8\frac{1}{2}\ \frac{1}{10}\ \frac{1}{25}$ or $8\frac{3}{5} + \frac{1}{5} \times \frac{1}{5}$', i.e. $8\frac{3}{5} + \frac{1}{25}$ (*ibid.* 12. 46). (In Hultsch's edition of Heron single accents were used to denote whole numbers and the numerators of fractions, while aliquot parts or denominators were represented by double accents; thus the last quoted expression was written η′ S ι″ κε″ ἤτοι μονάδες η′ ε″ ε″ γ′ καὶ ε″ τὸ ε″.)

But (2) the most convenient notation of all is that which is regularly employed by Diophantus, and occasionally in the *Metrica* of Heron. In this system the numerator of any fraction is written in the line, with the denominator *above* it, without accents or other marks (except where the numerator or denominator itself contains an accented fraction); the method is therefore simply the reverse of ours, but equally convenient. In Tannery's edition of Diophantus a line is put between the numerator below and the denominator above: thus $\frac{\iota\varsigma}{\rho\kappa\alpha} = \frac{121}{16}$. But it is better to omit the horizontal line (cf. $\overset{\rho\kappa\eta}{\rho} = \frac{100}{128}$ in Kenyon's Papyri ii, No. cclxv. 40, and the fractions in Schöne's edition of Heron's *Metrica*). A few more instances from Diophantus may be given: $\overset{\phi\iota\beta}{,\beta\upsilon\nu\varsigma} = \frac{2456}{512}$ (IV. 28); $\overset{\alpha\,.\,\sigma\alpha}{,\epsilon\tau\nu\eta} = \frac{5358}{10201}$ (V. 9); $\overset{\rho\nu\beta}{\tau\pi\theta\angle'} = \frac{389\frac{1}{2}}{152}$. The denominator is rarely found above the numerator, but to the right (like an exponent); e.g. $\overline{\iota\epsilon}^{\delta} = \frac{15}{4}$ (I. 39). Even in the case of a submultiple, where, as we have said, the orthodox method was to omit the numerator and simply write the denominator with an accent, Diophantus often follows the method applicable to other fractions, e.g. he writes $\overset{\phi\iota\beta}{\alpha}$ for $\frac{1}{512}$ (IV. 28). Numbers partly integral and partly fractional, where the fraction is a submultiple or expressed as the sum of submultiples, are written much as we write them, the fractions simply following the integer, e.g. $\alpha\ \gamma^{\chi} = 1\frac{1}{3}$; $\beta\ \angle'\ \varsigma^{\chi} = 2\frac{1}{2}\ \frac{1}{6}$ (Lemma to V. 8); $\tau o\ \angle'\ \iota\varsigma^{\chi} = 370\frac{1}{2}\ \frac{1}{16}$ (III. 11). Complicated fractions in which the numerator and denominator are algebraical expressions or large numbers are often expressed by writing the numerator first and separating it by *μορίου* or *ἐν μορίῳ* from the denominator; i.e. the fraction is expressed as the numerator *divided by* the denominator: thus $\overset{\Upsilon}{M}\rho\nu\,.\,,\zeta\,\text{ϡ}\pi\delta$ *μορίου* $\kappa\varsigma\,.\,,\beta\rho\mu\delta = 1507984/262144$ (IV. 28).

(γ) *Sexagesimal fractions.*

Great interest attaches to the system of sexagesimal fractions (Babylonian in its origin, as we have seen) which was used by the Greeks in astronomical calculations, and appears fully developed in the *Syntaxis* of Ptolemy. The circumference of a circle, and with it the four right angles subtended by it at the centre, were divided into 360 parts (τμήματα or μοῖραι), as we should say *degrees*, each μοῖρα into 60 parts called (πρῶτα) ἑξηκοστά, (*first*) *sixtieths* or *minutes* (λεπτά), each of these again into 60 δεύτερα ἑξηκοστά, *seconds*, and so on. In like manner, the diameter of the circle was divided into 120 τμήματα, *segments*, and each of these segments was divided into sixtieths, each sixtieth again into sixty parts, and so on. Thus a convenient fractional system was available for arithmetical calculations in general; for the unit could be chosen at will, and any mixed number could be expressed as so many of those units *plus* so many of the fractions which we should represent by $\frac{1}{60}$, so many of those which we should write $(\frac{1}{60})^2$, $(\frac{1}{60})^3$, and so on to any extent. The units, τμήματα or μοῖραι (the latter often denoted by the abbreviation μ°), were written first, with the ordinary numeral representing the number of them; then came a simple numeral with one accent representing that number of *first sixtieths*, or minutes, then a numeral with two accents representing that number of *second sixtieths*, or seconds, and so on. Thus $\mu^{\circ}\ \beta = 2^{\circ}$, μοιρῶν $\mu\zeta\ \mu\beta'\ \mu'' = 47^{\circ}\ 42'\ 40''$. Similarly, τμημάτων $\xi\zeta\ \delta'\ \nu\epsilon'' = 67^{p}\ 4'\ 55''$, where p denotes the *segment* (of the diameter). Where there was no unit, or no number of sixtieths, second sixtieths, &c., the symbol O, signifying οὐδεμία μοῖρα, οὐδὲν ἑξηκοστόν, and the like, was used; thus μοιρῶν $\mathrm{O}\ \alpha'\ \beta''\ \mathrm{O}''' = 0^{\circ} 1'\ 2''\ 0'''$. The system is parallel to our system of decimal fractions, with the difference that the submultiple is $\frac{1}{60}$ instead of $\frac{1}{10}$; nor is it much less easy to work with, while it furnishes a very speedy way of approximating to the values of quantities not expressible in whole numbers. For example, in his Table of Chords, Ptolemy says that the chord subtending an angle of 120° at the centre is (τμημάτων) $\rho\gamma\ \nu\epsilon'\ \kappa\gamma''$ or $103^{p}\ 55'\ 23''$; this is equivalent (since the radius of the circle is 60 τμήματα) to saying that

$$\sqrt{3} = 1 + \frac{43}{60} + \frac{55}{60^2} + \frac{23}{60^3}, \text{ and this works out to } 1{\cdot}7320509\ldots,$$

which is correct to the seventh decimal place, and exceeds the true value by 0·00000003 only.

Practical calculation.

(*a*) *The abacus.*

In practical calculation it was open to the Greeks to secure the advantages of a position-value system by using the abacus. The essence of the abacus was the arrangement of it in columns which might be vertical or horizontal, but were generally vertical, and pretty certainly so in Greece and Egypt; the columns were marked off by lines or in some other way and allocated to the successive denominations of the numerical system in use, i.e., in the case of the decimal system, the units, tens, hundreds, thousands, myriads, and so on. The number of units of each denomination was shown in each column by means of pebbles, pegs, or the like. When, in the process of addition or multiplication, the number of pebbles collected in one column becomes sufficient to make one or more units of the next higher denomination, the number of pebbles representing the complete number of the higher units is withdrawn from the column in question and the proper number of the higher units added to the next higher column. Similarly, in subtraction, when a number of units of one denomination has to be subtracted and there are not enough pebbles in the particular column to subtract from, one pebble from the next higher column is withdrawn and actually or mentally resolved into the number of the lower units equivalent in value; the latter number of additional pebbles increases the number already in the column to a number from which the number to be subtracted can actually be withdrawn. The details of the columns of the Greek abacus have unfortunately to be inferred from the corresponding details of the Roman abacus, for the only abaci which have been preserved and can with certainty be identified as such are Roman. There were two kinds; in one of these the marks were buttons or knobs which could be moved up and down in each column, but could not be taken out of it, while in the other kind they were pebbles which could also be moved from one column to another. Each column was in two parts, a shorter portion at the top containing one button only, which itself represented half the number of units necessary to make up one of the next higher units, and a longer portion below containing one less than half the same number. This arrangement of the columns in two parts enabled the total number of buttons to be economized. The columns represented, so far as integral numbers were concerned, units, tens, hundreds, thousands, &c., and in these cases the one button in the top portion of each column represented five units, and there were four buttons in the lower portion representing four units. But after the columns representing integers came columns representing fractions; the first contained buttons representing *unciae,* of which there were 12 to the unit, i.e. fractions of $\frac{1}{12}$th, and in this case the one button in the top portion represented 6 *unciae* or $\frac{6}{12}$ths, while there were 5 buttons in the lower portion (instead of 4), the buttons in the column thus representing in all 11 *unciae* or 12ths. After this column there were (in one specimen) three other shorter ones alongside the lower portions only of the columns for integers, the first representing fractions of $\frac{1}{24}$th (one button), the second fractions of $\frac{1}{48}$th (one button), and the third fractions of $\frac{1}{72}$nd (two buttons, which of course together made up $\frac{1}{36}$th).

The mediaeval writer of the so-called geometry of Boëtius describes another method of indicating in the various columns the number of units of each denomination.[1] According to him 'abacus' was a later name for what was previously called *mensa Pythagorea*, in honour of the Master who had taught its use. The method was to put in the columns, not the necessary number of pebbles or buttons, but the corresponding *numeral*, which might be written in sand spread over the surface (in the same way as Greek geometers are said to have drawn geometrical figures in sand strewn on boards similarly called ἄβαξ or ἀβάκιον). The figures put in the columns were called *apices*. The first variety of numerals mentioned by the writer are rough forms of the Indian figures (a fact which proves the late date of the composition); but other forms were (1) the first letters of the alphabet (which presumably mean the Greek alphabetic numerals) or (2) the ordinary Roman figures.

We should expect the arrangement of the Greek abacus to correspond to the Roman, but the actual evidence regarding its form and the extent to which it was used is so scanty that

[1] Boëtius, *De Inst. Ar.*, ed. Friedlein, pp. 396 sq.

we may well doubt whether any great use was made of it at all. But the use of pebbles to reckon with is attested by several writers. In Aristophanes (*Wasps*, 656–64) Bdelycleon tells his father to do an easy sum 'not with pebbles but with fingers', as much as to say, 'There is no need to use pebbles for this sum; you can do it on your fingers.' 'The income of the state', he says, 'is 2000 talents; the yearly payment to the 6000 dicasts is only 150 talents.' 'Why', answers the old man, 'we don't get a tenth of the revenue.' The calculation in this case amounted to multiplying 150 by 10 to show that the product is less than 2000. But more to the purpose are the following allusions. Herodotus says that, in reckoning with pebbles, as in writing, the Greeks move their hand from left to right, the Egyptians from right to left[1]; this indicates that the columns were vertical, facing the reckoner. Diogenes Laertius attributes to Solon a statement that those who had influence with tyrants were like the pebbles on a reckoning-board, because they sometimes stood for more and sometimes for less.[2] A character in a fourth-century comedy asks for an abacus and pebbles to do his accounts.[3] But most definite of all is a remark of Polybius that 'These men are really like the pebbles on reckoning-boards. For the latter, according to the pleasure of the reckoner, have the value, now of a χαλκοῦς ($\frac{1}{8}$th of an obol or $\frac{1}{48}$th of a drachma), and the next moment of a talent.'[4] The passages of Diogenes Laertius and Polybius both indicate that the pebbles were not fixed in the columns, but could be transferred from one to another, and the latter passage has some significance in relation to the Salaminian table presently to be mentioned, because the talent and the χαλκοῦς are actually the extreme denominations on one side of the table.

Two relics other than the Salaminian table may throw some light on the subject. First, the so-called Darius-vase found at Canosa (Canusium), south-west of Barletta, represents a collector of tribute of distressful countenance with a table in front of him having pebbles, or (as some maintain) coins, upon it and, on the right-hand edge, beginning on the side farthest away and written in the direction towards him, the letters

[1] Herodotus, ii. c. 36.
[2] Diog. L. i. 59.
[3] Alexis in Athenaeus, 117 c.
[4] Polybius, v. 26. 13.

ΜΨΗ ⊳ΓΟ<Τ, while in his left hand he holds a sort of book in which, presumably, he has to enter the receipts. Now Μ, Ψ (= Χ), Η, and ⊳ are of course the initial letters of the words for 10000, 1000, 100, and 10 respectively. Here therefore we have a purely decimal system, without the halfway numbers represented by Γ (= πέντε, 5) in combination with the other initial letters which we find in the 'Attic' system. The sign Γ after ⊳ seems to be wrongly written for Ρ, the older sign for a drachma, Ο stands for the obol, < for the $\frac{1}{2}$-obol, and Τ (τεταρτημόριον) for the $\frac{1}{4}$-obol.[1] Except that the fractions of the unit (here the drachma) are different from the fractions of the Roman unit, this scheme corresponds to the Roman, and so far might represent the abacus. Indeed, the decimal arrangement corresponds better to the abacus than does the Salaminian table with its intermediate 'Herodianic' signs for 500, 50, and 5 drachmas. Prof. David Eugene Smith is, however, clear that any one can see from a critical examination of the piece that what is represented is an ordinary money-changer or tax-receiver with coins on a table such as one might see anywhere in the East to-day, and that the table has no resemblance to an abacus.[2] On the other hand, it is to be observed that the open book held by the tax-receiver in his left hand has ΤΑΛΝ on one page and ΤΛΊΗ on the other, which would seem to indicate that he was entering totals in *talents* and must therefore presumably have been *adding* coins or pebbles on the table before him.

There is a second existing monument of the same sort, namely a so-called σήκωμα (or arrangement of measures) discovered about forty years ago[3]; it is a stone tablet with fluid measures and has, on the right-hand side, the numerals ΧΓΗΗΓΔΓΗΤΙC. The signs are the 'Herodianic', and they include those for 500, 50, and 5 drachmas; Ͱ is the sign for a drachma, Τ evidently stands for some number of obols making a fraction of the drachma, i.e. the τριώβολον or 3 obols, Ι for an obol, and C for a $\frac{1}{2}$-obol.

The famous Salaminian table was discovered by Rangabé, who gave a drawing and description of it immediately after-

[1] Keil in *Hermes*, 29, 1894, pp. 262–3.
[2] *Bibliotheca Mathematica*, ix_3, p. 193.
[3] Dumont in *Revue archéologique*, xxvi (1873), p. 43.

wards (1846).[1] The table, now broken into two unequal parts, is in the Epigraphical Museum at Athens. The facts with regard to it are stated, and a photograph of it is satisfactorily produced, by Wilhelm Kubitschek.[2] A representation of it is also given by Nagl [3] based on Rangabé's description, and the sketch of it here appended follows Nagl's drawing. The size and material of the table (according to Rangabé's measurements it is 1·5 metres long and 0·75 metre broad) show that it was no ordinary abacus; it may have been a fixture intended for quasi-public use, such as a banker's or money-changer's table, or again it may have been a scoring-table for some kind of game like *tric-trac* or backgammon. Opinion has from the first been divided between the two views; it has even been suggested that the table was intended for both purposes. But there can be no doubt that it was used for some kind of calculation and, if it was not actually an abacus, it may at least serve to give an idea of what the abacus was like. The difficulties connected with its interpretation are easily seen. The series of letters on the three sides are the same except that two of them go no higher than Χ (1000 drachmae), but the third has 𐅆 (5000 drachmae), and Τ (the talent or 6000 drachmae) in addition; ⊢ is the sign for a drachma, Ι for an obol ($\frac{1}{6}$th of the drachma), C for $\frac{1}{2}$-obol, Τ for $\frac{1}{4}$-obol (τεταρτημόριον, Boeckh's suggestion), not $\frac{1}{3}$-obol (τριτημόριον, Vincent), and Χ for $\frac{1}{8}$-obol (χαλκοῦς). It seems to be agreed that the four spaces provided between the five shorter lines were intended for the fractions of the drachma; the first space would require 5 pebbles (one less than the 6 obols making up a drachma), the others one each. The longer

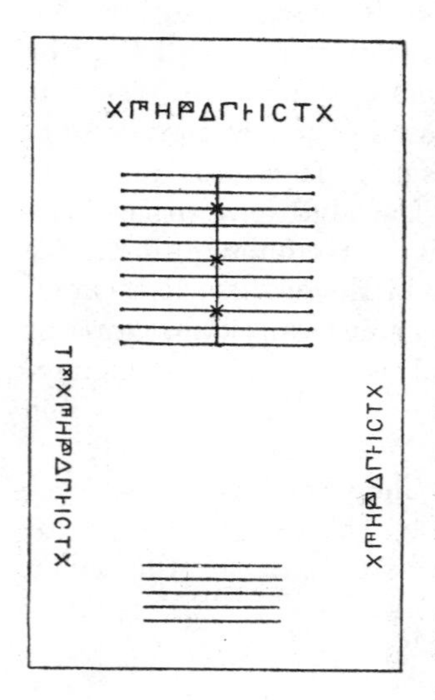

[1] *Revue archéologique*, iii. 1846.

[2] *Wiener numismatische Zeitschrift*, xxxi. 1899, pp. 393–8, with Plate xxiv.

[3] *Abh. zur Gesch. d. Math.* ix. 1899, plate after p. 357.

lines would provide the spaces for the drachmae and higher denominations. On the assumption that the cross line indicates the Roman method of having one pebble above it to represent 5, and four below it representing units, it is clear that, including denominations up to the talent (6000 drachmae), only five columns are necessary, namely one for the talent or 6000 drachmae, and four for 1000, 100, 10 drachmae, and 1 drachma respectively. But there are actually ten spaces provided by the eleven lines. On the theory of the game-board, five of the ten on one side (right or left) are supposed to belong to each of two players placed facing each other on the two longer sides of the table (but, if in playing they had to use the shorter columns for the fractions, it is not clear how they would make them suffice); the cross on the middle of the middle line might in that case serve to mark the separation between the lines belonging to the two players, or perhaps all the crosses may have the one object of helping the eye to distinguish all the columns from one another. On the assumption that the table is an abacus, a possible explanation of the *eleven* lines is to suppose that they really supply *five* columns only, the odd lines marking the divisions between the columns, and the even lines, one in the middle of each column, marking where the pebbles should be placed in rows; in this case, if the crosses are intended to mark divisions between the four pebbles representing units and the one pebble representing 5 in each column, the crosses are only required in the last three columns (for 100, 10, and 1), because, the highest denomination being 6000 drachmae, there was no need for a division of the 1000-column, which only required five unit-pebbles altogether. Nagl, a thorough-going supporter of the abacus-theory to the exclusion of the other, goes further and shows how the Salaminian table could have been used for the special purpose of carrying out a long multiplication; but this development seems far-fetched, and there is no evidence of such a use.

The Greeks in fact had little need of the abacus for calculations. With their alphabetic numerals they could work out their additions, subtractions, multiplications, and divisions without the help of any marked columns, in a form little less convenient than ours: examples of long multiplications, which include addition as the last step in each case, are found in Eutocius's commentary on Archimedes's *Measurement of a Circle.* We will take the four arithmetical operations separately.

(β) *Addition and Subtraction.*

There is no doubt that, in writing down numbers for the purpose of these operations, the Greeks would keep the several powers of 10 separate in a manner practically corresponding to our system of numerals, the hundreds, thousands, &c., being written in separate vertical rows. The following would be a typical example of a sum in addition:

$$\begin{array}{lcr}
\phantom{\overset{\alpha}{\mathrm{M}}}\,{}_{\prime}\alpha\;\upsilon\;\kappa\;\delta & = & 1424 \\
\phantom{\overset{\alpha}{\mathrm{M}}\,{}_{\prime}\alpha}\;\rho\;\;\;\gamma & & 103 \\
\overset{\alpha}{\mathrm{M}}\,{}_{\prime}\beta\sigma\pi\alpha & & 12281 \\
\overset{\gamma}{\mathrm{M}}\qquad\lambda & & 30030 \\
\hline
\overset{\delta}{\mathrm{M}}\,{}_{\prime}\gamma\omega\lambda\eta & & 43838
\end{array}$$

and the mental part of the work would be the same for the Greek as for us.

Similarly a subtraction would be represented as follows:

$$\begin{array}{lcr}
\overset{\theta}{\mathrm{M}}\,{}_{\prime}\gamma\chi\lambda\varsigma & = & 93636 \\
\overset{\beta}{\mathrm{M}}\,{}_{\prime}\gamma\upsilon\;\;\theta & & 23409 \\
\hline
\overset{\zeta}{\mathrm{M}}\quad\sigma\kappa\zeta & & 70227
\end{array}$$

(γ) *Multiplication.*

(i) The Egyptian method.

For carrying out multiplications two things were required. The first was a multiplication table. This the Greeks are certain to have had from very early times. The Egyptians, indeed, seem never to have had such a table. We know from the Papyrus Rhind that in order to multiply by any number the Egyptians began by successive doubling, thus obtaining twice, four times, eight times, sixteen times the multiplicand, and so on; they then added such sums of this series of multiples (including once the multiplicand) as were required. Thus, to multiply by 13, they did not take 10 times and 3 times the multiplicand respectively and add them, but they found 13 times the multiplicand by adding once and 4 times and 8 times it, which elements they had obtained by the doubling process; similarly they would find 25 times any number by adding once and 8 times and 16 times the number.[1] Division was performed by the Egyptians in an even more rudimentary fashion, namely by a tentative back-multiplication beginning with the same doubling process. But, as we have seen (p. 14), the scholiast to the *Charmides* says that the branches of λογιστική include the 'so-called Greek and Egyptian methods in multiplications and divisions'.

(ii) The Greek method.

The Egyptian method being what we have just described, it seems clear that the Greek method, which was different, depended on the direct use of a multiplication table. A fragment of such a multiplication table is preserved on a two-leaved wax tablet in the British Museum (Add. MS. 34186).

[1] I have been told that there is a method in use to-day (some say in Russia, but I have not been able to verify this), which is certainly attractive and looks original, but which will immediately be seen to amount simply to an elegant practical method of carrying out the Egyptian procedure. Write out side by side in successive lines, so as to form two columns, (1) the multiplier and multiplicand, (2) half the multiplier (or the nearest integer below it if the multiplier is odd) and twice the multiplicand, (3) half (or the nearest integer below the half) of the number in the first column of the preceding row and twice the number in the second column of the preceding row, and so on, until we have 1 in the first column. Then strike out all numbers in the second column which are opposite *even* numbers in the first column, and add all the numbers left in the second column. The sum will be the required product. Suppose e.g. that 157 is to be multiplied by 83. The rows and columns then are:

83	157
41	314
20	~~628~~
10	~~1256~~
5	2512
2	~~5024~~
1	10048
	13031 = 83 × 157

The explanation is, of course, that, where we take half the preceding number in the first column *less one*, we omit once the figure in the right-hand column, so that it must be left in that column to be added in at the end; and where we take the exact half of an even number, we omit nothing in the right-hand column, but the new line is the *exact* equivalent of the preceding one, which can therefore be struck out.

It is believed to date from the second century A.D., and it probably came from Alexandria or the vicinity. But the form of the characters and the mingling of capitals and small letters both allow of an earlier date; e.g. there is in the Museum a Greek papyrus assigned to the third century B.C. in which the numerals are very similar to those on the tablet.[1]

The second requirement is connected with the fact that the Greeks began their multiplications by taking the product of the highest constituents first, i.e. they proceeded as we should if we were to begin our long multiplications from the left instead of the right. The only difficulty would be to settle the denomination of the products of two high powers of ten. With such numbers as the Greeks usually had to multiply there would be no trouble; but if, say, the factors were unusually large numbers, e.g. millions multiplied by millions or billions, care would be required, and even some rule for settling the denomination, or determining the particular power or powers of 10 which the product would contain. This exceptional necessity was dealt with in the two special treatises, by Archimedes and Apollonius respectively, already mentioned. The former, the *Sand-reckoner*, proves that, if there be a series of numbers, $1, 10, 10^2, 10^3 \dots 10^m \dots 10^n \dots$, then, if 10^m, 10^n be any two terms of the series, their product $10^m \cdot 10^n$ will be a term in the same series and will be as many terms distant from 10^n as the term 10^m is distant from 1; also it will be distant from 1 by a number of terms less by one than the sum of the numbers of terms by which 10^m and 10^n respectively are distant from 1. This is easily seen to be equivalent to the fact that, 10^m being the $(m+1)$th term beginning with 1, and 10^n the $(n+1)$th term beginning with 1, the product of the two terms is the $(m+n+1)$th term beginning with 1, and is 10^{m+n}.

(iii) Apollonius's continued multiplications.

The system of Apollonius deserves a short description.[2] Its object is to give a handy method of finding the continued product of any number of factors, each of which is represented by a single letter in the Greek numeral notation. It does not

[1] David Eugene Smith in *Bibliotheca Mathematica*, ix_3, pp. 193–5.

[2] Our authority here is the *Synagoge* of Pappus, Book ii, pp. 2–28, Hultsch.

therefore show how to multiply two large numbers each of which contains a number of digits (in our notation), that is, a certain number of units, a certain number of tens, a certain number of hundreds, &c.; it is confined to the multiplication of any number of factors each of which is one or other of the following: (*a*) a number of units as 1, 2, 3, ... 9, (*b*) a number of even tens as 10, 20, 30, ... 90, (*c*) a number of even hundreds as 100, 200, 300, ... 900. It does not deal with factors above hundreds, e.g. 1000 or 4000; this is because the Greek numeral alphabet only went up to 900, the notation beginning again after that with ͵α, ͵β, ... for 1000, 2000, &c. The essence of the method is the separate multiplication (1) of the *bases*, πυθμένες, of the several factors, (2) of the powers of ten contained in the factors, that is, what we represent by the ciphers in each factor. Given a multiple of ten, say 30, 3 is the πυθμήν or base, being the same number of units as the number contains tens; similarly in a multiple of 100, say 800, 8 is the base. In multiplying three numbers such as 2, 30, 800, therefore, Apollonius first multiplies the bases, 2, 3, and 8, then finds separately the product of the ten and the hundred, and lastly multiplies the two products. The final product has to be expressed as a certain number of units less than a myriad, then a certain number of myriads, a certain number of 'double myriads' (myriads squared), 'triple myriads' (myriads cubed), &c., in other words in the form

$$A_0 + A_1 M + A_2 M^2 + \dots ,$$

where M is a myriad or 10^4 and A_0, A_1... respectively represent some number not exceeding 9999.

No special directions are given for carrying out the multiplication of the *bases* (digits), or for the multiplication of their product into the product of the tens, hundreds, &c., when separately found (directions for the latter multiplication may have been contained in propositions missing from the mutilated fragment in Pappus). But the method of dealing with the tens and hundreds (the ciphers in our notation) is made the subject of a considerable number of separate propositions. Thus in two propositions the factors are all of one sort (tens or hundreds), in another we have factors of two sorts (a number of factors containing units only multiplied by a number of multiples of ten, each less than 100, or by multiples of 100, each less than 1000), and so on. In the final proposition (25), with which the introductory lemmas close, the factors are of all three kinds, some containing units only, others being multiples of 10 (less than 100) and a third set being multiples of 100 (less than 1000 in each case). As Pappus frequently says, the proof is easy 'in numbers'; Apollonius himself seems to have proved the propositions by means of lines or a diagram in some form. The method is the equivalent of taking the indices of all the separate powers of ten included in the factors (in which process ten $=10^1$ counts as 1, and $100=10^2$ as 2), adding the indices together, and then dividing the sum by 4 to obtain the power of the myriad (10000) which the product contains. If the whole number in the quotient is n, the product contains $(10000)^n$ or the n-myriad in Apollonius's notation. There will in most cases be a remainder left after division by 4, namely 3, 2, or 1: the remainder then represents (in our notation) 3, 2, or 1 more ciphers, that is, the product is 1000, 100, or 10 times the n-myriad, or the 10000^n, as the case may be.

We cannot do better than illustrate by the main problem which Apollonius sets himself, namely that of multiplying together all the numbers represented by the separate letters in the hexameter:

Ἀρτέμιδος κλεῖτε κράτος ἔξοχον ἐννέα κοῦραι.

The number of letters, and therefore of factors, is 38, of which 10 are multiples of 100 less than 1000, namely ρ, τ, σ, τ, ρ, τ, σ, χ, υ, ρ (=100, 300, 200, 300, 100, 300, 200, 600, 400, 100), 17 are multiples of 10 less than 100, namely μ, ι, o, κ, λ, ι, κ, o, ξ, o, o, ν, ν, ν, κ, o, ι (=40, 10, 70, 20, 30, 10, 20, 70, 60, 70, 70, 50, 50, 50, 20, 70, 10), and 11 are numbers of units not exceeding 9, namely α, ϵ, δ, ϵ, ϵ, α, ϵ, ϵ, ϵ, α, α (=1, 5, 4, 5, 5, 1, 5, 5, 5, 1, 1). The sum of the indices of powers of ten contained in the factors is therefore $10 \cdot 2 + 17 \cdot 1 = 37$. This, when divided by 4, gives 9 with 1 as remainder. Hence the product of all the tens and hundreds, excluding the *bases* in each, is $10 \cdot 10000^9$.

We have now, as the second part of the operation, to multiply the numbers containing units only by the *bases* of all the other factors, i.e. (beginning with the *bases*, first of the hundreds, then of the tens) to multiply together the numbers:

1, 3, 2, 3, 1, 3, 2, 6, 4, 1,

4, 1, 7, 2, 3, 1, 2, 7, 6, 7, 7, 5, 5, 5, 2, 7, 1,

and 1, 5, 4, 5, 5, 1, 5, 5, 5, 1, 1.

The product is at once given in the text as 19 'quadruple myriads', 6036 'triple myriads', and 8480 'double myriads', or

$$19 . 10000^4 + 6036 . 10000^3 + 8480 . 10000^2.$$

(The detailed multiplication line by line, which is of course perfectly easy, is bracketed by Hultsch as interpolated.)

Lastly, says Pappus, this product multiplied by the other (the product of the tens and hundreds without the *bases*), namely $10 . 10000^9$, as above, gives

$$196 . 10000^{13} + 368 . 10000^{12} + 4800 . 10000^{11}.$$

(iv) Examples of ordinary multiplications.

I shall now illustrate, by examples taken from Eutocius, the Greek method of performing long multiplications. It will be seen that, as in the case of addition and subtraction, the working is essentially the same as ours. The multiplicand is written first, and below it is placed the multiplier preceded by ἐπί (= 'by' or 'into'). Then the term containing the highest power of 10 in the multiplier is taken and multiplied into all the terms in the multiplicand, one after the other, first into that containing the highest power of 10, then into that containing the next highest power of 10, and so on in descending order; after which the term containing the next highest power of 10 in the multiplier is multiplied into all the terms of the multiplicand in the same order; and so on. The same procedure is followed where either or both of the numbers to be multiplied contain fractions. Two examples from Eutocius will make the whole operation clear.

(1)						
,ατνα	1351					
ἐπὶ ,ατνα	× 1351					
$\overset{\rho}{M}\overset{\lambda}{M}\overset{\epsilon}{M}$,α	1000000	300000	50000	1000		
$\overset{\lambda}{M}\overset{\theta}{M}\overset{\alpha}{M}$,ετ	300000	90000	15000	300		
$\overset{\epsilon}{M}\overset{\alpha}{M}$,ε,βφν		50000	15000	2500	50	
,ατνα			1000	300	50	1
ὁμοῦ $\overset{\rho\pi\beta}{M}$,εσα	*together* 1825201.					

(2)	$,\gamma\iota\gamma\,∠'\delta'$		$3013\frac{1}{2}\,\frac{1}{4}$ [$=3013\frac{3}{4}$]				
$\dot{\epsilon}\pi\grave{\iota}$	$,\gamma\iota\gamma\,∠'\delta'$	×	$3013\frac{1}{2}\,\frac{1}{4}$				
	$\overset{ϡ}{\mathrm{M}}\overset{\gamma}{\mathrm{M}},\theta,\alpha\phi\psi\nu$		9000000	30000	9000	1500	750
	$\overset{\gamma}{\mathrm{M}}\rho\lambda\epsilon\beta∠'$		30000	100	30	5	$2\frac{1}{2}$
	$,\theta\lambda\theta\alpha∠'∠'\delta'$		9000	30	9	$1\frac{1}{2}$	$\frac{1}{2}\,\frac{1}{4}$
	$,\alpha\phi\epsilon\alpha∠'\delta'\eta'$		1500	5	$1\frac{1}{2}$	$\frac{1}{4}$	$\frac{1}{8}$
	$\psi\nu\beta∠'∠'\delta'\eta'\,\iota\varsigma'$		750	$2\frac{1}{2}$	$\frac{1}{2}\,\frac{1}{4}$	$\frac{1}{8}$	$\frac{1}{16}$
$\acute{o}\mu o\hat{\upsilon}$	$\overset{ϡ\eta}{\mathrm{M}},\beta\chi\pi\theta\,\iota\varsigma'$		*together* $9082689\frac{1}{16}$.				

The following is one among many instances in which Heron works out a multiplication of two numbers involving fractions. He has to multiply $4\frac{33}{64}$ by $7\frac{62}{64}$, which he effects as follows (*Geom.* 12. 68):

$$4.7 = 28,$$
$$4.\frac{62}{64} = \frac{248}{64},$$
$$\frac{33}{64}.7 = \frac{231}{64}$$
$$\frac{33}{64}\cdot\frac{62}{64} = \frac{2046}{64}\cdot\frac{1}{64} = \frac{31}{64} + \frac{62}{64}\cdot\frac{1}{64};$$

the result is therefore

$$28\,\frac{510}{64} + \frac{62}{64}\cdot\frac{1}{64} = 28 + 7\,\frac{62}{64} + \frac{62}{64}\cdot\frac{1}{64}$$
$$= 35\,\frac{62}{64} + \frac{62}{64}\cdot\frac{1}{64}.$$

The multiplication of $37^\circ\ 4'\ 55''$ (in the sexagesimal system) by itself is performed by Theon of Alexandria in his commentary on Ptolemy's *Syntaxis* in an exactly similar manner.

(δ) *Division.*

The operation of division depends on those of multiplication and subtraction, and was performed by the Greeks, *mutatis mutandis*, in the same way as we perform it to-day. Suppose, for example, that the process in the first of the above multiplications had to be reversed and $\overset{\rho\pi\beta}{\mathrm{M}},\epsilon\sigma\alpha$ (1825201) had to be divided by $,\alpha\tau\nu\alpha$ (1351). The terms involving the successive powers of 10 would be mentally kept separate, as in addition and subtraction, and the first question would be, how many times does one thousand go into one million, allowing for the fact that the one thousand has 351 behind it, while the one million has 825 thousands behind it. The answer is one thousand or $,\alpha$, and this multiplied by the divisor $,\alpha\tau\nu\alpha$ gives $\overset{\rho\lambda\epsilon}{\mathrm{M}},\alpha$ which, subtracted from $\overset{\rho\pi\beta}{\mathrm{M}},\epsilon\sigma\alpha$, leaves $\overset{\mu\varsigma}{\mathrm{M}},\delta\sigma\alpha$. This remainder (=474201) has now to be divided by $,\alpha\tau\nu\alpha$ (1351), and it would be seen that the latter would go into the former τ (300) times, but not υ (400) times. Multiplying $,\alpha\tau\nu\alpha$ by τ, we obtain $\overset{\mu}{\text{M}},\epsilon\tau$ (405300), which, when subtracted from $\overset{\mu\zeta}{\text{M}},\delta\sigma\alpha$ (474201), leaves $\overset{\varsigma}{\text{M}},\eta\text{ϡ}\alpha$ (68901). This has again to be divided by $,\alpha\tau\nu\alpha$ and goes ν (50) times; multiplying $,\alpha\tau\nu\alpha$ by ν, we have $\overset{\varsigma}{\text{M}},\zeta\phi\nu$ (67550), which, subtracted from $\overset{\varsigma}{\text{M}},\eta\text{ϡ}\alpha$ (68901), leaves $,\alpha\tau\nu\alpha$ (1351). The last quotient is therefore α (1), and the whole quotient is $,\alpha\tau\nu\alpha$ (1351).

An actual case of long division where both dividend and divisor contain sexagesimal fractions is described by Theon. The problem is to divide 1515 20′ 15″ by 25 12′ 10″, and Theon's account of the process amounts to the following:

Divisor.		Dividend.			Quotient.
25 12′ 10″		1515 20′ 15″			First term 60
	25 . 60 =	1500			
	Remainder 15 =	900′			
	Sum	920′			
	12′ . 60 =	720′			
	Remainder	200′			
	10″ . 60 =	10′			
	Remainder	190′			Second term 7′
	25 . 7′ =	175′			
		15′ =	900″		
	Sum		915″		
	12′ . 7′ =		84″		
	Remainder		831″		
	10″ . 7′ =		1″ 10‴		
	Remainder		829″ 50‴		Third term 33‴
	25 . 33″ =		825″		
	Remainder		4″ 50‴ =	290‴	
	12′ . 33″ =			396‴	
			(*too great by*)	106″	

Thus the quotient is something less than 60 7′ 33″. It will be observed that the difference between this operation of

Theon's and that of dividing $\overset{\rho\pi\beta}{\mathrm{M}}\,{,}\epsilon\sigma\alpha$ by ${,}\alpha\tau\nu\alpha$ as above is that Theon makes *three* subtractions for one term of the quotient, whereas the remainder was arrived at in the other case after *one* subtraction. The result is that, though Theon's method is quite clear, it is longer, and moreover makes it less easy to foresee what will be the proper figure to try in the quotient, so that more time would probably be lost in making unsuccessful trials.

(ϵ) *Extraction of the square root.*

We are now in a position to see how the problem of extracting the square root of a number would be attacked. First, as in the case of division, the given whole number would be separated into terms containing respectively such and such a number of units and of the separate powers of 10. Thus there would be so many units, so many tens, so many hundreds, &c., and it would have to be borne in mind that the squares of numbers from 1 to 9 lie between 1 and 99, the squares of numbers from 10 to 90 between 100 and 9900, and so on. Then the first term of the square root would be some number of tens or hundreds or thousands, and so on, and would have to be found in much the same way as the first term of a quotient in a long division, by trial if necessary. If A is the number the square root of which is required, while a represents the first term or denomination of the square root, and x the next term or denomination to be found, it would be necessary to use the identity $(a+x)^2 = a^2+2ax+x^2$ and to find x so that $2ax+x^2$ might be somewhat less than the remainder $A-a^2$, i.e. we have to divide $A-a^2$ by $2a$, allowing for the fact that not only must $2ax$ (where x is the quotient) but also $(2a+x)x$ be less than $A-a^2$. Thus, by trial, the highest possible value of x satisfying the condition would be easily found. If that value were b, the further quantity $2ab+b^2$ would have to be subtracted from the first remainder $A-a^2$, and from the second remainder thus left a third term or denomination of the square root would have to be found in like manner; and so on. That this was the actual procedure followed is clear from a simple case given by Theon of Alexandria in his commentary on the *Syntaxis*. Here the square root of 144 is in question, and it is obtained by means of

Eucl. II. 4. The highest possible denomination (i.e. power of 10) in the square root is 10; 10^2 subtracted from 144 leaves 44, and this must contain, not only twice the product of 10 and the next term of the square root, but also the square of the next term itself. Now twice 1 . 10 itself produces 20, and the division of 44 by 20 suggests 2 as the next term of the square root; this turns out to be the exact figure required, since $2 \cdot 20 + 2^2 = 44$.

The same procedure is illustrated by Theon's explanation of Ptolemy's method of extracting square roots according to the sexagesimal system of fractions. The problem is to find approximately the square root of 4500 μοῖραι or *degrees*, and

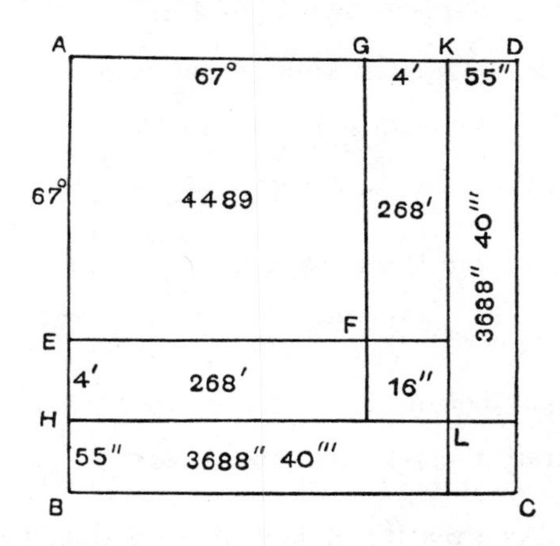

a geometrical figure is used which proves beyond doubt the essentially Euclidean basis of the whole method. The following arithmetical representation of the purport of the passage, when looked at in the light of the figure, will make the matter clear. Ptolemy has first found the integral part of $\sqrt{(4500)}$ to be 67. Now $67^2 = 4489$, so that the remainder is 11. Suppose now that the rest of the square root is expressed by means of sexagesimal fractions, and that we may therefore write

$$\sqrt{(4500)} = 67 + \frac{x}{60} + \frac{y}{60^2},$$

where x, y are yet to be found. Thus x must be such that $2 \cdot 67x/60$ is somewhat less than 11, or x must be somewhat less than $\frac{11.60}{2.67}$ or $\frac{330}{67}$, which is at the same time greater than 4. On trial it turns out that 4 will satisfy the conditions of the problem, namely that $\left(67+\frac{4}{60}\right)^2$ must be less than 4500, so that a remainder will be left by means of which y can be found.

Now this remainder is $11-\frac{2.67.4}{60}-\left(\frac{4}{60}\right)^2$, and this is equal to $\frac{11.60^2-2.67.4.60-16}{60^2}$ or $\frac{7424}{60^2}$.

Thus we must suppose that $2\left(67+\frac{4}{60}\right)\frac{y}{60^2}$ approximates to $\frac{7424}{60^2}$, or that $8048y$ is approximately equal to 7424.60. Therefore y is approximately equal to 55.

We have then to subtract $2\left(67+\frac{4}{60}\right)\frac{55}{60^2}+\left(\frac{55}{60^2}\right)^2$, or $\frac{442640}{60^3}+\frac{3025}{60^4}$, from the remainder $\frac{7424}{60^2}$ above found.

The subtraction of $\frac{442640}{60^3}$ from $\frac{7424}{60^2}$ gives $\frac{2800}{60^3}$ or $\frac{46}{60^2}+\frac{40}{60^3}$; but Theon does not go further and subtract the remaining $\frac{3025}{60^4}$; he merely remarks that the square of $\frac{55}{60^2}$ approximates to $\frac{46}{60^2}+\frac{40}{60^3}$. As a matter of fact, if we deduct the $\frac{3025}{60^4}$ from $\frac{2800}{60^3}$, so as to obtain the correct remainder, it is found to be $\frac{164975}{60^4}$.

Theon's plan does not work conveniently, so far as the determination of the first fractional term (the *first-sixtieths*) is concerned, unless the integral term in the square root is large relatively to $\frac{x}{60}$; if this is not the case, the term $\left(\frac{x}{60}\right)^2$ is not comparatively negligible, and the tentative ascertainment of x is more difficult. Take the case of $\sqrt{3}$, the value of which, in Ptolemy's Table of Chords, is equal to $1+\frac{43}{60}+\frac{55}{60^2}+\frac{23}{60^3}$.

If we first found the unit 1 and then tried to find the next term by trial, it would probably involve a troublesome amount of trials. An alternative method in such a case was to multiply the number by 60^2, thus reducing it to second-sixtieths, and then, taking the square root, to ascertain the number of first-sixtieths in it. Now $3 \cdot 60^2 = 10800$, and, as $103^2 = 10609$, the first element in the square root of 3 is found in this way to be $\frac{103}{60}$ $\left(= 1 + \frac{43}{60}\right)$. That this was the method in such cases is indicated by the fact that, in the Table of Chords, each chord is expressed as a certain number of first-sixtieths, followed by the second-sixtieths, &c., $\sqrt{3}$ being expressed as $\frac{103}{60} + \frac{55}{60^2} + \frac{23}{60^3}$. The same thing is indicated by the scholiast to Eucl., Book X, who begins the operation of finding the square root of 31 10′ 36″ by reducing this to second-sixtieths; the number of second-sixtieths is 112236, which gives, as the number of first-sixtieths in the square root, 335, while $\frac{335}{60} = 5\ 35'$. The second-sixtieths in the square root can then be found in the same way as in Theon's example. Or, as the scholiast says, we can obtain the square root as far as the second-sixtieths by reducing the original number to fourth-sixtieths, and so on. This would no doubt be the way in which the approximate value 2 49′ 42″ 20‴ 10⁗ given by the scholiast for $\sqrt{8}$ was obtained, and similarly with other approximations of his, such as $\sqrt{2} = 1\ 24'\ 51''$ and $\sqrt{(27)} \doteqdot 5\ 11'\ 46''\ 50'''$ (the 50‴ should be 10‴).

(ϛ) *Extraction of the cube root*

Our method of extracting the cube root of a number depends upon the formula $(a+x)^3 = a^3 + 3a^2x + 3ax^2 + x^3$, just as the extraction of the square root depends on the formula $(a+x)^2 = a^2 + 2ax + x^2$. As we have seen, the Greek method of extracting the square root was to use the latter (Euclidean) formula just as we do; but in no extant Greek writer do we find any description of the operation of extracting the cube root. It is possible that the Greeks had not much occasion for extracting cube roots, or that a table of cubes would suffice for most of their purposes. But that they had some method is clear from a passage of Heron, where he gives $4\frac{9}{14}$ as an approximation to $\sqrt[3]{(100)}$, and shows how he obtains it.[1] Heron merely gives the working dogmatically, in concrete numbers, without explaining its theoretical basis, and we cannot be quite certain as to the precise formula underlying the operation. The best suggestion which has been made on the subject will be given in its proper place, the chapter on Heron.

[1] Heron, *Metrica*, iii. c. 20.

III

PYTHAGOREAN ARITHMETIC

There is very little early evidence regarding Pythagoras's own achievements, and what there is does not touch his mathematics. The earliest philosophers and historians who refer to him would not be interested in this part of his work. Heraclitus speaks of his wide knowledge, but with disparagement: 'much learning does not teach wisdom; otherwise it would have taught Hesiod and Pythagoras, and again Xenophanes and Hecataeus'.[1] Herodotus alludes to Pythagoras and the Pythagoreans several times; he calls Pythagoras 'the most able philosopher among the Greeks' (*Ἑλλήνων οὐ τῷ ἀσθενεστάτῳ σοφιστῇ Πυθαγόρῃ*).[2] In Empedocles he had an enthusiastic admirer: 'But there was among them a man of prodigious knowledge who acquired the profoundest wealth of understanding and was the greatest master of skilled arts of every kind; for, whenever he willed with his whole heart, he could with ease discern each and every truth in his ten—nay, twenty—men's lives.'[3]

Pythagoras himself left no written exposition of his doctrines, nor did any of his immediate successors, not even Hippasus, about whom the different stories ran (1) that he was expelled from the school because he published doctrines of Pythagoras, and (2) that he was drowned at sea for revealing the construction of the dodecahedron in the sphere and claiming it as his own, or (as others have it) for making known the discovery of the irrational or incommensurable. Nor is the absence of any written record of Pythagorean

[1] Diog. L. ix. 1 (Fr. 40 in *Vorsokratiker*, i³, p. 86. 1–3).
[2] Herodotus, iv. 95.
[3] Diog. L. viii. 54 and Porph. *V. Pyth.* 30 (Fr. 129 in *Vors.* i³, p. 272. 15–20).

doctrines down to the time of Philolaus to be attributed to a pledge of secrecy binding the school; at all events, it did not apply to their mathematics or their physics; the supposed secrecy may even have been invented to explain the absence of documents. The fact appears to be that oral communication was the tradition of the school, while their doctrine would in the main be too abstruse to be understood by the generality of people outside.

In these circumstances it is difficult to disentangle the portions of the Pythagorean philosophy which can safely be attributed to the founder of the school. Aristotle evidently felt this difficulty; it is clear that he knew nothing for certain of any ethical or physical doctrines going back to Pythagoras himself; and when he speaks of the Pythagorean system, he always refers it to 'the Pythagoreans', sometimes even to 'the so-called Pythagoreans'.

The earliest direct testimony to the eminence of Pythagoras in mathematical studies seems to be that of Aristotle, who in his separate book *On the Pythagoreans*, now lost, wrote that

> 'Pythagoras, the son of Mnesarchus, first worked at mathematics and arithmetic, and afterwards, at one time, condescended to the wonder-working practised by Pherecydes.'[1]

In the *Metaphysics* he speaks in similar terms of the Pythagoreans:

> 'In the time of these philosophers (Leucippus and Democritus) and before them the so-called Pythagoreans applied themselves to the study of mathematics, and were the first to advance that science; insomuch that, having been brought up in it, they thought that its principles must be the principles of all existing things.'[2]

It is certain that the Theory of Numbers originated in the school of Pythagoras; and, with regard to Pythagoras himself, we are told by Aristoxenus that he 'seems to have attached supreme importance to the study of arithmetic, which he advanced and took out of the region of commercial utility'.[3]

[1] Apollonius, *Hist. mirabil.* 6 (*Vors.* i³, p. 29. 5).
[2] Arist. *Metaph.* A. 5, 985 b 23.
[3] Stobaeus, *Ecl.* i. proem. 6 (*Vors.* i³, p. 346. 12).

Numbers and the universe.

We know that Thales (about 624–547 B.C.) and Anaximander (born probably in 611/10 B.C.) occupied themselves with astronomical phenomena, and, even before their time, the principal constellations had been distinguished. Pythagoras (about 572–497 B.C. or a little later) seems to have been the first Greek to discover that the planets have an independent movement of their own from west to east, i.e. in a direction contrary to the daily rotation of the fixed stars; or he may have learnt what he knew of the planets from the Babylonians. Now any one who was in the habit of intently studying the heavens would naturally observe that each constellation has two characteristics, the number of the stars which compose it and the geometrical figure which they form. Here, as a recent writer has remarked,[1] we find, if not the origin, a striking illustration of the Pythagorean doctrine. And, just as the constellations have a number characteristic of them respectively, so all known objects have a number; as the formula of Philolaus states, 'all things which can be known have number; for it is not possible that without number anything can either be conceived or known'.[2]

This formula, however, does not yet express all the content of the Pythagorean doctrine. Not only do all things possess numbers; but, in addition, all things *are* numbers; 'these thinkers', says Aristotle, 'seem to consider that number is the principle both as matter for things and as constituting their attributes and permanent states'.[3] True, Aristotle seems to regard the theory as originally based on the analogy between the properties of things and of numbers.

'They thought they found in numbers, more than in fire, earth, or water, many resemblances to things which are and become; thus such and such an attribute of numbers is justice, another is soul and mind, another is opportunity, and so on; and again they saw in numbers the attributes and ratios of the musical scales. Since, then, all other things seemed in their whole nature to be assimilated to numbers, while numbers seemed to be the first things in the whole of nature,

[1] L. Brunschvicg, *Les étapes de la philosophie mathématique*, 1912, p. 33.
[2] Stob. *Ecl.* i. 21, 7[b] (*Vors.* i[3], p. 310. 8–10).
[3] Aristotle, *Metaph.* A. 5, 986 a 16.

they supposed the elements of numbers to be the elements of all things, and the whole heaven to be a musical scale and a number.'[1]

This passage, with its assertion of 'resemblances' and 'assimilation', suggests numbers as affections, states, or relations rather than as substances, and the same is implied by the remark that existing things exist by virtue of their *imitation* of numbers.[2] But again we are told that the numbers are not separable from the things, but that existing things, even perceptible substances, are made up of numbers; that the substance of all things is number, that things are numbers, that numbers are made up from the unit, and that the whole heaven is numbers.[3] Still more definite is the statement that the Pythagoreans 'construct the whole heaven out of numbers, but not of *monadic* numbers, since they suppose the units to have magnitude', and that, 'as we have said before, the Pythagoreans assume the numbers to have magnitude'.[4] Aristotle points out certain obvious difficulties. On the one hand the Pythagoreans speak of 'this number of which the heaven is composed'; on the other hand they speak of 'attributes of numbers' and of numbers as 'the *causes* of the things which exist and take place in the heaven both from the beginning and now'. Again, according to them, abstractions and immaterial things are also numbers, and they place them in different regions; for example, in one region they place opinion and opportunity, and in another, a little higher up or lower down, such things as injustice, sifting, or mixing. Is it this same 'number in the heaven' which we must assume each of these things to be, or a number other than this number?[5]

May we not infer from these scattered remarks of Aristotle about the Pythagorean doctrine that 'the number in the heaven' is the number of the visible stars, made up of units which are material points? And may this not be the origin of the theory that all things are numbers, a theory which of course would be confirmed when the further

[1] *Metaph.* A. 5, 985 b 27–986 a 2. [2] *Ib.* A. 5, 987 b 11.
[3] *Ib.* N. 3, 1090 a 22–23; M. 7, 1080 b 17; A. 5, 987 a 19, 987 b 27, 986 a 20.
[4] *Ib.* M. 7, 1080 b 18, 32. [5] *Ib.* A. 8, 990 a 18–29.

capital discovery was made that musical harmonies depend on numerical ratios, the octave representing the ratio 2 : 1 in length of string, the fifth 3 : 2 and the fourth 4 : 3 ?

The use by the Pythagoreans of visible points to represent the units of a number of a particular form is illustrated by the remark of Aristotle that

'Eurytus settled what is the number of what object (e.g. this is the number of a man, that of a horse) and imitated the shapes of living things by pebbles *after the manner of those who bring numbers into the forms of triangle or square*'.[1]

They treated the unit, which is a point without position (στιγμὴ ἄθετος), as a point, and a point as a unit having position (μονὰς θέσιν ἔχουσα).[2]

Definitions of the unit and of number.

Aristotle observes that the One is reasonably regarded as not being itself a number, because a measure is not the things measured, but the measure or the One is the beginning (or principle) of number.[3] This doctrine may be of Pythagorean origin; Nicomachus has it[4]; Euclid implies it when he says that a unit is that by virtue of which each of existing things is called one, while a number is 'the multitude made up of units'[5]; and the statement was generally accepted. According to Iamblichus,[6] Thymaridas (an ancient Pythagorean, probably not later than Plato's time) defined a unit as 'limiting quantity' (περαίνουσα ποσότης) or, as we might say, 'limit of fewness', while some Pythagoreans called it 'the confine between number and parts', i.e. that which separates multiples and submultiples. Chrysippus (third century B.C.) called it 'multitude one' (πλῆθος ἕν), a definition objected to by Iamblichus as a contradiction in terms, but important as an attempt to bring 1 into the conception of number.

The first definition of number is attributed to Thales, who defined it as a collection of units (μονάδων σύστημα), 'follow-

[1] *Metaph.* N. 5, 1092 b 10.
[2] *Ib.* M. 8, 1084 b 25; *De an.* i. 4, 409 a 6; Proclus on Eucl. I, p. 95. 21.
[3] *Metaph.* N. 1, 1088 a 6.
[4] Nicom. *Introd. arithm.* ii. 6. 3, 7. 3.
[5] Eucl. VII, Defs. 1, 2.
[6] Iambl. *in Nicom. ar. introd.*, p. 11. 2–10.

ing the Egyptian view'.[1] The Pythagoreans 'made number out of one'[2]; some of them called it 'a progression of multitude beginning from a unit and a regression ending in it'.[3] (Stobaeus credits Moderatus, a Neo-Pythagorean of the time of Nero, with this definition.[4]) Eudoxus defined number as a 'determinate multitude' (πλῆθος ὡρισμένον).[5] Nicomachus has yet another definition, 'a flow of quantity made up of units'[6] (ποσότητος χύμα ἐκ μονάδων συγκείμενον). Aristotle gives a number of definitions equivalent to one or other of those just mentioned, 'limited multitude',[7] 'multitude (or 'combination') of units',[8] 'multitude of indivisibles',[9] 'several ones' (ἕνα πλείω),[10] 'multitude measurable by one',[11] 'multitude measured', and 'multitude of measures'[12] (the measure being the unit).

Classification of numbers.

The distinction between *odd* (περισσός) and *even* (ἄρτιος) doubtless goes back to Pythagoras. A Philolaus fragment says that 'number is of two special kinds, odd and even, with a third, even-odd, arising from a mixture of the two; and of each kind there are many forms'.[13] According to Nicomachus, the Pythagorean definitions of odd and even were these:

'An *even* number is that which admits of being divided, by one and the same operation, into the greatest and the least parts, greatest in size but least in number (i. e. into *two halves*) . . ., while an *odd* number is that which cannot be so divided but is only divisible into two unequal parts.'[14]

Nicomachus gives another ancient definition to the effect that

'an *even* number is that which can be divided both into two equal parts and into two unequal parts (except the fundamental dyad which can only be divided into two equal parts), but, however it is divided, must have its two parts *of the same kind* without part in the other kind (i. e. the two parts are

[1] Iambl. *in Nicom. ar. introd.*, p. 10. 8–10.
[2] Arist. *Metaph.* A. 5, 986 a 20.
[3] Theon of Smyrna, p. 18. 3–5.
[4] Stob. *Ecl.* i. pr. 8.
[5] Iambl. *op. cit.*, p. 10. 17.
[6] Nicom. i. 7. 1.
[7] *Metaph.* Δ. 13, 1020 a 13.
[8] *Ib.* I. 1, 1053 a 30; Z. 13, 1039 a 12.
[9] *Ib.* M. 9, 1085 b 22.
[10] *Phys.* iii. 7, 207 b 7.
[11] *Metaph.* I. 6, 1057 a 3.
[12] *Ib.* N. 1, 1088 a 5.
[13] Stob. *Ecl.* i. 21. 7c (*Vors.* i³, p. 310. 11–14).
[14] Nicom. i. 7. 3.

both odd or both even); while an *odd* number is that which, however divided, must in any case fall into two unequal parts, and those parts always belonging to the two *different* kinds respectively (i.e. one being odd and one even).'[1]

In the latter definition we have a trace of the original conception of 2 (the dyad) as being, not a number at all, but the principle or beginning of the even, just as one was not a number but the principle or beginning of number; the definition implies that 2 was not originally regarded as an even number, the qualification made by Nicomachus with reference to the dyad being evidently a later addition to the original definition (Plato already speaks of two as even).[2]

With regard to the term 'odd-even', it is to be noted that, according to Aristotle, the Pythagoreans held that 'the One arises from both kinds (the odd and the even), for it is both even and odd'.[3] The explanation of this strange view might apparently be that the unit, being the principle of all number, even as well as odd, cannot itself be odd and must therefore be called even-odd. There is, however, another explanation, attributed by Theon of Smyrna to Aristotle, to the effect that the unit when added to an even number makes an odd number, but when added to an odd number makes an even number: which could not be the case if it did not partake of both species; Theon also mentions Archytas as being in agreement with this view.[4] But, inasmuch as the fragment of Philolaus speaks of 'many forms' of the species odd and even, and 'a third' (even-odd) obtained from a combination of them, it seems more natural to take 'even-odd' as there meaning, not the unit, but the product of an odd and an even number, while, if 'even' in the same passage excludes such a number, 'even' would appear to be confined to powers of 2, or 2^n.

We do not know how far the Pythagoreans advanced towards the later elaborate classification of the varieties of odd and even numbers. But they presumably had not got beyond the point of view of Plato and Euclid. In Plato we have the terms 'even-times even' (ἄρτια ἀρτιάκις), 'odd-times odd' (περιττὰ περιττάκις), 'odd-times even' (ἄρτια

[1] Nicom. i. 7. 4.
[2] Plato, *Parmenides*, 143 D.
[3] Arist. *Metaph.* A. 5, 986 a 19.
[4] Theon of Smyrna, p. 22. 5–10.

περιττάκις) and 'even-times odd' (περιττὰ ἀρτιάκις), which are evidently used in the simple sense of the products of even and even, odd and odd, odd and even, and even and odd factors respectively.[1] Euclid's classification does not go much beyond this; he does not attempt to make the four definitions mutually exclusive.[2] An 'odd-times odd' number is of course any odd number which is not prime; but 'even-times even' ('a number measured by an even number according to an even number') does not exclude 'even-times odd' ('a number measured by an even number according to an odd number'); e.g. 24, which is 6 times 4, or 4 times 6, is also 8 times 3. Euclid did not apparently distinguish, any more than Plato, between 'even-times odd' and 'odd-times even' (the definition of the latter in the texts of Euclid was probably interpolated). The Neo-Pythagoreans improved the classification thus. With them the 'even-times even' number is that which has its halves even, the halves of the halves even, and so on till unity is reached'[3]; in short, it is a number of the form 2^n. The 'even-odd' number (ἀρτιοπέριττος in one word) is such a number as, when once halved, leaves as quotient an odd number,[4] i.e. a number of the form $2(2m+1)$. The 'odd-even' number (περισσάρτιος) is a number such that it can be halved twice or more times successively, but the quotient left when it can no longer be halved is an odd number not unity,[5] i.e. it is a number of the form $2^{n+1}(2m+1)$. The 'odd-times odd' number is not defined as such by Nicomachus and Iamblichus, but Theon of Smyrna quotes a curious use of the term; he says that it was one of the names applied to prime numbers (excluding of course 2), for these have two odd factors, namely 1 and the number itself.[6]

Prime or *incomposite* numbers (πρῶτος καὶ ἀσύνθετος) and *secondary* or *composite* numbers (δεύτερος καὶ σύνθετος) are distinguished in a fragment of Speusippus based upon works of Philolaus.[7] We are told[8] that Thymaridas called a prime number *rectilinear* (εὐθυγραμμικός), the ground being that it can only be set out in one dimension[9] (since the only measure

[1] Plato, *Parmenides*, 143 E.
[2] See Eucl. VII. Defs. 8–10.
[3] Nicom. i. 8. 4.
[4] *Ib.* i. 9. 1.
[5] *Ib.* i. 10. 1.
[6] Theon of Smyrna, p. 23. 14–23.
[7] *Theol. Ar.* (Ast), p. 62 (*Vors.* i³, p. 304. 5).
[8] Iambl. *in Nicom.*, p. 27. 4.
[9] Cf. Arist. *Metaph.* Δ. 13, 1020 b 3, 4.

of it, excluding the number itself, is 1); Theon of Smyrna gives *euthymetric* and *linear* as alternative terms,[1] and the latter (γραμμικός) also occurs in the fragment of Speusippus. Strictly speaking, the prime number should have been called that which is rectilinear or linear *only*. As we have seen, 2 was not originally regarded as a prime number, or even as a number at all. But Aristotle speaks of the dyad as 'the only even number which is prime,'[2] showing that this divergence from early Pythagorean doctrine took place before Euclid's time. Euclid defined a prime number as 'that which is measured by a unit alone',[3] a composite number as 'that which is measured by some number',[4] while he adds definitions of numbers 'prime to one another' ('those which are measured by a unit alone as a common measure') and of numbers 'composite to one another' ('those which are measured by some number as a common measure').[5] Euclid then, as well as Aristotle, includes 2 among prime numbers. Theon of Smyrna says that even numbers are not measured by the unit alone, except 2, which therefore is odd-*like* without being prime.[6] The Neo-Pythagoreans, Nicomachus and Iamblichus, not only exclude 2 from prime numbers, but define composite numbers, numbers prime to one another, and numbers composite to one another as excluding all even numbers; they make all these categories subdivisions of *odd*.[7] Their object is to divide odd into three classes parallel to the three subdivisions of even, namely even-even $= 2^n$, even-odd $= 2(2m+1)$ and the quasi-intermediate odd-even $= 2^{n+1}(2m+1)$; accordingly they divide odd numbers into (*a*) the prime and incomposite, which are Euclid's primes excluding 2, (*b*) the secondary and composite, the factors of which must all be not only odd but prime numbers, (*c*) those which are 'secondary and composite in themselves but prime and incomposite to another number,' e.g. 9 and 25, which are both secondary and composite but have no common measure except 1. The inconvenience of the restriction in (*b*) is obvious, and there is the

[1] Theon of Smyrna, p. 23. 12.
[2] Arist. *Topics*, Θ. 2, 157 a 39.
[3] Eucl. VII. Def. 11.
[4] *Ib.* Def. 13.
[5] *Ib.* Defs. 12, 14.
[6] Theon of Smyrna, p. 24. 7.
[7] Nicom. i, cc. 11–13; Iambl. *in Nicom.*, pp. 26–8.

further objection that (*b*) and (*c*) overlap, in fact (*b*) includes the whole of (*c*).

'Perfect' and 'Friendly' numbers.

There is no trace in the fragments of Philolaus, in Plato or Aristotle, or anywhere before Euclid, of the *perfect* number (τέλειος) in the well-known sense of Euclid's definition (VII. Def. 22), a number, namely, which is 'equal to (the sum of) its own parts' (i.e. all its factors including 1), e.g. $6 = 1+2+3$; $28 = 1+2+4+7+14$;

$$496 = 1+2+4+8+16+31+62+124+248.$$

The law of the formation of these numbers is proved in Eucl. IX. 36, which is to the effect that, if the sum of any number of terms of the series $1, 2, 2^2, 2^3 \ldots . 2^{n-1} (= S_n)$ is prime, then $S_n . 2^{n-1}$ is a 'perfect' number. Theon of Smyrna[1] and Nicomachus[2] both define a 'perfect' number and explain the law of its formation; they further distinguish from it two other kinds of numbers, (1) *over-perfect* (ὑπερτελής or ὑπερτέλειος), so called because the sum of all its aliquot parts is greater than the number itself, e.g. 12, which is less than $1+2+3+4+6$, (2) *defective* (ἐλλιπής), so called because the sum of all its aliquot parts is less than the number itself, e.g. 8, which is greater than $1+2+4$. Of perfect numbers Nicomachus knew four (namely 6, 28, 496, 8128) but no more. He says they are formed in 'ordered' fashion, there being one among the units (i.e. less than 10), one among the tens (less than 100), one among the hundreds (less than 1000), and one among the thousands (less than a myriad); he adds that they terminate alternately in 6 or 8. They do all terminate in 6 or 8 (as we can easily prove by means of the formula $(2^n - 1)\, 2^{n-1}$), but not alternately, for the fifth and sixth perfect numbers both end in 6, and the seventh and eighth both end in 8. Iamblichus adds a tentative suggestion that there may (εἰ τύχοι) in like manner be one perfect number among the first myriads (less than 10000^2), one among the second myriads (less than 10000^3), and so on *ad infinitum*.[3] This is incorrect, for the next perfect numbers are as follows:[4]

[1] Theon of Smyrna, p. 45.
[2] Nicom. i. 16, 1–4.
[3] Iambl. *in Nicom.*, p. 33. 20–23.
[4] The fifth perfect number may have been known to Iamblichus,

fifth, $2^{12}(2^{13}-1) = 33\ 550\ 336$
sixth, $2^{16}(2^{17}-1) = 8\ 589\ 869\ 056$
seventh, $2^{18}(2^{19}-1) = 137\ 438\ 691\ 328$
eighth, $2^{30}(2^{31}-1) = 2\ 305\ 843\ 008\ 139\ 952\ 128$
ninth, $2^{60}(2^{61}-1) = 2\ 658\ 455\ 991\ 569\ 831\ 744\ 654\ 692\ 615\ 953\ 842\ 176$
tenth, $2^{88}(2^{89}-1)$.

With these 'perfect' numbers should be compared the so-called 'friendly numbers'. Two numbers are 'friendly' when each is the sum of all the aliquot parts of the other, e.g. 284 and 220 (for $284 = 1+2+4+5+10+11+20+22+44+55+110$, while $220 = 1+2+4+71+142$). Iamblichus attributes the discovery of such numbers to Pythagoras himself, who, being asked 'what is a friend?' said '*Alter ego*', and on this analogy applied the term 'friendly' to two numbers the aliquot parts of either of which make up the other.[1]

While for Euclid, Theon of Smyrna, and the Neo-Pythagoreans the 'perfect' number was the kind of number above described, we are told that the Pythagoreans made 10 the perfect number. Aristotle says that this was because they found within it such things as the void, proportion, oddness, and so on.[2] The reason is explained more in detail by Theon of Smyrna[3] and in the fragment of Speusippus. 10 is the sum of the numbers 1, 2, 3, 4 forming the τετρακτύς ('their greatest oath', alternatively called the 'principle of health'[4]). These numbers include the ratios corresponding to the musical intervals discovered by Pythagoras, namely 4 : 3 (the fourth),

though he does not give it; it was, however, known, with all its factors, in the fifteenth century, as appears from a tract written in German which was discovered by Curtze (Cod. lat. Monac. 14908). The first eight 'perfect' numbers were calculated by Jean Prestet (d. 1670); Fermat (1601-65) had stated, and Euler proved, that $2^{31}-1$ is prime. The ninth perfect number was found by P. Seelhoff, *Zeitschr. f. Math. u. Physik*, 1886, pp. 174 sq.) and verified by E. Lucas (*Mathésis*, vii, 1887, pp. 44-6). The tenth was found by R. E. Powers (*Bull. Amer. Math. Soc.*, 1912, p. 162).

[1] Iambl. *in Nicom.*, p. 35. 1-7. The subject of 'friendly' numbers was taken up by Euler, who discovered no less than sixty-one pairs of such numbers. Descartes and van Schooten had previously found three pairs but no more.

[2] Arist. *Metaph.* M. 8, 1084 a 32-4.

[3] Theon of Smyrna, p. 93. 17-94. 9 (*Vorsokratiker*, i³, pp. 303-4).

[4] Lucian, *De lapsu in salutando*, 5.

3 : 2 (the fifth), and 2 : 1 (the octave). Speusippus observes further that 10 contains in it the 'linear', 'plane' and 'solid' varieties of number; for 1 is a point, 2 is a line,[1] 3 a triangle, and 4 a pyramid.[2]

Figured numbers.

This brings us once more to the theory of figured numbers, which seems to go back to Pythagoras himself. A point or dot is used to represent 1; two dots placed apart represent 2, and at the same time define the straight line joining the two dots; three dots, representing 3, mark out the first rectilinear plane figure, a triangle; four dots, one of which is outside the plane containing the other three, represent 4 and also define the first rectilineal solid figure. It seems clear that the oldest Pythagoreans were acquainted with the formation of triangular and square numbers by means of pebbles or dots[3]; and we judge from the account of Speusippus's book, *On the Pythagorean Numbers*, which was based on works of Philolaus, that the latter dealt with linear numbers, polygonal numbers, and plane and solid numbers of all sorts, as well as with the five regular solid figures.[4] The varieties of plane numbers (triangular, square, oblong, pentagonal, hexagonal, and so on), solid numbers (cube, pyramidal, &c.) are all discussed, with the methods of their formation, by Nicomachus[5] and Theon of Smyrna.[6]

(α) *Triangular numbers.*

To begin with *triangular* numbers. It was probably Pythagoras who discovered that the sum of any number of successive terms of the series of natural numbers 1, 2, 3 . . . beginning from 1 makes a triangular number. This is obvious enough from the following arrangements of rows of points;

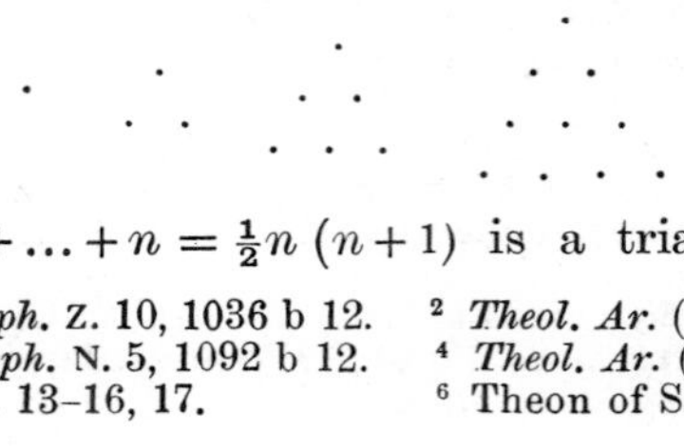

```
                              .
                   .         . .
 .      .         . .       . . .
       . .       . . .     . . . .
```

Thus $1+2+3+\ldots+n = \frac{1}{2}n(n+1)$ is a triangular number

[1] Cf. Arist. *Metaph.* Z. 10, 1036 b 12.
[2] *Theol. Ar.* (Ast), p. 62. 17–22.
[3] Cf. Arist. *Metaph.* N. 5, 1092 b 12.
[4] *Theol. Ar.* (Ast), p. 61.
[5] Nicom. i. 7–11, 13–16, 17.
[6] Theon of Smyrna, pp. 26–42.

of side n. The particular triangle which has 4 for its side is mentioned in a story of Pythagoras by Lucian. Pythagoras told some one to count. He said 1, 2, 3, 4, whereon Pythagoras interrupted, 'Do you see? What you take for 4 is 10, a perfect triangle and our oath'.[1] This connects the knowledge of triangular numbers with true Pythagorean ideas.

(β) *Square numbers and gnomons.*

We come now to *square* numbers. It is easy to see that, if we have a number of dots forming and filling up a square as in the accompanying figure representing 16, the square of 4, the next higher square, the square of 5, can be formed by adding a row of dots round two sides of the original square, as shown; the number of these dots is $2 \cdot 4 + 1$, or 9. This process of forming successive squares can be applied throughout, beginning from the first square number 1. The successive additions are shown in the annexed figure between the successive pairs of straight lines forming right angles; and the successive numbers added to the 1 are

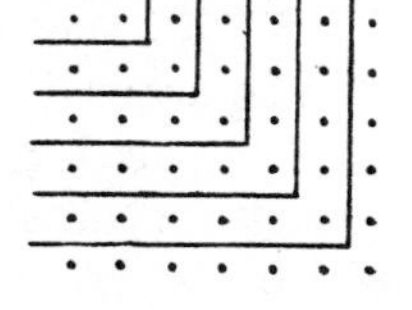

$$3, 5, 7 \dots (2n+1),$$

that is to say, the successive odd numbers. This method of formation shows that the sum of any number of successive terms of the series of odd numbers 1, 3, 5, 7 ... starting from 1 is a square number, that, if n^2 is any square number, the addition of the odd number $2n+1$ makes it into the next square, $(n+1)^2$, and that the sum of the series of odd numbers $1+3+5+7+\dots+(2n+1) = (n+1)^2$, while

$$1+3+5+7+\dots+(2n-1) = n^2.$$

All this was known to Pythagoras. The odd numbers successively added were called *gnomons*; this is clear from Aristotle's allusion to gnomons placed round 1 which now produce different figures every time (oblong figures, each dissimilar to the preceding one), now preserve one and the same figure (squares)[2]; the latter is the case with the gnomons now in question.

[1] Lucian, Βίων πρᾶσις, 4.

[2] Arist. *Phys.* iii. 4, 203 a 13–15.

(γ) *History of the term 'gnomon'.*

It will be noticed that the gnomons shown in the above figure correspond in shape to the geometrical gnomons with which Euclid, Book II, has made us familiar. The history of the word 'gnomon' is interesting. (1) It was originally an astronomical instrument for the measuring of time, and consisted of an upright stick which cast shadows on a plane or hemispherical surface. This instrument is said to have been introduced into Greece by Anaximander[1] and to have come from Babylon.[2] Following on this application of the word 'gnomon' (a 'marker' or 'pointer', a means of reading off and knowing something), we find Oenopides calling a perpendicular let fall on a straight line from an external point a straight line drawn '*gnomon-wise*' (κατὰ γνώμονα).[3] Next (2) we find the term used of an instrument for drawing right angles, which took the form shown in the annexed figure. This seems to be the meaning in Theognis 805, where it is said that the envoy sent to consult the oracle at Delphi should be 'straighter than the τόρνος (an instrument with a stretched string for drawing a circle), the στάθμη (a plumb-line), and the *gnomon*'. It was natural that, owing to its shape, the gnomon should then be used to describe (3) the figure which remained of a square when a smaller square was cut out of it (or the figure which, as Aristotle says, when added to a square, preserves the shape and makes up a larger square). The term is used in a fragment of Philolaus where he says that 'number makes all things knowable and mutually agreeing in the way characteristic of the *gnomon*'.[4] Presumably, as Boeckh says, the connexion between the gnomon and the square to which it is added was regarded as symbolical of union and agreement, and Philolaus used the idea to explain the knowledge of things, making the *knowing* embrace the *known* as the gnomon does the square.[5] (4) In Euclid the geometrical meaning of the word is further extended (II. Def. 2) to cover

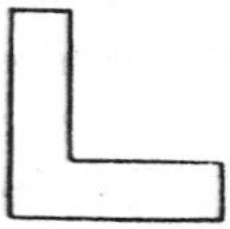

[1] Suidas, *s. v.*

[2] Herodotus, ii. 109.

[3] Proclus on Eucl. I, p. 283. 9.

[4] Boeckh, *Philolaos des Pythagoreers Lehren*, p. 141; *ib.*, p. 144; *Vors.* i[3], p. 313. 15.

[5] Cf. Scholium No. 11 to Book II in Euclid, ed. Heib., vol. v, p. 225.

the figure similarly related to any parallelogram, instead of a square; it is defined as made up of 'any one whatever of the parallelograms about the diameter (diagonal) with the two complements'. Later still (5) Heron of Alexandria defines a *gnomon* in general as that which, when added to anything, number or figure, makes the whole similar to that to which it is added.[1]

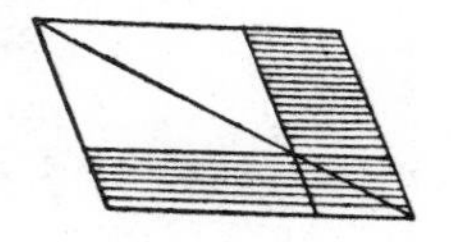

(δ) *Gnomons of the polygonal numbers.*

Theon of Smyrna uses the term in this general sense with reference to numbers: 'All the successive numbers which [by being successively added] produce triangles or squares *or polygons* are called gnomons.'[2] From the accompanying figures showing successive pentagonal and hexagonal numbers it will be seen that the outside rows or gnomons to be succes-

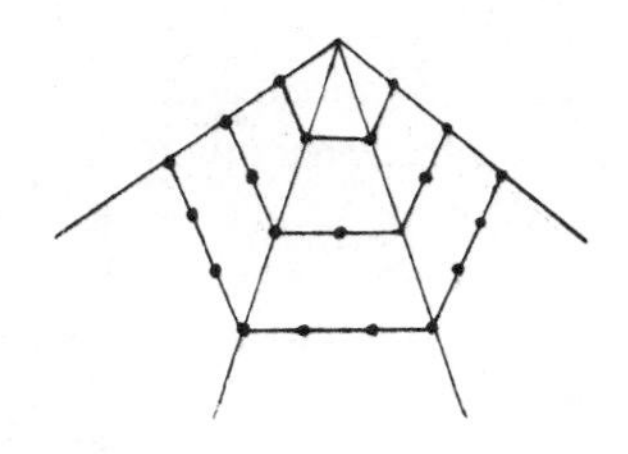

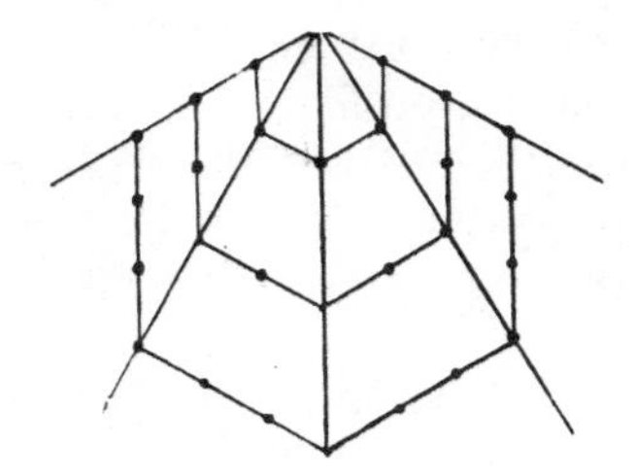

sively added after 1 (which is the first pentagon, hexagon, &c.) are in the case of the pentagon 4, 7, 10 , . . or the terms of an arithmetical progression beginning from 1 with common difference 3, and in the case of the hexagon 5, 9, 13 . . . . or the terms of an arithmetical progression beginning from 1 with common difference 4. In general the successive *gnomonic* numbers for any polygonal number, say of n sides, have $(n-2)$ for their common difference.[3]

(ε) *Right-angled triangles with sides in rational numbers.*

To return to Pythagoras. Whether he learnt the fact from Egypt or not, Pythagoras was certainly aware that, while $3^2+4^2=5^2$, any triangle with its sides in the ratio of the

[1] Heron, Def. 58 (Heron, vol. iv, Heib., p. 225).
[2] Theon of Smyrna, p. 37. 11–13. [3] *Ib.*, p. 34. 13–15.

numbers 3, 4, 5 is right angled. This fact could not but add strength to his conviction that all things were numbers, for it established a connexion between numbers and the *angles* of geometrical figures. It would also inevitably lead to an attempt to find other square numbers besides 5^2 which are the sum of two squares, or, in other words, to find other sets of three integral numbers which can be made the sides of right-angled triangles; and herein we have the beginning of the *indeterminate analysis* which reached so high a stage of development in Diophantus. In view of the fact that the sum of any number of successive terms of the series of odd numbers 1, 3, 5, 7 . . . beginning from 1 is a square, it was only necessary to pick out of this series the odd numbers which are themselves squares; for if we take one of these, say 9, the addition of this square to the square which is the sum of all the preceding odd numbers makes the square number which is the sum of the odd numbers up to the number (9) that we have taken. But it would be natural to seek a formula which should enable all the three numbers of a set to be immediately written down, and such a formula is actually attributed to Pythagoras.[1] This formula amounts to the statement that, if m be any odd number,

$$m^2 + \left(\frac{m^2-1}{2}\right)^2 = \left(\frac{m^2+1}{2}\right)^2.$$

Pythagoras would presumably arrive at this method of formation in the following way. Observing that the gnomon put round n^2 is $2n+1$, he would only have to make $2n+1$ a square.

If we suppose that $\quad 2n+1 = m^2,$

we obtain $\quad n = \frac{1}{2}(m^2-1),$

and therefore $\quad n+1 = \frac{1}{2}(m^2+1).$

It follows that

$$m^2 + \left(\frac{m^2-1}{2}\right)^2 = \left(\frac{m^2+1}{2}\right)^2.$$

[1] Proclus on Eucl. I, p. 487. 7–21.

Another formula, devised for the same purpose, is attributed to Plato,[1] namely

$$(2m)^2 + (m^2 - 1)^2 = (m^2 + 1)^2.$$

We could obtain this formula from that of Pythagoras by doubling the sides of each square in the latter; but it would be incomplete if so obtained, for in Pythagoras's formula m is necessarily odd, whereas in Plato's it need not be. As Pythagoras's formula was most probably obtained from the gnomons of dots, it is tempting to suppose that Plato's was similarly evolved. Consider the square with n dots in its side in relation to the next smaller square $(n-1)^2$ and the next larger $(n+1)^2$. Then n^2 exceeds $(n-1)^2$ by the gnomon $2n-1$, but falls short of $(n+1)^2$ by the gnomon $2n+1$. Therefore the square $(n+1)^2$ exceeds the square $(n-1)^2$ by the sum of the two gnomons $2n-1$ and $2n+1$, which is $4n$.

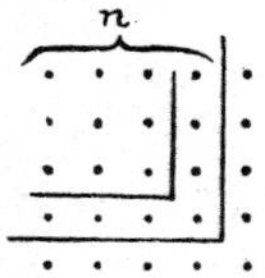

That is, $$4n + (n-1)^2 = (n+1)^2,$$
and, substituting m^2 for n in order to make $4n$ a square, we obtain the Platonic formula

$$(2m)^2 + (m^2 - 1)^2 = (m^2 + 1)^2.$$

The formulae of Pythagoras and Plato supplement each other. Euclid's solution (X, Lemma following Prop. 28) is more general, amounting to the following.

If AB be a straight line bisected at C and produced to D, then (Eucl. II. 6)

$$AD \,.\, DB + CB^2 = CD^2,$$

which we may write thus:

$$uv = c^2 - b^2,$$

where $$u = c + b, \quad v = c - b,$$

and consequently

$$c = \tfrac{1}{2}(u+v), \quad b = \tfrac{1}{2}(u-v).$$

In order that uv may be a square, says Euclid, u and v must, if they are not actually squares, be 'similar plane numbers', and further they must be either both odd or both even

[1] Proclus on Eucl. I, pp. 428. 21–429. 8.

in order that b (and c also) may be a whole number. 'Similar plane' numbers are of course numbers which are the product of two factors proportional in pairs, as $mp.np$ and $mq.nq$, or mnp^2 and mnq^2. Provided, then, that these numbers are both even or both odd,

$$m^2 n^2 p^2 q^2 + \left(\frac{mnp^2 - mnq^2}{2}\right)^2 = \left(\frac{mnp^2 + mnq^2}{2}\right)^2$$

is the solution, which includes both the Pythagorean and the Platonic formulae.

(ζ) *Oblong numbers.*

Pythagoras, or the earliest Pythagoreans, having discovered that, by adding any number of successive terms (beginning from 1) of the series $1+2+3+\ldots+n = \frac{1}{2}n(n+1)$, we obtain triangular numbers, and that by adding the successive odd numbers $1+3+5+\ldots+(2n-1) = n^2$ we obtain squares, it cannot be doubted that in like manner they summed the series of even numbers $2+4+6+\ldots+2n = n(n+1)$ and discovered accordingly that the sum of any number of successive terms of the series beginning with 2 was an 'oblong' number (ἑτερομήκης), with 'sides' or factors differing by 1. They would also see that the oblong number is double of a triangular number. These facts would be brought out by taking two dots representing 2 and then placing round them, gnomon-wise and successively, the even numbers 4, 6, &c., thus:

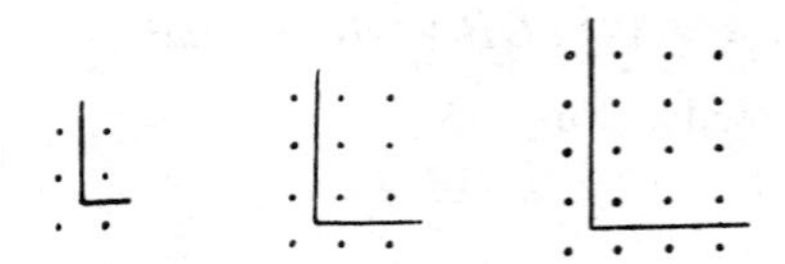

The successive oblong numbers are

$$2.3 = 6,\quad 3.4 = 12,\quad 4.5 = 20\ldots,\quad n(n+1)\ldots,$$

and it is clear that no two of these numbers are similar, for the ratio $n:(n+1)$ is different for all different values of n. We may have here an explanation of the Pythagorean identification of 'odd' with 'limit' or 'limited' and of 'even' with 'unlimited'[1] (cf. the Pythagorean scheme of ten pairs of opposites, where odd, limit and square in one set are opposed to even, unlimited and oblong respectively in the other).[2] For, while the adding of the successive odd numbers as gnomons round 1 gives only one form, the square, the addition of the successive even numbers to 2 gives a succession of 'oblong' numbers all dissimilar in form, that is to say, an infinity of forms. This seems to be indicated in the passage of Aristotle's *Physics* where, as an illustration of the view that the even is unlimited, he says that, where gnomons are put round 1, the resulting figures are in one case always different in species, while in the other they always preserve one form[3]; the one form is of course the square formed by adding the odd numbers as gnomons round 1; the words *καὶ χωρίς* ('and in the separate case', as we may perhaps translate) imperfectly describe the second case, since in that case even numbers are put round 2, not 1, but the meaning seems clear.[4] It is to be noted that the word *ἑτερομήκης* ('oblong') is in Theon of Smyrna and Nicomachus limited to numbers which are the product of two factors differing by unity, while they apply the term *προμήκης* ('prolate', as it were) to numbers which are the product of factors differing by two or more (Theon makes *προμήκης* include *ἑτερομήκης*). In Plato and Aristotle *ἑτερομήκης* has the wider sense of any non-square number with two unequal factors.

It is obvious that any 'oblong' number $n(n+1)$ is the

sum of two equal triangular numbers. Scarcely less obvious is the theorem of Theon that any square number is made up of two triangular numbers[5]; in this case, as is seen from the

[1] Arist. *Metaph.* A. 5, 986 a 17.
[2] *Ib.* A. 5, 986 a 23–26.
[3] Arist. *Phys.* iii. 4, 203 a 10–15.
[4] Cf. Plut. (?) Stob. *Ecl.* i. pr. 10, p. 22. 16 Wachsmuth.
[5] Theon of Smyrna, p. 41. 3–8.

figure, the sides of the triangles differ by unity, and of course

$$\tfrac{1}{2}n(n-1)+\tfrac{1}{2}n(n+1) = n^2.$$

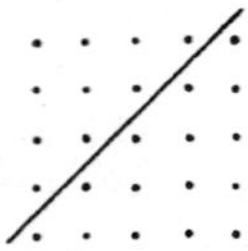

Another theorem connecting triangular numbers and squares, namely that 8 times any triangular number + 1 makes a square, may easily go back to the early Pythagoreans. It is quoted by Plutarch[1] and used by Diophantus,[2] and is equivalent to the formula

$$8 \,.\, \tfrac{1}{2}n(n+1)+1 = 4n(n+1)+1 = (2n+1)^2.$$

It may easily have been proved by means of a figure made up of dots in the usual way. Two equal triangles make up an oblong figure of the form $n(n+1)$, as above. Therefore we have to prove that four equal figures of this form with one more dot make up $(2n+1)^2$. The annexed figure representing 7^2 shows how it can be divided into four 'oblong' figures 3 . 4 leaving 1 over.

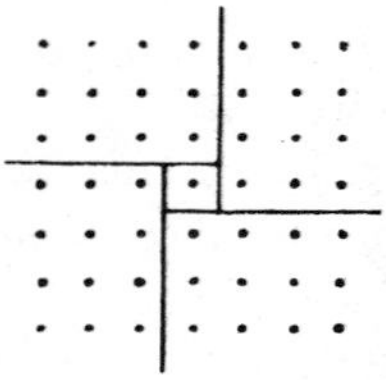

In addition to Speusippus, Philippus of Opus (fourth century), the editor of Plato's *Laws* and author of the *Epinomis*, is said to have written a work on polygonal numbers.[3] Hypsicles, who wrote about 170 B.C., is twice mentioned in Diophantus's *Polygonal Numbers* as the author of a 'definition' of a polygonal number.

The theory of proportion and means.

The 'summary' of Proclus (as to which see the beginning of Chapter IV) states (if Friedlein's reading is right) that Pythagoras discovered 'the theory of irrationals (τὴν τῶν ἀλόγων πραγματείαν) and the construction of the cosmic figures' (the five regular solids).[4] We are here concerned with the first part of this statement in so far as the reading ἀλόγων ('irrationals') is disputed. Fabricius seems to have been the first to record the variant ἀναλόγων, which is also noted by E. F. August[5]; Mullach adopted this reading from

[1] Plutarch, *Plat. Quaest.* v. 2. 4, 1003 F. [2] Dioph. IV. 38.
[3] Βιογράφοι, *Vitarum scriptores Graeci minores*, ed. Westermann, p. 446.
[4] Proclus on Eucl. I, p. 65. 19.
[5] In his edition of the Greek text of Euclid (1824–9), vol. i, p. 290.

Fabricius. ἀναλόγων is not the correct form of the word, but the meaning would be 'proportions' or 'proportionals', and the true reading may be either τῶν ἀναλογιῶν ('proportions'), or, more probably, τῶν ἀνὰ λόγον ('proportionals'); Diels reads τῶν ἀνὰ λόγον, and it would seem that there is now general agreement that ἀλόγων is wrong, and that the theory which Proclus meant to attribute to Pythagoras is the theory of *proportion* or *proportionals*, not of irrationals.

(α) *Arithmetic, geometric, and harmonic means.*

It is true that we have no positive evidence of the use by Pythagoras of proportions in geometry, although he must have been conversant with similar figures, which imply some theory of proportion. But he discovered the dependence of musical intervals on numerical ratios, and the theory of *means* was developed very early in his school with reference to the theory of music and arithmetic. We are told that in Pythagoras's time there were three means, the arithmetic, the geometric, and the subcontrary, and that the name of the third ('subcontrary') was changed by Archytas and Hippasus to 'harmonic'.[1] A fragment of Archytas's work *On Music* actually defines the three; we have the *arithmetic* mean when, of three terms, the first exceeds the second by the same amount as the second exceeds the third; the *geometric* mean when, of the three terms, the first is to the second as the second is to the third; the '*subcontrary, which we call harmonic*', when the three terms are such that 'by whatever part of itself the first exceeds the second, the second exceeds the third by the same part of the third'.[2] That is, if a, b, c are in harmonic progression, and $a = b + \frac{a}{n}$, we must have $b = c + \frac{c}{n}$, whence in fact

$$\frac{a}{c} = \frac{a-b}{b-c}, \quad \text{or} \quad \frac{1}{c} - \frac{1}{b} = \frac{1}{b} - \frac{1}{a}.$$

Nicomachus too says that the name 'harmonic mean' was adopted in accordance with the view of Philolaus about the 'geometrical harmony', a name applied to the cube because it has 12 edges, 8 angles, and 6 faces, and 8 is the mean

[1] Iambl. *in Nicom.*, p. 100. 19–24.
[2] Porph. *in Ptol. Harm.*, p. 267 (*Vors.* i³, p. 334. 17 sq.).

between 12 and 6 according to the theory of harmonics (κατὰ τὴν ἁρμονικήν).[1]

Iamblichus,[2] after Nicomachus,[3] mentions a special 'most perfect proportion' consisting of four terms and called 'musical', which, according to tradition, was discovered by the Babylonians and was first introduced into Greece by Pythagoras. It was used, he says, by many Pythagoreans, e. g. (among others) Aristaeus of Croton, Timaeus of Locri, Philolaus and Archytas of Tarentum, and finally by Plato in the *Timaeus*, where we are told that the double and triple intervals were filled up by two means, one of which exceeds and is exceeded by the same part of the extremes (the harmonic mean), and the other exceeds and is exceeded by the same numerical magnitude (the arithmetic mean).[4] The proportion is

$$a : \frac{a+b}{2} = \frac{2ab}{a+b} : b,$$

an example being $12 : 9 = 8 : 6$.

(β) *Seven other means distinguished.*

The theory of means was further developed in the school by the gradual addition of seven others to the first three, making ten in all. The accounts of the discovery of the fourth, fifth, and sixth are not quite consistent. In one place Iamblichus says they were added by Eudoxus[5]; in other places he says they were in use by the successors of Plato down to Eratosthenes, but that Archytas and Hippasus made a beginning with their discovery,[6] or that they were part of the Archytas and Hippasus tradition.[7] The remaining four means (the seventh to the tenth) are said to have been added by two later Pythagoreans, Myonides and Euphranor.[8] From a remark of Porphyry it would appear that one of the first seven means was discovered by Simus of Posidonia, but that the jealousy of other Pythagoreans would have robbed him of the credit.[9] The ten means are described by

[1] Nicom. ii. 26. 2.
[2] Iambl. *in Nicom.*, p. 118. 19 sq.
[3] Nicom. ii. 29.
[4] Plato, *Timaeus*, 36 A.
[5] Iambl. *in Nicom.*, p. 101. 1-5.
[6] *Ib.*, p. 116. 1-4.
[7] *Ib.*, p. 113, 16-18.
[8] *Ib.*, p. 116. 4-6.
[9] Porphyry, *Vit. Pyth.* 3; *Vors.* i³, p. 343. 12-15 and note.

Nicomachus[1] and Pappus[2]; their accounts only differ as regards one of the ten. If $a>b>c$, the formulae in the third column of the following table show the various means.

No. in Nicom.	No. in Pappus.	Formulae.	Equivalent.
1	1	$\frac{a-b}{b-c}=\frac{a}{a}=\frac{b}{b}=\frac{c}{c}$	$a+c=2b$ (arithmetic)
2	2	$\frac{a-b}{b-c}=\frac{a}{b}\left[=\frac{b}{c}\right]$	$ac=b^2$ (geometric)
3	3	$\frac{a-b}{b-c}=\frac{a}{c}$	$\frac{1}{a}+\frac{1}{c}=\frac{2}{b}$ (harmonic)
4	4	$\frac{a-b}{b-c}=\frac{c}{a}$	$\frac{a^2+c^2}{a+c}=b$ (subcontrary to harmonic)
5	5	$\frac{a-b}{b-c}=\frac{c}{b}$	$a=b+c-\frac{c^2}{b}$ (subcontrary to geometric)
6	6	$\frac{a-b}{b-c}=\frac{b}{a}$	$c=a+b-\frac{a^2}{b}$ (subcontrary to geometric)
7	(omitted)	$\frac{a-c}{b-c}=\frac{a}{c}$	$c^2=2ac-ab$
8	9	$\frac{a-c}{a-b}=\frac{a}{c}$	$a^2+c^2=a(b+c)$
9	10	$\frac{a-c}{b-c}=\frac{b}{c}$	$b^2+c^2=c(a+b)$
10	7	$\frac{a-c}{a-b}=\frac{b}{c}$	$a=b+c$
(omitted)	8	$\frac{a-c}{a-b}=\frac{a}{b}$	$a^2=2ab-bc$

The two lists together give *five* means in addition to the first six which are common to both; there would be six more (as Theon of Smyrna says[3]) were it not that $\frac{a-c}{b-c}=\frac{a}{b}$ is illusory, since it gives $a=b$. Tannery has remarked that

[1] Nicom. ii. 28.
[2] Pappus, iii, p. 102.
[3] Theon of Smyrna, p. 106. 15, p. 116. 3.

Nos. 4, 5, 6 of the above means give equations of the second degree, and he concludes that the geometrical and even the arithmetical solution of such equations was known to the discoverer of these means, say about the time of Plato[1]; Hippocrates of Chios, in fact, assumed the geometrical solution of a mixed quadratic equation in his quadrature of lunes.

Pappus has an interesting series of propositions with regard to eight out of the ten means defined by him.[2] He observes that if α, β, γ be three terms in geometrical progression, we can form from these terms three other terms a, b, c, being linear functions of α, β, γ which satisfy respectively eight of the above ten relations; that is to say, he gives a solution of eight problems in indeterminate analysis of the second degree. The solutions are as follows:

No. in Nicom.	No. in Pappus.	Formulae.	Solution in terms of α, β, γ.	Smallest solution.
2	2	$\frac{a-b}{b-c}=\frac{a}{b}=\frac{b}{c}$	$a = \alpha+2\beta+\gamma$	$a = 4$
			$b = \beta+\gamma$	$b = 2$
			$c = \gamma$	$c = 1$
3	3	$\frac{a-b}{b-c}=\frac{a}{c}$	$a = 2\alpha+3\beta+\gamma$	$a = 6$
			$b = 2\beta+\gamma$	$b = 3$
			$c = \beta+\gamma$	$c = 2$
4	4	$\frac{a-b}{b-c}=\frac{c}{a}$	$a = 2\alpha+3\beta+\gamma$	$a = 6$
			$b = 2\alpha+2\beta+\gamma$	$b = 5$
			$c = \beta+\gamma$	$c = 2$
5	5	$\frac{a-b}{b-c}=\frac{c}{b}$	$a = \alpha+3\beta+\gamma$	$a = 5$
			$b = \alpha+2\beta+\gamma$	$b = 4$
			$c = \beta+\gamma$	$c = 2$
6	6	$\frac{a-b}{b-c}=\frac{b}{a}$	$a = \alpha+3\beta+2\gamma$	$a = 6$
			$b = \alpha+2\beta+\gamma$	$b = 4$
			$c = \alpha+\beta-\gamma$	$c = 1$

[1] Tannery, *Mémoires scientifiques*, i, pp. 92-3.
[2] Pappus, iii, pp. 84-104.

No. in Nicom.	No. in Pappus.	Formulae.	Solution in terms of α, β, γ.	Smallest solution.
—	8	$\frac{a-c}{a-b} = \frac{a}{b}$	$a = 2\alpha+3\beta+\gamma$ $b = \alpha+2\beta+\gamma$ $c = 2\beta+\gamma$	$a = 6$ $b = 4$ $c = 3$
8	9	$\frac{a-c}{a-b} = \frac{a}{c}$	$a = \alpha+2\beta+\gamma$ $b = \alpha+\beta+\gamma$ $c = \beta+\gamma$	$a = 4$ $b = 3$ $c = 2$
9	10	$\frac{a-c}{b-c} = \frac{b}{c}$	$a = \alpha+\beta+\gamma$ $b = \beta+\gamma$ $c = \gamma$	$a = 3$ $b = 2$ $c = 1$

Pappus does not include a corresponding solution for his No. 1 and No. 7, and Tannery suggests as the reason for this that, the equations in these cases being already linear, there is no necessity to assume $\alpha\gamma = \beta^2$, and consequently there is one indeterminate too many.[1] Pappus does not so much prove as verify his results, by transforming the proportion $\frac{\alpha}{\beta} = \frac{\beta}{\gamma}$ in all sorts of ways, *componendo*, *dividendo*, &c.

(γ) *Plato on geometric means between two squares or two cubes.*

It is well known that the mathematics in Plato's *Timaeus* is essentially Pythagorean. It is therefore *a priori* probable that Plato πυθαγορίζει in the passage[2] where he says that between two *planes* one mean suffices, but to connect two *solids* two means are necessary. By *planes* and *solids* he really means square and cube numbers, and his remark is equivalent to stating that, if p^2, q^2 are two square numbers,

$$p^2 : pq = pq : q^2,$$

while, if p^3, q^3 are two cube numbers,

$$p^3 : p^2q = p^2q : pq^2 = pq^2 : q^3,$$

the means being of course means in continued geometric proportion. Euclid proves the properties for square and cube

[1] Tannery, *loc. cit.*, pp. 97–8.

[2] Plato, *Timaeus*, 32 A, B.

numbers in VIII. 11, 12, and for similar plane and solid numbers in VIII. 18, 19. Nicomachus quotes the substance of Plato's remark as a 'Platonic theorem', adding in explanation the equivalent of Eucl. VIII. 11, 12.[1]

(δ) A theorem of Archytas.

Another interesting theorem relative to geometric means evidently goes back to the Pythagoreans. If we have two numbers in the ratio known as ἐπιμόριος, or *superparticularis*, i.e. the ratio of $n+1$ to n, there can be no number which is a mean proportional between them. The theorem is Prop. 3 of Euclid's *Sectio Canonis*,[2] and Boëtius has preserved a proof of it by Archytas, which is substantially identical with that of Euclid.[3] The proof will be given later (pp. 215–16). So far as this chapter is concerned, the importance of the proposition lies in the fact that it implies the existence, at least as early as the date of Archytas (about 430–365 B.C.), of an *Elements of Arithmetic* in the form which we call Euclidean; and no doubt text-books of the sort existed even before Archytas, which probably Archytas himself and others after him improved and developed in their turn.

The 'irrational'.

We mentioned above the dictum of Proclus (if the reading ἀλόγων is right) that Pythagoras discovered the theory, or study, of *irrationals*. This subject was regarded by the Greeks as belonging to geometry rather than arithmetic. The irrationals in Euclid, Book X, are straight lines or areas, and Proclus mentions as special topics in geometry matters relating (1) to *positions* (for numbers have no position), (2) to *contacts* (for tangency is between *continuous* things), and (3) to *irrational straight lines* (for where there is division *ad infinitum*, there also is the irrational).[4] I shall therefore postpone to Chapter V on the Pythagorean geometry the question of the date of the discovery of the theory of irrationals. But it is certain that the incommensurability of the

[1] Nicom. ii. 24. 6, 7.

[2] *Musici Scriptores Graeci*, ed. Jan, pp. 148–66; Euclid, vol. viii, ed. Heiberg and Menge, p. 162.

[3] Boëtius, *De Inst. Musica*, iii. 11 (pp. 285–6, ed. Friedlein); see *Bibliotheca Mathematica*, vi_3, 1905/6, p. 227.

[4] Proclus on Eucl. I, p. 60. 12–16.

diagonal of a square with its side, that is, the 'irrationality' of $\sqrt{2}$, was discovered in the school of Pythagoras, and it is more appropriate to deal with this particular case here, both because the traditional proof of the fact depends on the elementary theory of numbers, and because the Pythagoreans invented a method of obtaining an infinite series of arithmetical ratios approaching more and more closely to the value of $\sqrt{2}$.

The actual method by which the Pythagoreans proved the fact that $\sqrt{2}$ is incommensurable with 1 was doubtless that indicated by Aristotle, a *reductio ad absurdum* showing that, if the diagonal of a square is commensurable with its side, it will follow that the same number is both odd and even.[1] This is evidently the proof interpolated in the texts of Euclid as X. 117, which is in substance as follows:

Suppose AC, the diagonal of a square, to be commensurable with AB, its side; let $\alpha : \beta$ be their ratio expressed in the smallest possible numbers.

Then $\alpha > \beta$, and therefore α is necessarily > 1.

Now $$AC^2 : AB^2 = \alpha^2 : \beta^2;$$
and, since $$AC^2 = 2\,AB^2, \quad \alpha^2 = 2\,\beta^2.$$

Hence α^2, and therefore α, is even.

Since $\alpha : \beta$ is in its lowest terms, it follows that β must be *odd*.

Let $\alpha = 2\,\gamma$; therefore $4\,\gamma^2 = 2\,\beta^2$, or $2\,\gamma^2 = \beta^2$, so that β^2, and therefore β, is *even*.

But β was also *odd*: which is impossible.

Therefore the diagonal AC cannot be commensurable with the side AB.

Algebraic equations.

(α) '*Side-*' *and* '*diameter-*' *numbers, giving successive approximations to* $\sqrt{2}$.

The Pythagorean method of finding any number of successive approximations to the value of $\sqrt{2}$ amounts to finding all the integral solutions of the indeterminate equations

$$2\,x^2 - y^2 = \pm 1,$$

the solutions being successive pairs of what were called *side-*

[1] Arist. *Anal. pr.* i. 23, 41 a 26–7.

and *diameter-* (diagonal-) *numbers* respectively. The law of formation of these numbers is explained by Theon of Smyrna, and is as follows.[1] The unit, being the beginning of all things, must be potentially both a side and a diameter. Consequently we begin with two units, the one being the first *side*, which we will call a_1, the other being the first *diameter*, which we will call d_1.

The second side and diameter (a_2, d_2) are formed from the first, the third side and diameter (a_3, d_3) from the second, and so on, as follows:

$$a_2 = a_1 + d_1, \quad d_2 = 2\,a_1 + d_1,$$
$$a_3 = a_2 + d_2, \quad d_3 = 2\,a_2 + d_2,$$
$$\cdots \cdots \cdots$$
$$a_{n+1} = a_n + d_n, \quad d_{n+1} = 2\,a_n + d_n.$$

Since $a_1 = d_1 = 1$, it follows that

$$a_2 = 1 + 1 = 2, \quad d_2 = 2\,.\,1 + 1 = 3,$$
$$a_3 = 2 + 3 = 5, \quad d_3 = 2\,.\,2 + 3 = 7,$$
$$a_4 = 5 + 7 = 12, \quad d_4 = 2\,.\,5 + 7 = 17,$$

and so on.

Theon states, with reference to these numbers, the general proposition that

$$d_n^{\,2} = 2\,a_n^{\,2} \pm 1,$$

and he observes (1) that the signs alternate as successive d's and a's are taken, $d_1^2 - 2\,a_1^2$ being equal to -1, $d_2^2 - 2\,a_2^2$ equal to $+1$, $d_3^2 - 2\,a_3^2$ equal to -1, and so on, while (2) the sum of the squares of *all* the d's will be double of the squares of *all* the a's. [If the number of successive terms in each series is finite, it is of course necessary that the number should be even.]

The properties stated depend on the truth of the following identity

$$(2\,x + y)^2 - 2\,(x + y)^2 = 2\,x^2 - y^2;$$

for, if x, y be numbers which satisfy one of the two equations

$$2\,x^2 - y^2 = \pm 1,$$

the formula (if true) gives us two higher numbers, $x + y$ and $2\,x + y$, which satisfy the other of the two equations.

Not only is the identity true, but we know from Proclus

[1] Theon of Smyrna, pp. 43, 44.

how it was proved.[1] Observing that 'it is proved by him (Euclid) graphically (γραμμικῶς) in the Second Book of the

A C B D

Elements', Proclus adds the enunciation of Eucl. II. 10. This proposition proves that, if AB is bisected at C and produced to D, then

$$AD^2 + DB^2 = 2\,AC^2 + 2\,CD^2;$$

and, if $AC = CB = x$ and $BD = y$, this gives

$$(2x+y)^2 + y^2 = 2\,x^2 + 2\,(x+y)^2,$$

or

$$(2x+y)^2 - 2\,(x+y)^2 = 2\,x^2 - y^2,$$

which is the formula required.

We can of course prove the property of consecutive side- and diameter- numbers algebraically thus:

$$\begin{aligned} d_n^{\,2} - 2\,a_n^{\,2} &= (2\,a_{n-1} + d_{n-1})^2 - 2\,(a_{n-1} + d_{n-1})^2 \\ &= 2a_{n-1}^{\,2} - d_{n-1}^{\,2} \\ &= -(d_{n-1}^{\,2} - 2\,a_{n-1}^{\,2}) \\ &= +(d_{n-2}^{\,2} - 2\,a_{n-2}^{\,2}), \text{ in like manner;} \end{aligned}$$

and so on.

In the famous passage of the *Republic* (546 c) dealing with the geometrical number Plato distinguishes between the 'irrational diameter of 5', i.e. the diagonal of a square having 5 for its side, or $\sqrt{(50)}$, and what he calls the 'rational diameter' of 5. The square of the 'rational diameter' is less by 1 than the square of the 'irrational diameter', and is therefore 49, so that the 'rational diameter' is 7; that is, Plato refers to the fact that $2 \,.\, 5^2 - 7^2 = 1$, and he has in mind the particular pair of side- and diameter- numbers, 5 and 7, which must therefore have been known before his time. As the proof of the property of these numbers in general is found, as Proclus says, in the geometrical theorem of Eucl. II. 10, it is a fair inference that that theorem is Pythagorean, and was probably invented for the special purpose.

[1] Proclus, *Comm. on Rep. of Plato*, ed. Kroll, vol. ii, 1901, cc. 23 and 27, pp. 24, 25, and 27–9.

(β) *The ἐπάνθημα ('bloom') of Thymaridas.*

Thymaridas of Paros, an ancient Pythagorean already mentioned (p. 69), was the author of a rule for solving a certain set of n simultaneous simple equations connecting n unknown quantities. The rule was evidently well known, for it was called by the special name of ἐπάνθημα, the 'flower' or 'bloom' of Thymaridas.[1] (The term ἐπάνθημα is not, however, confined to the particular proposition now in question; Iamblichus speaks of ἐπανθήματα of the *Introductio arithmetica*, 'arithmetical ἐπανθήματα' and ἐπανθήματα of particular numbers.) The rule is stated in general terms and no symbols are used, but the content is pure algebra. The known or determined quantities (ὡρισμένον) are distinguished from the undetermined or unknown (ἀόριστον), the term for the latter being the very word used by Diophantus in the expression πλῆθος μονάδων ἀόριστον, 'an undefined or undetermined number of units', by which he describes his ἀριθμός or unknown quantity ($=x$). The rule is very obscurely worded, but it states in effect that, if we have the following n equations connecting n unknown quantities $x, x_1, x_2 \ldots x_{n-1}$, namely

$$\begin{aligned} x+x_1+x_2+\ldots+x_{n-1} &= s, \\ x+x_1 &= a_1, \\ x+x_2 &= a_2 \\ \cdot \quad \cdot \quad &\cdot \quad \cdot \\ x+x_{n-1} &= a_{n-1}, \end{aligned}$$

the solution is given by

$$x = \frac{(a_1+a_2+\ldots+a_{n-1})-s}{n-2}.$$

Iamblichus, our informant on this subject, goes on to show that other types of equations can be reduced to this, so that the rule does not 'leave us in the lurch' in those cases either.[2] He gives as an instance the indeterminate problem represented by the following three linear equations between four unknown quantities:

$$\begin{aligned} x+y &= a(z+u), \\ x+z &= b(u+y), \\ x+u &= c(y+z). \end{aligned}$$

[1] Iambl. *in Nicom.*, p. 62. 18 sq.

[2] *Ib.*, p. 63. 16.

From these equations we obtain

$$x+y+z+u = (a+1)(z+u) = (b+1)(u+y) = (c+1)(y+z).$$

If now x, y, z, u are all to be integers, $x+y+z+u$ must contain $a+1, b+1, c+1$ as factors. If L be the least common multiple of $a+1$, $b+1$, $c+1$, we can put $x+y+z+u=L$, and we obtain from the above equations in pairs

$$x+y = \frac{a}{a+1}L,$$

$$x+z = \frac{b}{b+1}L,$$

$$x+u = \frac{c}{c+1}L,$$

while $$x+y+z+u = L.$$

These equations are of the type to which Thymaridas's rule applies, and, since the number of unknown quantities (and equations) is 4, $n-2$ is in this case 2, and

$$x = \frac{L\left(\frac{a}{a+1}+\frac{b}{b+1}+\frac{c}{c+1}\right)-L}{2}$$

The numerator is integral, but it may be an odd number, in which case, in order that x may be integral, we must take $2L$ instead of L as the value of $x+y+z+u$.

Iamblichus has the particular case where $a=2$, $b=3$, $c=4$. L is thus $3.4.5 = 60$, and the numerator of the expression for x becomes $133-60$, or 73, an odd number; he has therefore to put $2L$ or 120 in place of L, and so obtains $x=73$, $y=7$, $z=17$, $u=23$.

Iamblichus goes on to apply the method to the equations

$$x+y = \frac{3}{2}(z+u),$$

$$x+z = \frac{4}{3}(u+y),$$

$$x+u = \frac{5}{4}(y+z),$$

which give

$$x+y+z+u=\frac{5}{2}(z+u)=\frac{7}{3}(u+y)=\frac{9}{4}(y+z).$$

Therefore

$$x+y+z+u=\frac{5}{3}(x+y)=\frac{7}{4}(x+z)=\frac{9}{5}(x+u).$$

In this case we take L, the least common multiple of 5, 7, 9, or 315, and put

$$x+y+z+u= \quad L=315,$$

$$x+y=\frac{3}{5}L=189,$$

$$x+z=\frac{4}{7}L=180,$$

$$x+u=\frac{5}{9}L=175,$$

whence $$x=\frac{544-315}{2}=\frac{229}{2}.$$

In order that x may be integral, we have to take $2L$, or 630, instead of L, or 315, and the solution is $x=229$, $y=149$, $z=131$, $u=121$.

(γ) *Area of rectangles in relation to perimeter.*

Sluse,[1] in letters to Huygens dated Oct. 4, 1657, and Oct. 25, 1658, alludes to a property of the numbers 16 and 18 of which he had read somewhere in Plutarch that it was known to the Pythagoreans, namely that each of these numbers represents the perimeter as well as the area of a rectangle; for $4.4=2.4+2.4$ and $3.6=2.3+2.6$. I have not found the passage of Plutarch, but the property of 16 is mentioned in the *Theologumena Arithmetices*, where it is said that 16 is the only square the area of which is equal to its perimeter, the perimeter of smaller squares being greater, and that of all larger squares being less, than the area.[2] We do not know whether the Pythagoreans proved that 16 and 18 were the only numbers having the property in question; but it is likely enough that they did, for the proof amounts to finding the integral

[1] *Œuvres complètes de C. Huygens*, pp. 64, 260.
[2] *Theol. Ar.*, pp. 10, 23 (Ast).

solutions of $xy = 2(x+y)$. This is easy, for the equation is equivalent to $(x-2)(y-2) = 4$, and we have only to equate $x-2$ and $y-2$ to the respective factors of 4. Since 4 is only divisible into integral factors in two ways, as 2 . 2 or as 1 . 4, we get, as the only possible solutions for x, y, (4, 4) or (3, 6).

Systematic treatises on arithmetic (theory of numbers).

It will be convenient to include in this chapter some account of the arithmetic of the later Pythagoreans, beginning with NICOMACHUS. If any systematic treatises on arithmetic were ever written between Euclid (Books VII–IX) and Nicomachus, none have survived. Nicomachus, of Gerasa, probably the Gerasa in Judaea east of the river Jordan, flourished about 100 A.D., for, on the one hand, in a work of his entitled the *Enchiridion Harmonices* there is an allusion to Thrasyllus, who arranged the Platonic dialogues, wrote on music, and was the astrologer-friend of Tiberius; on the other hand, the *Introductio Arithmetica* of Nicomachus was translated into Latin by Apuleius of Madaura under the Antonines. Besides the Ἀριθμητικὴ εἰσαγωγή, Nicomachus is said to have written another treatise on the theology or the mystic properties of numbers, called Θεολογούμενα ἀριθμητικῆς, in two Books. The curious farrago which has come down to us under that title and which was edited by Ast[1] is, however, certainly not by Nicomachus; for among the authors from whom it gives extracts is Anatolius, Bishop of Laodicaea (A.D. 270); but it contains quotations from Nicomachus which appear to come from the genuine work. It is possible that Nicomachus also wrote an *Introduction to Geometry*, since in one place he says, with regard to certain solid numbers, that they have been specially treated 'in the geometrical introduction, being more appropriate to the theory of magnitude'[2]; but this geometrical introduction may not necessarily have been a work of his own.

It is a very far cry from Euclid to Nicomachus. In the

[1] *Theologumena arithmeticae. Accedit Nicomachi Geraseni Institutio arithmetica*, ed. Ast, Leipzig, 1817.

[2] Nicom. *Arithm.* ii. 6. 1.

Introductio arithmetica we find the form of exposition entirely changed. Numbers are represented in Euclid by straight lines with letters attached, a system which has the advantage that, as in algebraical notation, we can work with numbers in general without the necessity of giving them specific values; in Nicomachus numbers are no longer denoted by straight lines, so that, when different undetermined numbers have to be distinguished, this has to be done by circumlocution, which makes the propositions cumbrous and hard to follow, and it is necessary, after each proposition has been stated, to illustrate it by examples in concrete numbers. Further, there are no longer any proofs in the proper sense of the word; when a general proposition has been enunciated, Nicomachus regards it as sufficient to show that it is true in particular instances; sometimes we are left to infer the general proposition by induction from particular cases which are alone given. Occasionally the author makes a quite absurd remark through failure to distinguish between the general and the particular case, as when, after he has defined the mean which is 'subcontrary to the harmonic' as being determined by the relation $\frac{a-b}{b-c}=\frac{c}{a}$, where $a>b>c$, and has given 6, 5, 3 as an illustration, he goes on to observe that it is a property peculiar to this mean that the product of the greatest and middle terms is double of the product of the middle and least,[1] simply because this happens to be true in the particular case! Probably Nicomachus, who was not really a mathematician, intended his *Introduction* to be, not a scientific treatise, but a popular treatment of the subject calculated to awaken in the beginner an interest in the theory of numbers by making him acquainted with the most noteworthy results obtained up to date; for proofs of most of his propositions he could refer to Euclid and doubtless to other treatises now lost. The style of the book confirms this hypothesis; it is rhetorical and highly coloured; the properties of numbers are made to appear marvellous and even miraculous; the most obvious relations between them are stated in turgid language very tiresome to read. It was the mystic rather than the mathematical side of the theory of numbers that

[1] Nicom. ii. 28. 3.

interested Nicomachus. If the verbiage is eliminated, the mathematical content can be stated in quite a small compass. Little or nothing in the book is original, and, except for certain definitions and refinements of classification, the essence of it evidently goes back to the early Pythagoreans. Its success is difficult to explain except on the hypothesis that it was at first read by philosophers rather than mathematicians (Pappus evidently despised it), and afterwards became generally popular at a time when there were no mathematicians left, but only philosophers who incidentally took an interest in mathematics. But a success it undoubtedly was; this is proved by the number of versions or commentaries which appeared in ancient times. Besides the Latin translation by Apuleius of Madaura (born about A.D. 125), of which no trace remains, there was the version of Boëtius (born about 480, died 524 A.D.); and the commentators include Iamblichus (fourth century), Heronas,[1] Asclepius of Tralles (sixth century), Joannes Philoponus, Proclus.[2] The commentary of Iamblichus has been published,[3] as also that of Philoponus,[4] while that of Asclepius is said to be extant in MSS. When (the pseudo-) Lucian in his *Philopatris* (c. 12) makes Critias say to Triephon 'you calculate like Nicomachus', we have an indication that the book was well known, although the remark may be less a compliment than a laugh at Pythagorean subtleties.[5]

Book I of the *Introductio*, after a philosophical prelude (cc. 1–6), consists principally of definitions and laws of formation. Numbers, odd and even, are first dealt with (c. 7); then comes the subdivision of even into three kinds (1) evenly-even, of the form 2^n, (2) even-odd, of the form $2(2n+1)$, and (3) odd-even, of the form $2^{m+1}(2n+1)$, the last-named occupying a sort of intermediate position in that it partakes of the character of both the others. The odd is next divided into three kinds: (1) 'prime and incomposite', (2) 'secondary and

[1] *v.* Eutoc. *in Archim.* (ed. Heib. iii, p. 120. 22). [2] *v.* Suidas.

[3] The latest edition is Pistelli's (Teubner, 1894).

[4] Ed. Hoche, Heft 1, Leipzig, 1864, Heft 2, Berlin, 1867.

[5] Triephon tells Critias to swear by the Trinity ('One (proceeding) from Three and Three from One'), and Critias replies, 'You would have me learn to calculate, for your oath is mere arithmetic and you calculate like Nicomachus of Gerasa. I do not know what you mean by your "One-Three and Three-One"; I suppose you don't mean the τετρακτύς of Pythagoras or the ὀγδοάς or the τριακάς?'

composite', a product of prime factors (excluding 2, which is even and not regarded as prime), and (3) 'that which is in itself secondary and composite but in relation to another is prime and incomposite', e.g. 9 in relation to 25, which again is a sort of intermediate class between the two others (cc. 11–13); the defects of this classification have already been noted (pp. 73–4). In c. 13 we have these different classes of odd numbers exhibited in a description of Eratosthenes's 'sieve' (κόσκινον), an appropriately named device for finding prime numbers. The method is this. We set out the series of odd numbers beginning from 3.

$$3, 5, 7, 9, 11, 13, 15, 17, 19, 21, 23, 25, 27, 29, 31, \ldots\ldots$$

Now 3 is a prime number, but multiples of 3 are not; these multiples, 9, 15 ... are got by passing over two numbers at a time beginning from 3; we therefore strike out these numbers as not being prime. Similarly 5 is a prime number, but by passing over four numbers at a time, beginning from 5, we get multiples of 5, namely 15, 25 . . .; we accordingly strike out all these multiples of 5. In general, if n be a prime number, its multiples appearing in the series are found by passing over $n-1$ terms at a time, beginning from n; and we can strike out all these multiples. When we have gone far enough with this process, the numbers which are still left will be primes. Clearly, however, in order to make sure that the odd number $2n+1$ in the series is prime, we should have to try all the prime divisors between 3 and $\sqrt{(2n+1)}$; it is obvious, therefore, that this primitive empirical method would be hopeless as a practical means of obtaining prime numbers of any considerable size.

The same c. 13 contains the rule for finding whether two given numbers are prime to one another; it is the method of Eucl. VII. 1, equivalent to our rule for finding the greatest common measure, but Nicomachus expresses the whole thing in words, making no use of any straight lines or symbols to represent the numbers. If there is a common measure greater than unity, the process gives it; if there is none, i.e. if 1 is left as the last remainder, the numbers are prime to one another.

The next chapters (cc. 14–16) are on *over-perfect* (ὑπερτελής), *deficient* (ἐλλιπής), and *perfect* (τέλειος) numbers respectively. The definitions, the law of formation of perfect numbers, and Nicomachus's observations thereon have been given above (p. 74).

Next comes (cc. 17–23) the elaborate classification of numerical ratios greater than unity, with their counterparts which are less than unity. There are five categories of each, and under each category there is (*a*) the general name, (*b*) the particular names corresponding to the particular numbers taken.

The enumeration is tedious, but, for purposes of reference, is given in the following table:—

RATIOS GREATER THAN UNITY	RATIOS LESS THAN UNITY
1. (a) General πολλαπλάσιος, multiple (multiplex)	1. (a) General ὑποπολλαπλάσιος, submultiple (submultiplex)
(b) Particular διπλάσιος, double (duplus) τριπλάσιος, triple (triplus) &c.	(b) Particular ὑποδιπλάσιος, one half (subduplus) ὑποτριπλάσιος, one third (subtriplus) &c.
2. (a) General ἐπιμόριος (superparticularis) a number which is of the form $1+\frac{1}{n}$ or $\frac{n+1}{n}$, where n is any integer.	2. (a) General ὑπεπιμόριος (subsuperparticularis) the fraction $\frac{n}{n+1}$, where n is any integer.
(b) Particular According to the value of n, we have the names ἡμιόλιος (sesquialter) $= 1\frac{1}{2}$ ἐπίτριτος) (sesquitertius) $= 1\frac{1}{3}$ ἐπιτέταρτος (sesquiquartus) $= 1\frac{1}{4}$ &c.	(b) Particular ὑφημιόλιος (subsesquialter) $= \frac{2}{3}$ ὑπεπίτριτος (subsesquitertius) $= \frac{3}{4}$ ὑπεπιτέταρτος (subsesquiquartus) $= \frac{4}{5}$ &c.

RATIOS GREATER THAN UNITY	RATIOS LESS THAN UNITY
3. (a) General ἐπιμερής (superpartiens) which exceeds 1 by twice, thrice, or more times a submultiple, and which therefore may be represented by $1+\frac{m}{m+n}$ or $\frac{2m+n}{m+n}$.	3. (a) General ὑπεπιμερής (subsuperpartiens) which is of the form $\frac{m+n}{2m+n}$.
(b) Particular The formation of the names for the series of particular *superpartientes* follows three different plans. Thus, of numbers of the form $1+\frac{m}{m+1}$, $1\ \frac{2}{3}$ { ἐπιδιμερής (superbipartiens) or ἐπιδίτριτος (superbitertius) or δισεπίτριτος $1\frac{3}{4}$ { ἐπιτριμερής (supertripartiens) or ἐπιτριτέταρτος (supertriquartus) or τρισεπιτέταρτος $1\frac{4}{5}$ is { ἐπιτετραμερής (superquadripartiens) or ἐπιτετράπεμπτος (superquadriquintus) or τετρακισεπίπεμπτος &c. As regards the first name in each case we note that, with ἐπιδιμερής we must understand τρίτων; with ἐπιτριμερής, τετάρτων, and so on.	The corresponding names are not specified in Nicomachus.

RATIOS GREATER THAN UNITY	RATIOS LESS THAN UNITY
Where the more general form $1+\frac{m}{m+n}$, instead of $1+\frac{m}{m+1}$, has to be expressed, Nicomachus uses terms following the *third* plan of formation above, e.g. $1\frac{3}{5}$ = τρισεπίπεμπτος $1\frac{4}{7}$ = τετρακισεφέβδομος $1\frac{5}{9}$ = πεντακισεπένατος and so on, although he might have used the second and called these ratios ἐπιτρίπεμπτος, &c.	
4. (a) General πολλαπλασιεπιμόριος (multiplex superparticularis) This contains a certain *multiple* plus a certain submultiple (instead of 1 plus a submultiple) and is therefore of the form $m+\frac{1}{n}$ (instead of the $1+\frac{1}{n}$ of the ἐπιμόριος) or $\frac{mn+1}{n}$.	4. (a) General ὑποπολλαπλασιεπιμόριος (submultiplex superparticularis) of the form $\frac{n}{mn+1}$.
(b) Particular $2\frac{1}{2}$ = διπλασιεφήμισυς (duplex sesquialter) $2\frac{1}{3}$ = διπλασιεπίτριτος (duplex sesquitertius) $3\frac{1}{5}$ = τριπλασιεπίπεμπτος (triplex sesquiquintus) &c.	The corresponding particular names do not seem to occur in Nicomachus, but Boëtius has them, e.g. subduplex sesquialter, subduplex sesquiquartus.
5. (a) General πολλαπλασιεπιμερής (multiplex superpartiens). This is related to ἐπιμερής [(3) above] in the same way as πολλαπλασιεπιμόριος to ἐπιμόριος; that is to say, it is of the form $p+\frac{m}{m+n}$ or $\frac{(p+1)m+n}{m+n}$.	5. (a) General ὑποπολλαπλασιεπιμερής (submultiplex superpartiens), a fraction of the form $\frac{m+n}{(p+1)m+n}$.

RATIOS GREATER THAN UNITY	RATIOS LESS THAN UNITY
(b) Particular These names are only given for cases where $n = 1$; they follow the first form of the names for particular ἐπιμερεῖς, e. g. $2\frac{2}{3}$ = διπλασιεπιδιμερής (duplex superbipartiens) &c.	Corresponding names not found in Nicomachus; but Boëtius has *subduplex superbipartiens*, &c.

In c. 23 Nicomachus shows how these various ratios can be got from one another by means of a certain rule. Suppose that

$$a, b, c$$

are three numbers such that $a:b = b:c =$ one of the ratios described; we form the three numbers

$$a, \quad a+b, \quad a+2b+c$$

and also the three numbers

$$c, \quad c+b, \quad c+2b+a$$

Two illustrations may be given. If $a = b = c = 1$, repeated application of the first formula gives (1, 2, 4), then (1, 3, 9), then (1, 4, 16), and so on, showing the successive multiples. Applying the second formula to (1, 2, 4), we get (4, 6, 9) where the ratio is $\frac{3}{2}$; similarly from (1, 3, 9) we get (9, 12, 16) where the ratio is $\frac{4}{3}$, and so on; that is, from the πολλαπλάσιοι we get the ἐπιμόριοι. Again from (9, 6, 4), where the ratio is of the latter kind, we get by the first formula (9, 15, 25), giving the ratio $1\frac{2}{3}$, an ἐπιμερής, and by the second formula (4, 10, 25), giving the ratio $2\frac{1}{2}$, a πολλαπλασιεπιμόριος. And so on.

Book II begins with two chapters showing how, by a converse process, three terms in continued proportion with any one of the above forms as common ratio can be reduced to three equal terms. If

$$a, b, c$$

are the original terms, a being the smallest, we take three terms of the form

$$a, \qquad b-a, \qquad \{c-a-2(b-a)\} = c+a-2b,$$

then apply the same rule to these three, and so on.

In cc. 3–4 it is pointed out that, if

$$1, r, r^2 \ldots, r^n \ldots$$

be a geometrical progression, and if

$$\rho_n = r^{n-1} + r^n,$$

then $$\frac{\rho_n}{r^n} = \frac{r+1}{r}, \quad \text{an } \acute{\epsilon}\pi\iota\mu\acute{o}\rho\iota o\varsigma \text{ ratio,}$$

and similarly, if $$\rho'_n = \rho_{n-1} + \rho_n,$$

$$\frac{\rho'_n}{\rho_n} = \frac{r+1}{r};$$

and so on.

If we set out in rows numbers formed in this way,

$$\begin{array}{lllll} 1, r, & r^2, & r^3 \ldots & & r^n \\ & r+1, \; r^2+r, & r^3+r^2 \ldots & & r^n + r^{n-1} \\ & & r^2+2r+1, \; r^3+2r^2+r \ldots & & r^n + 2r^{n-1} + r^{n-2} \\ & & & r^3+3r^2+3r+1 \ldots & r^n + 3r^{n-1} + 3r^{n-2} + r^{n-3} \\ & & & & \vdots \\ & & & & r^n + nr^{n-1} + \frac{n(n-1)}{2} r^{n-2} + \ldots + 1, \end{array}$$

the vertical rows are successive numbers in the ratio $r/(r+1)$, while diagonally we have the geometrical series $1, r+1, (r+1)^2, (r+1)^3 \ldots$.

Next follows the theory of polygonal numbers. It is prefaced by an explanation of the quasi-geometrical way of representing numbers by means of dots or α's. Any number from 2 onwards can be represented as a *line*; the *plane* numbers begin with 3, which is the first number that can be represented in the form of a *triangle*; after triangles follow squares, pentagons, hexagons, &c. (c. 7). Triangles (c. 8) arise by adding any number of successive terms, beginning with 1, of the series of natural numbers

$$1, 2, 3, \ldots n, \ldots.$$

The *gnomons* of triangles are therefore the successive natural numbers. Squares (c. 9) are obtained by adding any number of successive terms of the series of odd numbers, beginning with 1, or

$$1, 3, 5, \ldots 2n-1, \ldots.$$

The *gnomons* of squares are the successive odd numbers. Similarly the *gnomons* of pentagonal numbers (c. 10) are the numbers forming an arithmetical progression with 3 as common difference, or

$$1, 4, 7, \ldots 1+(n-1)3, \ldots;$$

and generally (c. 11) the gnomons of polygonal numbers of a sides are

$$1, \quad 1+(a-2), \quad 1+2(a-2), \ldots 1+(r-1)(a-2), \ldots$$

and the a-gonal number with side n is

$$1+1+(a-2)+1+2(a-2)+\ldots+1+(n-1)(a-2)$$
$$= n+\tfrac{1}{2}n(n-1)(a-2).$$

The general formula is not given by Nicomachus, who contents himself with writing down a certain number of polygonal numbers of each species up to heptagons.

After mentioning (c. 12) that any square is the sum of two successive triangular numbers, i.e.

$$n^2 = \tfrac{1}{2}(n-1)n+\tfrac{1}{2}n(n+1),$$

and that an a-gonal number of side n is the sum of an $(a-1)$-gonal number of side n plus a triangular number of side $n-1$, i.e.

$$n+\tfrac{1}{2}n(n-1)(a-2) = n+\tfrac{1}{2}n(n-1)(a-3)+\tfrac{1}{2}n(n-1),$$

he passes (c. 13) to the first *solid* number, the *pyramid.* The base of the pyramid may be a triangular, a square, or any polygonal number. If the base has the side n, the pyramid is formed by similar and similarly situated polygons placed successively upon it, each of which has 1 less in its side than that which precedes it; it ends of course in a unit at the top, the unit being 'potentially' any polygonal number. Nicomachus mentions the first triangular pyramids as being 1, 4, 10, 20, 35, 56, 84, and (c. 14) explains the formation of the series of pyramids with square bases, but he gives no general formula or summation. An a-gonal number with n in its side being

$$n+\tfrac{1}{2}n(n-1)(a-2),$$

it follows that the pyramid with that polygonal number for base is

$$1+2+3+\ldots+n+\tfrac{1}{2}(a-2)\{1.2+2.3+\ldots+(n-1)n\}$$
$$=\frac{n(n+1)}{2}+\frac{a-2}{2}\cdot\frac{(n-1)n(n+1)}{3}.$$

A pyramid is κόλουρος, *truncated*, when the unit is cut off the top, δικόλουρος, *twice-truncated*, when the unit and the next layer is cut off, τρικόλουρος, *thrice-truncated*, when three layers are cut off, and so on (c. 14).

Other solid numbers are then classified (cc. 15–17): *cubes*, which are the product of three equal numbers; *scalene* numbers, which are the product of three numbers all unequal, and which are alternatively called *wedges* (σφηνίσκοι), *stakes* (σφηκίσκοι), or *altars* (βωμίσκοι). The latter three names are in reality inappropriate to mere products of three unequal factors, since the figure which could properly be called by these names should *taper*, i.e. should have the plane face at the top less than the base. We shall find when we come to the chapter on Heron's mensuration that true (geometrical) βωμίσκοι and σφηνίσκοι have there to be measured in which the top rectangular face is in fact smaller than the rectangular base parallel to it. Iamblichus too indicates the true nature of βωμίσκοι and σφηνίσκοι when he says that they have not only their dimensions but also their faces and angles unequal, and that, while the πλινθίς or δοκίς corresponds to the parallelogram, the σφηνίσκος corresponds to the trapezium.[1] The use, therefore, of the terms in question as alternatives to *scalene* appears to be due to a misapprehension. Other varieties of solid numbers are *parallelepipeds*, in which there are faces which are ἑτερομήκεις (oblong) or of the form $n(n+1)$, so that two factors differ by unity; *beams* (δοκίδες) or *columns* (στηλίδες, Iamblichus) of the form $m^2(m+n)$; *tiles* (πλινθίδες) of the form $m^2(m-n)$. Cubes, the last digit (the units) of which are the same as the last digit in the side, are *spherical*

[1] Iambl. *in Nicom.*, p. 93. 18, 94. 1–3.

(σφαιρικοί) or *recurring* (ἀποκαταστατικοί); these sides and cubes end in 1, 5, or 6, and, as the squares end in the same digits, the squares are called *circular* (κυκλικοί).

Oblong numbers (ἑτερομήκεις) are, as we have seen, of the form $m(m+1)$; *prolate* numbers (προμήκεις) of the form $m(m+n)$ where $n>1$ (c. 18). Some simple relations between oblong numbers, squares, and triangular numbers are given (cc. 19–20). If h_n represents the oblong number $n(n+1)$, and t_n the triangular number $\frac{1}{2}n(n+1)$ of side n, we have, for example,

$$h_n/n^2 = (n+1)/n, \quad h_n - n^2 = n, \quad n^2/h_{n-1} = n/(n-1),$$
$$n^2/h_n = h_n/(n+1)^2, \quad n^2+(n+1)^2+2h_n = (2n+1)^2,$$
$$n^2+h_n = t_{2n}, \quad h_n+(n+1)^2 = t_{2n+1},$$
$$n^2 \pm n = \left.\begin{matrix} h_n \\ h_{n-1} \end{matrix}\right\},$$

all of which formulae are easily verified.

Sum of series of cube numbers.

C. 20 ends with an interesting statement about cubes. If, says Nicomachus, we set out the series of odd numbers

$$1,\ 3,\ 5,\ 7,\ 9,\ 11,\ 13,\ 15,\ 17,\ 19, \ldots$$

the first (1) is a cube, the sum of the next *two* $(3+5)$ is a cube, the sum of the next *three* $(7+9+11)$ is a cube, and so on. We can prove this law by assuming that n^3 is equal to the sum of n odd numbers beginning with $2x+1$ and ending with $2x+2n-1$. The sum is $(2x+n)n$; since therefore $(2x+n)n = n^3$,

$$x = \tfrac{1}{2}(n^2-n),$$

and the formula is

$$(n^2-n+1)+(n^2-n+3)+\ldots+(n^2+n-1) = n^3.$$

By putting successively $n = 1, 2, 3 \ldots r$, &c., in this formula and adding the results we find that

$$1^3+2^3+3^3+\ldots+r^3 = 1+(3+5)+(7+9+11)+\ldots+(\ldots r^2+r-1).$$

The number of terms in this series of odd numbers is clearly

$$1+2+3+\ldots+r \quad \text{or} \quad \tfrac{1}{2}r(r+1).$$

Therefore
$$\begin{aligned} 1^3+2^3+3^3+\ldots+r^3 &= \tfrac{1}{4}r(r+1)(1+r^2+r-1) \\ &= \{\tfrac{1}{2}r(r+1)\}^2. \end{aligned}$$

Nicomachus does not give this formula, but it was known to the Roman *agrimensores*, and it would be strange if Nicomachus was not aware of it. It may have been discovered by the same mathematician who found out the proposition actually stated by Nicomachus, which probably belongs to a much earlier time. For the Greeks were from the time of the early Pythagoreans accustomed to summing the series of odd numbers by placing 3, 5, 7, &c., successively as gnomons round 1; they knew that the result, whatever the number of gnomons, was always a square, and that, if the number of gnomons added to 1 is (say) r, the sum (including the 1) is $(r+1)^2$. Hence, when it was once discovered that the first cube after 1, i.e. 2^3, is $3+5$, the second, or 3^3, is $7+9+11$, the third, or 4^3, is $13+15+17+19$, and so on, they were in a position to sum the series $1^3+2^3+3^3+\ldots+r^3$; for it was only necessary to find out how many terms of the series $1+3+5+\ldots$ this sum of cubes includes. The number of terms being clearly $1+2+3+\ldots+r$, the number of gnomons (including the 1 itself) is $\frac{1}{2}r(r+1)$; hence the sum of them all (including the 1), which is equal to

$$1^3+2^3+3^3+\ldots+r^3,$$

is $\{\frac{1}{2}r(r+1)\}^2$. Fortunately we possess a piece of evidence which makes it highly probable that the Greeks actually dealt with the problem in this way. Alkarkhī, the Arabian algebraist of the tenth-eleventh century, wrote an algebra under the title *Al-Fakhrī*. It would seem that there were at the time two schools in Arabia which were opposed to one another in that one favoured Greek, and the other Indian, methods. Alkarkhī was one of those who followed Greek models almost exclusively, and he has a proof of the theorem now in question by means of a figure with gnomons drawn in it, furnishing an excellent example of the geometrical algebra which is so distinctively Greek.

Let AB be the side of a square AC; let

$$AB = 1+2+\ldots+n = \tfrac{1}{2}n(n+1),$$

and suppose $BB' = n$, $B'B'' = n-1$, $B''B''' = n-2$, and so on. Draw the squares on AB', $AB''\ldots$ forming the gnomons shown in the figure.

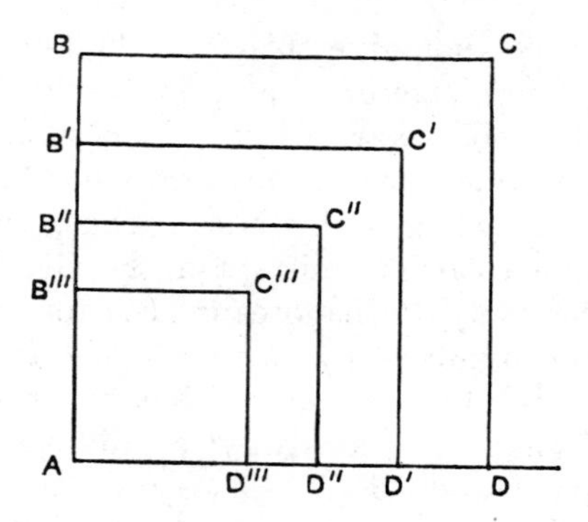

Then the gnomon

$$BC'D = BB' \,.\, BC + DD' \,.\, C'D'$$
$$= BB'(BC + C'D').$$

Now $BC = \frac{1}{2}n(n+1)$,

$$C'D' = 1+2+3+\ldots+(n-1) = \tfrac{1}{2}n(n-1), \qquad BB' = n;$$

therefore (gnomon $BC'D$) $= n \,.\, n^2 = n^3$.

Similarly (gnomon $B'C''D'$) $= (n-1)^3$, and so on.

Therefore $1^3+2^3+\ldots+n^3$ = the sum of the gnomons round the small square at A which has 1 for its side *plus* that small square; that is,

$$1^3+2^3+3^3+\ldots+n^3 = \text{square } AC = \{\tfrac{1}{2}n(n+1)\}^2.$$

It is easy to see that the first gnomon about the small square at A is $3+5 = 2^3$, the next gnomon is $7+9+11 = 3^3$, and so on.

The demonstration therefore hangs together with the theorem stated by Nicomachus. Two alternatives are possible. Alkarkhī may have devised the proof himself in the Greek manner, following the hint supplied by Nicomachus's theorem. Or he may have found the whole proof set out in some Greek treatise now lost and reproduced it. Whichever alternative is the true one, we can hardly doubt the Greek origin of the summation of the series of cubes.

Nicomachus passes to the theory of arithmetical proportion and the various *means* (cc. 21–9), a description of which has already been given (p. 87 above). There are a few more propositions to be mentioned under this head. If $a-b = b-c$, so that a, b, c are in arithmetical progression, then (c. 23. 6)

$$b^2 - ac = (a-b)^2 = (b-c)^2,$$

a fact which, according to Nicomachus, was not generally known. Boëtius[1] mentions this proposition which, if we take $a+d$, a, $a-d$ as the three terms in arithmetical progression, may be written $a^2 = (a+d)(a-d)+d^2$. This is presumably the origin of the *regula Nicomachi* quoted by one Ocreatus (? O'Creat), the author of a tract, *Prologus in Helceph*, written in the twelfth or thirteenth century[2] ('Helceph' or 'Helcep' is evidently equivalent to *Algorismus*; may it perhaps be meant for the *Al-Kāfī* of Alkarkhī ?). The object of the *regula* is to find the square of a number containing a single digit. If $d = 10-a$, or $a+d = 10$, the rule is represented by the formula

$$a^2 = 10(a-d)+d^2,$$

so that the calculation of a^2 is made to depend on that of d^2 which is easier to evaluate if $d<a$.

Again (c. 24. 3, 4), if a, b, c be three terms in descending geometrical progression, r being the common ratio (a/b or b/c), then

$$\frac{a-b}{b-c} = \frac{a}{b} = \frac{b}{c}$$

and

$$(a-b) = (r-1)b, \quad (b-c) = (r-1)c,$$
$$(a-b)-(b-c) = (r-1)(b-c).$$

It follows that

$$b = a-b(r-1) = c+c(r-1).$$

This is the property of three terms in geometrical progression which corresponds to the property of three terms a, b, c of a harmonical progression

$$b = a-\frac{a}{n} = c+\frac{c}{n},$$

from which we derive

$$n = (a+c)/(a-c).$$

If a, b, c are in descending order, Nicomachus observes (c. 25) that $\frac{a}{b} <=> \frac{b}{c}$ according as a, b, c are in arithmetical, geometrical, or harmonical progression.

[1] Boëtius, *Inst. Ar.* ii. c. 43.
[2] See *Abh. zur Gesch. d. Math.* 3, 1880, p. 134.

The 'Platonic theorem' (c. 24. 6) about the number of possible means (geometric) between two square numbers and between two cube numbers respectively has already been mentioned (pp. 89, 90), as also the 'most perfect proportion' (p. 86).

THEON OF SMYRNA was the author of a book purporting to be a manual of mathematical subjects such as a student would require to enable him to understand Plato. A fuller account of this work will be given later; at present we are only concerned with the arithmetical portion. This gives the elementary theory of numbers on much the same lines as we find it in Nicomachus, though less systematically. We can here pass over the things which are common to Theon and Nicomachus and confine ourselves to what is peculiar to the former. The important things are two. One is the theory of side- and diameter-numbers invented by the Pythagoreans for the purpose of finding the successive integral solutions of the equations $2x^2-y^2 = \pm 1$; as to this see pp. 91–3 above. The other is an explanation of the limited number of forms which square numbers may have.[1] If m^2 is a square number, says Theon, either m^2 or m^2-1 is divisible by 3, and again either m^2 or m^2-1 is divisible by 4: which is equivalent to saying that a square number cannot be of any of the following forms, $3n+2$, $4n+2$, $4n+3$. Again, he says, for any square number m^2, *one* of the following alternatives must hold:

(1) $\dfrac{m^2-1}{3}$, $\dfrac{m^2}{4}$ both integral (e.g. $m^2 = 4$),

(2) $\dfrac{m^2-1}{4}$, $\dfrac{m^2}{3}$ both integral (e.g. $m^2 = 9$),

(3) $\dfrac{m^2}{3}$, $\dfrac{m^2}{4}$ both integral (e.g. $m^2 = 36$),

(4) $\dfrac{m^2-1}{3}$, $\dfrac{m^2-1}{4}$ both integral (e.g. $m^2 = 25$).

[1] Theon of Smyrna, p. 35. 17–36. 2.

Iamblichus states the same facts in a slightly different form.[1] The truth of these statements can be seen in the following way.[2] Since any number m must have one of the following forms

$$6k, \quad 6k \pm 1, \quad 6k \pm 2, \quad 6k \pm 3,$$

any square m^2 must have one or other of the forms

$$36k^2, \quad 36k^2 \pm 12k + 1, \quad 36k^2 \pm 24k + 4, \quad 36k^2 \pm 36k + 9.$$

For squares of the first type $\frac{m^2}{3}$ and $\frac{m^2}{4}$ are both integral, for those of the second type $\frac{m^2-1}{3}$, $\frac{m^2-1}{4}$ are both integral, for those of the third type $\frac{m^2-1}{3}$ and $\frac{m^2}{4}$ are both integral, and for those of the fourth type $\frac{m^2}{3}$ and $\frac{m^2-1}{4}$ are both integral; which agrees with Theon's statement. Again, if the four forms of squares be divided by 3 or 4, the remainder is always either 0 or 1; so that, as Theon says, no square can be of the form $3n+2$, $4n+2$, or $4n+3$. We can hardly doubt that these discoveries were also Pythagorean.

IAMBLICHUS, born at Chalcis in Coele-Syria, was a pupil of Anatolius and Porphyry, and belongs to the first half of the fourth century A.D. He wrote nine Books on the Pythagorean Sect, the titles of which were as follows: I. On the Life of Pythagoras; II. Exhortation to philosophy (Προτρεπτικὸς ἐπὶ φιλοσοφίαν); III. On mathematical science in general; IV. On Nicomachus's *Introductio Arithmetica*; V. On arithmetical science in physics; VI. On arithmetical science in ethics; VII. On arithmetical science in theology; VIII. On the Pythagorean geometry; IX. On the Pythagorean music. The first four of these books survive and are accessible in modern editions; the other five are lost, though extracts from VII. are doubtless contained in the *Theologumena arithmetices*. Book IV. on Nicomachus's *Introductio* is that which concerns us here; and the few things requiring notice are the following. The first is the view of a square number

[1] Iambl. *in Nicom.*, p. 90. 6–11.

[2] Cf. Loria, *Le scienze esatte nell' antica Grecia*, p. 834.

as a race-course (δίαυλος)[1] formed of successive numbers from 1 (as *start*, ὕσπληξ) up to n, the side of the square, which is the turning-point (καμπτήρ), and then back again through $(n-1)$, $(n-2)$, &c., to 1 (the *goal*, νύσσα), thus:

$$\begin{array}{l} 1+2+3+4 \ldots \qquad\qquad (n-1)+_{n} \\ 1+2+3+4 \ldots (n-2)+(n-1)+ \end{array}.$$

This is of course equivalent to the proposition that n^2 is the sum of the two triangular numbers $\frac{1}{2}n(n+1)$ and $\frac{1}{2}(n-1)n$ with sides n and $n-1$ respectively. Similarly Iamblichus points out[2] that the *oblong* number

$$n(n-1) = (1+2+3+\ldots+n)+(n-2+n-3+\ldots+3+2).$$

He observes that it was on this principle that, after 10, which was called the *unit of the second course* (δευτερωδουμένη μονάς), the Pythagoreans regarded $100 = 10.10$ as the *unit of the third course* (τριωδουμένη μονάς), $1000 = 10^3$ as the *unit of the fourth course* (τετρωδουμένη μονάς), and so on,[3] since

$$1+2+3+\ldots+10+9+8+\ldots+2+1 = 10.10,$$
$$10+20+30+\ldots+100+90+80+\ldots+20+10 = 10^3,$$
$$100+200+300+\ldots+1000+900+\ldots+200+100 = 10^4,$$

and so on. Iamblichus sees herein the special virtue of 10: but of course the same formulae would hold in any scale of notation as well as the decimal.

In connexion with this Pythagorean decimal terminology Iamblichus gives a proposition of the greatest interest.[4] Suppose we have any three consecutive numbers the greatest of which is divisible by 3. Take the sum of the three numbers; this will consist of a certain number of units, a certain number of tens, a certain number of hundreds, and so on. Now take the units in the said sum as they are, then as many units as there are tens in the sum, as many units as there are hundreds, and so on, and add all the units so obtained together (i.e. add the *digits* of the sum expressed in our decimal notation). Apply the same procedure to the

[1] Iambl. *in Nicom.*, p. 75. 25–77. 4.
[2] *Ib.*, pp. 77. 4–80. 9.
[3] *Ib.*, pp. 88. 15–90. 2.
[4] *Ib.*, pp. 103. 10–104. 13.

result, and so on. Then, says Iamblichus, *the final result will be the number* 6. E.g. take the numbers 10, 11, 12; the sum is 33. Add the digits, and the result is 6. Take 994, 995, 996: the sum is 2985; the sum of the digits is 24; and the sum of the digits of 24 is again 6. The truth of the general proposition is seen in this way.[1]

Let $$N = n_0 + 10n_1 + 10^2 n_2 + \dots$$

be a number written in the decimal notation. Let $S(N)$ represent the sum of its digits, $S^{(2)}(N)$ the sum of the digits of $S(N)$ and so on.

Now $$N - S(N) = 9\,(n_1 + 11n_2 + 111n_3 + \dots),$$

whence $$N \equiv S(N) \pmod{9}.$$

Similarly $$S(N) \equiv S^{(2)}N \pmod{9}.$$
$$\vdots$$

Let $$S^{(k-1)}(N) \equiv S^{(k)}N \pmod{9}$$

be the last possible relation of this kind; $S^{(k)}N$ will be a number $N' \leqq 9$.

Adding the congruences, we obtain

$$N \equiv N' \pmod{9}, \text{ while } N' \leqq 9.$$

Now, if we have three consecutive numbers the greatest of which is divisible by 3, we can put for their sum

$$N = (3p+1) + (3p+2) + (3p+3) = 9p + 6,$$

and the above congruence becomes

$$9p + 6 \equiv N' \pmod{9},$$

so that $$N' \equiv 6 \pmod{9};$$

and, since $N' \leqq 9$, N' can only be equal to 6.

This addition of the digits of a number expressed in our notation has an important parallel in a passage of the *Refutation of all Heresies* by saint Hippolytus,[2] where there is a description of a method of foretelling future events called the 'Pythagorean calculus'. Those, he says, who claim to predict events by means of calculations with numbers, letters and names use the principle of the *pythmen* or *base*,

[1] Loria, *op. cit.*, pp. 841–2.
[2] Hippolytus, *Refut.* iv, c. 14.

that is, what we call a digit of a number expressed in our decimal notation; for the Greeks, in the case of any number above 9, the *pythmen* was the same number of units as the alphabetical numeral contains tens, hundreds, thousands, &c. Thus the *pythmen* of 700 (ψ in Greek) is 7 (ζ); that of ͵ϛ (6000) is ϛ (6), and so on. The method then proceeded to find the *pythmen* of a certain name, say Ἀγαμέμνων. Taking the *pythmenes* of all the letters and adding them, we have

$$1+3+1+4+5+4+5+8+5 = 36.$$

Take the *pythmenes* of 36, namely 3 and 6, and their sum is 9. The *pythmen* of Ἀγαμέμνων is therefore 9. Next take the name Ἕκτωρ; the *pythmenes* are 5, 2, 3, 8, 1, the sum of which is 19; the *pythmenes* of 19 are 1, 9; the sum of 1 and 9 is 10, the pythmen of which is 1. The *pythmen* of Ἕκτωρ is therefore 1. 'It is easier', says Hippolytus, 'to proceed thus. Finding the *pythmenes* of the letters, we obtain, in the case of Ἕκτωρ, 19 as their sum. Divide this by 9 and note the remainder: thus, if I divide 19 by 9, the remainder is 1, for nine times 2 is 18, and 1 is left, which will accordingly be the *pythmen* of the name Ἕκτωρ.' Again, take the name Πάτροκλος. The sum of the *pythmenes* is

$$8+1+3+1+7+2+3+7+2 = 34:$$

and $3+4 = 7$, so that 7 is the *pythmen* of Πάτροκλος. 'Those then who calculate by the *rule of nine* take one-ninth of the sum of the *pythmenes* and then determine the sum of the *pythmenes* in the remainder. Those on the other hand who follow the "rule of seven" divide by 7. Thus the sum of the *pythmenes* in Πάτροκλος was found to be 34. This, divided by 7, gives 4, and since 7 times 4 is 28, the remainder is 6. . . .' 'It is necessary to observe that, if the division gives an integral quotient (without remainder), . . . the *pythmen* is the number 9 itself' (that is, if the *rule of nine* is followed). And so on.

Two things emerge from this fragment. (1) The use of the *pythmen* was not appearing for the first time when Apollonius framed his system for expressing and multiplying large numbers; it originated much earlier, with the Pythagoreans.

(2) The method of calculating the *pythmen* is like the operation of 'casting out nines' in the proof which goes by that name, where we take the sum of the digits of a number and divide by 9 to get the remainder. The method of verification by 'casting out nines' came to us from the Arabs, who may, as Avicenna and Maximus Planudes tell us, have got it from the Indians; but the above evidence shows that, at all events, the elements from which it was built up lay ready to hand in the Pythagorean arithmetic.

IV

THE EARLIEST GREEK GEOMETRY. THALES

The 'Summary' of Proclus.

WE shall often, in the course of this history, have occasion to quote from the so-called 'Summary' of Proclus, which has already been cited in the preceding chapter. Occupying a few pages (65–70) of Proclus's *Commentary on Euclid*, Book I, it reviews, in the briefest possible outline, the course of Greek geometry from the earliest times to Euclid, with special reference to the evolution of the Elements. At one time it was often called the 'Eudemian summary', on the assumption that it was an extract from the great *History of Geometry* in four Books by Eudemus, the pupil of Aristotle. But a perusal of the summary itself is sufficient to show that it cannot have been written by Eudemus; the most that can be said is that, down to a certain sentence, it was probably based, more or less directly, upon data appearing in Eudemus's *History*. At the sentence in question there is a break in the narrative, as follows:

'Those who have compiled histories bring the development of this science up to this point. Not much younger than these is Euclid, who put together the Elements, collecting many of the theorems of Eudoxus, perfecting many others by Theaetetus, and bringing to irrefragable demonstration the propositions which had only been somewhat loosely proved by his predecessors.'

Since Euclid was later than Eudemus, it is impossible that Eudemus can have written this; while the description 'those who have compiled histories', and who by implication were a little older than Euclid, suits Eudemus excellently. Yet the style of the summary after the break does not show any such change from that of the earlier portion as to suggest different authorship. The author of the earlier portion frequently refers to the question of the origin of the Elements of Geometry in a way in which no one would be likely to write who was not later than Euclid; and it seems to be the same hand which, in the second portion, connects the Elements of Euclid with the work of Eudoxus and Theaetetus. Indeed the author, whoever he was, seems to have compiled the summary with one main object in view, namely, to trace the origin and growth of the Elements of Geometry; consequently he omits to refer to certain famous discoveries in geometry such as the solutions of the problem of the duplication of the cube, doubtless because they did not belong to the Elements. In two cases he alludes to such discoveries, as it were in parenthesis, in order to recall to the mind of the reader a current association of the name of a particular geometer with a particular discovery. Thus he mentions Hippocrates of Chios as a famous geometer for the particular reason that he was the first to write Elements, and he adds to his name, for the purpose of identification, 'the discoverer of the quadrature of the lune'. Similarly, when he says of Pythagoras '(he it was) who' (ὃς δὴ ...) 'discovered the theory of irrationals [or "proportions"] and the construction of the cosmic figures', he seems to be alluding, entirely on his own account, to a popular tradition to that effect. If the summary is the work of one author, who was it? Tannery answers that it was Geminus; but this seems highly improbable, for the extracts from Geminus's work which we possess suggest that the subjects therein discussed were of a different kind; they seem rather to have been general questions relating to the philosophy and content of mathematics, and even Tannery admits that historical details could only have come incidentally into the work.

Could the author have been Proclus himself? This again seems, on the whole, improbable. In favour of the authorship of Proclus are the facts (1) that the question of the origin of the Elements is kept prominent and (2) that there is no mention of Democritus, whom Eudemus would not have ignored, while a follower of Plato such as Proclus might have done him this injustice, following the example of Plato himself, who was an opponent of Democritus, never once mentions him, and is said to have wished to burn all his writings. On the other hand (1) the style of the summary is not such as to point to Proclus as the author; (2) if he wrote it, it is hardly conceivable that he would have passed over in silence the discovery of the analytical method, 'the finest', as he says elsewhere, of the traditional methods in geometry, 'which Plato is said to have communicated to Leodamas'. Nor (3) is it easy to suppose that Proclus would have spoken in the detached way that the author does of Euclid whose *Elements* was the subject of his whole commentary: 'Not much younger than these is Euclid, who compiled the Elements . . .'. 'This man lived in the time of the first Ptolemy . . .'. On the whole, therefore, it would seem probable that the body of the summary was taken by Proclus from a compendium made by some writer later than Eudemus, though the earlier portion was based, directly or indirectly, upon notices in Eudemus's *History*. But the prelude with which the summary is introduced may well have been written, or at all events expanded, by Proclus himself, for it is in his manner to bring in 'the inspired Aristotle' (ὁ δαιμόνιος Ἀριστοτέλης)—as he calls him here and elsewhere—and the transition to the story of the Egyptian origin of geometry may also be his:

'Since, then, we have to consider the beginnings of the arts and sciences with reference to the particular cycle [of the series postulated by Aristotle] through which the universe is at present passing, *we say* that, according to most accounts, geometry was first discovered in Egypt, having had its origin in the measurement of areas. For this was a necessity for the Egyptians owing to the rising of the Nile which effaced the proper boundaries of everybody's lands.'

The next sentences also may well be due to Proclus:

'And it is in no way surprising that the discovery of this as well as the other sciences had its beginning in practical needs, seeing that everything that is in the course of becoming progresses from the imperfect to the perfect. Thus the transition from sensation to reasoning and from reasoning to understanding is only natural.'

These sentences look like reflections by Proclus, and the transition to the summary proper follows, in the words:

'Accordingly, just as exact arithmetic began among the

Phoenicians owing to its use in commerce and contracts, so geometry was discovered in Egypt for the reason aforesaid.'

Tradition as to the origin of geometry.

Many Greek writers besides Proclus give a similar account of the origin of geometry. Herodotus says that Sesostris (Ramses II, *circa* 1300 B.C.) distributed the land among all the Egyptians in equal rectangular plots, on which he levied an annual tax; when therefore the river swept away a portion of a plot and the owner applied for a corresponding reduction in the tax, surveyors had to be sent down to certify what the reduction in the area had been. 'This, in my opinion (δοκέει μοι)', he continues, 'was the origin of geometry, which then passed into Greece.'[1] The same story, a little amplified, is repeated by other writers, Heron of Alexandria,[2] Diodorus Siculus,[3] and Strabo.[4] True, all these statements (even if that in Proclus was taken directly from Eudemus's *History of Geometry*) may all be founded on the passage of Herodotus, and Herodotus may have stated as his own inference what he was told in Egypt; for Diodorus gives it as an Egyptian tradition that geometry and astronomy were the discoveries of Egypt, and says that the Egyptian priests claimed Solon, Pythagoras, Plato, Democritus, Oenopides of Chios, and Eudoxus as their pupils. But the Egyptian claim to the discoveries was never disputed by the Greeks. In Plato's *Phaedrus* Socrates is made to say that he had heard that the Egyptian god Theuth was the first to invent arithmetic, the science of calculation, geometry, and astronomy.[5] Similarly Aristotle says that the mathematical arts first took shape in Egypt, though he gives as the reason, not the practical need which arose for a scientific method of measuring land, but the fact that in Egypt there was a leisured class, the priests, who could spare time for such things.[6] Democritus boasted that no one of his time had excelled him 'in making lines into figures and proving their properties, not even the so-called *Harpedonaptae* in Egypt'.[7] This word, compounded of two Greek words, ἁρπεδόνη and ἅπτειν, means 'rope-stretchers' or 'rope-

[1] Herodotus ii. 109.
[2] Heron, *Geom.* c. 2, p. 176, Heib.
[3] Diod. Sic. i. 69, 81.
[4] Strabo xvii. c. 3.
[5] Plato, *Phaedrus* 274 C.
[6] Arist. *Metaph.* A. 1, 981 b 23.
[7] Clem. *Strom.* i. 15. 69 (*Vorsokratiker*, ii³, p. 123. 5–7).

fasteners'; and, while it is clear from the passage that the persons referred to were clever geometers, the word reveals a characteristic *modus operandi*. The Egyptians were extremely careful about the orientation of their temples, and the use of ropes and pegs for marking out the limits, e.g. corners, of the sacred precincts is portrayed in all pictures of the laying of foundation stones of temples.[1] The operation of 'rope-stretching' is mentioned in an inscription on leather in the Berlin Museum as having been in use as early as Amenemhat I (say 2300 B.C.).[2] Now it was the practice of ancient Indian and probably also of Chinese geometers to make, for instance, a right angle by stretching a rope divided into three lengths in the ratio of the sides of a right-angled triangle in rational numbers, e.g. 3, 4, 5, in such a way that the three portions formed a triangle, when of course a right angle would be formed at the point where the two smaller sides meet. There seems to be no doubt that the Egyptians knew that the triangle (3, 4, 5), the sides of which are so related that the square on the greatest side is equal to the sum of the squares on the other two, is right-angled; if this is so, they were acquainted with at least one case of the famous proposition of Pythagoras.

Egyptian geometry, i.e. mensuration.

We might suppose, from Aristotle's remark about the Egyptian priests being the first to cultivate mathematics because they had leisure, that their geometry would have advanced beyond the purely practical stage to something more like a theory or science of geometry. But the documents which have survived do not give any ground for this supposition; the art of geometry in the hands of the priests never seems to have advanced beyond mere routine. The most important available source of information about Egyptian mathematics is the Papyrus Rhind, written probably about 1700 B.C. but copied from an original of the time of King Amenemhat III (Twelfth Dynasty), say 2200 B.C. The geometry in this 'guide for calculation, a means of ascertaining everything, of elucidating all obscurities, all mysteries, all

[1] Brugsch, *Steininschrift und Bibelwort*, 2nd ed., p. 36.
[2] Dümichen, *Denderatempel*, p. 33.

difficulties', as it calls itself, is rough *mensuration.* The following are the cases dealt with which concern us here. (1) There is the *rectangle,* the area of which is of course obtained by multiplying together the numbers representing the sides. (2) The measure of a *triangle* is given as the product of half the base into the *side.* And here there is a difference of opinion as to the kind of triangle measured. Eisenlohr and Cantor, taking the diagram to represent an *isosceles* triangle rather inaccurately drawn, have to assume error on the part of the writer in making the area $\frac{1}{2}ab$ instead of $\frac{1}{2}a\sqrt{(b^2-\frac{1}{4}a^2)}$ where a is the base and b the 'side', an error which of course becomes less serious as a becomes smaller relatively to b (in the case taken $a = 4, b = 10$, and the area as given according to the rule, i.e. 20, is not greatly different from the true value 19·5959). But other authorities take the triangle to be *right-angled* and b to be the side perpendicular to the base, their argument being that the triangle as drawn is not a worse representation of a right-angled triangle than other triangles purporting to be right-angled which are found in other manuscripts, and indeed is a better representation of a right-angled triangle than it is of an isosceles triangle, while the number representing the side is shown in the figure alongside one only of the sides, namely that adjacent to the angle which the more nearly represents a right angle. The advantage of this interpretation is that the rule is then correct instead of being more inaccurate than one would expect from a people who had expert land surveyors to measure land for the purpose of assessing it to tax. The same doubt arises with reference to (3) the formula for the area of a trapezium, namely $\frac{1}{2}(a+c)\times b$, where a, c are the base and the opposite parallel side respectively, while b is the 'side', i.e. one of the non-parallel sides. In this case the figure seems to have been intended to be isosceles, whereas the formula is only accurate if b, one of the non-parallel sides, is at right angles to the base, in which case of course the side opposite to b is not at right angles to the base. As the parallel sides (6, 4) in the case taken are short relatively to the 'side' (20), the angles at the base are not far short of being right angles, and it is possible that one of them, adjacent to the particular side which is marked 20, was intended to be right. The hypothesis that the triangles and trapezia are isosceles, and that the formulae are therefore crude and inaccurate, was thought to be confirmed by the evidence of inscriptions on the Temple of Horus at Edfu. This temple was planned out in 237 B.C.; the inscriptions which refer to the assignment of plots of ground to the priests belong to the reign of Ptolemy XI, Alexander I (107–88 B.C.). From so much of these inscriptions as were published by Lepsius[1] we gather that $\frac{1}{2}(a+c) \cdot \frac{1}{2}(b+d)$ was a formula for the area of a quadrilateral the sides of which in order are a, b, c, d. Some of the quadrilateral figures are evidently trapezia with the non-parallel sides equal; others are not, although they are commonly not far from being rectangles or isosceles trapezia. Examples are '16 to 15 and 4 to $3\frac{1}{2}$ make $58\frac{1}{8}$' (i.e. $\frac{1}{2}(16+15) \times \frac{1}{2}(4+3\frac{1}{2}) = 58\frac{1}{8}$); '$9\frac{1}{2}$ to $10\frac{1}{2}$ and $24\frac{1}{2}\ \frac{1}{8}$ to $22\frac{1}{2}\ \frac{1}{8}$ make $236\frac{1}{4}$'; '22 to 23 and 4 to 4 make 90', and so on. Triangles are not made the subject of a separate formula, but are regarded as cases of quadrilaterals in which the length of one side is zero. Thus the triangle 5, 17, 17 is described as a figure with sides '0 to 5 and 17 to 17', the area being accordingly $\frac{1}{2}(0+5) \cdot \frac{1}{2}(17+17)$ or $42\frac{1}{2}$; 0 is expressed by hieroglyphs meaning the word Nen. It is remarkable enough that the use of a formula so inaccurate should have lasted till 200 years or so after Euclid had lived and taught in Egypt; there is also a case of its use in the *Liber Geeponicus* formerly attributed to Heron,[2] the quadrilateral having two opposite sides parallel and the pairs of opposite sides being (32, 30) and (18, 16). But it is right to add that, in the rest of the Edfu inscriptions published later by Brugsch, there are cases where the inaccurate formula is not used, and it is suggested that what is being attempted in these cases is an approximation to the square root of a non-square number.[3]

We come now (4) to the mensuration of circles as found in the Papyrus Rhind. If d is the diameter, the area is given as $\{(1-\frac{1}{9})d\}^2$ or $\frac{64}{81}d^2$. As this is the corresponding figure to $\frac{1}{4}\pi d^2$, it follows that the value of π is taken as $\frac{256}{81} = (\frac{16}{9})^2$, or 3·16, very nearly. A somewhat different value for π has been inferred from measurements of certain

[1] 'Ueber eine hieroglyphische Inschrift am Tempel von Edfu' (*Abh. der Berliner Akad.*, 1855, pp. 69–114).

[2] Heron, ed. Hultsch, p. 212. 15–20 (Heron, *Geom.* c. 6. 2, Heib.).

[3] M. Simon, *Gesch. d. Math. im Altertum*, p. 48.

heaps of grain or of spaces which they fill. Unfortunately the shape of these spaces or heaps cannot be determined with certainty. The word in the Papyrus Rhind is *ṣhaȧ*; it is evident that it ordinarily means a rectangular parallelepiped, but it can also be applied to a figure with a circular base, e.g. a cylinder, or a figure resembling a thimble, i.e. with a rounded top. There is a measurement of a mass of corn apparently of the latter sort in one of the Kahuṇ papyri.[1] The figure shows a circle with $1365\frac{1}{3}$ as the content of the heap written within it, and with 12 and 8 written above and to the left of the circle respectively. The calculation is done in this way. 12 is taken and $\frac{1}{3}$ of it added; this gives 16; 16 is squared, which gives 256, and finally 256 is multiplied by $\frac{2}{3}$ of 8, which gives $1365\frac{1}{3}$. If for the original figures 12 and 8 we write h and k respectively, the formula used for the content is $(\frac{4}{3}h)^2 . \frac{2}{3}k$. Griffith took 12 to be the height of the figure and 8 to be the diameter of the base. But according to another interpretation,[2] 12 is simply $\frac{3}{2}$ of 8, and the figure to be measured is a hemisphere with diameter 8 ells. If this is so, the formula makes the content of a hemisphere of diameter k to be $(\frac{4}{3} . \frac{3}{2}k)^2 . \frac{2}{3}k$ or $\frac{8}{3}k^3$. Comparing this with the true volume of the hemisphere, $\frac{2}{3} . \frac{1}{8}\pi k^3$ or $\frac{1}{12}\pi k^3 = 134{\cdot}041$ cubic ells, we see that the result $1365\frac{1}{3}$ obtained by the formula must be expressed in $\frac{1}{10}$ths of a cubic ell: consequently for $\frac{1}{12}\pi$ the formula substitutes $\frac{8}{30}$, so that the formula gives $3{\cdot}2$ in place of π, a value different from the $3{\cdot}16$ of Ahmes. Borchardt suggests that the formula for the measurement of a hemisphere was got by repeated practical measurements of heaps of corn built up as nearly as possible in that form, in which case the inaccuracy in the figure for π is not surprising. With this problem from the Kahuṇ papyri must be compared No. 43 from the Papyrus Rhind. A curious feature in the measurements of stores or heaps of corn in the Papyrus Rhind is the fact, not as yet satisfactorily explained, that the area of the base (square or circular) is first found and is then regularly multiplied, not into the 'height' itself, but into $\frac{3}{2}$ times the height. But in No. 43 the calculation is different and more parallel to the case in the Kahuṇ papyrus. The problem is to find the content of a space round

[1] Griffith, *Kahuṇ Papyri*, Pt. I, Plate 8.

[2] Simon, *l. c.*

in form '9 in height and 6 in breadth'. The word *qa*, here translated 'height', is apparently used in other documents for 'length' or 'greatest dimension', and must in this case mean the diameter of the base, while the 'breadth' is the height in our sense. If we denote the diameter of the circular base by k, and the height by h, the formula used in this problem for finding the volume is $(\frac{4}{3} . \frac{8}{9}k)^2 . \frac{2}{3}h$. Here it is not $\frac{3}{2}h$, but $\frac{2}{3}h$, which is taken as the last factor of the product. Eisenlohr suggests that the analogy of the formula for a hemisphere, $\pi r^2 . \frac{2}{3}r$, may have operated to make the calculator take $\frac{2}{3}$ of the height, although the height is not in the particular case the same as the radius of the base, but different. But there remains the difficulty that $(\frac{4}{3})^2$ or $\frac{16}{9}$ times the area of the circle of diameter k is taken instead of the area itself. As to this Eisenlohr can only suggest that the circle of diameter k which was accessible for measurement was not the real or mean circular section, and that allowance had to be made for this, or that the base was not a circle of diameter k but an *ellipse* with $\frac{16}{9}k$ and k as major and minor axes. But such explanations can hardly be applied to the factor $(\frac{4}{3})^2$ in the Kahuṇ case *if* the latter is really the case of a hemispherical space as suggested. Whatever the true explanation may be, it is clear that these rules of measurement must have been empirical and that there was little or no geometry about them.

Much more important geometrically are certain calculations with reference to the proportions of pyramids (Nos. 56–9 of the Papyrus Rhind) and a monument (No. 60). In the case of the pyramid two lines in the figure are distinguished, (1) *ukha-thebt*, which is evidently some line in the base, and (2) *pir-em-us* or *per-em-us* ('height'), a word from which the name πυραμίς may have been derived.[1] The object of

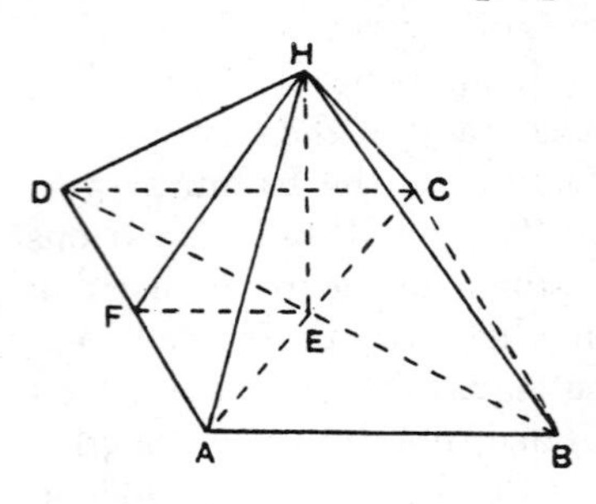

[1] Another view is that the words πυραμίς and πυραμοῦς, meaning a kind of cake made from roasted wheat and honey, are derived from πυροί, 'wheat', and are thus of purely Greek origin.

the problems is to find a certain relation called *se-qeṭ*, literally 'that which makes the nature', i.e. that which determines the proportions of the pyramid. The relation $se\text{-}qe\d{t} = \dfrac{\frac{1}{2}\,ukha\text{-}thebt}{piremus}$. In the case of the monument we have two other names for lines in the figure, (1) *senti*, 'foundation', or base, (2) *qay en ḥeru*, 'vertical length', or height; the same term *se-qeṭ* is used for the relation $\dfrac{\frac{1}{2}\,senti}{qay\ en\ \d{h}eru}$ or the same inverted. Eisenlohr and Cantor took the lines (1) and (2) in the case of the pyramid to be different from the lines (1) and (2) called by different names in the monument. Suppose $ABCD$ to be the square base of a pyramid, E its centre, H the vertex, and F the middle point of the side AD of the base. According to Eisenlohr and Cantor the *ukha-thebt* is the diagonal, say AC, of the base, and the *pir-em-us* is the *edge*, as AH. On this assumption the *se-qeṭ*

$$= \frac{AE}{AH} = \cos HAE.$$

In the case of the monument they took the *senti* to be the side of the base, as AB, the *qay en ḥeru* to be the height of the pyramid EH, and the *se-qeṭ* to be the ratio of EH to $\frac{1}{2}AB$ or of EH to EF, i.e. the *tangent* of the angle HFE which is the slope of the faces of the pyramid. According to Eisenlohr and Cantor, therefore, the one term *se-qeṭ* was used in two different senses, namely, in Nos. 56–9 for $\cos HAE$ and in No. 60 for $\tan HFE$. Borchardt has, however, proved that the *se-qeṭ* in all the cases has one meaning, and represents the *cotangent* of the slope of the faces of the pyramid, i.e. $\cot HFE$ or the ratio of FE to EH. There is no difficulty in the use of the different words *ukha-thebt* and *senti* to express the same thing, namely, the side of the base, and of the different words *per-em-us* and *qay en ḥeru* in the same sense of 'height'; such synonyms are common in Egypt, and, moreover, the word *mer* used of the pyramids is different from the word *ȧn* for the monument. Again, it is clear that, while the *slope*, the angle HFE, is what the builder would want to know, the cosine of the angle HAE, formed by the *edge* with the plane of the base, would be of no direct use to him. But, lastly, the *se-qeṭ* in No. 56 is $\frac{18}{25}$ and, if *se-qeṭ* is taken in the sense of $\cot HFE$, this gives for the angle HFE the value of $54° 14' 16''$, which is *precisely*, to the seconds, the slope of the lower half of the southern stone pyramid of Dakshūr; in Nos. 57–9 the *se-qeṭ*, $\frac{3}{4}$, is the cotangent of an angle of $53° 7' 48''$, which again is exactly the slope of the second pyramid of Gizeh as measured by Flinders Petrie; and the *se-qeṭ* in No. 60, which is $\frac{1}{4}$, is the cotangent of an angle of $75° 57' 50''$, corresponding exactly to the slope of the Mastaba-tombs of the Ancient Empire and of the sides of the Mēdūm pyramid.[1]

These measurements of *se-qeṭ* indicate at all events a rule-of-thumb use of geometrical proportion, and connect themselves naturally enough with the story of Thales's method of measuring the heights of pyramids.

The beginnings of Greek geometry.

At the beginning of the summary of Proclus we are told that Thales (624–547 B.C.)

'first went to Egypt and thence introduced this study (geometry) into Greece. He discovered many propositions himself, and instructed his successors in the principles underlying many others, his method of attack being in some cases more general (i.e. more theoretical or scientific), in others more empirical (*αἰσθητικώτερον*, more in the nature of simple inspection or observation).'[2]

With Thales, therefore, geometry first becomes a deductive science depending on general propositions; this agrees with what Plutarch says of him as one of the Seven Wise Men:

'he was apparently the only one of these whose wisdom stepped, in speculation, beyond the limits of practical utility: the rest acquired the reputation of wisdom in politics.'[3]

(Not that Thales was inferior to the others in political wisdom. Two stories illustrate the contrary. He tried to save Ionia by urging the separate states to form a federation

[1] Flinders Petrie, *Pyramids and Temples of Gizeh*, p. 162.
[2] Proclus on Eucl. I, p. 65. 7–11.
[3] Plutarch, *Solon*, c. 3.

with a capital at Teos, that being the most central place in Ionia. And when Croesus sent envoys to Miletus to propose an alliance, Thales dissuaded his fellow-citizens from accepting the proposal, with the result that, when Cyrus conquered, the city was saved.)

(α) *Measurement of height of pyramid.*

The accounts of Thales's method of measuring the heights of pyramids vary. The earliest and simplest version is that of Hieronymus, a pupil of Aristotle, quoted by Diogenes Laertius:

'Hieronymus says that he even succeeded in measuring the pyramids by observation of the length of their shadow at the moment when our shadows are equal to our own height.'[1]

Pliny says that

'Thales discovered how to obtain the height of pyramids and all other similar objects, namely, by measuring the shadow of the object at the time when a body and its shadow are equal in length.'[2]

Plutarch embellishes the story by making Niloxenus say to Thales:

'Among other feats of yours, he (Amasis) was particularly pleased with your measurement of the pyramid, when, without trouble or the assistance of any instrument, you merely set up a stick at the extremity of the shadow cast by the pyramid and, having thus made two triangles by the impact of the sun's rays, you showed that the pyramid has to the stick the same ratio which the shadow has to the shadow.'[3]

The first of these versions is evidently the original one and, as the procedure assumed in it is more elementary than the more general method indicated by Plutarch, the first version seems to be the more probable. Thales could not have failed to observe that, at the time when the shadow of a particular object is equal to its height, the same relation holds for all other objects casting a shadow; this he would probably infer by induction, after making actual measurements in a

[1] Diog. L. i. 27.

[2] *N. H.* xxxvi. 12 (17).

[3] Plut. *Conv. sept. sap.* 2, p. 147 A.

considerable number of cases at a time when he found the length of the shadow of one object to be equal to its height. But, even if Thales used the more general method indicated by Plutarch, that method does not, any more than the Egyptian *se-qet* calculations, imply any general theory of similar triangles or proportions; the solution is itself a *se-qeṭ* calculation, just like that in No. 57 of Ahmes's handbook. In the latter problem the base and the *se-qeṭ* are given, and we have to find the height. So in Thales's problem we get a certain *se-qeṭ* by dividing the measured length of the shadow of the stick by the length of the stick itself; we then only require to know the distance between the point of the shadow corresponding to the apex of the pyramid and the centre of the base of the pyramid in order to determine the height; the only difficulty would be to measure or estimate the distance from the apex of the shadow to the centre of the base.

(β) *Geometrical theorems attributed to Thales.*

The following are the general theorems in elementary geometry attributed to Thales.

(1) He is said to have been the first to demonstrate that a circle is bisected by its diameter.[1]

(2) Tradition credited him with the first statement of the theorem (Eucl. I. 5) that the angles at the base of any isosceles triangle are equal, although he used the more archaic term 'similar' instead of 'equal'.[2]

(3) The proposition (Eucl. I. 15) that, if two straight lines cut one another, the vertical and opposite angles are equal was discovered, though not scientifically proved, by Thales. Eudemus is quoted as the authority for this.[3]

(4) Eudemus in his History of Geometry referred to Thales the theorem of Eucl. I. 26 that, if two triangles have two angles and one side respectively equal, the triangles are equal in all respects.

'For he (Eudemus) says that the method by which Thales showed how to find the distances of ships from the shore necessarily involves the use of this theorem.'[4]

[1] Proclus on Eucl. I, p. 157. 10.
[2] *Ib.*, pp. 250. 20–251. 2.
[3] *Ib.*, p. 299. 1–5.
[4] *Ib.*, p. 352. 14–18.

(5) 'Pamphile says that Thales, who learnt geometry from the Egyptians, was the first to describe on a circle a triangle (which shall be) right-angled (καταγράψαι κύκλου τὸ τρίγωνον ὀρθογώνιον), and that he sacrificed an ox (on the strength of the discovery). Others, however, including Apollodorus the calculator, say that it was Pythagoras.'[1]

The natural interpretation of Pamphile's words is to suppose that she attributed to Thales the discovery that the angle in a semicircle is a right angle.

Taking these propositions in order, we may observe that, when Thales is said to have 'demonstrated' (ἀποδεῖξαι) that a circle is bisected by its diameter, whereas he only 'stated' the theorem about the isosceles triangle and 'discovered', without scientifically proving, the equality of vertically opposite angles, the word 'demonstrated' must not be taken too literally. Even Euclid did not 'demonstrate' that a circle is bisected by its diameter, but merely stated the fact in I. Def. 17. Thales therefore probably observed rather than proved the property; and it may, as Cantor says, have been suggested by the appearance of certain figures of circles divided into a number of equal sectors by 2, 4, or 6 diameters such as are found on Egyptian monuments or represented on vessels brought by Asiatic tributary kings in the time of the eighteenth dynasty.[2]

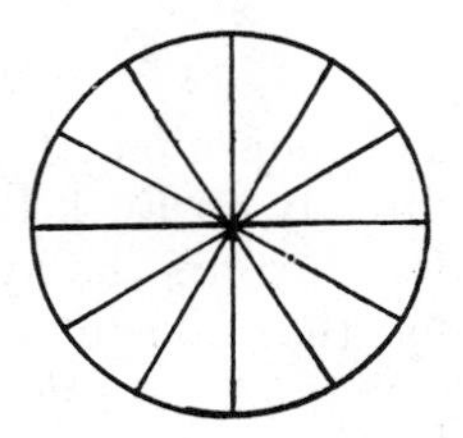

It has been suggested that the use of the word 'similar' to describe the equal angles of an isosceles triangle indicates that Thales did not yet conceive of an angle as a magnitude, but as a *figure* having a certain *shape*, a view which would agree closely with the idea of the Egyptian *se-qeṭ*, 'that which makes the nature', in the sense of determining a similar or the same inclination in the faces of pyramids.

With regard to (4), the theorem of Eucl. I. 26, it will be observed that Eudemus only inferred that this theorem was known to Thales from the fact that it is necessary to Thales's determination of the distance of a ship from the shore. Unfortunately the method used can only be conjectured.

[1] Diog. L. i. 24, 25.

[2] Cantor, *Gesch. d. Math.* i³, pp. 109, 140.

The most usual supposition is that Thales, observing the ship from the top of a tower on the sea-shore, used the practical equivalent of the proportionality of the sides of two similar right-angled triangles, one small and one large. Suppose B to be the base of the tower, C the ship. It was only necessary for a man standing at the top of the tower to have an instrument with two legs forming a right angle, to place it with one leg DA vertical and in a straight line with B, and the other leg DE in the direction of the ship, to take any point A on DA, and then to mark on DE the point E where the line of sight from A to C cuts the leg DE. Then AD ($= l$, say) and DE ($= m$, say) can be actually measured, as also the height BD ($= h$, say) from D to the foot of the tower, and, by similar triangles,

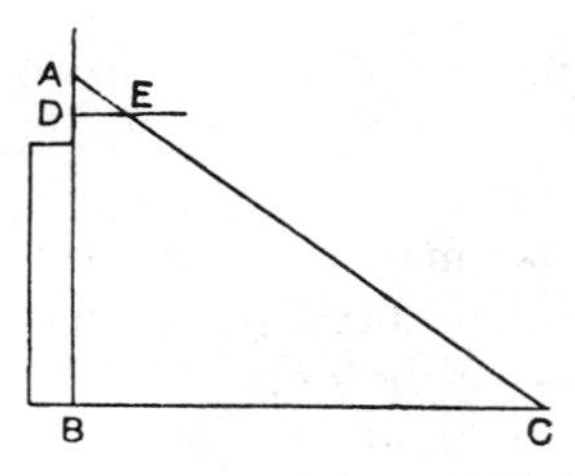

$$BC = (h + l) \cdot \frac{m}{l}.$$

The objection to this solution is that it does not depend directly on Eucl. I. 26, as Eudemus implies. Tannery [1] therefore favours the hypothesis of a solution on the lines followed by the Roman agrimensor Marcus Junius Nipsus in his *fluminis varatio*.—To find the distance from A to an inaccessible point B. Measure from A, along a straight line at right angles to AB, a distance AC, and bisect it at D. From C, on the side of AC remote from B, draw CE at right angles to AC, and let E be the point on it which is in a straight line with B and D. Then clearly, by Eucl. I. 26, CE is equal to AB; and CE can be measured, so that AB is known.

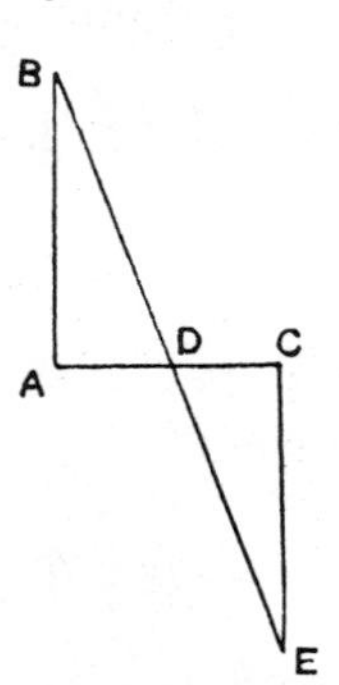

This hypothesis is open to a different objection, namely that, as a rule, it would be difficult, in the supposed case, to get a sufficient amount of free and level space for the construction and measurements.

I have elsewhere [2] suggested a still simpler method free

[1] Tannery, *La géométrie grecque*, pp. 90–1.

[2] *The Thirteen Books of Euclid's Elements*, vol. i, p. 305.

from this objection, and depending equally directly on Eucl. I. 26. If the observer was placed on the top of a tower, he had only to use a rough instrument made of a straight stick and a cross-piece fastened to it so as to be capable of turning about the fastening (say a nail) so that it could form any angle with the stick and would remain where it was put. Then the natural thing would be to fix the stick upright (by means of a plumb-line) and direct the cross-piece towards the ship. Next, leaving the cross-piece at the angle so found, he would turn the stick round, while keeping it vertical, until the cross-piece pointed to some visible object on the shore, which would be mentally noted; after this it would only be necessary to measure the distance of the object from the foot of the tower, which distance would, by Eucl. I. 26, be equal to the distance of the ship. It appears that this precise method is found in so many practical geometries of the first century of printing that it must be assumed to have long been a common expedient. There is a story that one of Napoleon's engineers won the Imperial favour by quickly measuring, in precisely this way, the width of a stream that blocked the progress of the army.[1]

There is even more difficulty about the dictum of Pamphile implying that Thales first discovered the fact that the angle in a semicircle is a right angle. Pamphile lived in the reign of Nero (A. D. 54–68), and is therefore a late authority. The date of Apollodorus the 'calculator' or arithmetician is not known, but he is given as only one of several authorities who attributed the proposition to Pythagoras. Again, the story of the sacrifice of an ox by Thales on the occasion of his discovery is suspiciously like that told in the distich of Apollodorus 'when Pythagoras discovered that famous proposition, on the strength of which he offered a splendid sacrifice of oxen'. But, in quoting the distich of Apollodorus, Plutarch expresses doubt whether the discovery so celebrated was that of the theorem of the square of the hypotenuse or the solution of the problem of 'application of areas'[2]; there is nothing about the discovery of the fact of the angle in a semicircle being a right angle. It may therefore be that

[1] David Eugene Smith, *The Teaching of Geometry*, pp. 172–3.

[2] Plutarch, *Non posse suaviter vivi secundum Epicurum*, c. 11, p. 1094 B.

Diogenes Laertius was mistaken in bringing Apollodorus into the story now in question at all; the mere mention of the sacrifice in Pamphile's account would naturally recall Apollodorus's lines about Pythagoras, and Diogenes may have forgotten that they referred to a different proposition.

But, even if the story of Pamphile is accepted, there are difficulties of substance. As Allman pointed out, if Thales knew that the angle in a semicircle is a right angle, he was in a position at once to infer that the sum of the angles of any *right-angled* triangle is equal to two right angles. For suppose that BC is the diameter of the semicircle, O the centre, and A a point on the semicircle; we are then supposed to know that the angle BAC is a right angle. Joining OA, we form two isosceles triangles OAB, OAC; and Thales knows that the base angles in each of these triangles are equal. Consequently the sum of the angles OAB, OAC is equal to the sum of the angles OBA, OCA. The former sum is known to be a right angle; therefore the second sum is also a right angle, and the three angles of the triangle ABC are together equal to twice the said sum, i.e. to two right angles.

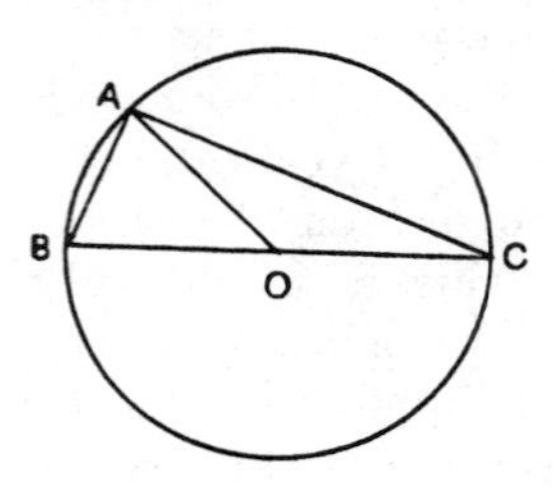

Next it would easily be seen that *any* triangle can be divided into two right-angled triangles by drawing a perpendicular AD from a vertex A to the opposite side BC. Then the three angles of each of the right-angled triangles ABD, ADC are together equal to two right angles. By adding together the three angles of both triangles we find that the sum of the three angles of the triangle ABC together with the angles ADB, ADC is equal to four right angles; and, the sum of the latter two angles being two right angles, it follows that the sum of the remaining angles, the angles at A, B, C, is equal to two right angles. And ABC is *any* triangle.

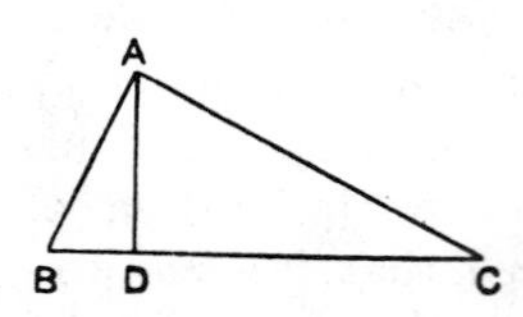

Now Euclid in III. 31 proves that the angle in a semicircle is a right angle by means of the general theorem of I. 32

that the sum of the angles of any triangle is equal to two right angles; but if Thales was aware of the truth of the latter general proposition and proved the proposition about the semicircle in this way, by means of it, how did Eudemus come to credit the Pythagoreans, not only with the general proof, but with the *discovery*, of the theorem that the angles of any triangle are together equal to two right angles? [1]

Cantor, who supposes that Thales proved his proposition after the manner of Euclid III. 31, i.e. by means of the general theorem of I. 32, suggests that Thales arrived at the truth of the latter, not by a general proof like that attributed by Eudemus to the Pythagoreans, but by an argument following the steps indicated by Geminus. Geminus says that

'the *ancients* investigated the theorem of the two right angles in each individual species of triangle, first in the equilateral, then in the isosceles, and afterwards in the scalene triangle, but later geometers demonstrated the general theorem that in *any* triangle the three interior angles are equal to two right angles'. [2]

The 'later geometers' being the Pythagoreans, it is assumed that the 'ancients' may be Thales and his contemporaries. As regards the equilateral triangle, the fact might be suggested by the observation that six such triangles arranged round one point as common vertex would fill up the space round that point; whence it follows that each angle is one-sixth of four right angles, and three such angles make up two right angles. Again, suppose that in either an equilateral or an isosceles

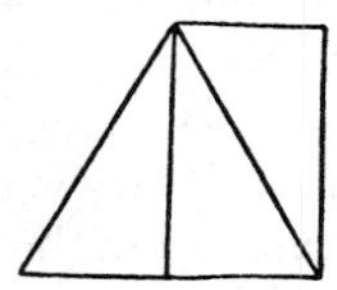
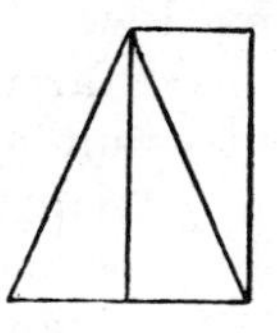

triangle the vertical angle is bisected by a straight line meeting the base, and that the rectangle of which the bisector and one half of the base are adjacent sides is completed; the rectangle is double of the half of the original triangle, and the angles of the half-triangle are together equal to half the sum

[1] Proclus on Eucl. I, p. 379. 2–5.
[2] See Eutocius, Comm. on *Conics* of Apollonius (vol. ii, p. 170, Heib.).

of the angles of the rectangle, i.e. are equal to two right angles; and it immediately follows that the sum of the angles of the original equilateral or isosceles triangle is equal to two right angles. The same thing is easily proved of any triangle by dividing it into two right-angled triangles and completing the rectangles which are their doubles respectively, as in the figure. But the fact that a proof on these lines is just as easy in the case of the general triangle as it is for the equilateral and isosceles triangles throws doubt on the whole procedure; and we are led to question whether there is any foundation for Geminus's account at all. Aristotle has a remark that

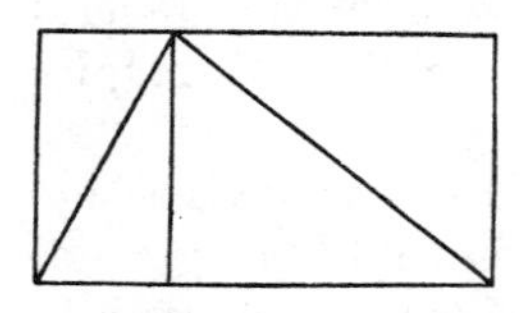

'even if one should prove, with reference to each (sort of) triangle, the equilateral, scalene, and isosceles, separately, that each has its angles equal to two right angles, either by one proof or by different proofs, he does not yet know that *the triangle*, i.e. the triangle *in general*, has its angles equal to two right angles, except in a sophistical sense, even though there exists no triangle other than triangles of the kinds mentioned. For he knows it not *quâ* triangle, nor of *every* triangle, except in a numerical sense; he does not know it *notionally* of every triangle, even though there be actually no triangle which he does not know'.[1]

It may well be that Geminus was misled into taking for a historical fact what Aristotle gives only as a hypothetical illustration, and that the exact stages by which the proposition was first proved were not those indicated by Geminus.

Could Thales have arrived at his proposition about the semicircle without assuming, or even knowing, that the sum of the angles of *any* triangle is equal to two right angles? It seems possible, and in the following way. Many propositions were doubtless first discovered by drawing all sorts of figures and lines in them, and observing *apparent* relations of equality, &c., between parts. It would, for example, be very natural to draw a rectangle, a figure with four right angles (which, it

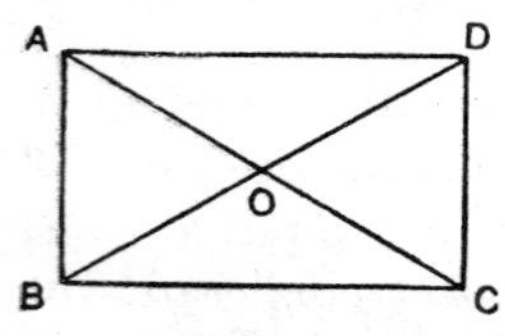

[1] Arist. *Anal. Post.* i. 5, 74 a 25 sq.

would be found, could be drawn in practice), and to put in the two diagonals. The equality of the opposite sides would doubtless, in the first beginnings of geometry, be assumed as obvious, or verified by measurement. If then it was *assumed* that a rectangle is a figure with all its angles right angles and each side equal to its opposite, it would be easy to deduce certain consequences. Take first the two triangles ADC, BCD. Since by hypothesis $AD = BC$ and CD is common, the two triangles have the sides AD, DC respectively equal to the sides BC, CD, and the included angles, being right angles, are equal; therefore the triangles ADC, BCD are equal in all respects (cf. Eucl. I. 4), and accordingly the angles ACD (i.e. OCD) and BDC (i.e. ODC) are equal, whence (by the converse of Eucl. I. 5, known to Thales) $OD = OC$. Similarly by means of the equality of AB, CD we prove the equality of OB, OC. Consequently OB, OC, OD (and OA) are all equal. It follows that a circle with centre O and radius OA passes through B, C, D also; since AO, OC are in a straight line, AC is a diameter of the circle, and the angle ABC, by hypothesis a right angle, is an 'angle in a semicircle'. It would then appear that, given any right angle as ABC standing on AC as base, it was only necessary to bisect AC at O, and O would then be the centre of a semicircle on AC as diameter and passing through B. The construction indicated would be the construction of a circle about the right-angled triangle ABC, which seems to correspond well enough to Pamphile's phrase about 'describing on (i.e. in) a circle a triangle (which shall be) right angled'.

(γ) *Thales as astronomer.*

Thales was also the first Greek astronomer. Every one knows the story of his falling into a well when star-gazing, and being rallied by 'a clever and pretty maidservant from Thrace' for being so eager to know what goes on in the heavens that he could not see what was straight in front of him, nay, at his very feet. But he was not merely a stargazer. There is good evidence that he predicted a solar eclipse which took place on May 28, 585 B.C. We can conjecture the basis of this prediction. The Babylonians, as the result of observations continued through centuries, had discovered the period of 223 lunations after which eclipses recur; and this period was doubtless known to Thales, either directly or through the Egyptians as intermediaries. Thales, however, cannot have known the *cause* of eclipses; he could not have given the true explanation of *lunar* eclipses (as the *Doxographi* say he did) because he held that the earth is a circular disc floating on the water like a log; and, if he had correctly accounted for *solar* eclipses, it is impossible that all the succeeding Ionian philosophers should, one after another, have put forward the fanciful explanations which we find recorded.

Thales's other achievements in astronomy can be very shortly stated. Eudemus attributed to him the discovery of 'the fact that the period of the sun with reference to the solstices is not always the same'[1]; the vague phrase seems to mean that he discovered the inequality of the length of the four astronomical seasons, that is, the four parts of the 'tropical' year as divided by the solstices and equinoxes. Eudemus presumably referred to the written works by Thales *On the Solstice* and *On the Equinoxes* mentioned by Diogenes Laertius.[2] He knew of the division of the year into 365 days, which he probably learnt from Egypt.

Thales observed of the Hyades that there were two of them, one north and the other south. He used the Little Bear as a means of finding the pole, and advised the Greeks to sail by the Little Bear, as the Phoenicians did, in preference to their own practice of sailing by the Great Bear. This instruction was probably noted in the handbook under the title of *Nautical Astronomy*, attributed by some to Thales and by others to Phocus of Samos.

It became the habit of the *Doxographi* to assign to Thales, in common with other astronomers in each case, a number of discoveries not made till later. The following is the list, with the names of the astronomers to whom the respective discoveries may with most certainty be attributed: (1) the fact that the moon takes its light from the sun (Anaxagoras and possibly Parmenides); (2) the sphericity of the earth (Pythagoras); (3) the division of the heavenly sphere into five zones (Pythagoras and Parmenides); (4) the obliquity of the ecliptic (Oenopides of Chios); (5) the estimate of the

[1] See Theon of Smyrna, p. 198. 17.

[2] Diog. L. i. 23.

sun's diameter as 1/720th part of the sun's circle (Aristarchus of Samos).

From Thales to Pythagoras.

We are completely in the dark as to the progress of geometry between the times of Thales and Pythagoras. ANAXIMANDER (born about 611/10 B.C.) put forward some daring and original hypotheses in astronomy. According to him the earth is a short cylinder with two bases (on one of which we live) and of depth equal to one-third of the diameter of either base. It is suspended freely in the middle of the universe without support, being kept there in equilibrium by virtue of its equidistance from the extremities and from the other heavenly bodies all round. The sun, moon, and stars are enclosed in opaque rings of compressed air concentric with the earth and filled with fire; what we see is the fire shining through vents (like gas-jets, as it were). The sun's ring is 27 or 28 times, the moon's ring 19 times, as large as the earth, i.e. the sun's and moon's distances are estimated in terms (as we may suppose) of the radius of the circular face of the earth; the fixed stars and the planets are nearer to the earth than the sun and moon. This is the first speculation on record about sizes and distances. Anaximander is also said to have introduced the *gnomon* (or sun-dial with a vertical needle) into Greece and to have shown on it the solstices, the times, the seasons, and the equinox[1] (according to Herodotus[2] the Greeks learnt the use of the *gnomon* from the Babylonians). He is also credited, like Thales before him, with having constructed a sphere to represent the heavens.[3] But Anaximander has yet another claim to undying fame. He was the first who ventured to draw a map of the inhabited earth. The Egyptians had drawn maps before, but only of particular districts; Anaximander boldly planned out the whole world with 'the circumference of the earth and sea'.[4] This work involved of course an attempt to estimate the dimensions of the earth, though we have no information as to his results. It is clear, therefore, that Anaximander was something of

[1] Euseb. *Praep. Evang.* x. 14. 11 (*Vors.* i³, p. 14. 28).
[2] Hdt. ii. 109.
[3] Diog. L. ii. 2.
[4] Diog. L. *l. c.*

a mathematician; but whether he contributed anything to geometry as such is uncertain. True, Suidas says that he 'introduced the gnomon and generally set forth a sketch or outline of geometry' (ὅλως γεωμετρίας ὑποτύπωσιν ἔδειξεν); but it may be that 'geometry' is here used in its literal sense of earth-measurement, and that the reference is only to the famous map.

'Next to Thales, Ameristus, a brother of the poet Stesichorus, is mentioned as having engaged in the study of geometry; and from what Hippias of Elis says it appears that he acquired a reputation for geometry.'[1]

Stesichorus the poet lived about 630–550 B.C. The brother therefore would probably be nearly contemporary with Thales. We know nothing of him except from the passage of Proclus, and even his name is uncertain. In Friedlein's edition of Proclus it is given as Mamercus, after a later hand in cod. Monac. 427; Suidas has it as Mamertinus (*s.v.* Stesichorus); Heiberg in his edition of Heron's *Definitions* writes Mamertius, noting Μαρμέτιος as the reading of Cod. Paris. Gr. 2385.

[1] Proclus on Eucl. I, p. 65. 11–15.

V

PYTHAGOREAN GEOMETRY

The special service rendered by PYTHAGORAS to geometry is thus described in the Proclus summary:

'After these (Thales and Ameristus or Mamercus) Pythagoras transformed the study of geometry into a liberal education, examining the principles of the science from the beginning and probing the theorems in an immaterial and intellectual manner: he it was who discovered the theory of irrationals' (or 'proportions') 'and the construction of the cosmic figures'.[1]

These supposed discoveries will claim our attention presently; the rest of the description agrees with another passage about the Pythagoreans:

'Herein', says Proclus, 'I emulate the Pythagoreans who even had a conventional phrase to express what I mean, "a figure and a platform, not a figure and sixpence", by which they implied that the geometry which is deserving of study is that which, at each new theorem, sets up a platform to ascend by, and lifts the soul on high instead of allowing it to go down among sensible objects and so become subservient to the common needs of this mortal life'.[2]

In like manner we are told that 'Pythagoras used definitions on account of the mathematical nature of the subject',[3] which again implies that he took the first steps towards the systematization of geometry as a subject in itself.

A comparatively early authority, Callimachus (about 250 B.C.), is quoted by Diodorus as having said that Pythagoras discovered some geometrical problems himself and was the first to introduce others from Egypt into Greece.[4] Diodorus gives what appear to be five verses of Callimachus *minus* a few words;

[1] Proclus on Eucl. I, p. 65. 15–21.
[2] *Ib.*, p. 84. 15–22.
[3] Favorinus in Diog. L. viii. 25.
[4] Diodorus x. 6. 4 (*Vors.* i³, p. 346. 23).

a longer fragment including the same passage is now available (though the text is still deficient) in the Oxyrhynchus Papyri.[1] The story is that one Bathycles, an Arcadian, bequeathed a cup to be given to the best of the Seven Wise Men. The cup first went to Thales, and then, after going the round of the others, was given to him a second time. We are told that Bathycles's son brought the cup to Thales, and that (presumably on the occasion of the first presentation)

'by a happy chance he found . . . the old man scraping the ground and drawing the figure discovered by the Phrygian Euphorbus (= Pythagoras), who was the first of men to draw even scalene triangles and a circle . . ., and who prescribed abstinence from animal food'.

Notwithstanding the anachronism, the 'figure discovered by Euphorbus' is presumably the famous proposition about the squares on the sides of a right-angled triangle. In Diodorus's quotation the words after 'scalene triangles' are *κύκλον ἑπταμήκη* (*ἑπταμήκε'* Hunt), which seems unintelligible unless the 'seven-lengthed circle' can be taken as meaning the 'lengths of seven circles' (in the sense of the seven independent orbits of the sun, moon, and planets) or the circle (the zodiac) comprehending them all.[2]

But it is time to pass on to the propositions in geometry which are definitely attributed to the Pythagoreans.

[1] *Oxyrhynchus Papyri*, Pt. vii, p. 33 (Hunt).

[2] The papyrus has an accent over the ε and to the right of the accent, above the uncertain π, the appearance of a λ in dark ink, thus καικυκλονέπ̣ (with λ above), a reading which is not yet satisfactorily explained. Diels (*Vorsokratiker*, i[3], p. 7) considers that the accent over the ε is fatal to the reading ἑπταμήκη, and conjectures καὶ κύκλον ἕλ⟨ικα⟩ κἠδίδαξε νηστεύειν instead of Hunt's καὶ κύκλον ἑπ[ταμήκε', ἠδὲ νηστεύειν] and Diodorus's καὶ κύκλον ἑπταμήκη δίδαξε νηστεύειν. But κύκλον ἕλικα, 'twisted (or curved) circle', is very indefinite. It may have been suggested to Diels by Hermesianax's lines (Athenaeus xiii. 599 A) attributing to Pythagoras the 'refinements of the geometry of spirals' (ἑλίκων κομψὰ γεωμετρίης). One naturally thinks of Plato's dictum (*Timaeus* 39 A, B) about the circles of the sun, moon, and planets being twisted into spirals by the combination of their own motion with that of the daily rotation; but this can hardly be the meaning here. A more satisfactory sense would be secured if we could imagine the circle to be the circle described about the 'scalene' (right-angled) triangle, i.e. if we could take the reference to be to the discovery of the fact that the angle in a semicircle is a right angle, a discovery which, as we have seen, was alternatively ascribed to Thales and Pythagoras.

Discoveries attributed to the Pythagoreans.

(α) *Equality of the sum of the three angles of a triangle to two right angles.*

We have seen that Thales, if he really discovered that the angle in a semicircle is a right angle, was in a position, first, to show that in any right-angled triangle the sum of the three angles is equal to two right angles, and then, by drawing the perpendicular from a vertex of any triangle to the opposite side and so dividing the triangle into two right-angled triangles, to prove that the sum of the three angles of any triangle whatever is equal to two right angles. If this method of passing from the particular case of a right-angled triangle to that of any triangle did not occur to Thales, it is at any rate hardly likely to have escaped Pythagoras. But all that we know for certain is that Eudemus referred to the Pythagoreans the discovery of the general theorem that in any triangle the sum of the interior angles is equal to two right angles.[1] Eudemus goes on to tell us how they proved it. The method differs slightly from that of Euclid, but depends, equally with Euclid's proof, on the properties of parallels; it can therefore only have been evolved at a time when those properties were already known.

Let ABC be any triangle; through A draw DE parallel to BC.

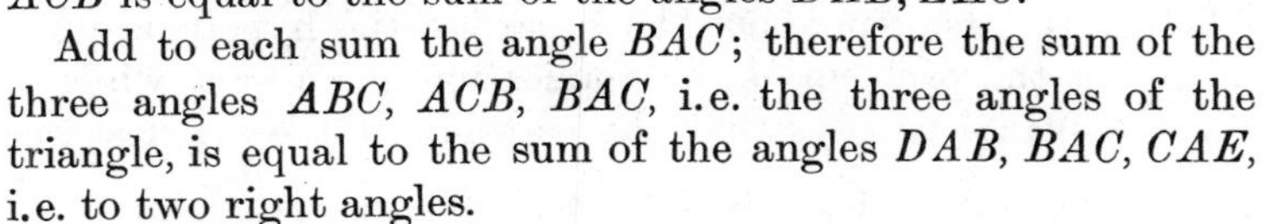

Then, since BC, DE are parallel, the alternate angles DAB, ABC are equal.

Similarly the alternate angles EAC, ACB are equal.

Therefore the sum of the angles ABC, ACB is equal to the sum of the angles DAB, EAC.

Add to each sum the angle BAC; therefore the sum of the three angles ABC, ACB, BAC, i.e. the three angles of the triangle, is equal to the sum of the angles DAB, BAC, CAE, i.e. to two right angles.

We need not hesitate to credit the Pythagoreans with the more general propositions about the angles of any polygon,

[1] Proclus on Eucl. I, p. 397. 2.

namely (1) that, if n be the number of the sides or angles, the interior angles of the polygon are together equal to $2n-4$ right angles, and (2) that the exterior angles of the polygon (being the supplements of the interior angles respectively) are together equal to four right angles. The propositions are interdependent, and Aristotle twice quotes the latter.[1] The Pythagoreans also discovered that the only three regular polygons the angles of which, if placed together round a common point as vertex, just fill up the space (four right angles) round the point are the equilateral triangle, the square, and the regular hexagon.

(β) *The 'Theorem of Pythagoras'* (= Eucl. I. 47).

Though this is the proposition universally associated by tradition with the name of Pythagoras, no really trustworthy evidence exists that it was actually discovered by him. The comparatively late writers who attribute it to him add the story that he sacrificed an ox to celebrate his discovery. Plutarch[2] (born about A.D. 46), Athenaeus[3] (about A.D. 200), and Diogenes Laertius[4] (A.D. 200 or later) all quote the verses of Apollodorus the 'calculator' already referred to (p. 133). But Apollodorus speaks of the 'famous theorem', or perhaps 'figure' (γράμμα), the discovery of which was the occasion of the sacrifice, without saying what the theorem was. Apollodorus is otherwise unknown; he may have been earlier than Cicero, for Cicero[5] tells the story in the same form without specifying what geometrical discovery was meant, and merely adds that he does not believe in the sacrifice, because the Pythagorean ritual forbade sacrifices in which blood was shed. Vitruvius[6] (first century B.C.) connects the sacrifice with the discovery of the property of the particular triangle 3, 4, 5. Plutarch, in quoting Apollodorus, questions whether the theorem about the square of the hypotenuse was meant, or the problem of the application of an area, while in another place[7] he says that the occasion of the sacrifice was

[1] *An. Post.* i. 24, 85 b 38; *ib.* ii. 17, 99 a 19.
[2] Plutarch, *Non posse suaviter vivi secundum Epicurum*, c. 11, p. 1094 B.
[3] Athenaeus x. 418 F.
[4] Diog. L. viii. 12, i. 25.
[5] Cicero, *De nat. deor.* iii. 36, 88.
[6] Vitruvius, *De architectura*, ix. pref.
[7] Plutarch, *Quaest. conviv.* viii. 2, 4, p. 720 A.

the solution of the problem, 'given two figures, to *apply* a third which shall be equal to the one and similar to the other', and he adds that this problem is unquestionably finer than the theorem about the square on the hypotenuse. But Athenaeus and Porphyry[1] (A.D. 233–304) connect the sacrifice with the latter proposition; so does Diogenes Laertius in one place. We come lastly to Proclus, who is very cautious, mentioning the story but declining to commit himself to the view that it was Pythagoras or even any single person who made the discovery:

'If we listen to those who wish to recount ancient history, we may find some of them referring this theorem to Pythagoras, and saying that he sacrificed an ox in honour of his discovery. But for my part, while I admire *those who* first observed the truth of this theorem, I marvel more at the writer of the Elements, not only because he made it fast by a most lucid demonstration, but because he compelled assent to the still more general theorem by the irrefutable arguments of science in the sixth book.'

It is possible that all these authorities may have built upon the verses of Apollodorus; but it is remarkable that, although in the verses themselves the particular theorem is not specified, there is practical unanimity in attributing to Pythagoras the theorem of Eucl. I. 47. Even in Plutarch's observations expressing doubt about the particular occasion of the sacrifice there is nothing to suggest that he had any hesitation in accepting as discoveries of Pythagoras *both* the theorem of the square on the hypotenuse and the problem of the application of an area. Like Hankel,[2] therefore, I would not go so far as to deny to Pythagoras the credit of the discovery of our proposition; nay, I like to believe that tradition is right, and that it was really his.

True, the discovery is also claimed for India.[3] The work relied on is the *Āpastamba-Śulba-Sūtra*, the date of which is put at least as early as the fifth or fourth century B.C., while it is remarked that the matter of it must have been much

[1] Porphyry, *Vit. Pyth.* 36.
[2] Hankel, *Zur Geschichte der Math. in Alterthum und Mittelalter*, p. 97.
[3] Bürk in the *Zeitschrift der morgenländ. Gesellschaft*, lv, 1901, pp. 543–91; lvi, 1902, pp. 327–91.

older than the book itself; thus one of the constructions for right angles, using cords of lengths 15, 36, 39 (= 5, 12, 13), was known at the time of the *Tāittirīya Saṃhitā* and the *Śatapatha Brāhmana,* still older works belonging to the eighth century B.C. at latest. A feature of the *Āpastamba-Śulba-Sūtra* is the construction of right angles in this way by means of cords of lengths equal to the three sides of certain rational right-angled triangles (or, as Āpastamba calls them, rational rectangles, i.e. those in which the diagonals as well as the sides are rational). The rational right-angled triangles actually used are (3, 4, 5), (5, 12, 13), (8, 15, 17), (12, 35, 37). There is a proposition stating the theorem of Eucl. I. 47 as a fact in general terms, but without proof, and there are rules based upon it for constructing a square equal to (1) the sum of two given squares and (2) the difference of two squares. But certain considerations suggest doubts as to whether the proposition had been established by any proof applicable to all cases. Thus Āpastamba mentions only seven rational right-angled triangles, really reducible to the above-mentioned four (one other, 7, 24, 25, appears, it is true, in the Bāudhāyana Ś. S., supposed to be older than Āpastamba); he had no general rule such as that attributed to Pythagoras for forming any number of rational right-angled triangles; he refers to his seven in the words 'so many *recognizable* constructions are there', implying that he knew of no other such triangles. On the other hand, the truth of the theorem was recognized in the case of the isosceles right-angled triangle; there is even a construction for $\sqrt{2}$, or the length of the diagonal of a square with side unity, which is constructed as $\left(1+\frac{1}{3}+\frac{1}{3.4}-\frac{1}{3.4.34}\right)$ of the side, and is then used with the side for the purpose of drawing the square on the side: the length taken is of course an approximation to $\sqrt{2}$ derived from the consideration that $2.12^2 = 288 = 17^2-1$; but the author does not say anything which suggests any knowledge on his part that the approximate value is not exact. Having drawn by means of the approximate value of the diagonal an inaccurate square, he proceeds to use it to construct a square with area equal to three times the original square, or, in other words, to construct $\sqrt{3}$, which is therefore only approximately found.

Thus the theorem is enunciated and used as if it were of general application; there is, however, no sign of any general proof; there is nothing in fact to show that the assumption of its universal truth was founded on anything better than an imperfect induction from a certain number of cases, discovered empirically, of triangles with sides in the ratios of whole numbers in which the property (1) that the square on the longest side is equal to the sum of the squares on the other two was found to be always accompanied by the property (2) that the latter two sides include a right angle. But, even if the Indians had actually attained to a scientific proof of the general theorem, there is no evidence or probability that the Greeks obtained it from India; the subject was doubtless developed quite independently in the two countries.

The next question is, how was the theorem proved by Pythagoras or the Pythagoreans? Vitruvius says that Pythagoras first discovered the triangle (3, 4, 5), and doubtless the theorem was first suggested by the discovery that this triangle is right-angled; but this discovery probably came to Greece from Egypt. Then a very simple construction would show that the theorem is true of an *isosceles* right-angled triangle. Two possible lines are suggested on which the general proof may have been developed. One is that of decomposing square and rectangular areas into squares, rectangles and triangles, and piecing them together again after the manner of Eucl., Book II; the isosceles right-angled triangle gives the most obvious case of this method. The other line is one depending upon proportions; and we have good reason for supposing that Pythagoras developed a theory of proportion. That theory was applicable to commensurable magnitudes only; but this would not be any obstacle to the use of the method so long as the existence of the incommensurable or irrational remained undiscovered. From Proclus's remark that, while he admired those who first noticed the truth of the theorem, he admired Euclid still more for his most clear proof of it and for the irrefutable demonstration of the extension of the theorem in Book VI, it is natural to conclude that Euclid's proof in I. 47 was new, though this is not quite certain. Now VI. 31 could be proved at once by using I. 47 along with VI. 22; but Euclid proves it independently of I. 47 by means of proportions. This seems to suggest that he proved I. 47 by the methods of Book I instead of by proportions in order to get the proposition into Book I instead of Book VI, to which it must have been relegated if the proof by proportions had been used. If, on the other hand, Pythagoras had proved it by means of the methods of Books I and II, it would hardly have been necessary for Euclid to devise a new proof of I. 47. Hence it would appear most probable that Pythagoras would prove the proposition by means of his (imperfect) theory of proportions. The proof may have taken one of three different shapes.

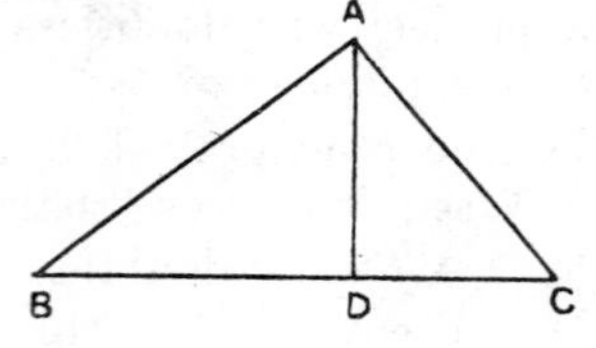

(1) If ABC is a triangle right-angled at A, and AD is perpendicular to BC, the triangles DBA, DAC are both similar to the triangle ABC.

It follows from the theorems of Eucl. VI. 4 and 17 that

$$BA^2 = BD \,.\, BC,$$
$$AC^2 = CD \,.\, BC,$$

whence, by addition, $BA^2 + AC^2 = BC^2$.

It will be observed that this proof is *in substance* identical with that of Eucl. I. 47, the difference being that the latter uses the relations between parallelograms and triangles on the same base and between the same parallels instead of proportions. The probability is that it was this particular proof by proportions which suggested to Euclid the method of I. 47; but the transformation of the proof depending on proportions into one based on Book I only (which was absolutely required under Euclid's arrangement of the *Elements*) was a stroke of genius.

(2) It would be observed that, in the similar triangles DBA, DAC, ABC, the corresponding sides opposite to the right angle in each case are BA, AC, BC.

The triangles therefore are in the duplicate ratios of these sides, and so are the squares on the latter.

But of the triangles two, namely DBA, DAC, make up the third, ABC.

The same must therefore be the case with the squares, or

$$BA^2 + AC^2 = BC^2.$$

(3) The method of VI. 31 might have been followed exactly, with squares taking the place of any similar rectilineal figures. Since the triangles DBA, ABC are similar,

$$BD : AB = AB : BC,$$

or BD, AB, BC are three proportionals, whence

$$AB^2 : BC^2 = BD^2 : AB^2 = BD : BC.$$

Similarly, $\quad AC^2 : BC^2 = CD : BC.$

Therefore $\quad (BA^2 + AC^2) : BC^2 = (BD + DC) : BC. \quad$ [V. 24]

$$= 1.$$

If, on the other hand, the proposition was originally proved by the methods of Euclid, Books I, II alone (which, as I have said, seems the less probable supposition), the suggestion of

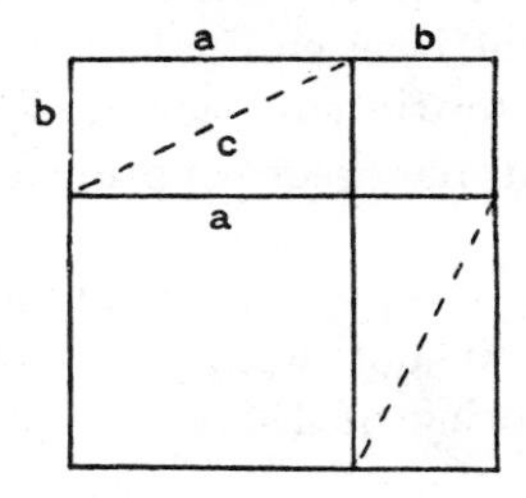

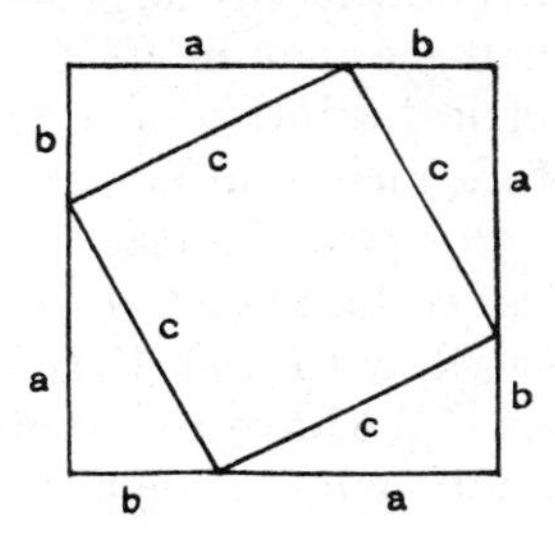

Bretschneider and Hankel seems to be the best. According to this we are to suppose, first, a figure like that of Eucl. II. 4, representing a larger square, of side $(a + b)$, divided into two smaller squares of sides a, b respectively, and two complements, being two equal rectangles with a, b as sides.

Then, dividing each complementary rectangle into two equal triangles, we dispose the four triangles round another square of side $a + b$ in the manner shown in the second figure.

Deducting the four triangles from the original square in each case we get, in the first figure, two squares a^2 and b^2 and, in the second figure, one square on c, the diagonal of the rectangle (a, b) or the hypotenuse of the right-angled triangle in which a, b are the sides about the right angle. It follows that $a^2 + b^2 = c^2$.

(γ) *Application of areas and geometrical algebra.*

We have seen that, in connexion with the story of the sacrifice of an ox, Plutarch attributes to Pythagoras himself the discovery of the problem of the application of an area or, as he says in another place, the problem 'Given two figures, to "apply" a third figure which shall be equal to the one, and similar to the other (of the given figures).' The latter problem (= Eucl. VI. 25) is, strictly speaking, not so much a case of *applying* an area as of *constructing* a figure, because the base is not given in length; but it depends directly upon the simplest case of 'application of areas', namely the problem, solved in Eucl. I. 44, 45, of applying to a given straight line as base a parallelogram containing a given angle and equal in area to a given triangle or rectilineal figure. The method of application of areas is fundamental in Greek geometry and requires detailed notice. We shall see that in its general form it is equivalent to the geometrical solution of a mixed quadratic equation, and it is therefore an essential part of what has been appropriately called *geometrical algebra.*

It is certain that the theory of application of areas originated with the Pythagoreans, if not with Pythagoras himself. We have this on the authority of Eudemus, quoted in the following passage of Proclus:

'These things, says Eudemus, are ancient, being discoveries of the Muse of the Pythagoreans, I mean the *application of areas* (παραβολὴ τῶν χωρίων), their *exceeding* (ὑπερβολή) and their *falling short* (ἔλλειψις). It was from the Pythagoreans that later geometers [i. e. Apollonius of Perga] took the names, which they then transferred to the so-called *conic* lines (curves), calling one of these a *parabola* (application), another a *hyperbola* (exceeding), and the third an *ellipse* (falling short), whereas those god-like men of old saw the things signified by these names in the construction, in a plane, of areas upon a given finite straight line. For, when you have a straight line set out, and lay the given area exactly alongside the whole of the straight line, they say that you *apply* the said area; when, however, you make the length of the area greater than the straight line, it is said to *exceed,* and, when you make it less, in which case after the area has been drawn there is some part of the straight line extending beyond it, it is said to *fall short*. Euclid, too, in the sixth book speaks in this way both of exceeding and falling short; but in this place (I. 44) he needed the *application* simply, as he sought to apply to a given straight line an area equal to a given triangle, in order that we might have in our power, not only the *construction* (σύστασις) of a parallelogram equal to a given triangle, but also the application of it to a limited straight line.'[1]

The general form of the problem involving *application* with *exceeding* or *falling short* is the following:

'To apply to a given straight line a rectangle (or, more generally, a parallelogram) equal to a given rectilineal figure, and (1) *exceeding* or (2) *falling short* by a square figure (or, in the more general case, by a parallelogram similar to a given parallelogram).'

The most general form, shown by the words in brackets, is found in Eucl. VI. 28, 29, which are equivalent to the geometrical solution of the quadratic equations

$$ax \pm \frac{b}{c}x^2 = \frac{C}{m},$$

and VI. 27 gives the condition of possibility of a solution when the sign is negative and the parallelogram *falls short*. This general case of course requires the use of proportions; but the simpler case where the area applied is a rectangle, and the form of the portion which overlaps or falls short is a square, can be solved by means of Book II only. The proposition II. 11 is the geometrical solution of the particular quadratic equation

$$a(a-x) = x^2,$$

or

$$x^2 + ax = a^2.$$

The propositions II. 5 and 6 are in the form of theorems. Taking, e.g., the figure of the former proposition, and supposing $AB = a$, $BD = x$, we have

$$\begin{aligned} ax - x^2 &= \text{rectangle } AH \\ &= \text{gnomon } NOP. \end{aligned}$$

If, then, the area of the gnomon is given ($= b^2$, say, for any area can be transformed into the equivalent square by means of the problems of Eucl. I. 45 and II. 14), the solution of the equation

$$ax - x^2 = b^2$$

[1] Proclus on Eucl. I, pp. 419. 15–420. 12.

would be, in the language of application of areas, 'To a given straight line (a) to apply a rectangle which shall be equal to a given square (b^2) and shall fall short by a square figure.'

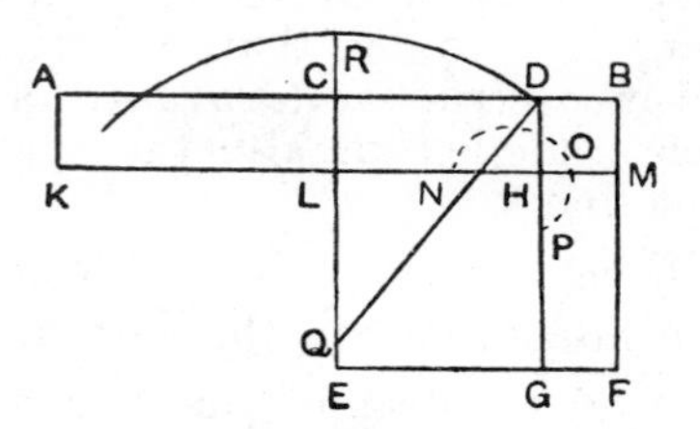

As the Pythagoreans solved the somewhat similar equation in II. 11, they cannot have failed to solve this one, as well as the equations corresponding to II. 6. For in the present case it is only necessary to draw CQ at right angles to AB from its middle point C, to make CQ equal to b, and then, with centre Q and radius equal to CB, or $\frac{1}{2}a$, to draw a circle cutting QC produced in R and CB in D (b^2 must be not greater than $\frac{1}{2}a^2$; otherwise a solution is impossible).

Then the determination of the point D constitutes the solution of the quadratic.

For, by the proposition II. 5,

$$AD.DB + CD^2 = CB^2$$
$$= QD^2 = QC^2 + CD^2;$$

therefore $$AD.DB = QC^2,$$

or $$ax - x^2 = b^2.$$

Similarly II. 6 enables us to solve the equations

$$ax + x^2 = b^2,$$

and $$x^2 - ax = b^2;$$

A C B D

the first equation corresponding to $AB = a$, $BD = x$ and the second to $AB = a$, $AD = x$, in the figure of the proposition.

The application of the theory to conics by Apollonius will be described when we come to deal with his treatise.

One great feature of Book II of Euclid's *Elements* is the use of the *gnomon* (Props. 5 to 8), which is undoubtedly Pythagorean and is connected, as we have seen, with the application of areas. The whole of Book II, with the latter section of Book I from Prop. 42 onwards, may be said to deal with the transformation of areas into equivalent areas of different shape or composition by means of 'application' and the use of the theorem of I. 47. Eucl. II. 9 and 10 are special cases which are very useful in geometry generally, but were also employed by the Pythagoreans for the specific purpose of proving the property of 'side-' and 'diameter-' numbers, the object of which was clearly to develop a series of closer and closer approximations to the value of $\sqrt{2}$ (see p. 93 *ante*).

The *geometrical algebra*, therefore, as we find it in Euclid, Books I and II, was Pythagorean. It was of course confined to problems not involving expressions above the second degree. Subject to this, it was an effective substitute for modern algebra. The product of two linear factors was a rectangle, and Book II of Euclid made it possible to *multiply* two factors with any number of linear terms in each; the compression of the result into a single product (rectangle) followed by means of the *application*-theorem (Eucl. I. 44). That theorem itself corresponds to *dividing* the product of any two linear factors by a third linear expression. To transform any area into a square, we have only to turn the area into a rectangle (as in Eucl. I. 45), and then find a square equal to that rectangle by the method of Eucl. II. 14; the latter problem then is equivalent to the *extraction of the square root*. And we have seen that the theorems of Eucl. II. 5, 6 enable mixed quadratic equations of certain types to be solved so far as their roots are real. In cases where a quadratic equation has one or both roots negative, the Greeks would transform it into one having a positive root or roots (by the equivalent of substituting $-x$ for x); thus, where one root is positive and one negative, they would solve the problem in two parts by taking two cases.

The other great engine of the Greek geometrical algebra, namely the method of proportions, was not in its full extent available to the Pythagoreans because their theory of proportion was only applicable to commensurable magnitudes (Eudoxus was the first to establish the general theory, applicable to commensurables and incommensurables alike, which we find in Eucl. V, VI). Yet it cannot be doubted that they used the method quite freely before the discovery of the irrational showed them that they were building on an insecure and inadequate foundation.

(δ) *The irrational.*

To return to the sentence about Pythagoras in the summary of Proclus already quoted more than once (pp. 84, 90, 141). Even if the reading ἀλόγων were right and Proclus really meant to attribute to Pythagoras the discovery of 'the theory, or study, of irrationals', it would be necessary to consider the authority for this statement, and how far it is supported by other evidence. We note that it occurs in a relative sentence ὃς δὴ . . . , which has the appearance of being inserted in parenthesis by the compiler of the summary rather than copied from his original source; and the shortened form of the first part of the same summary published in the *Variae collectiones* of Hultsch's Heron, and now included by Heiberg in Heron's *Definitions*,[1] contains no such parenthesis. Other authorities attribute the discovery of the theory of the irrational not to Pythagoras but to the Pythagoreans. A scholium to Euclid, Book X, says that

'the Pythagoreans were the first to address themselves to the investigation of commensurability, having discovered it as the result of their observation of numbers; for, while the unit is a common measure of all numbers, they were unable to find a common measure of all magnitudes, . . . because all magnitudes are divisible *ad infinitum* and never leave a magnitude which is too small to admit of further division, but that remainder is equally divisible *ad infinitum*,'

and so on. The scholiast adds the legend that

'the first of the Pythagoreans who made public the investigation of these matters perished in a shipwreck'.[2]

Another commentary on Eucl. X discovered by Woepcke in an Arabic translation and believed, with good reason, to be part of the commentary of Pappus, says that the theory of irrational magnitudes 'had its origin in the school of Pythagoras'. Again, it is impossible that Pythagoras himself should have discovered a 'theory' or 'study' of irrationals in any

[1] Heron, vol. iv, ed. Heib., p. 108.
[2] Euclid, ed. Heib., vol. v, pp. 415, 417.

proper sense. We are told in the *Theaetetus*[1] that Theodorus of Cyrene (a pupil of Protagoras and the teacher of Plato) proved the irrationality of $\sqrt{3}$, $\sqrt{5}$, &c., up to $\sqrt{17}$, and this must have been at a date not much, if anything, earlier than 400 B.C.; while it was Theaetetus who, inspired by Theodorus's investigation of these particular 'roots' (or surds), was the first to generalize the theory, seeking terms to cover all such incommensurables; this is confirmed by the continuation of the passage from Pappus's commentary, which says that the theory was

> 'considerably developed by Theaetetus the Athenian, who gave proof, in this part of mathematics as in others, of ability which has been justly admired . . . As for the exact distinctions of the above-named magnitudes and the rigorous demonstrations of the propositions to which this theory gives rise, I believe that they were chiefly established by this mathematician'.

It follows from all this that, if Pythagoras discovered anything about irrationals, it was not any 'theory' of irrationals but, at the most, some particular case of incommensurability. Now the passage which states that Theodorus proved that $\sqrt{3}$, $\sqrt{5}$, &c. are incommensurable says nothing of $\sqrt{2}$. The reason is, no doubt, that the incommensurability of $\sqrt{2}$ had been proved earlier, and everything points to the probability that this was the first case to be discovered. But, if Pythagoras discovered even this, it is difficult to see how the theory that number is the essence of all existing things, or that all things are made of number, could have held its ground for any length of time. The evidence suggests the conclusion that geometry developed itself for some time on the basis of the numerical theory of proportion which was inapplicable to any but commensurable magnitudes, and that it received an unexpected blow later by reason of the discovery of the irrational. The inconvenience of this state of things, which involved the restriction or abandonment of the use of proportions as a method pending the discovery of the generalized theory by Eudoxus, may account for the idea of the existence of the irrational having been kept secret, and of punishment having overtaken the first person who divulged it.

[1] Plato, *Theaetetus*, 147 D sq.

If then it was not Pythagoras but some Pythagorean who discovered the irrationality of $\sqrt{2}$, at what date are we to suppose the discovery to have been made? A recent writer [1] on the subject holds that it was the *later* Pythagoreans who made the discovery, not much before 410 B.C. It is impossible, he argues, that fifty or a hundred years would elapse between the discovery of the irrationality of $\sqrt{2}$ and the like discovery by Theodorus (about 410 or 400 B.C.) about the other surds $\sqrt{3}$, $\sqrt{5}$, &c. It is difficult to meet this argument except by the supposition that, in the interval, the thoughts of geometers had been taken up by other famous problems, such as the quadrature of the circle and the duplication of the cube (itself equivalent to finding $\sqrt[3]{2}$). Another argument is based on the passage in the *Laws* where the Athenian stranger speaks of the shameful ignorance of the generality of Greeks, who are not aware that it is not all geometrical magnitudes that are commensurable with one another; the speaker adds that it was only 'late' (ὀψέ ποτε) that he himself learnt the truth.[2] Even if we knew for certain whether 'late' means 'late in the day' or 'late in life', the expression would not help much towards determining the date of the first discovery of the irrationality of $\sqrt{2}$; for the language of the passage is that of rhetorical exaggeration (Plato speaks of men who are unacquainted with the existence of the irrational as more comparable to swine than to human beings). Moreover, the irrational appears in the *Republic* as something well known, and precisely with reference to $\sqrt{2}$; for the expressions 'the rational diameter of (the square the side of which is) 5' [= the approximation $\sqrt{(49)}$ or 7] and the 'irrational (ἄρρητος) diameter of 5' [= $\sqrt{(50)}$] are used without any word of explanation.[3]

Further, we have a well-authenticated title of a work by Democritus (born 470 or 460 B.C.), περὶ ἀλόγων γραμμῶν καὶ ναστῶν αβ, 'two books on irrational lines and solids' (ναστόν is πλῆρες, 'full', as opposed to κενόν. 'void', and Democritus called his 'first bodies' ναστά). Of the contents of this work we are not informed; the recent writer already mentioned

[1] H. Vogt in *Bibliotheca mathematica*, x_3, 1910, pp. 97–155 (cf. ix_3, p. 190 sq.).

[2] Plato, *Laws*, 819 D–820 C.

[3] Plato, *Republic*, vii. 546 D.

suggests that ἄλογος does not here mean irrational or incommensurable at all, but that the book was an attempt to connect the atomic theory with continuous magnitudes (lines) through 'indivisible lines' (cf. the Aristotelian treatise *On indivisible lines*), and that Democritus meant to say that, since any two lines are alike made up of an infinite number of the (indivisible) elements, they cannot be said to have any expressible ratio to one another, that is, he would regard them as 'having no ratio'! It is, however, impossible to suppose that a mathematician of the calibre of Democritus could have denied that any two lines can have a ratio to one another; moreover, on this view, since no two straight lines would have a ratio to one another, ἄλογοι γραμμαί would not be a *class* of lines, but *all* lines, and the title would lose all point. But indeed, as we shall see, it is also on other grounds inconceivable that Democritus should have been an upholder of 'indivisible lines' at all. I do not attach any importance to the further argument used in support of the interpretation in question, namely that ἄλογος in the sense of 'irrational' is not found in any other writer before Aristotle, and that Plato uses the words ἄρρητος and ἀσύμμετρος only. The latter statement is not even strictly true, for Plato does in fact use the word ἄλογοι specifically of γραμμαί in the passage of the *Republic* where he speaks of youths not being ἄλογοι ὥσπερ γραμμαί, 'irrational like lines'.[1] Poor as the joke is, it proves that ἄλογοι γραμμαί was a recognized technical term, and the remark looks like a sly reference to the very treatise of Democritus of which we are speaking. I think there is no reason to doubt that the book was on 'irrationals' in the technical sense. We know from other sources that Democritus was already on the track of infinitesimals in geometry; and nothing is more likely than that he would write on the kindred subject of irrationals.

I see therefore no reason to doubt that the irrationality of $\sqrt{2}$ was discovered by some Pythagorean at a date appreciably earlier than that of Democritus; and indeed the simple proof of it indicated by Aristotle and set out in the proposition interpolated at the end of Euclid's Book X seems appropriate to an early stage in the development of geometry.

[1] Plato, *Republic*, 534 D.

(ε) *The five regular solids.*

The same parenthetical sentence in Proclus which attributes to Pythagoras the discovery of the theory of irrationals (or proportions) also states that he discovered the 'putting together (σύστασις) of the cosmic figures' (the five regular solids). As usual, there has been controversy as to the sense in which this phrase is to be taken, and as to the possibility of Pythagoras having done what is attributed to him, in any sense of the words. I do not attach importance to the argument that, whereas Plato, presumably 'Pythagorizing', assigns the first four solids to the four elements, earth, fire, air, and water, Empedocles and not Pythagoras was the first to declare these four elements to be the material principles from which the universe was evolved; nor do I think it follows that, because the elements are four, only the first four solids had been discovered at the time when the four elements came to be recognized, and that the dodecahedron must therefore have been discovered later. I see no reason why all five should not have been discovered by the early Pythagoreans before any question of identifying them with the elements arose. The fragment of Philolaus, indeed, says that

'there are five bodies in the sphere, the fire, water, earth, and air in the sphere, and the vessel of the sphere itself making the fifth',[1]

but as this is only to be understood of the *elements* in the sphere of the universe, not of the solid figures, in accordance with Diels's translation, it would appear that Plato in the *Timaeus*[2] is the earliest authority for the allocation, and it may very well be due to Plato himself (were not the solids called the 'Platonic figures'?), although put into the mouth of a Pythagorean. At the same time, the fact that the *Timaeus* is fundamentally Pythagorean may have induced Aëtius's authority (probably Theophrastus) to conclude too

[1] Stobaeus, *Ecl.* I, proem. 3 (p. 18. 5 Wachsmuth); Diels, *Vors.* i³, p. 314. The Greek of the last phrase is καὶ ὁ τᾶς σφαίρας ὁλκάς, πέμπτον, but ὁλκάς is scarcely an appropriate word, and von Wilamowitz (*Platon*, vol. ii, 1919, pp. 91–2) proposes ὁ τᾶς σφαίρας ὁλκός, taking ὁλκός (which implies 'winding') as *volumen*. We might then translate by 'the spherical envelope'.

[2] *Timaeus*, 53 C–55 C.

hastily that 'here, too, Plato Pythagorizes', and to say dogmatically on the faith of this that

'*Pythagoras*, seeing that there are five solid figures, which are also called the mathematical figures, says that the earth arose from the cube, fire from the pyramid, air from the octahedron, water from the icosahedron, and the sphere of the universe from the dodecahedron.'[1]

It may, I think, be conceded that Pythagoras or the early Pythagoreans would hardly be able to 'construct' the five regular solids in the sense of a complete theoretical construction such as we find in Eucl. XIII; and it is possible that Theaetetus was the first to give these constructions, whether ἔγραψε in Suidas's notice means that 'he was the first to *construct*' or 'to *write upon* the five solids so called'. But there is no reason why the Pythagoreans should not have 'put together' the five figures in the manner in which Plato puts them together in the *Timaeus*, namely, by bringing a certain number of angles of equilateral triangles, squares, or pentagons severally together at one point so as to make a solid angle, and then completing all the solid angles in that way. That the early Pythagoreans should have discovered the five regular solids in this elementary way agrees well with what we know of their having put angles of certain regular figures round a point and shown that only three kinds of such angles would fill up the space in one plane round the point.[2] How elementary the construction still was in Plato's hands may be inferred from the fact that he argues that only three of the elements are transformable into one another because only three of the solids are made from equilateral triangles; these triangles, when present in sufficient numbers in given regular solids, can be separated again and redistributed so as to form regular solids of a different number of faces, as if the solids were really hollow shells bounded by the triangular faces as planes or laminae (Aristotle criticizes this in *De caelo*, iii. 1)! We may indeed treat Plato's elementary method as an indication that this was actually the method employed by the earliest Pythagoreans.

[1] Aët. ii. 6. 5 (*Vors*. i³, p. 306. 3–7).
[2] Proclus on Eucl. I, pp. 304. 11–305. 3.

Putting together squares three by three, forming eight solid angles, and equilateral triangles three by three, four by four, or five by five, forming four, six, or twelve solid angles respectively, we readily form a cube, a tetrahedron, an octahedron, or an icosahedron, but the fifth regular solid, the dodecahedron, requires a new element, the regular pentagon. True, if we form the angle of an icosahedron by putting together five equilateral triangles, the bases of those triangles when put together form a regular pentagon; but Pythagoras or the Pythagoreans would require a theoretical construction. What is the evidence that the early Pythagoreans could have constructed and did construct pentagons? That they did construct them seems established by the story of Hippasus,

'who was a Pythagorean but, owing to his being the first to publish and write down the (construction of the) sphere with (ἐκ, from) the twelve pentagons, perished by shipwreck for his impiety, but received credit for the discovery, whereas it really belonged to HIM (ἐκείνου τοῦ ἀνδρός), for it is thus that they refer to Pythagoras, and they do not call him by his name.'[1]

The connexion of Hippasus's name with the subject can hardly be an invention, and the story probably points to a positive achievement by him, while of course the Pythagoreans' jealousy for the Master accounts for the reflection upon Hippasus and the moral. Besides, there is evidence for the very early existence of dodecahedra in actual fact. In 1885 there was discovered on Monte Loffa (Colli Euganei, near Padua) a regular dodecahedron of Etruscan origin, which is held to date from the first half of the first millennium B.C.[2] Again, it appears that there are extant no less than twenty-six objects of dodecahedral form which are of Celtic origin.[3] It may therefore be that Pythagoras or the Pythagoreans had seen dodecahedra of this kind, and that their merit was to have treated them as mathematical objects and brought them into their theoretical geometry. Could they then have

[1] Iambl. *Vit. Pyth.* 88, *de c. math. scient.* c. 25, p. 77. 18–24.

[2] F. Lindemann, 'Zur Geschichte der Polyeder und der Zahlzeichen' (*Sitzungsber. der K. Bay. Akad. der Wiss.* xxvi. 1897, pp. 625–768).

[3] L. Hugo in *Comptes rendus* of the Paris Acad. of Sciences, lxiii, 1873, pp. 420–1; lxvii, 1875, pp. 433, 472; lxxxi, 1879, p. 332.

constructed the regular pentagon? The answer must, I think, be yes. If $ABCDE$ be a regular pentagon, and AC, AD, CE be joined, it is easy to prove, from the (Pythagorean) propositions about the sum of the internal angles of a polygon and

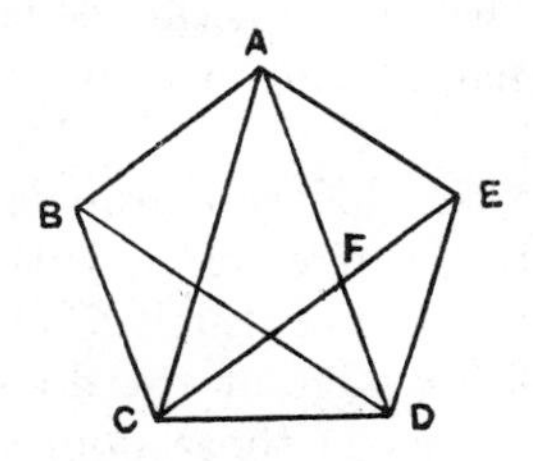

the sum of the angles of a triangle, that each of the angles BAC, DAE, ECD is $\frac{2}{5}$ths of a right angle, whence, in the triangle ACD, the angle CAD is $\frac{2}{5}$ths of a right angle, and each of the base angles ACD, ADC is $\frac{4}{5}$ths of a right angle or double of the vertical angle CAD; and from these facts it easily follows that, if CE and AD meet in F, CDF is an isosceles triangle equiangular, and therefore similar, to ACD, and also that $AF = FC = CD$. Now, since the triangles ACD, CDF are similar,

$$AC : CD = CD : DF,$$

or

$$AD : AF = AF : FD;$$

that is, if AD is given, the length of AF, or CD, is found by dividing AD at F in 'extreme and mean ratio' by Eucl. II. 11. This last problem is a particular case of the problem of 'application of areas', and therefore was obviously within the power of the Pythagoreans. This method of constructing a pentagon is, of course, that taught in Eucl. IV. 10, 11. If further evidence is wanted of the interest of the early Pythagoreans in the regular pentagon, it is furnished by the fact, attested by Lucian and the scholiast to the *Clouds* of Aristophanes, that the 'triple interwoven triangle, the pentagram', i.e. the star-pentagon, was used by the Pythagoreans as a symbol of recognition between the members of the same school, and was called by them Health.[1] Now it will be seen from the separate diagram of the star-pentagon above that it actually

[1] Lucian, *Pro lapsu in salut.* § 5 (vol. i, pp. 447–8, Jacobitz); schol. on *Clouds* 609.

shows the equal sides of the five isosceles triangles of the type referred to and also the points at which they are divided in extreme and mean ratio. (I should perhaps add that the pentagram is said to be found on the vase of Aristonophus found at Caere and supposed to belong to the seventh century B.C., while the finds at Mycenae include ornaments of pentagonal form.)

It would be easy to conclude that the dodecahedron is inscribable in a sphere, and to find the centre of it, without constructing both in the elaborate manner of Eucl. XIII. 17 and working out the relation between an edge of the dodecahedron and the radius of the sphere, as is there done: an investigation probably due to Theaetetus. It is right to mention here the remark in scholium No. 1 to Eucl. XIII that the book is about

'the five so-called Platonic figures, which, however, do not belong to Plato, three of the five being due to the Pythagoreans, namely the cube, the pyramid, and the dodecahedron, while the octahedron and icosahedron are due to Theaetetus'.[1]

This statement (taken probably from Geminus) may perhaps rest on the fact that Theaetetus was the first to write at any length about the two last-mentioned solids, as he was probably the first to construct all five theoretically and investigate fully their relations to one another and the circumscribing spheres.

(ζ) *Pythagorean astronomy.*

Pythagoras and the Pythagoreans occupy an important place in the history of astronomy. (1) Pythagoras was one of the first to maintain that the universe and the earth are spherical in form. It is uncertain what led Pythagoras to conclude that the earth is a sphere. One suggestion is that he inferred it from the roundness of the shadow cast by the earth in eclipses of the moon. But it is certain that Anaxagoras was the first to suggest this, the true, explanation of eclipses. The most likely supposition is that Pythagoras's ground was purely mathematical, or mathematico-aesthetical; that is, he

[1] Heiberg's Euclid, vol. v, p. 654.

attributed spherical shape to the earth (as to the universe) for the simple reason that the sphere is the most beautiful of solid figures. For the same reason Pythagoras would surely hold that the sun, the moon, and the other heavenly bodies are also spherical in shape. (2) Pythagoras is credited with having observed the identity of the Morning and the Evening Stars. (3) It is probable that he was the first to state the view (attributed to Alcmaeon and 'some of the mathematicians') that the planets as well as the sun and moon have a motion of their own from west to east opposite to and independent of the daily rotation of the sphere of the fixed stars from east to west.[1] Hermesianax, one of the older generation of Alexandrine poets (about 300 B.C.), is quoted as saying:

'What inspiration laid forceful hold on Pythagoras when he discovered the subtle geometry of (the heavenly) spirals and compressed in a small sphere the whole of the circle which the aether embraces.'[2]

This would seem to imply the construction of a sphere on which were represented the circles described by the sun, moon and planets together with the daily revolution of the heavenly sphere; but of course Hermesianax is not altogether a trustworthy authority.

It is improbable that Pythagoras himself was responsible for the astronomical system known as the Pythagorean, in which the earth was deposed from its place at rest in the centre of the universe, and became a 'planet', like the sun, the moon and the other planets, revolving about the central fire. For Pythagoras the earth was still at the centre, while about it there moved (*a*) the sphere of the fixed stars revolving daily from east to west, the axis of rotation being a straight line through the centre of the earth, (*b*) the sun, moon and planets moving in independent circular orbits in a sense opposite to that of the daily rotation, i.e. from west to east.

The later Pythagorean system is attributed by Aëtius (probably on the authority of Theophrastus) to Philolaus, and

[1] Aët. ii. 16. 2, 3 (*Vors.* i[3], p. 132. 15).
[2] See Athenaeus, xiii. 599 A.

may be described thus. The universe is spherical in shape and finite in size. Outside it is infinite void which enables the universe to breathe, as it were. At the centre is the central fire, the Hearth of the Universe, called by various names, the Tower or Watch-tower of Zeus, the Throne of Zeus, the House of Zeus, the Mother of the Gods, the Altar, Bond and Measure of Nature. In this central fire is located the governing principle, the force which directs the movement and activity of the universe. In the universe there revolve in circles about the central fire the following bodies. Nearest to the central fire revolves the counter-earth, which always accompanies the earth, the orbit of the earth coming next to that of the counter-earth; next to the earth, reckoning in order from the centre outwards, comes the moon, next to the moon the sun, next to the sun the five planets, and last of all, outside the orbits of the five planets, the sphere of the fixed stars. The counter-earth, which accompanies the earth and revolves in a smaller orbit, is not seen by us because the hemisphere of the earth on which we live is turned away from the counter-earth (the analogy of the moon which always turns one side towards us may have suggested this); this involves, incidentally, a rotation of the earth about its axis completed in the same time as it takes the earth to complete a revolution about the central fire. As the latter revolution of the earth was held to produce day and night, it is a natural inference that the earth was supposed to complete one revolution round the central fire in a day and a night, or in twenty-four hours. This motion on the part of the earth with our hemisphere always turned outwards would, of course, be equivalent, as an explanation of phenomena, to a rotation of the earth about a fixed axis, but for the parallax consequent on the earth describing a circle in space with radius greater than its own radius; this parallax, if we may trust Aristotle,[1] the Pythagoreans boldly asserted to be negligible. The superfluous thing in this system is the introduction of the counter-earth. Aristotle says in one place that its object was to bring up the number of the moving bodies to ten, the perfect number according to

[1] Arist. *De caelo*, ii. 13, 293 b 25–30.

the Pythagoreans[1]; but he hints at the truer explanation in another passage where he says that eclipses of the moon were considered to be due sometimes to the interposition of the earth, sometimes to the interposition of the counter-earth (to say nothing of other bodies of the same sort assumed by 'some' in order to explain why there appear to be more lunar eclipses than solar)[2]; we may therefore take it that the counter-earth was invented for the purpose of explaining eclipses of the moon and their frequency.

Recapitulation.

The astronomical systems of Pythagoras and the Pythagoreans illustrate the purely mathematical character of their physical speculations; the heavenly bodies are all spheres, the most perfect of solid figures, and they move in circles; there is no question raised of *forces* causing the respective movements; astronomy is pure mathematics, it is geometry, combined with arithmetic and harmony. The capital discovery by Pythagoras of the dependence of musical intervals on numerical proportions led, with his successors, to the doctrine of the 'harmony of the spheres'. As the ratio 2 : 1 between the lengths of strings of the same substance and at the same tension corresponds to the octave, the ratio 3 : 2 to the fifth, and the ratio 4 : 3 to the fourth, it was held that bodies moving in space produce sounds, that those which move more quickly give a higher note than those which move more slowly, while those move most quickly which move at the greatest distance; the sounds therefore produced by the heavenly bodies, depending on their distances (i.e. the size of their orbits), combine to produce a harmony; 'the whole heaven is number and harmony'.[3]

We have seen too how, with the Pythagoreans, the theory of numbers, or 'arithmetic', goes hand in hand with geometry; numbers are represented by dots or lines forming geometrical figures; the species of numbers often take their names from their geometrical analogues, while their properties are proved by geometry. The Pythagorean mathematics, therefore, is all one science, and their science is all mathematics.

[1] Arist. *Metaph.* A. 5, 986 a 8–12.

[2] Arist. *De caelo,* ii. 13, 293 b 21–5.

[3] Arist. *Metaph.* A. 5, 986 a 2.

It is this identification of mathematics (and of geometry in particular) with science in general, and their pursuit of it for its own sake, which led to the extraordinary advance of the subject in the Pythagorean school. It was the great merit of Pythagoras himself (apart from any particular geometrical or arithmetical theorems which he discovered) that he was the first to take this view of mathematics; it is characteristic of him that, as we are told, 'geometry was called by Pythagoras *inquiry* or *science*' (ἐκαλεῖτο δὲ ἡ γεωμετρία πρὸς Πυθαγόρου ἱστορία).[1] Not only did he make geometry a liberal education; he was the first to attempt to explore it down to its first principles; as part of the scientific basis which he sought to lay down he 'used definitions'. A point was, according to the Pythagoreans, a 'unit having position'[2]; and, if their method of regarding a line, a surface, a solid, and an angle does not amount to a definition, it at least shows that they had reached a clear idea of the *differentiae*, as when they said that 1 was a point, 2 a line, 3 a triangle, and 4 a pyramid. A surface they called χροιά, 'colour'; this was their way of describing the superficial appearance, the idea being, as Aristotle says, that the colour is either in the limiting surface (πέρας) or is the πέρας,[3] so that the meaning intended to be conveyed is precisely that intended by Euclid's definition (XI. Def. 2) that 'the limit of a solid is a surface'. An angle they called γλωχίς, a 'point' (as of an arrow) made by a line broken or bent back at one point.[4]

The positive achievements of the Pythagorean school in geometry, and the immense advance made by them, will be seen from the following summary.

1. They were acquainted with the properties of parallel lines, which they used for the purpose of establishing by a general proof the proposition that the sum of the three angles of any triangle is equal to two right angles. This latter proposition they again used to establish the well-known theorems about the sums of the exterior and interior angles, respectively, of any polygon.

2. They originated the subject of equivalent areas, the transformation of an area of one form into another of different

[1] Iambl. *Vit. Pyth.* 89.
[2] Proclus on Eucl. I, p. 95. 21.
[3] Arist. *De sensu*, 3, 439 a 31.
[4] Heron, Def. 15.

form and, in particular, the whole method of *application of areas*, constituting a *geometrical algebra*, whereby they effected the equivalent of the algebraical processes of addition, subtraction, multiplication, division, squaring, extraction of the square root, and finally the complete solution of the mixed quadratic equation $x^2 \pm px \pm q = 0$, so far as its roots are real. Expressed in terms of Euclid, this means the whole content of Book I. 35–48 and Book II. The method of *application of areas* is one of the most fundamental in the whole of later Greek geometry; it takes its place by the side of the powerful method of proportions; moreover, it is the starting point of Apollonius's theory of conics, and the three fundamental terms, *parabole, ellipsis,* and *hyperbole* used to describe the three separate problems in 'application' were actually employed by Apollonius to denote the three conics, names which, of course, are those which we use to-day. Nor was the use of the geometrical algebra for solving *numerical* problems unknown to the Pythagoreans; this is proved by the fact that the theorems of Eucl. II. 9, 10 were invented for the purpose of finding successive integral solutions of the indeterminate equations

$$2x^2 - y^2 = \pm 1.$$

3. They had a theory of proportion pretty fully developed. We know nothing of the form in which it was expounded; all we know is that it took no account of incommensurable magnitudes. Hence we conclude that it was a numerical theory, a theory on the same lines as that contained in Book VII of Euclid's *Elements*.

They were aware of the properties of similar figures. This is clear from the fact that they must be assumed to have solved the problem, which was, according to Plutarch, attributed to Pythagoras himself, of describing a figure which shall be similar to one given figure and equal in area to another given figure This implies a knowledge of the proposition that similar figures (triangles or polygons) are to one another in the duplicate ratio of corresponding sides (Eucl. VI. 19, 20). As the problem is solved in Eucl. VI. 25, we assume that, subject to the qualification that their theorems about similarity, &c., were only established of figures in which corresponding elements are commensurable, they had theorems corresponding to a great part of Eucl., Book VI.

Again, they knew how to cut a straight line in extreme and mean ratio (Eucl. VI. 30); this problem was presumably solved by the method used in Eucl. II. 11, rather than by that of Eucl. VI. 30, which depends on the solution of a problem in the application of areas more general than the methods of Book II enable us to solve, the problem namely of Eucl. VI. 29.

4. They had discovered, or were aware of the existence of, the five regular solids. These they may have constructed empirically by putting together squares, equilateral triangles, and pentagons. This implies that they could construct a regular pentagon and, as this construction depends upon the construction of an isosceles triangle in which each of the base angles is double of the vertical angle, and this again on the cutting of a line in extreme and mean ratio, we may fairly assume that this was the way in which the construction of the regular pentagon was actually evolved. It would follow that the solution of problems by *analysis* was already practised by the Pythagoreans, notwithstanding that the discovery of the analytical method is attributed by Proclus to Plato. As the particular construction is practically given in Eucl. IV. 10, 11, we may assume that the content of Eucl. IV was also partly Pythagorean.

5. They discovered the existence of the irrational in the sense that they proved the incommensurability of the diagonal of a square with reference to its side; in other words, they proved the irrationality of $\sqrt{2}$. As a proof of this is referred to by Aristotle in terms which correspond to the method used in a proposition interpolated in Euclid, Book X, we may conclude that this proof is ancient, and therefore that it was probably the proof used by the discoverers of the proposition. The method is to prove that, if the diagonal of a square is commensurable with the side, then the same number must be both odd and even; here then we probably have an early Pythagorean use of the method of *reductio ad absurdum*.

Not only did the Pythagoreans discover the irrationality of $\sqrt{2}$; they showed, as we have seen, how to approximate as closely as we please to its numerical value.

After the discovery of this one case of irrationality, it would be obvious that propositions theretofore proved by means of the numerical theory of proportion, which was inapplicable to incommensurable magnitudes, were only partially proved. Accordingly, pending the discovery of a theory of proportion applicable to incommensurable as well as commensurable magnitudes, there would be an inducement to substitute, where possible, for proofs employing the theory of proportions other proofs independent of that theory. This substitution is carried rather far in Euclid, Books I–IV; it does not follow that the Pythagoreans remodelled their proofs to the same extent as Euclid felt bound to do.

VI

PROGRESS IN THE ELEMENTS DOWN TO PLATO'S TIME

In tracing the further progress in the Elements which took place down to the time of Plato, we do not get much assistance from the summary of Proclus. The passage in which he states the succession of geometers from Pythagoras to Plato and his contemporaries runs as follows:

'After him [Pythagoras] Anaxagoras of Clazomenae dealt with many questions in geometry, and so did Oenopides of Chios, who was a little younger than Anaxagoras; Plato himself alludes, in the *Rivals*, to both of them as having acquired a reputation for mathematics. After them came Hippocrates of Chios, the discoverer of the quadrature of the lune, and Theodorus of Cyrene, both of whom became distinguished geometers; Hippocrates indeed was the first of whom it is recorded that he actually compiled Elements. Plato, who came next to them, caused mathematics in general and geometry in particular to make a very great advance, owing to his own zeal for these studies; for every one knows that he even filled his writings with mathematical discourses and strove on every occasion to arouse enthusiasm for mathematics in those who took up philosophy. At this time too lived Leodamas of Thasos, Archytas of Taras, and Theaetetus of Athens, by whom the number of theorems was increased and a further advance was made towards a more scientific grouping of them.'[1]

It will be seen that we have here little more than a list of names of persons who advanced, or were distinguished in, geometry. There is no mention of specific discoveries made by particular geometers, except that the work of Hippocrates on the squaring of certain lunes is incidentally alluded to, rather as a means of identifying Hippocrates than as a detail relevant to the subject in hand. It would appear that

[1] Proclus on Eucl. I, p. 65. 21-66. 18.

the whole summary was directed to the one object of tracing progress in the Elements, particularly with reference to improvements of method in the direction of greater generality and more scientific order and treatment; hence only those writers are here mentioned who contributed to this development. Hippocrates comes into the list, not because of his lunes, but because he was a distinguished geometer and was the first to write Elements. Hippias of Elis, on the other hand, though he belongs to the period covered by the extract, is omitted, presumably because his great discovery, that of the curve known as the *quadratrix*, does not belong to elementary geometry; Hippias is, however, mentioned in two other places by Proclus in connexion with the quadratrix,[1] and once more as authority for the geometrical achievements of Ameristus (or Mamercus or Mamertius).[2] Less justice is done to Democritus, who is neither mentioned here nor elsewhere in the commentary; the omission here of the name of Democritus is one of the arguments for the view that this part of the summary is not quoted from the *History of Geometry* by Eudemus (who would not have been likely to omit so accomplished a mathematician as Democritus), but is the work either of an intermediary or of Proclus himself, based indeed upon data from Eudemus's history, but limited to particulars relevant to the object of the commentary, that is to say, the elucidation of Euclid and the story of the growth of the Elements.

There are, it is true, elsewhere in Proclus's commentary a very few cases in which particular propositions in Euclid, Book I, are attributed to individual geometers, e.g. those which Thales is said to have discovered. Two propositions presently to be mentioned are in like manner put to the account of Oenopides; but except for these details about Oenopides we have to look elsewhere for evidence of the growth of the Elements in the period now under notice. Fortunately we possess a document of capital importance, from this point of view, in the fragment of Eudemus on Hippocrates's quadrature of lunes preserved in Simplicius's commentary on the *Physics* of Aristotle.[3] This fragment will

[1] Proclus on Eucl. I, p. 272. 7, p. 356. 11. [2] *Ib.*, p. 65. 14.
[3] Simpl. *in Arist. Phys.* pp. 54–69 Diels.

be described below. Meantime we will take the names mentioned by Proclus in their order.

ANAXAGORAS (about 500–428 B.C.) was born at Clazomenae in the neighbourhood of Smyrna. He neglected his possessions, which were considerable, in order to devote himself to science. Some one once asked him what was the object of being born, to which he replied, 'The investigation of sun, moon and heaven.' He was apparently the first philosopher to take up his abode at Athens, where he enjoyed the friendship of Pericles. When Pericles became unpopular shortly before the outbreak of the Peloponnesian War, he was attacked through his friends, and Anaxagoras was accused of impiety for holding that the sun was a red-hot stone and the moon earth. According to one account he was fined five talents and banished; another account says that he was kept in prison and that it was intended to put him to death, but that Pericles obtained his release; he went and lived at Lampsacus till his death.

Little or nothing is known of Anaxagoras's achievements in mathematics proper, though it is credible enough that he was a good mathematician. But in astronomy he made one epoch-making discovery, besides putting forward some remarkably original theories about the evolution of the universe. We owe to him the first clear recognition of the fact that the moon does not shine by its own light but receives its light from the sun; this discovery enabled him to give the true explanation of lunar and solar eclipses, though as regards the former (perhaps in order to explain their greater frequency) he erroneously supposed that there were other opaque and invisible bodies 'below the moon' which, as well as the earth, sometimes by their interposition caused eclipses of the moon. A word should be added about his cosmology on account of the fruitful ideas which it contained. According to him the formation of the world began with a vortex set up, in a portion of the mixed mass in which 'all things were together', by Mind (*νοῦς*). This rotatory movement began in the centre and then gradually spread, taking in wider and wider circles. The first effect was to separate two great masses, one consisting of the rare, hot, light, dry, called the 'aether', the other of the opposite categories and called 'air'. The aether took the outer, the air the inner place. From the air were next separated clouds, water, earth and stones. The dense, the moist, the dark and cold, and all the heaviest things, collected in the centre as the result of the circular motion, and it was from these elements when consolidated that the earth was formed; but after this, in consequence of the violence of the whirling motion, the surrounding fiery aether tore stones away from the earth and kindled them into stars. Taking this in conjunction with the remark that stones 'rush outwards more than water', we see that Anaxagoras conceived the idea of a *centrifugal* force as well as that of concentration brought about by the motion of the vortex, and that he assumed a series of projections or 'whirlings-off' of precisely the same kind as the theory of Kant and Laplace assumed for the formation of the solar system. At the same time he held that one of the heavenly bodies might break away and fall (this may account for the story that he prophesied the fall of the meteoric stone at Aegospotami in 468/7 B.C.), a *centripetal* tendency being here recognized.

In mathematics we are told that Anaxagoras 'while in prison wrote (or drew, ἔγραφε) the squaring of the circle'.[1] But we have no means of judging what this amounted to. Rudio translates ἔγραφε as 'zeichnete', 'drew', observing that he probably knew the Egyptian rule for squaring, and simply drew on the sand a square as nearly as he could equal to the area of a circle.[2] It is clear to me that this cannot be right, but that the word means 'wrote upon' in the sense that he tried to work out theoretically the problem in question. For the same word occurs (in the passive) in the extract from Eudemus about Hippocrates: 'The squarings of the lunes . . . were first written (or proved) by Hippocrates and were found to be correctly expounded',[3] where the context shows that ἐγράφησαν cannot merely mean 'were drawn'. Besides, τετραγωνισμός, *squaring*, is a process or operation, and you cannot, properly speaking, 'draw' a process, though you can 'describe' it or prove its correctness.

[1] Plutarch, *De exil.* 17, 607 F.

[2] Rudio, *Der Bericht des Simplicius über die Quadraturen des Antiphon und Hippokrates*, 1907, p. 92, 93.

[3] Simpl. *in Phys.*, p. 61. 1–3 Diels; Rudio, *op. cit.*, pp. 46. 22–48. 4.

Vitruvius tells us that one Agatharchus was the first to paint stage-scenes at Athens, at the time when Aeschylus was having his tragedies performed, and that he left a treatise on the subject which was afterwards a guide to Democritus and Anaxagoras, who discussed the same problem, namely that of painting objects on a plane surface in such a way as to make some of the things depicted appear to be in the background while others appeared to stand out in the foreground, so that you seemed, e.g., to have real buildings before you; in other words, Anaxagoras and Democritus both wrote treatises on perspective.[1]

There is not much to be gathered from the passage in the *Rivals* to which Proclus refers. Socrates, on entering the school of Dionysius, finds two lads disputing a certain point, something about Anaxagoras or Oenopides, he was not certain which; but they appeared to be drawing circles, and to be imitating certain inclinations by placing their hands at an angle.[2] Now this description suggests that what the lads were trying to represent was the circles of the equator and the zodiac or ecliptic; and we know that in fact Eudemus in his *History of Astronomy* attributed to Oenopides the discovery of 'the cincture of the zodiac circle',[3] which must mean the discovery of the obliquity of the ecliptic. It would probably be unsafe to conclude that Anaxagoras was also credited with the same discovery, but it certainly seems to be suggested that Anaxagoras had to some extent touched the mathematics of astronomy.

OENOPIDES OF CHIOS was primarily an astronomer. This is shown not only by the reference of Eudemus just cited, but by a remark of Proclus in connexion with one of two propositions in elementary geometry attributed to him.[4] Eudemus is quoted as saying that he not only discovered the obliquity of the ecliptic, but also the period of a Great Year. According to Diodorus the Egyptian priests claimed that it was from them that Oenopides learned that the sun moves in an inclined orbit and in a sense opposite to the motion of the fixed stars. It does not appear that Oenopides made any measurement of

[1] Vitruvius, *De architectura*, vii. praef. 11.
[2] Plato, *Erastae* 132 A, B.
[3] Theon of Smyrna, p. 198. 14.
[4] Proclus on Eucl. I, p. 283. 7-8.

the obliquity of the ecliptic. The duration of the Great Year he is said to have put at 59 years, while he made the length of the year itself to be $365\frac{22}{59}$ days. His Great Year clearly had reference to the sun and moon only; he merely sought to find the least integral number of complete years which would contain an exact number of lunar months. Starting, probably, with 365 days as the length of a year and $29\frac{1}{2}$ days as the length of a lunar month, approximate values known before his time, he would see that twice $29\frac{1}{2}$, or 59, years would contain twice 365, or 730, lunar months. He may then, from his knowledge of the calendar, have obtained 21,557 as the number of days in 730 months, for 21,557 when divided by 59 gives $365\frac{22}{59}$ as the number of days in the year.

Of Oenopides's geometry we have no details, except that Proclus attributes to him two propositions in Eucl. Bk. I. Of I. 12 ('to draw a perpendicular to a given straight line from a point outside it') Proclus says:

'This problem was first investigated by Oenopides, who thought it useful for astronomy. He, however, calls the perpendicular in the archaic manner (a straight line drawn) *gnomon-wise* (κατὰ γνώμονα), because the gnomon is also at right angles to the horizon.'[1]

On I. 23 ('on a given straight line and at a given point on it to construct a rectilineal angle equal to a given rectilineal angle') Proclus remarks that this problem is 'rather the discovery of Oenopides, as Eudemus says'.[2] It is clear that the geometrical reputation of Oenopides could not have rested on the mere solution of such simple problems as these. Nor, of course, could he have been the first to draw a perpendicular in practice; the point may be that he was the first to solve the problem by means of the ruler and compasses only, whereas presumably, in earlier days, perpendiculars would be drawn by means of a set square or a right-angled triangle originally constructed, say, with sides proportional to 3, 4, 5. Similarly Oenopides may have been the first to give the theoretical, rather than the practical, construction for the problem of I. 23 which we find in Euclid. It may therefore be that Oenopides's significance lay in improvements of method from the point of view of theory; he may, for example, have been the first to

[1] Proclus on Eucl. I, p. 283. 7–8. [2] Proclus on Eucl. I, p. 333. 5.

lay down the restriction of the means permissible in constructions to the ruler and compasses which became a canon of Greek geometry for all 'plane' constructions, i.e. for all problems involving the equivalent of the solution of algebraical equations of degree not higher than the second.

DEMOCRITUS, as mathematician, may be said to have at last come into his own. In the *Method* of Archimedes, happily discovered in 1906, we are told that Democritus was the first to state the important propositions that the volume of a cone is one third of that of a cylinder having the same base and equal height, and that the volume of a pyramid is one third of that of a prism having the same base and equal height; that is to say, Democritus enunciated these propositions some fifty years or more before they were first scientifically proved by Eudoxus.

Democritus came from Abdera, and, according to his own account, was young when Anaxagoras was old. Apollodorus placed his birth in Ol. 80 (= 460–457 B.C.), while according to Thrasyllus he was born in Ol. 77. 3 (= 470/69 B.C.), being one year older than Socrates. He lived to a great age, 90 according to Diodorus, 104, 108, 109 according to other authorities. He was indeed, as Thrasyllus called him, πένταθλος in philosophy[1]; there was no subject to which he did not notably contribute, from mathematics and physics on the one hand to ethics and poetics on the other; he even went by the name of 'Wisdom' (*Σοφία*).[2] Plato, of course, ignores him throughout his dialogues, and is said to have wished to burn all his works; Aristotle, on the other hand, pays handsome tribute to his genius, observing, e.g., that on the subject of change and growth no one save Democritus had observed anything except superficially; whereas Democritus seemed to have thought of everything.[3] He could say of himself (the fragment is, it is true, considered by Diels to be spurious, while Gomperz held it to be genuine), 'Of all my contemporaries I have covered the most ground in my travels, making the most exhaustive inquiries the while; I have seen the most climates and countries and listened to

[1] Diog. L. ix. 37 (*Vors.* ii³, p. 11. 24–30).
[2] Clem. *Strom.* vi. 32 (*Vors.* ii³, p. 16. 28).
[3] Arist. *De gen. et corr.* i. 2, 315 a 35.

the greatest number of learned men '.[1] His travels lasted for five years, and he is said to have visited Egypt, Persia and Babylon, where he consorted with the priests and magi; some say that he went to India and Aethiopia also. Well might he undertake the compilation of a geographical survey of the earth as, after Anaximander, Hecataeus of Miletus and Damastes of Sigeum had done. In his lifetime his fame was far from world-wide: 'I came to Athens', he says, 'and no one knew me.'[2]

A long list of his writings is preserved in Diogenes Laertius, the authority being Thrasyllus. In astronomy he wrote, among other works, a book *On the Planets*, and another *On the Great Year or Astronomy* including a *parapegma*[3] (or calendar). Democritus made the order of the heavenly bodies, reckoning outwards from the earth, the following: Moon, Venus, Sun, the other planets, the fixed stars. Lucretius[4] has preserved an interesting explanation which he gave of the reason why the sun takes a year to describe the full circle of the zodiac, while the moon completes its circle in a month. The nearer any body is to the earth (and therefore the farther from the sphere of the fixed stars) the less swiftly can it be carried round by the revolution of the heaven. Now the moon is nearer than the sun, and the sun than the signs of the zodiac; therefore the moon seems to get round faster than the sun because, while the sun, being lower and therefore slower than the signs, is left behind by them, the moon, being still lower and therefore slower still, is still more left behind. Democritus's Great Year is described by Censorinus[5] as 82 (LXXXII) years including 28 intercalary months, the latter number being the same as that included by Callippus in his cycle of 76 years; it is therefore probable that LXXXII is an incorrect reading for LXXVII (77).

As regards his mathematics we have first the statement in

[1] Clement, *Strom.* i. 15, 69 (*Vors.* ii³, p. 123. 3).

[2] Diog. L. ix. 36 (*Vors.* ii³, p. 11. 22).

[3] The *parapegma* was a posted record, a kind of almanac, giving, for a series of years, the movements of the sun, the dates of the phases of the moon, the risings and settings of certain stars, besides ἐπισημασίαι or weather indications; many details from Democritus's *parapegma* are preserved in the Calendar at the end of Geminus's *Isagoge* and in Ptolemy.

[4] Lucretius, v. 621 sqq.

[5] *De die natali*, 18. 8.

the continuation of the fragment of doubtful authenticity already quoted that

'in the putting together of lines, with the necessary proof, no one has yet surpassed me, not even the so-called *harpedonaptae* (rope-stretchers) of Egypt'.

This does not tell us much, except that it indicates that the 'rope-stretchers', whose original function was land-measuring or practical geometry, had by Democritus's time advanced some way in theoretical geometry (a fact which the surviving documents, such as the book of Ahmes, with their merely practical rules, would not have enabled us to infer). However, there is no reasonable doubt that in geometry Democritus was fully abreast of the knowledge of his day; this is fully confirmed by the titles of treatises by him and from other sources. The titles of the works classed as mathematical are (besides the astronomical works above mentioned):

1. *On a difference of opinion* (*γνώμης*: *v. l.* *γνώμονος*, gnomon), *or on the contact of a circle and a sphere;*
2. *On Geometry;*
3. *Geometricorum* (? I, II);
4. *Numbers;*
5. *On irrational lines and solids* (*ναστῶν*, atoms ?);
6. *'Εκπετάσματα.*

As regards the first of these works I think that the attempts to extract a sense out of Cobet's reading *γνώμονος* (on a difference of a gnomon) have failed, and that *γνώμης* (Diels) is better. But 'On a difference of opinion' seems scarcely determinative enough, if this was really an alternative title to the book. We know that there were controversies in ancient times about the nature of the 'angle of contact' (the 'angle' formed, at the point of contact, between an arc of a circle and the tangent to it, which angle was called by the special name *hornlike*, *κερατοειδής*), and the 'angle' complementary to it (the 'angle of a semicircle').[1] The question was whether the 'hornlike angle' was a magnitude comparable with the rectilineal angle, i.e. whether by being multiplied a sufficient number of times it could be made to exceed a

[1] Proclus on Eucl. I, pp. 121. 24–122. 6.

given rectilineal angle. Euclid proved (in III. 16) that the 'angle of contact' is less than any rectilineal angle, thereby setting the question at rest. This is the only reference in Euclid to this angle and the 'angle *of* a semicircle', although he defines the 'angle *of* a segment' in III, Def. 7, and has statements about the angles *of* segments in III. 31. But we know from a passage of Aristotle that before his time 'angles *of* segments' came into geometrical text-books as elements in figures which could be used in the proofs of propositions[1]; thus e.g. the equality of the two angles *of* a segment (assumed as known) was used to prove the theorem of Eucl. I. 5. Euclid abandoned the use of all such angles in proofs, and the references to them above mentioned are only survivals. The controversies doubtless arose long before his time, and such a question as the nature of the contact of a circle with its tangent would probably have a fascination for Democritus, who, as we shall see, broached other questions involving infinitesimals. As, therefore, the questions of the nature of the contact of a circle with its tangent and of the character of the 'hornlike' angle are obviously connected, I prefer to read γωνίης ('of an angle') instead of γνώμης; this would give the perfectly comprehensible title, '*On a difference in an angle, or on the contact of a circle and a sphere*'. We know from Aristotle that Protagoras, who wrote a book on mathematics, περὶ τῶν μαθημάτων, used against the geometers the argument that no such straight lines and circles as they assume exist in nature, and that (e.g.) a material circle does not in actual fact touch a ruler at one point only[2]; and it seems probable that Democritus's work was directed against this sort of attack on geometry.

We know nothing of the contents of Democritus's book *On Geometry* or of his *Geometrica*. One or other of these works may possibly have contained the famous dilemma about sections of a cone parallel to the base and very close together, which Plutarch gives on the authority of Chrysippus.[3]

'If', said Democritus, 'a cone were cut by a plane parallel to the base [by which is clearly meant a plane indefinitely

[1] Arist. *Anal. Pr.* i. 24, 41 b 13–22.
[2] Arist. *Metaph.* B. 2, 998 a 2.
[3] Plutarch, *De comm. not. adv. Stoicos*, xxxix. 3.

near to the base], what must we think of the surfaces forming the sections? Are they equal or unequal? For, if they are unequal, they will make the cone irregular as having many indentations, like steps, and unevennesses; but, if they are equal, the sections will be equal, and the cone will appear to have the property of the cylinder and to be made up of equal, not unequal, circles, which is very absurd.'

The phrase '*made up* of equal . . . circles' shows that Democritus already had the idea of a solid being the sum of an infinite number of parallel planes, or indefinitely thin laminae, indefinitely near together: a most important anticipation of the same thought which led to such fruitful results in Archimedes. This idea may be at the root of the argument by which Democritus satisfied himself of the truth of the two propositions attributed to him by Archimedes, namely that a cone is one third part of the cylinder, and a pyramid one third of the prism, which has the same base and equal height. For it seems probable that Democritus would notice that, if two pyramids having the same height and equal triangular bases are respectively cut by planes parallel to the base and dividing the heights in the same ratio, the corresponding sections of the two pyramids are equal, whence he would infer that the pyramids are equal as being the sum of the same infinite number of equal plane sections or indefinitely thin laminae. (This would be a particular anticipation of Cavalieri's proposition that the areal or solid content of two figures is equal if two sections of them taken at the same height, whatever the height may be, always give equal straight lines or equal surfaces respectively.) And Democritus would of course see that the three pyramids into which a prism on the same base and of equal height with the original pyramid is divided (as in Eucl. XII. 7) satisfy this test of equality, so that the pyramid would be one third part of the prism. The extension to a pyramid with a polygonal base would be easy. And Democritus may have stated the proposition for the cone (of course without an absolute proof) as a natural inference from the result of increasing indefinitely the number of sides in a regular polygon forming the base of a pyramid.

Tannery notes the interesting fact that the order in the list of Democritus's works of the treatises *On Geometry*, *Geometrica*, *Numbers*, and *On irrational lines and solids* corresponds to the order of the separate sections of Euclid's *Elements*, Books I–VI (plane geometry), Books VII–IX (on numbers), and Book X (on irrationals). With regard to the work *On irrational lines and solids* it is to be observed that, inasmuch as his investigation of the cone had brought Democritus consciously face to face with infinitesimals, there is nothing surprising in his having written on irrationals; on the contrary, the subject is one in which he would be likely to take special interest. It is useless to speculate on what the treatise actually contained; but of one thing we may be sure, namely that the ἄλογοι γραμμαί, 'irrational lines', were not ἄτομοι γραμμαί, '*indivisible* lines'.[1] Democritus was too good a mathematician to have anything to do with such a theory. We do not know what answer he gave to his puzzle about the cone; but his statement of the dilemma shows that he was fully alive to the difficulties connected with the conception of the continuous as illustrated by the particular case, and he cannot have solved it, in a sense analogous to his physical theory of atoms, by assuming indivisible lines, for this would have involved the inference that the consecutive parallel sections of the cone are *unequal*, in which case the surface would (as he said) be discontinuous, forming steps, as it were. Besides, we are told by Simplicius that, according to Democritus himself, his atoms were, in a mathematical sense divisible further and in fact *ad infinitum*,[2] while the scholia to Aristotle's *De caelo* implicitly deny to Democritus any theory of indivisible lines: 'of those who have maintained the existence of indivisibles, some, as for example Leucippus and Democritus, believe in indivisible bodies, others, like Xenocrates, in indivisible lines'.[3]

With reference to the Ἐκπετάσματα it is to be noted that this word is explained in Ptolemy's *Geography* as the projection of the armillary sphere upon a plane.[4] This work and that *On irrational lines* would hardly belong to elementary geometry.

[1] On this cf. O. Apelt, *Beiträge zur Geschichte der griechischen Philosophie*, 1891, p. 265 sq.

[2] Simpl. *in Phys.*, p. 83. 5.

[3] Scholia in Arist., p. 469 b 14, Brandis.

[4] Ptolemy, *Geogr.* vii. 7.

HIPPIAS OF ELIS, the famous sophist already mentioned (pp. 2, 23-4), was nearly contemporary with Socrates and Prodicus, and was probably born about 460 B.C. Chronologically, therefore, his place would be here, but the only particular discovery attributed to him is that of the curve afterwards known as the *quadratrix*, and the *quadratrix* does not come within the scope of the *Elements*. It was used first for trisecting any rectilineal angle or, more generally, for dividing it in any ratio whatever, and secondly for squaring the circle, or rather for finding the length of any arc of a circle; and these problems are not what the Greeks called 'plane' problems, i.e. they cannot be solved by means of the ruler and compasses. It is true that some have denied that the Hippias who invented the *quadratrix* can have been Hippias of Elis; Blass [1] and Apelt [2] were of this opinion, Apelt arguing that at the time of Hippias geometry had not got far beyond the theorem of Pythagoras. To show how wide of the mark this last statement is we have only to think of the achievements of Democritus. We know, too, that Hippias the sophist specialized in mathematics, and I agree with Cantor and Tannery that there is no reason to doubt that it was he who discovered the *quadratrix*. This curve will be best described when we come to deal with the problem of squaring the circle (Chapter VII); here we need only remark that it implies the proposition that the lengths of arcs in a circle are proportional to the angles subtended by them at the centre (Eucl. VI. 33).

The most important name from the point of view of this chapter is HIPPOCRATES OF CHIOS. He is indeed the first person of whom it is recorded that he compiled a book of Elements. This is lost, but Simplicius has preserved in his commentary on the *Physics* of Aristotle a fragment from Eudemus's *History of Geometry* giving an account of Hippocrates's quadratures of certain 'lunules' or lunes.[3] This is one of the most precious sources for the history of Greek geometry before Euclid; and, as the methods, with one slight apparent exception, are those of the straight line and circle, we can form a good idea of the progress which had been made in the Elements up to Hippocrates's time.

[1] Fleckeisen's *Jahrbuch*, cv, p. 28.
[2] *Beiträge zur Gesch. d. gr. Philosophie*, p. 379.
[3] Simpl. *in Phys.*, pp. 60. 22-68. 32, Diels.

It would appear that Hippocrates was in Athens during a considerable portion of the second half of the fifth century, perhaps from 450 to 430 B.C. We have quoted the story that what brought him there was a suit to recover a large sum which he had lost, in the course of his trading operations, through falling in with pirates; he is said to have remained in Athens on this account a long time, during which he consorted with the philosophers and reached such a degree of proficiency in geometry that he tried to discover a method of squaring the circle.[1] This is of course an allusion to the quadratures of lunes.

Another important discovery is attributed to Hippocrates. He was the first to observe that the problem of doubling the cube is reducible to that of finding two mean proportionals in continued proportion between two straight lines.[2] The effect of this was, as Proclus says, that thenceforward people addressed themselves (exclusively) to the equivalent problem of finding two mean proportionals between two straight lines.[3]

(α) *Hippocrates's quadrature of lunes.*

I will now give the details of the extract from Eudemus on the subject of Hippocrates's quadrature of lunes, which (as I have indicated) I place in this chapter because it belongs to elementary 'plane' geometry. Simplicius says he will quote Eudemus 'word for word' (κατὰ λέξιν) except for a few additions taken from Euclid's *Elements*, which he will insert for clearness' sake, and which are indeed necessitated by the summary (memorandum-like) style of Eudemus, whose form of statement is condensed, 'in accordance with ancient practice'. We have therefore in the first place to distinguish between what is textually quoted from Eudemus and what Simplicius has added. To Bretschneider[4] belongs the credit of having called attention to the importance of the passage of Simplicius to the historian of mathematics; Allman[5] was the first to attempt the task of distinguishing between the actual

[1] Philop. *in Phys.*, p. 31. 3, Vitelli.

[2] Pseudo-Eratosthenes to King Ptolemy in Eutoc. on Archimedes (vol. iii, p. 88, Heib.).

[3] Proclus on Eucl. I, p. 213. 5.

[4] Bretschneider, *Die Geometrie und die Geometer vor Euklides*, 1870, pp. 100–21.

[5] *Hermathena*, iv, pp. 180–228; *Greek Geometry from Thales to Euclid*, pp. 64–75.

extracts from Eudemus and Simplicius's amplifications; then came the critical text of Simplicius's commentary on the *Physics* edited by Diels (1882), who, with the help of Usener, separated out, and marked by spacing, the portions which they regarded as Eudemus's own. Tannery,[1] who had contributed to the preface of Diels some critical observations, edited (in 1883), with a translation and notes, what he judged to be Eudemian (omitting the rest). Heiberg[2] reviewed the whole question in 1884; and finally Rudio,[3] after giving in the *Bibliotheca Mathematica* of 1902 a translation of the whole passage of Simplicius with elaborate notes, which again he followed up by other articles in the same journal and elsewhere in 1903 and 1905, has edited the Greek text, with a translation, introduction, notes, and appendices, and summed up the whole controversy.

The occasion of the whole disquisition in Simplicius's commentary is a remark by Aristotle that there is no obligation on the part of the exponent of a particular subject to refute a fallacy connected with it unless the author of the fallacy has based his argument on the admitted principles lying at the root of the subject in question. 'Thus', he says, 'it is for the geometer to refute the (supposed) quadrature of a circle by means of segments (τμημάτων), but it is not the business of the geometer to refute the argument of Antiphon.'[4] Alexander took the remark to refer to Hippocrates's attempted quadrature by means of *lunes* (although in that case τμῆμα is used by Aristotle, not in the technical sense of a *segment*, but with the non-technical meaning of any portion cut out of a figure). This, probable enough in itself (for in another place Aristotle uses the same word τμῆμα to denote a *sector* of a circle[5]), is made practically certain by two other allusions in Aristotle, one to a proof that a circle together with certain lunes is equal to a rectilineal figure,[6] and the other to 'the (fallacy) of Hippocrates or the quadrature by means of the lunes'.[7] The

[1] Tannery, *Mémoires scientifiques*, vol. i, 1912, pp. 339–70, esp. pp. 347–66.

[2] *Philologus*, 43, pp. 336–44.

[3] Rudio, *Der Bericht des Simplicius über die Quadraturen des Antiphon und Hippokrates* (Teubner, 1907).

[4] Arist. *Phys.* i. 2, 185 a 14–17.

[5] Arist. *De caelo*, ii. 8, 290 a 4.

[6] *Anal. Pr.* ii. 25, 69 a 32.

[7] *Soph. El.* 11, 171 b 15.

two expressions separated by 'or' may no doubt refer not to one but to two different fallacies. But if 'the quadrature by means of lunes' is different from Hippocrates's quadratures of lunes, it must apparently be some quadrature like the second quoted by Alexander (not by Eudemus), and the fallacy attributed to Hippocrates must be the quadrature of a certain lune *plus* a circle (which in itself contains no fallacy at all). It seems more likely that the two expressions refer to one thing, and that this is the argument of Hippocrates's tract taken as a whole.

The passage of Alexander which Simplicius reproduces before passing to the extract from Eudemus contains two simple cases of quadrature, of a lune, and of lunes *plus* a semicircle respectively, with an erroneous inference from these cases that a circle is thereby squared. It is evident that this account does not represent Hippocrates's own argument, for he would not have been capable of committing so obvious an error; Alexander must have drawn his information, not from Eudemus, but from some other source. Simplicius recognizes this, for, after giving the alternative account extracted from Eudemus, he says that we must trust Eudemus's account rather than the other, since Eudemus was 'nearer the times' (of Hippocrates).

The two quadratures given by Alexander are as follows.

1. Suppose that AB is the diameter of a circle, D its centre, and AC, CB sides of a square inscribed in it.

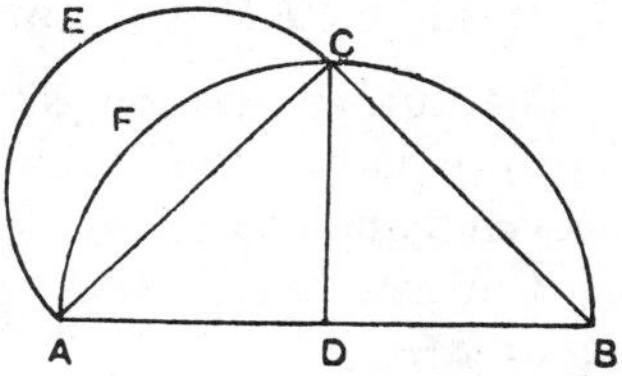

On AC as diameter describe the semicircle AEC. Join CD.

Now, since

$$AB^2 = 2\,AC^2,$$

and circles (and therefore semicircles) are to one another as the squares on their diameters,

$$(\text{semicircle } ACB) = 2(\text{semicircle } AEC).$$

But $\qquad (\text{semicircle } ACB) = 2(\text{quadrant } ADC)$;

therefore $\qquad (\text{semicircle } AEC) = (\text{quadrant } ADC)$.

If now we subtract the common part, the segment AFC, we have $\qquad (\text{lune } AECF) = \Delta\, ADC$,

and the lune is 'squared'.

2. Next take three consecutive sides *CE, EF, FD* of a regular hexagon inscribed in a circle of diameter *CD*. Also take *AB* equal to the radius of the circle and therefore equal to each of the sides.

On *AB, CE, EF, FD* as diameters describe semicircles (in the last three cases outwards with reference to the circle).

Then, since

$$CD^2 = 4AB^2 = AB^2 + CE^2 + EF^2 + FD^2,$$

and circles are to one another as the squares on their diameters,

semicircle *CEFD*) = 4 (semicircle *ALB*)

= (sum of semicircles *ALB, CGE, EHF, FKD*).

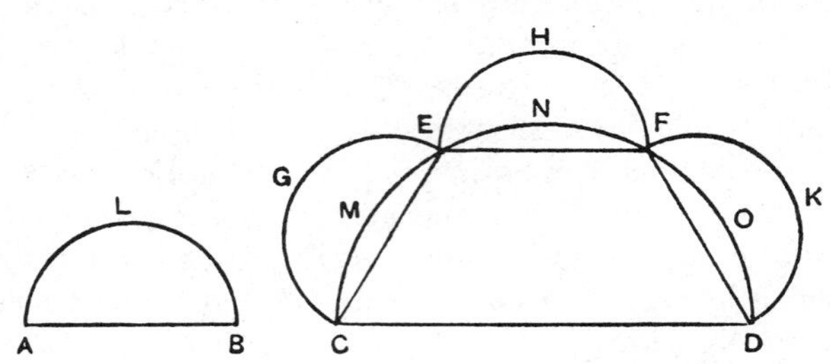

Subtracting from each side the sum of the small segments on *CE, EF, FD*, we have

(trapezium *CEFD*) = (sum of three lunes) + (semicircle *ALB*).

The author goes on to say that, subtracting the rectilineal figure equal to the three lunes ('for a rectilineal figure was proved equal to a lune'), we get a rectilineal figure equal to the semicircle *ALB*, 'and so the circle will have been squared'.

This conclusion is obviously false, and, as Alexander says, the fallacy is in taking what was proved only of the lune on the side of the inscribed square, namely that it can be squared, to be true of the lunes on the sides of an inscribed regular hexagon. It is impossible that Hippocrates (one of the ablest of geometers) could have made such a blunder. We turn therefore to Eudemus's account, which has every appearance of beginning at the beginning of Hippocrates's work and proceeding in his order.

It is important from the point of view of this chapter to preserve the phraseology of Eudemus, which throws light on the question how far the technical terms of Euclidean geometry were already used by Eudemus (if not by Hippocrates) in their technical sense. I shall therefore translate literally so much as can safely be attributed to Eudemus himself, except in purely geometrical work, where I shall use modern symbols.

'The quadratures of lunes, which were considered to belong to an uncommon class of propositions on account of the close relation (of lunes) to the circle, were first investigated by Hippocrates, and his exposition was thought to be in correct form[1]; we will therefore deal with them at length and describe them. He started with, and laid down as the first of the theorems useful for his purpose, the proposition that similar segments of circles have the same ratio to one another as the squares on their bases have [lit. as their bases in square, δυνάμει]. And this he proved by first showing that the squares on the diameters have the same ratio as the circles. [For, as the circles are to one another, so also are similar segments of them. For similar segments are those which are the same part of the circles respectively, as for instance a semicircle is similar to a semicircle, and a third part of a circle to a third part [here, Rudio argues, the word *segments*, *τμήματα*, would seem to be used in the sense of *sectors*]. It is for this reason also (διὸ καὶ) that similar segments contain equal angles [here 'segments' are certainly segments in the usual sense]. The angles of all semicircles are right, those of segments greater than a semicircle are less than right angles and are less in proportion as the segments are greater than semicircles, while those of segments less than a semicircle are greater than right angles and are greater in proportion as the segments are less than semicircles.']

I have put the last sentences of this quotation in dotted brackets because it is matter of controversy whether they belong to the original extract from Eudemus or were added by Simplicius.

I think I shall bring out the issues arising out of this passage into the clearest relief if I take as my starting-point the interpretation of it by Rudio, the editor of the latest

[1] κατὰ τρόπον ('werthvolle Abhandlung', Heib.).

edition of the whole extract. Whereas Diels, Usener, Tannery, and Heiberg had all seen in the sentences 'For, as the circles are to one another . . . less than semicircles' an addition by Simplicius, like the phrase just preceding (not quoted above), 'a proposition which Euclid placed second in his twelfth book with the enunciation "Circles are to one another as the squares on their diameters"', Rudio maintains that the sentences are wholly Eudemian, because 'For, as the circles are to one another, so are the similar segments' is obviously connected with the proposition that similar segments are as the squares on their bases a few lines back. Assuming, then, that the sentences are Eudemian, Rudio bases his next argument on the sentence defining similar segments, 'For similar segments are those which are the same part of the circles: thus a semicircle is similar to a semicircle, and a third part (of one circle) to a third part (of another circle)'. He argues that a 'segment' in the proper sense which is one third, one fourth, &c., of the circle is not a conception likely to have been introduced into Hippocrates's discussion, because it cannot be visualized by actual construction, and so would not have conveyed any clear idea. On the other hand, if we divide the four right angles about the centre of a circle into 3, 4, or n equal parts by means of 3, 4, or n radii, we have an obvious division of the circle into equal parts which would occur to any one; that is, any one would understand the expression one third or one fourth part of a circle if the parts were *sectors* and not segments. (The use of the word τμῆμα in the sense of sector is not impossible in itself at a date when mathematical terminology was not finally fixed; indeed it means 'sector' in one passage of Aristotle.[1]) Hence Rudio will have it that 'similar segments' in the second and third places in our passage are 'similar *sectors*'. But the 'similar segments' in the fundamental proposition of Hippocrates enunciated just before are certainly segments in the proper sense; so are those in the next sentence which says that similar segments contain equal angles. There is, therefore, the very great difficulty that, under Rudio's interpretation, the word τμήματα used in successive sentences means, first segments, then sectors, and then segments again. However, assuming this to be so, Rudio

[1] Arist. *De caelo*, ii. 8, 290 a 4.

is able to make the argument hang together, in the following way. The next sentence says, 'For this reason also (διὸ καὶ) similar segments contain equal angles'; therefore this must be inferred from the fact that similar sectors are the same part of the respective circles. The intermediate steps are not given in the text; but, since the similar sectors are the same part of the circles, they contain equal angles, and it follows that the angles in the segments which form part of the sectors are equal, since they are the supplements of the halves of the angles of the sectors respectively (this inference presupposes that Hippocrates knew the theorems of Eucl. III. 20–22, which is indeed clear from other passages in the Eudemus extract). Assuming this to be the line of argument, Rudio infers that in Hippocrates's time similar segments were not defined as in Euclid (namely as segments containing equal angles) but were regarded as the segments belonging to 'similar *sectors*', which would thus be the prior conception. Similar sectors would be sectors having their angles equal. The sequence of ideas, then, leading up to Hippocrates's proposition would be this. Circles are to one another as the squares on their diameters or radii. Similar sectors, having their angles equal, are to one another as the whole circles to which they belong. (Euclid has not this proposition, but it is included in Theon's addition to VI. 33, and would be known long before Euclid's time.) Hence similar sectors are as the squares on the radii. But so are the triangles formed by joining the extremities of the bounding radii in each sector. Therefore (cf. Eucl. V. 19) the differences between the sectors and the corresponding triangles respectively, i.e. the corresponding *segments*, are in the same ratio as (1) the similar sectors, or (2) the similar triangles, and therefore are as the squares on the radii.

We could no doubt accept this version subject to three *ifs*, (1) if the passage is Eudemian, (2) if we could suppose τμήματα to be used in different senses in consecutive sentences without a word of explanation, (3) if the omission of the step between the definition of similar 'segments' and the inference that the angles in similar segments are equal could be put down to Eudemus's 'summary' style. The second of these *ifs* is the crucial one; and, after full reflection, I feel bound to agree with the great scholars who have held that this hypothesis is impossible; indeed the canons of literary criticism seem to exclude it altogether. If this is so, the whole of Rudio's elaborate structure falls to the ground.

We can now consider the whole question *ab initio*. First, are the sentences in question the words of Eudemus or of Simplicius? On the one hand, I think the whole paragraph would be much more like the 'summary' manner of Eudemus if it stopped at 'have the same ratio as the circles', i.e. if the sentences were not there at all. Taken together, they are long and yet obscurely argued, while the last sentence is really otiose, and, I should have said, quite unworthy of Eudemus. On the other hand, I do not see that Simplicius had any sufficient motive for interpolating such an explanation: he might have added the words 'for, as the circles are to one another, so also are similar segments of them', but there was no need for him to define similar segments; *he* must have been familiar enough with the term and its meaning to take it for granted that his readers would know them too. I think, therefore, that the sentences, down to 'the same part of the circles respectively' at any rate, may be from Eudemus. In these sentences, then, can 'segments' mean segments in the proper sense (and not sectors) after all? The argument that it cannot rests on the assumption that the Greeks of Hippocrates's day would not be likely to speak of a segment which was one third of the whole circle if they did not see their way to visualize it by actual construction. But, though the idea would be of no use to *us*, it does not follow that their point of view would be the same as ours. On the contrary, I agree with Zeuthen that Hippocrates may well have said, of segments of circles which are in the same ratio as the circles, that they are 'the same part' of the circles respectively, for this is (in an incomplete form, it is true) the language of the definition of proportion in the only theory of proportion (the numerical) then known (cf. Eucl. VII. Def. 20, 'Numbers are proportional when the first is the same multiple, or the same part, or the same parts, of the second that the third is of the fourth', i.e. the two equal ratios are of one of the following forms m, $\frac{1}{n}$ or $\frac{m}{n}$ where m, n are integers); the illustrations, namely the semicircles and the segments which are one third of the circles respectively, are from this point of view quite harmless.

Only the transition to the view of similar segments as segments 'containing equal angles' remains to be explained. And here we are in the dark, because we do not know how, for instance, Hippocrates would have *drawn* a segment in one given circle which should be 'the same part' of that circle that a given segment of another given circle is of that circle. (If e.g. he had used the proportionality of the parts into which the bases of the two similar segments divide the diameters of the circles which bisect them perpendicularly, he could, by means of the sectors to which the segments belong, have proved that the segments, like the sectors, are in the ratio of the circles, just as Rudio supposes him to have done; and the equality of the angles in the segments would have followed as in Rudio's proof.)

As it is, I cannot feel certain that the sentence διὸ καὶ κτλ. 'this is the reason why similar segments contain equal angles' is not an addition by Simplicius. Although Hippocrates was fully aware of the fact, he need not have stated it in this place, and Simplicius may have inserted the sentence in order to bring Hippocrates's view of similar segments into relation with Euclid's definition. The sentence which follows about 'angles of' semicircles and 'angles of' segments, greater or less than semicircles, is out of place, to say the least, and can hardly come from Eudemus.

We resume Eudemus's account.

'After proving this, he proceeded to show in what way it was possible to square a lune the outer circumference of which is that of a semicircle. This he effected by circumscribing a semicircle about an isosceles right-angled triangle and (circumscribing) about the base [= describing on the base] a segment of a circle similar to those cut off by the sides.' [This is the problem of Eucl. III. 33, and involves the knowledge that similar segments contain equal angles.]

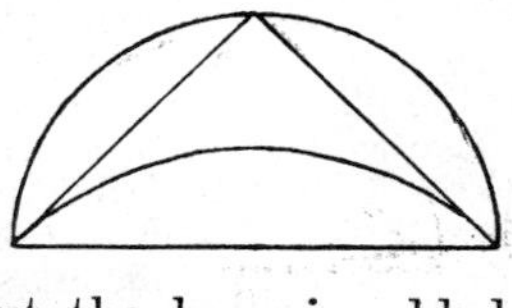

'Then, since the segment about the base is equal to the sum of those about the sides, it follows that, when the part of the triangle above the segment about the base is added to both alike, the lune will be equal to the triangle.

'Therefore the lune, having been proved equal to the triangle, can be squared.

'In this way, assuming that the outer circumference of the lune is that of a semicircle, Hippocrates easily squared the lune.

'Next after this he assumes (an outer circumference) greater than a semicircle (obtained) by constructing a trapezium in which three sides are equal to one another, while one, the greater of the parallel sides, is such that the square on it is triple of the square on each one of the other sides, and then comprehending the trapezium in a circle and circumscribing about (= describing on) its greatest side a segment similar to those cut off from the circle by the three equal sides.'

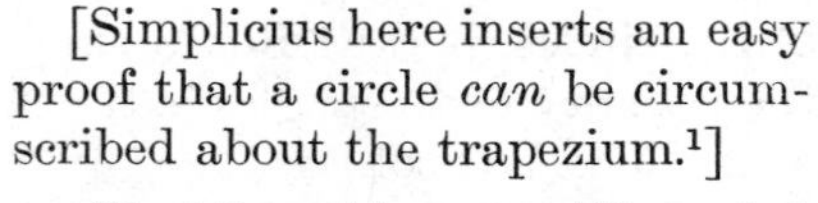

[Simplicius here inserts an easy proof that a circle *can* be circumscribed about the trapezium.[1]]

'That the said segment [bounded by the outer circumference $BACD$ in the figure] is greater than a semicircle is clear, if a diagonal be drawn in the trapezium.

'For this diagonal [say BC], subtending two sides [BA, AC] of the trapezium, is such that the square on it is greater than double the square on one of the remaining sides.'

[This follows from the fact that, AC being parallel to BD but less than it, BA and DC will meet, if produced, in a point F. Then, in the isosceles triangle FAC, the angle FAC is less than a right angle, so that the angle BAC is obtuse.]

'Therefore the square on [BD] the greatest side of the trapezium [$= 3\,CD^2$ by hypothesis] is less than the sum of the squares on the diagonal [BC] and that one of the other sides

[1] Heiberg (*Philologus*, 43, p. 340) thinks that the words καὶ ὅτι μὲν περιληφθήσεται κύκλῳ τὸ τραπέζιον δείξεις [οὕτως] διχοτομήσας τὰς τοῦ τραπεζίου γωνίας ('Now, that the trapezium can be comprehended in a circle you can prove by bisecting the angles of the trapezium') *may* (without οὕτως—F omits it) be Eudemus's own. For ὅτι μὲν . . . forms a natural contrast to ὅτι δὲ μεῖζον . . . in the next paragraph. Also cf. p. 65. 9 Diels, τούτων οὖν οὕτως ἐχόντων τὸ τραπέζιόν φημι ἐφ᾽ οὗ ΕΚΒΗ περιλήψεται κύκλος.

[CD] which is subtended[1] by the said (greatest) side [BD] together with the diagonal [BC]' [i.e. $BD^2 < BC^2 + CD^2$].

'Therefore the angle standing on the greater side of the trapezium [$\angle BCD$] is acute.

'Therefore the segment in which the said angle is is greater than a semicircle. And this (segment) is the outer circumference of the lune.'

[Simplicius observes that Eudemus has omitted the actual squaring of the lune, presumably as being obvious. We have only to supply the following.

Since $$BD^2 = 3\,BA^2,$$

$$(\text{segment on } BD) = 3\,(\text{segment on } BA)$$
$$= (\text{sum of segments on } BA,\ AC,\ CD).$$

Add to each side the area between BA, AC, CD, and the circumference of the segment on BD, and we have

(trapezium $ABDC$) = (lune bounded by the two circumferences).]

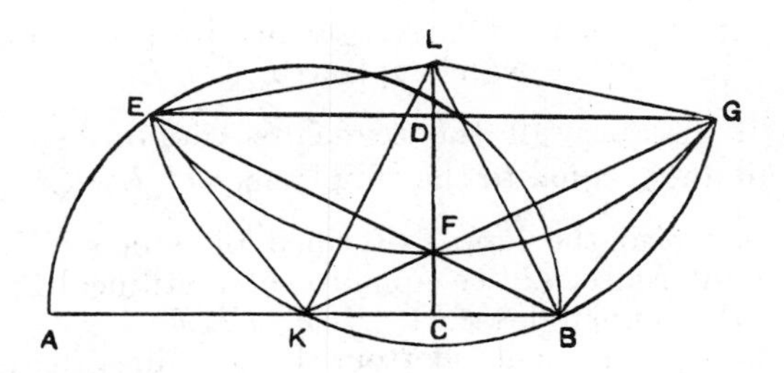

'A case too where the outer circumference is less than a semicircle was solved by Hippocrates,[2] who gave the following preliminary construction.

'*Let there be a circle with diameter AB, and let its centre be K.*

'*Let CD bisect BK at right angles; and let the straight line EF be so placed between CD and the circumference that it verges towards B* [i.e. will, if produced, pass through B], *while its length is also such that the square on it is* $1\frac{1}{2}$ *times the square on (one of) the radii.*

[1] Observe the curious use of ὑποτείνειν, stretch under, subtend. The third side of a triangle is said to be 'subtended' by the other two together.

[2] Literally 'If (the outer circumference) were less than a semicircle, Hippocrates solved (κατεσκεύασεν, constructed) this (case).'

'*Let EG be drawn parallel to AB, and let (straight lines) be drawn joining K to E and F.*

'*Let the straight line* [KF] *joined to F and produced meet EG in G, and again let (straight lines) be drawn joining B to F, G.*

'*It is then manifest that BF produced will pass through* ["fall on"] E [for by hypothesis EF verges towards B], *and BG will be equal to EK.*'

[Simplicius proves this at length. The proof is easy. The triangles FKC, FBC are equal in all respects [Eucl. I. 4]. Therefore, EG being parallel to KB, the triangles EDF, GDF are equal in all respects [Eucl. I. 15, 29, **26**]. Hence the trapezium is isosceles, and $BG = EK$.

'*This being so, I say that the trapezium EKBG can be comprehended in a circle.*'

[Let the segment $EKBG$ circumscribe it.]

'Next let a segment of a circle be circumscribed about the triangle EFG also;
then manifestly each of the segments [on] EF, FG will be similar to each of the segments [on] EK, KB, BG.'

[This is because all the segments contain equal angles, namely an angle equal to the supplement of EGK.]

'This being so, the lune so formed, of which $EKBG$ is the outer circumference, will be equal to the rectilineal figure made up of the three triangles BFG, BFK, EKF.

'For the segments cut off from the rectilineal figure, on the inner side of the lune, by the straight lines EF, FG, are (together) equal to the segments outside the rectilineal figure cut off by the straight lines EK, KB, BG, since each of the inner segments is $1\frac{1}{2}$ times each of the outer, because, by hypothesis,

$$EF^2 \,(= FG^2) = \tfrac{3}{2}\, EK^2$$

[i.e.

$$2\,EF^2 = 3\,EK^2, \quad = EK^2 + KB^2 + BG^2].$$

'If then

$$\text{(lune)} = \text{(the three segmts.)} + \{\text{(rect. fig.)} - \text{(the two segmts.)}\},$$

the trapezium including the two segments but not the three, while the (sum of the) two segments is equal to the (sum of the) three, it follows that

$$\text{(lune)} = \text{(rectilineal figure)}.$$

'The fact that this lune (is one which) has its outer circumference less than a semicircle he proves by means of the fact that the angle [EKG] in the outer segment is obtuse.

'And the fact that the angle EKG is obtuse he proves as follows.'

[This proof is supposed to have been given by Eudemus in Hippocrates's own words, but unfortunately the text is confused. The argument seems to have been substantially as follows.

By hypothesis, $EF^2 = \frac{3}{2} EK^2$.

Also $BK^2 > 2\,BF^2$ (this is assumed: we shall consider the ground later);

or $EK^2 > 2\,KF^2$.

Therefore $EF^2 = EK^2 + \frac{1}{2} EK^2$

$$> EK^2 + KF^2,$$

so that the angle EKF is obtuse, and the segment is less than a semicircle.

How did Hippocrates prove that $BK^2 > 2\,BF^2$? The manuscripts have the phrase 'because the angle at F is greater' (where presumably we should supply ὀρθῆς, 'than a right angle'). But, if Hippocrates proved this, he must evidently have proved it by means of his hypothesis $EF^2 = \frac{3}{2}EK^2$, and this hypothesis leads more directly to the consequence that $BK^2 > 2\,KF^2$ than to the fact that the angle at F is greater than a right angle.

We may supply the proof thus.

By hypothesis, $EF^2 = \frac{3}{2} KB^2$.

Also, since A, E, F, C are concyclic,

$$EB \,.\, BF = AB \,.\, BC$$
$$= KB^2,$$

or

$$EF \,.\, FB + BF^2 = KB^2$$
$$= \frac{2}{3} EF^2.$$

It follows from the last relations that $EF > FB$, and that

$$KB^2 > 2\,BF^2.$$

The most remarkable feature in the above proof is the assumption of the solution of the problem '*to place a straight line* [EF] *of length such that the square on it is* $1\frac{1}{2}$ *times the square on* AK *between the circumference of the semicircle and* CD *in such a way that it will verge* (νεύειν) *towards* B' [i.e. if produced, will pass through B]. This is a problem of a type which the Greeks called νεύσεις, *inclinationes* or *vergings*. Theoretically it may be regarded as the problem of finding a length (x) such that, if F be so taken on CD that $BF = x$, BF produced will intercept between CD and the circumference of the semicircle a length EF equal to $\sqrt{\frac{3}{2}}\,.\,AK$.

If we suppose it done, we have

$$EB\,.\,BF = AB\,.\,BC = AK^2;$$

or

$$x\,(x + \sqrt{\tfrac{3}{2}}\,.\,a) = a^2 \quad (\text{where } AK = a).$$

That is, the problem is equivalent to the solution of the quadratic equation

$$x^2 + \sqrt{\tfrac{3}{2}}\,.\,ax = a^2.$$

This again is the problem of 'applying to a straight line of length $\sqrt{\frac{3}{2}}\,.\,a$ a rectangle exceeding by a square figure and equal in area to a^2', and would theoretically be solved by the Pythagorean method based on the theorem of Eucl. II. 6. Undoubtedly Hippocrates could have solved the problem by this theoretical method; but he may, on this occasion, have used the purely mechanical method of marking on a ruler or straight edge a length equal to $\sqrt{\frac{3}{2}}\,.\,AK$, and then moving it till the points marked lay on the circumference and on CD respectively, while the straight edge also passed through B. This method is perhaps indicated by the fact that he first *places* EF (without producing it to B) and afterwards *joins* BF.

We come now to the last of Hippocrates's quadratures. Eudemus proceeds:]

'Thus Hippocrates squared every[1] (sort of) lune, seeing that[1] (he squared) not only (1) the lune which has for its outer

[1] Tannery brackets πάντα and εἴπερ καί. Heiberg thinks (*l.c*, p. 343) the *wording* is that of Simplicius reproducing the *content* of Eudemus. The wording of the sentence is important with reference to the questions (1) What was the paralogism with which Aristotle actually charged Hippocrates? and (2) What, if any, was the justification for the charge? Now the four quadratures as given by Eudemus are clever, and contain in themselves no fallacy at all. The supposed fallacy, then, can only have consisted in an assumption on the part of Hippocrates that, because he

circumference the arc of a semicircle, but also (2) the lune in which the outer circumference is greater, and (3) the lune in which it is less, than a semicircle.

'But he also squared the sum of a lune and a circle in the following manner.

'*Let there be two circles about K as centre, such that the square on the diameter of the outer is 6 times the square on that of the inner.*

'*Let a (regular) hexagon ABCDEF be inscribed in the inner circle, and let KA, KB, KC be joined from the centre and produced as far as the circumference of the outer circle. Let GH, HI, GI be joined.*'

[Then clearly GH, HI are sides of a hexagon inscribed in the outer circle.]

'*About GI* [i.e. on GI] *let a segment be circumscribed similar to the segment cut off by GH.*

'*Then* $$GI^2 = 3\,GH^2,$$

for $$GI^2 + (\text{side of outer hexagon})^2 = (\text{diam. of outer circle})^2 = 4\,GH^2.$$

[The original states this in words without the help of the letters of the figure.]

'*Also* $$GH^2 = 6\,AB^2.$$

had squared one particular lune of each of three types, namely those which have for their outer circumferences respectively (1) a semicircle, (2) an arc greater than a semicircle, (3) an arc less than a semicircle, he had squared all possible lunes, and therefore also the lune included in his last quadrature, the squaring of which (had it been possible) would actually have enabled him to square the circle. The question is, did Hippocrates so delude himself? Heiberg thinks that, in the then state of logic, he may have done so. But it seems impossible to believe this of so good a mathematician; moreover, if Hippocrates had really thought that he had squared the circle, it is inconceivable that he would not have said so in express terms at the end of his fourth quadrature.

Another recent view is that of Björnbo (in Pauly-Wissowa, *Real-Encyclopädie*, xvi, pp. 1787–99), who holds that Hippocrates realized perfectly the limits of what he had been able to do and knew that he had not squared the circle, but that he deliberately used language which, without being actually untrue, was calculated to mislead any one who read him into the belief that he had really solved the problem. This, too, seems incredible; for surely Hippocrates must have known that the first expert who read his tract would detect the fallacy at once, and that he was risking his reputation as a mathematician for no purpose. I prefer to think that he was merely trying to put what he had discovered in the most favourable light; but it must be admitted that the effect of his language was only to bring upon himself a charge which he might easily have avoided.

'*Therefore*

segment on GI [= 2(segmt. on GH) + 6(segmt. on AB)]
= (*segmts. on* GH, HI) + (*all segmts. in inner circle*).

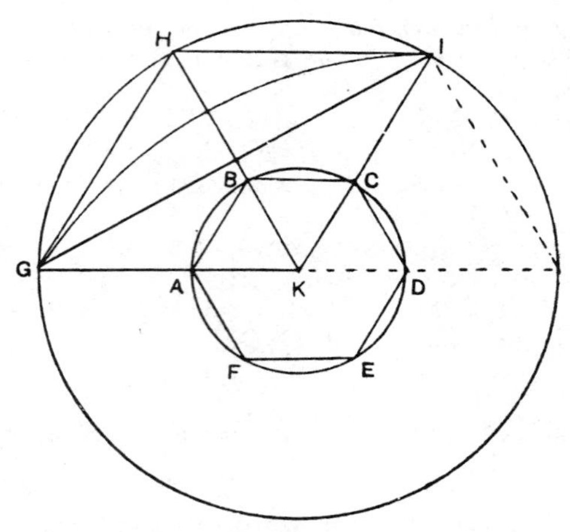

['Add to each side the area bounded by GH, HI and the arc GI;]

therefore ($\Delta\, GHI$) = (*lune* GHI) + (*all segmts. in inner circle*).

Adding to both sides the hexagon in the inner circle, we have

($\Delta\, GHI$) + (inner hexagon) = (lune GHI) + (inner circle).

'Since, then, the sum of the two rectilineal figures can be squared, so can the sum of the circle and the lune in question.'

Simplicius adds the following observations:

'Now, so far as Hippocrates is concerned, we must allow that Eudemus was in a better position to know the facts, since he was nearer the times, being a pupil of Aristotle. But, as regards the "squaring of the circle by means of segments" which Aristotle reflected on as containing a fallacy, there are three possibilities, (1) that it indicates the squaring by means of lunes (Alexander was quite right in expressing the doubt implied by his words, "if it is the same as the squaring by means of lunes"), (2) that it refers, not to the proofs of Hippocrates, but some others, one of which Alexander actually reproduced, or (3) that it is intended to reflect on the squaring by Hippocrates of the circle *plus* the lune, which Hippocrates did in fact prove "by means of segments", namely the three (in the greater circle) and those in the lesser circle. . . . On this third hypothesis the fallacy would lie in the fact that the sum of the circle and the lune is squared, and not the circle alone.'

If, however, the reference of Aristotle was really to Hippocrates's last quadrature alone, Hippocrates was obviously misjudged; there is no fallacy in it, nor is Hippocrates likely to have deceived himself as to what his proof actually amounted to.

In the above reproduction of the extract from Eudemus I have marked by italics the passages where the writer follows the ancient fashion of describing points, lines, angles, &c., with reference to the letters in the figure: the ancient practice was to write τὸ σημεῖον ἐφ' ᾧ (or ἐφ' οὗ) K, the (point) *on which* (is) the letter K, instead of the shorter form τὸ K σημεῖον, the point K, used by Euclid and later geometers; ἡ ἐφ' ᾗ AB (εὐθεῖα), the straight line *on which* (are the letters AB, for ἡ AB (εὐθεῖα), the straight line AB; τὸ τρίγωνον τὸ ἐφ' οὗ EZH, the triangle *on which* (are the letters) EFG, instead of τὸ EZH τρίγωνον, the triangle EFG; and so on. Some have assumed that, where the longer archaic form, instead of the shorter Euclidean, is used, Eudemus must be quoting Hippocrates *verbatim*; but this is not a safe criterion, because, e.g., Aristotle himself uses both forms of expression, and there are, on the other hand, some relics of the archaic form even in Archimedes.

Trigonometry enables us readily to find all the types of Hippocratean lunes that can be squared by means of the straight line and circle. Let ACB be the external circumference, ADB the internal circumference of such a lune, r, r' the radii, and O, O' the centres of the two arcs, θ, θ' the halves of the angles subtended by the arcs at the centres respectively.

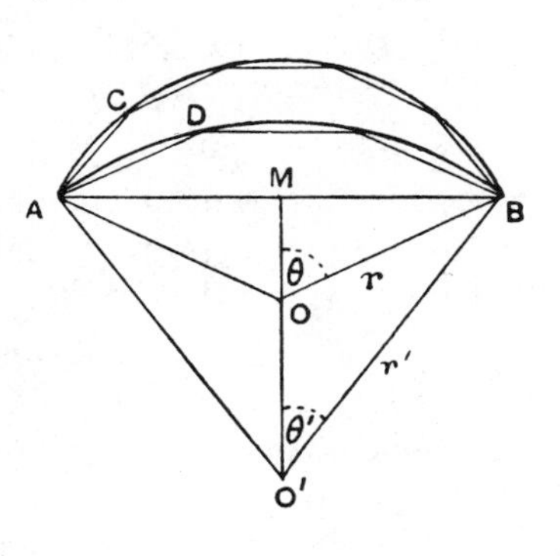

Now (area of lune)

$$= (\text{difference of segments } ACB, ADB)$$
$$= (\text{sector } OACB - \Delta AOB) - (\text{sector } O'ADB - \Delta AO'B)$$
$$= r^2\theta - r'^2\theta' + \tfrac{1}{2}(r'^2 \sin 2\theta' - r^2 \sin 2\theta).$$

We also have

$$r \sin \theta = \tfrac{1}{2} AB = r' \sin \theta' \quad . \quad . \quad . \quad . \quad . \quad . \quad (1)$$

In order that the lune may be squareable, we must have, in the first place, $$r^2 \theta = r'^2 \theta'.$$

Suppose that $\theta = m\theta'$, and it follows that

$$r' = \sqrt{m} \,.\, r.$$

Accordingly the area becomes

$$\tfrac{1}{2} r^2 (m \sin 2\theta' - \sin 2m\theta');$$

and it remains only to solve the equation (1) above, which becomes $$\sin m\theta' = \sqrt{m} \,.\, \sin \theta'.$$

This reduces to a quadratic equation only when m has one of the values $$2, \quad 3, \quad \tfrac{3}{2}, \quad 5, \quad \tfrac{5}{3}.$$

The solutions of Hippocrates correspond to the first three values of m. But the lune is squareable by 'plane' methods in the other two cases also. Clausen (1840) gave the last four cases of the problem as new[1] (it was not then known that Hippocrates had solved more than the first); but, according to M. Simon[2], all five cases were given much earlier in a dissertation by Martin Johan Wallenius of Åbo (Abveae, 1766). As early as 1687 Tschirnhausen noted the existence of an infinite number of squareable portions of the first of Hippocrates's lunes. Vieta[3] discussed the case in which $m = 4$, which of course leads to a cubic equation.

(β) *Reduction of the problem of doubling the cube to the finding of two mean proportionals.*

We have already alluded to Hippocrates's discovery of the reduction of the problem of duplicating the cube to that of finding two mean proportionals in continued proportion. That is, he discovered that, if

$$a : x = x : y = y : b,$$

then $a^3 : x^3 = a : b$. This shows that he could work with compound ratios, although for him the theory of proportion must still have been the incomplete, *numerical*, theory developed by the Pythagoreans. It has been suggested that

[1] Crelle, xxi, 1840, pp. 375–6.
[2] *Geschichte der Math. im Altertum*, p. 174.
[3] Vieta, *Variorum de rebus mathematicis responsorum* lib. viii, 1593.

the idea of the reduction of the problem of duplication may have occurred to him through analogy. The problem of doubling a square is included in that of finding *one* mean proportional between two lines; he might therefore have thought of what would be the effect of finding two mean proportionals. Alternatively he may have got the idea from the theory of numbers. Plato in the *Timaeus* has the propositions that between two square numbers there is one mean proportional number, but that two cube numbers are connected, not by one, but by two mean numbers in continued proportion.[1] These are the theorems of Eucl. VIII. 11, 12, the latter of which is thus enunciated: 'Between two cube numbers there are two mean proportional numbers, and the cube has to the cube the ratio triplicate of that which the side has to the side.' If this proposition was really Pythagorean, as seems probable enough, Hippocrates had only to give the geometrical adaptation of it.

(γ) *The Elements as known to Hippocrates.*

We can now take stock of the advances made in the Elements up to the time when Hippocrates compiled a work under that title. We have seen that the Pythagorean geometry already contained the substance of Euclid's Books I and II, part of Book IV, and theorems corresponding to a great part of Book VI; but there is no evidence that the Pythagoreans paid much attention to the geometry of the circle as we find it, e.g., in Eucl., Book III. But, by the time of Hippocrates, the main propositions of Book III were also known and used, as we see from Eudemus's account of the quadratures of lunes. Thus it is assumed that 'similar' segments contain equal angles, and, as Hippocrates assumes that two segments of circles are similar when the obvious thing about the figure is that the angles at the circumferences which are the supplements of the angles in the segments are one and the same, we may clearly infer, as above stated, that Hippocrates knew the theorems of Eucl. III. 20–2. Further, he assumes the construction on a given straight line of a segment similar to another given segment (cf. Eucl. III. 33). The theorems of Eucl. III. 26–9 would obviously be known to Hippocrates,

[1] Plato, *Timaeus*, 32 A, B.

as was that of III. 31 (that the angle in a semicircle is a right angle, and that, according as a segment is less or greater than a semicircle, the angle in it is obtuse or acute). He assumes the solution of the problem of circumscribing a circle about a triangle (Eucl. IV. 5), and the theorem that the side of a regular hexagon inscribed in a circle is equal to the radius (Eucl. IV. 15).

But the most remarkable fact of all is that, according to Eudemus, Hippocrates actually proved the theorem of Eucl. XII. 2, that *circles are to one another as the squares on their diameters*, afterwards using this proposition to prove that *similar segments are to one another as the squares on their bases.* Euclid of course proves XII. 2 by the *method of exhaustion*, the invention of which is attributed to Eudoxus on the ground of notices in Archimedes.[1] This method depends on the use of a certain lemma known as the Axiom of Archimedes, or, alternatively, a lemma similar to it. The lemma used by Euclid is his proposition X. 1, which is closely related to Archimedes's lemma in that the latter is practically used in the proof of it. Unfortunately we have no information as to the nature of Hippocrates's proof; if, however, it amounted to a genuine proof, as Eudemus seems to imply, it is difficult to see how it could have been effected otherwise than by some anticipation in essence of the method of exhaustion.

THEODORUS OF CYRENE, who is mentioned by Proclus along with Hippocrates as a celebrated geometer and is claimed by Iamblichus as a Pythagorean,[2] is only known to us from Plato's *Theaetetus*. He is said to have been Plato's teacher in mathematics,[3] and it is likely enough that Plato, while on his way to or from Egypt, spent some time with Theodorus at Cyrene,[4] though, as we gather from the *Theaetetus*, Theodorus had also been in Athens in the time of Socrates. We learn from the same dialogue that he was a pupil of Protagoras, and was distinguished not only in geometry but in astronomy, arithmetic, music, and all educational subjects.[5] The one notice

[1] Prefaces to *On the Sphere and Cylinder*, i, and *Quadrature of the Parabola*.

[2] Iambl. *Vit. Pyth.* c. 36.

[3] Diog. L. ii. 103.

[4] Cf. Diog. L. iii. 6.

[5] Plato, *Theaetetus*, 161 B, 162 A; *ib.* 145 A, C, D.

which we have of a particular achievement of his suggests that it was he who first carried the theory of irrationals beyond the first step, namely the discovery by the Pythagoreans of the irrationality of $\sqrt{2}$. According to the *Theaetetus*,[1] Theodorus

'was proving[2] to us a certain thing about square roots (δυνάμεις), I mean (the square roots, i.e. sides) of three square feet and of five square feet, namely that these roots are not commensurable in length with the foot-length, and he went on in this way, taking all the separate cases up to the root of 17 square feet, at which point, for some reason, he stopped'.

That is, he proved the irrationality of $\sqrt{3}$, $\sqrt{5}$ ... up to $\sqrt{17}$. It does not appear, however, that he had reached any definition of a surd in general or proved any general proposition about all surds, for Theaetetus goes on to say:

'The idea occurred to the two of us (Theaetetus and the younger Socrates), seeing that these square roots appeared

[1] *Theaetetus*, 147 D sq.

[2] Περὶ δυνάμεών τι ἡμῖν Θεόδωρος ὅδε ἔγραφε, τῆς τε τρίποδος πέρι καὶ πεντέποδος [ἀποφαίνων] ὅτι μήκει οὐ σύμμετροι τῇ ποδιαίᾳ. Certain writers (H. Vogt in particular) persist in taking ἔγραφε in this sentence to mean *drew* or *constructed*. The idea is that Theodorus's exposition must have included two things, first the *construction* of straight lines representing $\sqrt{3}$, $\sqrt{5}$ ... (of course by means of the Pythagorean theorem, Eucl. I. 47), in order to show that these straight lines exist, and secondly the *proof* that each of them is incommensurable with 1; therefore, it is argued, ἔγραφε must indicate the construction and ἀποφαίνων the proof. But in the first place it is impossible that ἔγραφέ τι περί, 'he wrote *something about*' (roots), should mean '*constructed* each of the roots'. Moreover, if ἀποφαίνων is bracketed (as it is by Burnet), the supposed contrast between ἔγραφε and ἀποφαίνων disappears, and ἔγραφε *must* mean 'proved', in accordance with the natural meaning of ἔγραφέ τι, because there is nothing else to govern ὅτι μήκει, κτλ. ('that they are not commensurable in length ...'), which phrase is of course a closer description of τι. There are plenty of instances of γράφειν in the sense of 'prove'. Aristotle says (*Topics*, Θ. 3, 158 b 29) 'It would appear that in mathematics too some things are difficult to prove (οὐ ῥᾳδίως γράφεσθαι) owing to the want of a definition, e.g. that a straight line parallel to the side and cutting a plane figure (parallelogram) divides the straight line (side) and the area similarly'. Cf. Archimedes, *On the Sphere and Cylinder*, ii, Pref., 'It happens that most of them are proved (γράφεσθαι) by means of the theorems ...'; 'Such of the theorems and problems as are proved (γράφεται) by means of these theorems I have proved (or written out, γράψας) and send you in this book'; *Quadrature of a Parabola*, Pref., 'I have proved (ἔγραφον) that every cone is one third of the cylinder with the same base and equal height by assuming a lemma similar to that aforesaid.'

I do not deny that Theodorus *constructed* his 'roots'; I have no doubt that he did; but this is not what ἔγραφέ τι means.

to be unlimited in multitude, to try to arrive at one collective term by which we could designate all these roots . . . We divided number in general into two classes. The number which can be expressed as equal multiplied by equal (*ἴσον ἰσάκις*) we likened to a square in form, and we called it square and equilateral (*ἰσόπλευρον*) . . . The intermediate number, such as three, five, and any number which cannot be expressed as equal multiplied by equal, but is either less times more or more times less, so that it is always contained by a greater and a less side, we likened to an oblong figure (*προμήκει σχήματι*) and called an oblong number. . . . Such straight lines then as square the equilateral and plane number we defined as *length* (*μῆκος*), and such as square the oblong (we called) *square roots* (*δυνάμεις*) as not being commensurable with the others in length but only in the plane areas to which their squares are equal. And there is another distinction of the same sort with regard to solids.'

Plato gives no hint as to how Theodorus proved the propositions attributed to him, namely that $\sqrt{3}$, $\sqrt{5}$ ... $\sqrt{17}$ are all incommensurable with 1; there is therefore a wide field open for speculation, and several conjectures have been put forward.

(1) Hultsch, in a paper on Archimedes's approximations to square roots, suggested that Theodorus took the line of seeking successive approximations. Just as $\frac{7}{5}$, the first approximation to $\sqrt{2}$, was obtained by putting $2 = \frac{50}{25}$, Theodorus might have started from $3 = \frac{48}{16}$, and found $\frac{7}{4}$ or $1\frac{1}{2}\frac{1}{4}$ as a first approximation, and then, seeing that $1\frac{1}{2}\frac{1}{4} > \sqrt{3} > 1\frac{1}{2}$, might (by successive trials, probably) have found that

$$1\tfrac{1}{2}\tfrac{1}{8}\tfrac{1}{16}\tfrac{1}{32}\tfrac{1}{64} > \sqrt{3} > 1\tfrac{1}{2}\tfrac{1}{8}\tfrac{1}{16}\tfrac{1}{32}\tfrac{1}{128}.$$

But the method of finding closer and closer approximations, although it might afford a presumption that the true value cannot be exactly expressed in fractions, would leave Theodorus as far as ever from *proving* that $\sqrt{3}$ is incommensurable.

(2) There is no mention of $\sqrt{2}$ in our passage, and Theodorus probably omitted this case because the incommensurability of $\sqrt{2}$ and the traditional method of proving it were already known. The traditional proof was, as we have seen, a *reductio ad absurdum* showing that, if $\sqrt{2}$ is commensurable with 1, it will follow that the same number is both even and odd, i.e. both divisible and not divisible by 2. The same method

of proof can be adapted to the cases of $\sqrt{3}$, $\sqrt{5}$, &c., if 3, 5 ... are substituted for 2 in the proof; e.g. we can prove that, if $\sqrt{3}$ is commensurable with 1, then the same number will be both divisible and not divisible by 3. One suggestion, therefore, is that Theodorus may have applied this method to all the cases from $\sqrt{3}$ to $\sqrt{17}$. We can put the proof quite generally thus. Suppose that N is a non-square number such as 3, 5 ..., and, if possible, let $\sqrt{N} = m/n$, where m, n are integers prime to one another.

Therefore $$m^2 = N \,.\, n^2;$$

therefore m^2 is divisible by N, so that m also is a multiple of N.

Let $$m = \mu \,.\, N, \qquad (1)$$

and consequently $$n^2 = N \,.\, \mu^2.$$

Then in the same way we can prove that n is a multiple of N.

Let $$n = \nu \,.\, N \qquad (2)$$

It follows from (1) and (2) that $m/n = \mu/\nu$, where $\mu < m$ and $\nu < n$; therefore m/n is not in its lowest terms, which is contrary to the hypothesis.

The objection to this conjecture as to the nature of Theodorus's proof is that it is so easy an adaptation of the traditional proof regarding $\sqrt{2}$ that it would hardly be important enough to mention as a new discovery. Also it would be quite unnecessary to repeat the proof for every case up to $\sqrt{17}$; for it would be clear, long before $\sqrt{17}$ was reached, that it is generally applicable. The latter objection seems to me to have force. The former objection may or may not; for I do not feel sure that Plato is necessarily attributing any important new discovery to Theodorus. The object of the whole context is to show that a definition by mere enumeration is no definition; e.g. it is no definition of ἐπιστήμη to enumerate particular ἐπιστῆμαι (as shoemaking, carpentering, and the like); this is to put the cart before the horse, the general definition of ἐπιστήμη being logically prior. Hence it was probably Theaetetus's generalization of the procedure of Theodorus which impressed Plato as being original and important rather than Theodorus's proofs themselves.

(3) The third hypothesis is that of Zeuthen.[1] He starts with the assumptions (*a*) that the method of proof used by Theodorus must have been original enough to call for special notice from Plato, and (*b*) that it must have been of such a kind that the application of it to each surd required to be set out separately in consequence of the variations in the numbers entering into the proofs. Neither of these conditions is satisfied by the hypothesis of a mere adaptation to $\sqrt{3}$, $\sqrt{5}$ ... of the traditional proof with regard to $\sqrt{2}$. Zeuthen therefore suggests another hypothesis as satisfying both conditions, namely that Theodorus used the criterion furnished by the process of finding the greatest common measure as stated in the theorem of Eucl. X. 2. 'If, when the lesser of two unequal magnitudes is continually subtracted in turn from the greater [this includes the subtraction from any term of the highest multiple of another that it contains], that which is left never measures the one before it, the magnitudes will be incommensurable'; that is, if two magnitudes are such that the process of finding their G. C. M. never comes to an end, the two magnitudes are incommensurable. True, the proposition Eucl. X. 2 depends on the famous X. 1 (Given two unequal magnitudes, if from the greater there be subtracted more than the half (or the half), from the remainder more than the half (or the half), and so on, there will be left, ultimately, some magnitude less than the lesser of the original magnitudes), which is based on the famous postulate of Eudoxus (= Eucl. V, Def. 4), and therefore belongs to a later date. Zeuthen gets over this objection by pointing out that the necessity of X. 1 for a rigorous demonstration of X. 2 may not have been noticed at the time; Theodorus may have proceeded by intuition, or he may even have postulated the truth proved in X. 1.

The most obvious case in which incommensurability can be proved by using the process of finding the greatest common measure is that of the two segments of a straight line divided in extreme and mean ratio. For, if AB is divided in this way at C, we have only to mark off along CA (the greater segment)

[1] Zeuthen, 'Sur la constitution des livres arithmétiques des Éléments d'Euclide et leur rapport à la question de l'irrationalité' in *Oversigt over det kgl. Danske videnskabernes Selskabs Forhandlinger*, 1915, pp. 422 sq.

a length CD equal to CB (the lesser segment), and CA is then divided at D in extreme and mean ratio, CD being the greater segment. (Eucl. XIII. 5 is the equivalent of this

A D E C B

proposition.) Similarly, DC is so divided if we set off DE along it equal to DA; and so on. This is precisely the process of finding the greatest common measure of AC, CB, the quotient being always unity; and the process never comes to an end. Therefore AC, CB are incommensurable. What is proved in this case is the irrationality of $\frac{1}{2}(\sqrt{5}-1)$. This of course shows incidentally that $\sqrt{5}$ is incommensurable with 1. It has been suggested, in view of the easiness of the above proof, that the irrational may first have been discovered with reference to the segments of a straight line cut in extreme and mean ratio, rather than with reference to the diagonal of a square in relation to its side. But this seems, on the whole, improbable.

Theodorus would, of course, give a geometrical form to the process of finding the G. C. M., after he had represented in a figure the particular surd which he was investigating. Zeuthen illustrates by two cases, $\sqrt{5}$ and $\sqrt{3}$.

We will take the former, which is the easier. The process of finding the G. C. M. (if any) of $\sqrt{5}$ and 1 is as follows:

$$
\begin{array}{l}
1\,)\ \sqrt{5}\ (\,2 \\
\quad\ \ 2 \\
\hline
\sqrt{5}-2\,)\ 1 \qquad\qquad (\,4 \\
\qquad\qquad 4\,(\sqrt{5}-2) \\
\hline
\qquad\qquad (\sqrt{5}-2)^2
\end{array}
$$

[The explanation of the second division is this:

$$1 = (\sqrt{5}-2)(\sqrt{5}+2) = 4\,(\sqrt{5}-2) + (\sqrt{5}-2)^2.]$$

Since, then, the ratio of the last term $(\sqrt{5}-2)^2$ to the preceding one, $\sqrt{5}-2$, is the same as the ratio of $\sqrt{5}-2$ to 1, the process will never end.

Zeuthen has a geometrical proof which is not difficult; but I think the following proof is neater and easier.

Let ABC be a triangle right-angled at B, such that $AB = 1$, $BC = 2$, and therefore $AC = \sqrt{5}$.

Cut off CD from CA equal to CB, and draw DE at right angles to CA. Then $DE = EB$.

Now $AD = \sqrt{5} - 2$, and by similar triangles

$$DE = 2\,AD = 2(\sqrt{5} - 2).$$

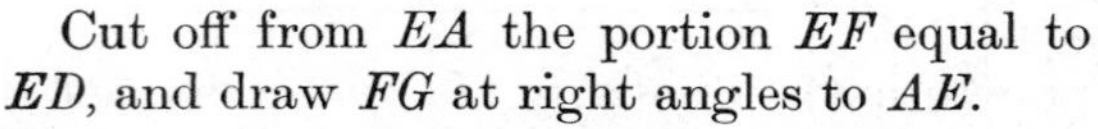

Cut off from EA the portion EF equal to ED, and draw FG at right angles to AE.

Then
$$\begin{aligned} AF = AB - BF &= AB - 2\,DE \\ &= 1 - 4(\sqrt{5} - 2) \\ &= (\sqrt{5} - 2)^2. \end{aligned}$$

Therefore ABC, ADE, AFG are diminishing similar triangles such that

$$AB : AD : AF = 1 : (\sqrt{5} - 2) : (\sqrt{5} - 2)^2,$$

and so on.

Also $AB > FB$, i.e. $2\,DE$ or $4\,AD$.

Therefore the side of each triangle in the series is less than $\frac{1}{4}$ of the corresponding side of the preceding triangle.

In the case of $\sqrt{3}$ the process of finding the G. C. M. of $\sqrt{3}$ and 1 gives

$$\begin{array}{l} 1\,)\ \sqrt{3}\quad(1 \\ \quad\ \ \underline{1} \\ \sqrt{3}-1\,)\ 1\quad(1 \\ \qquad\quad \underline{\sqrt{3}-1} \\ \qquad \tfrac{1}{2}(\sqrt{3}-1)^2\,)\ \sqrt{3}-1\quad(2 \\ \qquad\qquad\qquad \underline{(\sqrt{3}-1)^2} \\ \qquad\qquad\qquad \tfrac{1}{2}(\sqrt{3}-1)^3 \end{array}$$

the ratio of $\frac{1}{2}(\sqrt{3} - 1)^2$ to $\frac{1}{2}(\sqrt{3} - 1)^3$ being the same as that of 1 to $(\sqrt{3} - 1)$.

This case is more difficult to show in geometrical form because we have to make one more division before recurrence takes place.

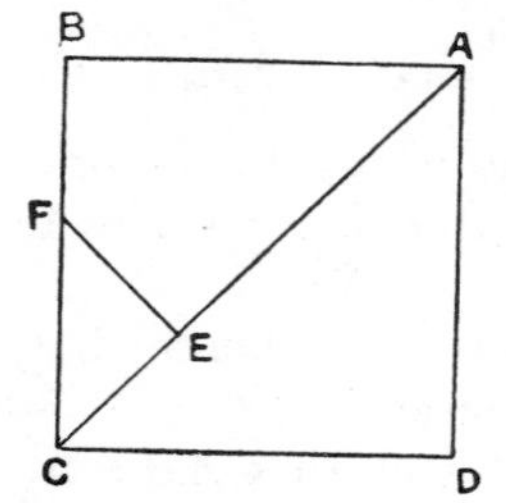

The cases $\sqrt{10}$ and $\sqrt{17}$ are exactly similar to that of $\sqrt{5}$.

The irrationality of $\sqrt{2}$ can, of course, be proved by the same method. If $ABCD$ is a square, we mark off along the diagonal AC a length AE equal to AB and draw EF at right angles to AC. The same thing is then done with the triangle CEF

as with the triangle ABC, and so on. This could not have escaped Theodorus if his proof in the cases of $\sqrt{3}$, $\sqrt{5}$... took the form suggested by Zeuthen; but he was presumably content to accept the traditional proof with regard to $\sqrt{2}$.

The conjecture of Zeuthen is very ingenious, but, as he admits, it necessarily remains a hypothesis.

THEAETETUS[1] (about 415–369 B.C.) made important contributions to the body of the Elements. These related to two subjects in particular, (*a*) the theory of irrationals, and (*b*) the five regular solids.

That Theaetetus actually succeeded in generalizing the theory of irrationals on the lines indicated in the second part of the passage from Plato's dialogue is confirmed by other evidence. The commentary on Eucl. X, which has survived in Arabic and is attributed to Pappus, says (in the passage partly quoted above, p. 155) that the theory of irrationals

'had its origin in the school of Pythagoras. It was considerably developed by Theaetetus the Athenian, who gave proof in this part of mathematics, as in others, of ability which has been justly admired. . . . As for the exact distinctions of the above-named magnitudes and the rigorous demonstrations of the propositions to which this theory gives rise, I believe that they were chiefly established by this mathematician. For Theaetetus had distinguished square roots[2] commensurable in length from those which are incommensurable, and had divided the well-known species of irrational lines after the different means, assigning the *medial* to geometry, the *binomial* to arithmetic, and the *apotome* to harmony, as is stated by Eudemus the Peripatetic.'[3]

[1] On Theaetetus the reader may consult a recent dissertation, *De Theaeteto Atheniensi mathematico*, by Eva Sachs (Berlin, 1914).

[2] 'Square roots'. The word in Woepcke's translation is 'puissances', which indicates that the original word was δυνάμεις. This word is always ambiguous; it might mean 'squares', but I have translated it 'square roots' because the δύναμις of Theaetetus's definition is undoubtedly the square root of a non-square number, a surd. The distinction in that case would appear to be between 'square roots' commensurable in length and square roots commensurable in square only; thus $\sqrt{3}$ and $\sqrt{12}$ are commensurable in length, while $\sqrt{3}$ and $\sqrt{7}$ are commensurable in square only. I do not see how δυνάμεις could here mean squares; for 'squares commensurable in length' is not an intelligible phrase, and it does not seem legitimate to expand it into 'squares ⟨on straight lines⟩ commensurable in length'.

[3] For an explanation of this see *The Thirteen Books of Euclid's Elements* vol. iii, p. 4.

The irrationals called by the names here italicized are described in Eucl. X. 21, 36 and 73 respectively.

Again, a scholiast[1] on Eucl. X. 9 (containing the general theorem that squares which have not to one another the ratio of a square number to a square number have their sides incommensurable in length) definitely attributes the discovery of this theorem to Theaetetus. But, in accordance with the traditional practice in Greek geometry, it was necessary to prove the existence of such incommensurable ratios, and this is done in the porism to Eucl. X. 6 by a geometrical construction; the porism first states that, given a straight line a and any two numbers m, n, we can find a straight line x such that $a:x = m:n$; next it is shown that, if y be taken a mean proportional between a and x, then

$$a^2:y^2 = a:x = m:n;$$

if, therefore, the ratio $m:n$ is not a ratio of a square to a square, we have constructed an irrational straight line $a\sqrt{(n/m)}$ and therefore shown that such a straight line exists.

The proof of Eucl. X. 9 formally depends on VIII. 11 alone (to the effect that between two square numbers there is one mean proportional number, and the square has to the square the duplicate ratio of that which the side has to the side); and VIII. 11 again depends on VII. 17 and 18 (to the effect that $ab:ac = b:c$, and $a:b = ac:bc$, propositions which are not identical). But Zeuthen points out that these propositions are an inseparable part of a whole theory established in Book VII and the early part of Book VIII, and that the real demonstration of X. 9 is rather contained in propositions of these Books which give a rigorous proof of the necessary and sufficient conditions for the rationality of the square roots of numerical fractions and integral numbers, notably VII. 27 and the propositions leading up to it, as well as VIII. 2. He therefore suggests that the theory established in the early part of Book VII was not due to the Pythagoreans, but was an innovation made by Theaetetus with the direct object of laying down a scientific basis for his theory of irrationals, and that this, rather than the mere formulation

[1] X, No. 62 (Heiberg's Euclid, vol. v, p. 450).

of the theorem of Eucl. X. 9, was the achievement which Plato intended to hold up to admiration.

This conjecture is of great interest, but it is, so far as I know, without any positive confirmation. On the other hand, there are circumstances which suggest doubts. For example, Zeuthen himself admits that Hippocrates, who reduced the duplication of the cube to the finding of two mean proportionals, must have had a proposition corresponding to the very proposition VIII. 11 on which X. 9 formally depends. Secondly, in the extract from Simplicius about the squaring of lunes by Hippocrates, we have seen that the proportionality of similar segments of circles to the circles of which they form part is explained by the statement that 'similar segments are those which are *the same part* of the circles'; and if we may take this to be a quotation by Eudemus from Hippocrates's own argument, the inference is that Hippocrates had a definition of numerical proportion which was at all events near to that of Eucl. VII, Def. 20. Thirdly, there is the proof (presently to be given) by Archytas of the proposition that there can be no number which is a (geometric) mean between two consecutive integral numbers, in which proof it will be seen that several propositions of Eucl., Book VII, are pre-supposed; but Archytas lived (say) 430–365 B.C., and Theaetetus was some years younger. I am not, therefore, prepared to give up the view, which has hitherto found general acceptance, that the Pythagoreans already had a theory of proportion of a numerical kind on the lines, though not necessarily or even probably with anything like the fullness and elaboration, of Eucl., Book VII.

While Pappus, in the commentary quoted, says that Theaetetus distinguished the well-known species of irrationals, and in particular the *medial*, the *binomial*, and the *apotome*, he proceeds thus:

'As for Euclid, he set himself to give rigorous rules, which he established, relative to commensurability and incommensurability in general; he made precise the definitions and distinctions between rational and irrational magnitudes, he set out a great number of orders of irrational magnitudes, and finally he made clear their whole extent.'

As Euclid proves that there are thirteen irrational straight lines in all, we may perhaps assume that the subdivision of the three species of irrationals distinguished by Theaetetus into thirteen was due to Euclid himself, while the last words of the quotation seem to refer to Eucl. X. 115, where it is proved that from the *medial* straight line an unlimited number of other irrationals can be derived which are all different from it and from one another.

It will be remembered that, at the end of the passage of the *Theaetetus* containing the definition of 'square roots' or surds, Theaetetus says that 'there is a similar distinction in the case of solids'. We know nothing of any further development of a theory of irrationals arising from solids; but Theaetetus doubtless had in mind a distinction related to VIII. 12 (the theorem that between two cube numbers there are two mean proportional numbers) in the same way as the definition of a 'square root' or surd is related to VIII. 11; that is to say, he referred to the incommensurable cube root of a non-cube number which is the product of three factors.

Besides laying the foundation of the theory of irrationals as we find it in Eucl., Book X, Theaetetus contributed no less substantially to another portion of the *Elements*, namely Book XIII, which is devoted (after twelve introductory propositions) to constructing the five regular solids, circumscribing spheres about them, and finding the relation between the dimensions of the respective solids and the circumscribing spheres. We have already mentioned (pp. 159, 162) the traditions that Theaetetus was the first to 'construct' or 'write upon' the five regular solids,[1] and that his name was specially associated with the octahedron and the icosahedron.[2] There can be little doubt that Theaetetus's 'construction' of, or treatise upon, the regular solids gave the theoretical constructions much as we find them in Euclid.

Of the mathematicians of Plato's time, two others are mentioned with Theaetetus as having increased the number of theorems in geometry and made a further advance towards a scientific grouping of them, LEODAMAS OF THASOS and ARCHYTAS OF TARAS. With regard to the former we are

[1] Suidas, *s. v.* Θεαίτητος.
[2] Schol. 1 to Eucl. XIII (Euclid, ed. Heiberg, vol. v, p. 654).

told that Plato 'explained (εἰσηγήσατο) to Leodamas of Thasos the method of inquiry by analysis'[1]; Proclus's account is fuller, stating that the finest method for discovering lemmas in geometry is that 'which by means of *analysis* carries the thing sought up to an acknowledged principle, a method which Plato, as they say, communicated to Leodamas, and by which the latter too is said to have discovered many things in geometry'.[2] Nothing more than this is known of Leodamas, but the passages are noteworthy as having given rise to the idea that Plato *invented* the method of mathematical analysis, an idea which, as we shall see later on, seems nevertheless to be based on a misapprehension.

Archytas of Taras, a Pythagorean, the friend of Plato, flourished in the first half of the fourth century, say 400 to 365 B.C. Plato made his acquaintance when staying in Magna Graecia, and he is said, by means of a letter, to have saved Plato from death at the hands of Dionysius. Statesman and philosopher, he was famed for every sort of accomplishment. He was general of the forces of his city-state for seven years, though ordinarily the law forbade any one to hold the post for more than a year; and he was never beaten. He is said to have been the first to write a systematic treatise on *mechanics* based on mathematical principles.[3] Vitruvius mentions that, like Archimedes, Ctesibius, Nymphodorus, and Philo of Byzantium, Archytas wrote on machines[4]; two mechanical devices in particular are attributed to him, one a mechanical dove made of wood which would fly,[5] the other a rattle which, according to Aristotle, was found useful to 'give to children to occupy them, and so prevent them from breaking things about the house (for the young are incapable of keeping still)'.[6]

We have already seen Archytas distinguishing the four mathematical sciences, geometry, arithmetic, sphaeric (or astronomy), and music, comparing the art of calculation with geometry in respect of its relative efficiency and conclusiveness, and defining the three means in music, the arithmetic,

[1] Diog. L. iii. 24.
[2] Proclus on Eucl. I, p. 211. 19–23.
[3] Diog. L. viii. 79–83.
[4] Vitruvius, *De architectura*, Praef. vii. 14.
[5] Gellius, x. 12. 8, after Favorinus (*Vors.* i³, p. 325. 21–9).
[6] Aristotle, *Politics*, E (Θ). 6, 1340 b 26.

the geometric, and the harmonic (a name substituted by Archytas and Hippasus for the older name 'sub-contrary').

From his mention of *sphaeric* in connexion with his statement that 'the mathematicians have given us clear knowledge about the speed of the heavenly bodies and their risings and settings' we gather that in Archytas's time astronomy was already treated mathematically, the properties of the sphere being studied so far as necessary to explain the movements in the celestial sphere. He discussed too the question whether the universe is unlimited in extent, using the following argument.

'If I were at the outside, say at the heaven of the fixed stars, could I stretch my hand or my stick outwards or not? To suppose that I could not is absurd; and if I can stretch it out, that which is outside must be either body or space (it makes no difference which it is, as we shall see). We may then in the same way get to the outside of that again, and so on, asking on arrival at each new limit the same question; and if there is always a new place to which the stick may be held out, this clearly involves extension without limit. If now what so extends is body, the proposition is proved; but even if it is space, then, since space is that in which body is or can be, and in the case of eternal things we must treat that which potentially is as being, it follows equally that there must be body and space (extending) without limit.'[1]

In *geometry*, while Archytas doubtless increased the number of theorems (as Proclus says), only one fragment of his has survived, namely the solution of the problem of finding two mean proportionals (equivalent to the duplication of the cube) by a remarkable theoretical construction in three dimensions. As this, however, belongs to higher geometry and not to the Elements, the description of it will come more appropriately in another place (pp. 246–9).

In *music* he gave the numerical ratios representing the intervals of the tetrachord on three scales, the anharmonic, the chromatic, and the diatonic.[2] He held that sound is due to impact, and that higher tones correspond to quicker motion communicated to the air, and lower tones to slower motion.[3]

[1] Simplicius *in Phys.*, p. 467. 26. [2] Ptol. *harm.* i. 13, p. 31 Wall.
[3] Porph. *in Ptol. harm.*, p. 236 (*Vors.* i³, p. 232–3); Theon of Smyrna, p. 61. 11–17.

Of the fragments of Archytas handed down to us the most interesting from the point of view of this chapter is a proof of the proposition that there can be no number which is a (geometric) mean between two numbers in the ratio known as ἐπιμόριος or *superparticularis*, that is, $(n+1):n$. This proof is preserved by Boetius [1], and the noteworthy fact about it is that it is substantially identical with the proof of the same theorem in Prop. 3 of Euclid's tract on the *Sectio canonis*.[2] I will quote Archytas's proof in full, in order to show the slight differences from the Euclidean form and notation.

Let A, B be the given 'superparticularis proportio' (ἐπιμόριον διάστημα in Euclid). [Archytas writes the smaller number first (instead of second, as Euclid does); we are then to suppose that A, B are integral numbers in the ratio of n to $(n+1)$.]

Take C, DE the smallest numbers which are in the ratio of A to B. [Here DE means $D+E$; in this respect the notation differs from that of Euclid, who, as usual, takes a straight line DF divided into two parts at G, the parts DG, GF corresponding to the D and E respectively in Archytas's proof. The step of finding C, DE the smallest numbers in the same ratio as that of A to B presupposes Eucl. VII. 33 applied to two numbers.]

Then DE exceeds C by an aliquot part of itself and of C [cf. the definition of ἐπιμόριος ἀριθμός in Nicomachus, i. 19. 1].

Let D be the excess [i.e. we suppose E equal to C].

I say that D is not a number, but a unit.

For, if D is a number and an aliquot part of DE, it measures DE; therefore it measures E, that is, C.

Thus D measures both C and DE: which is impossible, since the smallest numbers which are in the same ratio as any numbers are prime to one another. [This presupposes Eucl. VII. 22.]

Therefore D is a unit; that is, DE exceeds C by a unit.

Hence no number can be found which is a mean between the two numbers C, DE [for there is no *integer* intervening].

[1] Boetius, *De inst. mus.* iii. 11, pp. 285–6 Friedlein.

[2] *Musici scriptores Graeci*, ed. Jan, p. 14; Heiberg and Menge's Euclid, vol. viii, p. 162.

Therefore neither can any number be a mean between the original numbers A, B, which are in the same ratio as C, DE [cf. the more general proposition, Eucl. VIII. 8; the particular inference is a consequence of Eucl. VII. 20, to the effect that the least numbers of those which have the same ratio with them measure the latter the same number of times, the greater the greater and the less the less].

Since this proof cites as known several propositions corresponding to propositions in Euclid, Book VII, it affords a strong presumption that there already existed, at least as early as the time of Archytas, a treatise of some sort on the Elements of Arithmetic in a form similar to the Euclidean, and containing many of the propositions afterwards embodied by Euclid in his arithmetical books.

Summary.

We are now in a position to form an idea of the scope of the Elements at the stage which they had reached in Plato's time. The substance of Eucl. I–IV was practically complete. Book V was of course missing, because the theory of proportion elaborated in that book was the creation of Eudoxus. The Pythagoreans had a theory of proportion applicable to commensurable magnitudes only; this was probably a numerical theory on lines similar to those of Eucl., Book VII. But the theorems of Eucl., Book VI, in general, albeit insufficiently established in so far as they depended on the numerical theory of proportion, were known and used by the Pythagoreans. We have seen reason to suppose that there existed Elements of Arithmetic partly (at all events) on the lines of Eucl., Book VII, while some propositions of Book VIII (e.g. Props. 11 and 12) were also common property. The Pythagoreans, too, conceived the idea of perfect numbers (numbers equal to the sum of all their divisors) if they had not actually shown (as Euclid does in IX. 36) how they are evolved. There can also be little doubt that many of the properties of plane and solid numbers and of similar numbers of both classes proved in Euclid, Books VIII and IX, were known before Plato's time.

We come next to Book X, and it is plain that the foundation of the whole had been well and truly laid by Theaetetus, and the main varieties of irrationals distinguished, though their classification was not carried so far as in Euclid.

The substance of Book XI. 1–19 must already have been included in the Elements (e.g. Eucl. XI. 19 is assumed in Archytas's construction for the two mean proportionals), and the whole theory of the section of Book XI in question would be required for Theaetetus's work on the five regular solids: XI. 21 must have been known to the Pythagoreans: while there is nothing in the latter portion of the book about parallelepipedal solids which (subject to the want of a rigorous theory of proportion) was not within the powers of those who were familiar with the theory of plane and solid numbers.

Book XII employs throughout the *method of exhaustion*, the orthodox form of which is attributed to Eudoxus, who grounded it upon a lemma known as Archimedes's Axiom or its equivalent (Eucl. X. 1). Yet even XII. 2, to the effect that circles are to one another as the square of their diameters, had already been anticipated by Hippocrates of Chios, while Democritus had discovered the truth of the theorems of XII. 7, Por., about the volume of a pyramid, and XII. 10, about the volume of a cone.

As in the case of Book X, it would appear that Euclid was indebted to Theaetetus for much of the substance of Book XIII, the latter part of which (Props. 12–18) is devoted to the construction of the five regular solids, and the inscribing of them in spheres.

There is therefore probably little in the whole compass of the *Elements* of Euclid, except the new theory of proportion due to Eudoxus and its consequences, which was not in substance included in the recognized content of geometry and arithmetic by Plato's time, although the form and arrangement of the subject-matter and the methods employed in particular cases were different from what we find in Euclid.

VII

SPECIAL PROBLEMS

SIMULTANEOUSLY with the gradual evolution of the Elements, the Greeks were occupying themselves with problems in higher geometry; three problems in particular, the squaring of the circle, the doubling of the cube, and the trisection of any given angle, were rallying-points for mathematicians during three centuries at least, and the whole course of Greek geometry was profoundly influenced by the character of the specialized investigations which had their origin in the attempts to solve these problems. In illustration we need only refer to the subject of conic sections which began with the use made of two of the curves for the finding of two mean proportionals.

The Greeks classified problems according to the means by which they were solved. The ancients, says Pappus, divided them into three classes, which they called *plane*, *solid*, and *linear* respectively. Problems were *plane* if they could be solved by means of the straight line and circle only, *solid* if they could be solved by means of one or more conic sections, and *linear* if their solution required the use of other curves still more complicated and difficult to construct, such as spirals, *quadratrices*, cochloids (conchoids) and cissoids, or again the various curves included in the class of 'loci on surfaces' (τόποι πρὸς ἐπιφανείαις), as they were called.[1] There was a corresponding distinction between loci: *plane* loci are straight lines or circles; *solid* loci are, according to the most strict classification, conics only, which arise from the sections of certain solids, namely cones; while *linear* loci include all

[1] Pappus, iii, pp. 54-6, iv, pp. 270-2.

higher curves.[1] Another classification of loci divides them into *loci on lines* (τόποι πρὸς γραμμαῖς) and *loci on surfaces* (τόποι πρὸς ἐπιφανείαις).[2] The former term is found in Proclus, and seems to be used in the sense both of loci which *are* lines (including of course curves) and of loci which are spaces bounded by lines; e.g. Proclus speaks of 'the whole space between the parallels' in Eucl. I. 35 as being the locus of the (equal) parallelograms 'on the same base and in the same parallels'.[3] Similarly *loci on surfaces* in Proclus may be loci which *are* surfaces; but Pappus, who gives lemmas to the two books of Euclid under that title, seems to imply that they were curves drawn on surfaces, e.g. the cylindrical helix.[4]

It is evident that the Greek geometers came very early to the conclusion that the three problems in question were not *plane*, but required for their solution either higher curves than circles or constructions more mechanical in character than the mere use of the ruler and compasses in the sense of Euclid's Postulates 1–3. It was probably about 420 B.C. that Hippias of Elis invented the curve known as the *quadratrix* for the purpose of trisecting any angle, and it was in the first half of the fourth century that Archytas used for the duplication of the cube a solid construction involving the revolution of plane figures in space, one of which made a *tore* or anchor-ring with internal diameter *nil.* There are very few records of illusory attempts to do the impossible in these cases. It is practically only in the case of the squaring of the circle that we read of abortive efforts made by 'plane' methods, and none of these (with the possible exception of Bryson's, if the accounts of his argument are correct) involved any real fallacy. On the other hand, the bold pronouncement of Antiphon the Sophist that by inscribing in a circle a series of regular polygons each of which has twice as many sides as the preceding one, we shall use up or exhaust the area of the circle, though it was in advance of his time and was condemned as a fallacy on the technical ground that a straight line cannot coincide with an arc of a circle however short its length, contained an idea destined to be fruitful in the

[1] Cf. Pappus, vii, p. 662, 10–15.
[2] Proclus on Eucl. I, p. 394. 19.
[3] *Ib.*, p. 395. 5.
[4] Pappus, iv, p. 258 sq.

hands of later and abler geometers, since it gives a method of approximating, with any desired degree of accuracy, to the area of a circle, and lies at the root of the *method of exhaustion* as established by Eudoxus. As regards Hippocrates's quadrature of lunes, we must, notwithstanding the criticism of Aristotle charging him with a paralogism, decline to believe that he was under any illusion as to the limits of what his method could accomplish, or thought that he had actually squared the circle.

The squaring of the circle.

There is presumably no problem which has exercised such a fascination throughout the ages as that of rectifying or squaring the circle; and it is a curious fact that its attraction has been no less (perhaps even greater) for the non-mathematician than for the mathematician. It was naturally the kind of problem which the Greeks, of all people, would take up with zest the moment that its difficulty was realized. The first name connected with the problem is Anaxagoras, who is said to have occupied himself with it when in prison.[1] The Pythagoreans claimed that it was solved in their school, 'as is clear from the demonstrations of Sextus the Pythagorean, who got his method of demonstration from early tradition'[2]; but Sextus, or rather Sextius, lived in the reign of Augustus or Tiberius, and, for the usual reasons, no value can be attached to the statement.

The first serious attempts to solve the problem belong to the second half of the fifth century B.C. A passage of Aristophanes's *Birds* is quoted as evidence of the popularity of the problem at the time (414 B.C.) of its first representation. Aristophanes introduces Meton, the astronomer and discoverer of the Metonic cycle of 19 years, who brings with him a ruler and compasses, and makes a certain construction 'in order that your circle may become square'.[3] This is a play upon words, because what Meton really does is to divide a circle into four quadrants by two diameters at right angles to one another; the idea is of streets radiating from the agora in the centre

[1] Plutarch, *De exil.* 17, p. 607 F.
[2] Iambl. ap. Simpl. *in Categ.*, p. 192, 16–19 K., 64 b 11 Brandis.
[3] Aristophanes, *Birds* 1005.

of a town; the word τετράγωνος then really means 'with four (right) angles' (at the centre), and not 'square', but the word conveys a laughing allusion to the problem of squaring all the same.

We have already given an account of Hippocrates's quadratures of lunes. These formed a sort of *prolusio*, and clearly did not purport to be a solution of the problem; Hippocrates was aware that 'plane' methods would not solve it, but, as a matter of interest, he wished to show that, if circles could not be squared by these methods, they could be employed to find the area of *some* figures bounded by arcs of circles, namely certain lunes, and even of the sum of a certain circle and a certain lune.

ANTIPHON of Athens, the Sophist and a contemporary of Socrates, is the next person to claim attention. We owe to Aristotle and his commentators our knowledge of Antiphon's method. Aristotle observes that a geometer is only concerned to refute any fallacious arguments that may be propounded in his subject if they are based upon the admitted principles of geometry; if they are not so based, he is not concerned to refute them:

'thus it is the geometer's business to refute the quadrature by means of segments, but it is not his business to refute that of Antiphon'.[1]

As we have seen, the quadrature 'by means of segments' is probably Hippocrates's quadrature of lunes. Antiphon's method is indicated by Themistius[2] and Simplicius.[3] Suppose there is any regular polygon inscribed in a circle, e.g. a square or an equilateral triangle. (According to Themistius, Antiphon began with an equilateral triangle, and this seems to be the authentic version; Simplicius says he inscribed some one of the regular polygons which can be inscribed

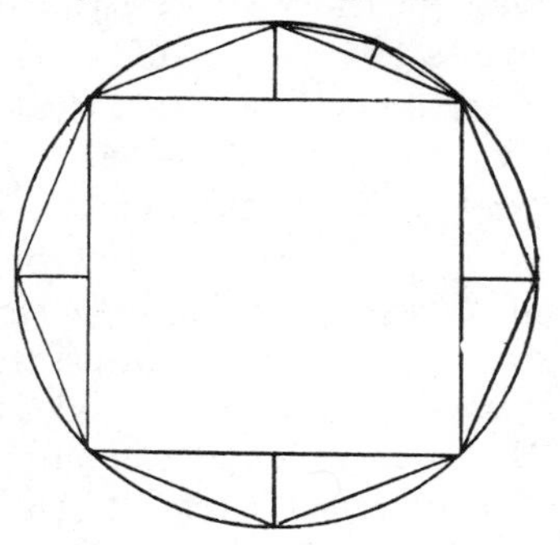

[1] Arist. *Phys.* i. 2, 185 a 14–17.
[2] Them. *in Phys.*, p. 4. 2 sq., Schenkl.
[3] Simpl. *in Phys.*, p. 54. 20–55. 24, Diels.

in a circle, 'suppose, if it so happen, that the inscribed polygon is a square'.) On each side of the inscribed triangle or square as base describe an isosceles triangle with its vertex on the arc of the smaller segment of the circle subtended by the side. This gives a regular inscribed polygon with double the number of sides. Repeat the construction with the new polygon, and we have an inscribed polygon with four times as many sides as the original polygon had. Continuing the process,

'Antiphon thought that in this way the area (of the circle) would be used up, and we should some time have a polygon inscribed in the circle the sides of which would, owing to their smallness, coincide with the circumference of the circle. And, as we can make a square equal to any polygon . . . we shall be in a position to make a square equal to a circle.'

Simplicius tells us that, while according to Alexander the geometrical principle hereby infringed is the truth that a circle touches a straight line in one point (only), Eudemus more correctly said it was the principle that magnitudes are divisible without limit; for, if the area of the circle is divisible without limit, the process described by Antiphon will never result in using up the whole area, or in making the sides of the polygon take the position of the actual circumference of the circle. But the objection to Antiphon's statement is really no more than verbal; Euclid uses exactly the same construction in XII. 2, only he expresses the conclusion in a different way, saying that, if the process be continued far enough, the small segments left over will be together less than any assigned area. Antiphon in effect said the same thing, which again we express by saying that the circle is the *limit* of such an inscribed polygon when the number of its sides is indefinitely increased. Antiphon therefore deserves an honourable place in the history of geometry as having originated the idea of *exhausting* an area by means of inscribed regular polygons with an ever increasing number of sides, an idea upon which, as we said, Eudoxus founded his epoch-making *method of exhaustion.* The practical value of Antiphon's construction is illustrated by Archimedes's treatise on the *Measurement of a Circle,* where, by constructing inscribed and circumscribed regular polygons with 96 sides, Archimedes proves that $3\frac{1}{7} > \pi > 3\frac{10}{71}$, the lower limit, $\pi > 3\frac{10}{71}$, being obtained by calculating the perimeter of the *inscribed* polygon of 96 sides, which is constructed in Antiphon's manner from an inscribed equilateral triangle. The same construction starting from a square was likewise the basis of Vieta's expression for $2/\pi$, namely

$$\frac{2}{\pi} = \cos\frac{\pi}{4}.\cos\frac{\pi}{8}.\cos\frac{\pi}{16}\dots$$
$$= \sqrt{\tfrac{1}{2}}.\sqrt{\tfrac{1}{2}(1+\sqrt{\tfrac{1}{2}})}.\sqrt{\tfrac{1}{2}(1+\sqrt{\tfrac{1}{2}(1+\sqrt{\tfrac{1}{2}})})}\dots \textit{(ad inf.)}$$

Bryson, who came a generation later than Antiphon, being a pupil of Socrates or of Euclid of Megara, was the author of another attempted quadrature which is criticized by Aristotle as 'sophistic' and 'eristic' on the ground that it was based on principles not special to geometry but applicable equally to other subjects.[1] The commentators give accounts of Bryson's argument which are substantially the same, except that Alexander speaks of *squares* inscribed and circumscribed to a circle [2], while Themistius and Philoponus speak of any polygons.[3] According to Alexander, Bryson inscribed a square in a circle and circumscribed another about it, while he also took a square intermediate between them (Alexander does not say how constructed); then he argued that, as the intermediate square is less than the outer and greater than the inner, while the circle is also less than the outer square and greater than the inner, and as *things which are greater and less than the same things respectively are equal*, it follows that the circle is equal to the intermediate square: upon which Alexander remarks that not only is the thing assumed applicable to other things besides geometrical magnitudes, e.g. to numbers, times, depths of colour, degrees of heat or cold, &c., but it is also false because (for instance) 8 and 9 are both less than 10 and greater than 7 and yet they are not equal. As regards the intermediate square (or polygon), some have assumed that it was the arithmetic mean between the inscribed and circumscribed figures, and others that it was the geometric mean. Both assumptions seem to be due to misunderstanding [4]; for

[1] Arist. *An. Post.* i. 9, 75 b 40.

[2] Alexander on *Soph. El.*, p. 90. 10–21, Wallies, 306 b 24 sq., Brandis.

[3] Them. on *An. Post.*, p. 19. 11–20, Wallies, 211 b 19, Brandis; Philop. on *An. Post.*, p. 111. 20–114. 17 W., 211 b 30, Brandis.

[4] Psellus (11th cent. A.D.) says, 'there are different opinions as to the

the ancient commentators do not attribute to Bryson any such statement, and indeed, to judge by their discussions of different interpretations, it would seem that tradition was by no means clear as to what Bryson actually did say. But it seems important to note that Themistius states (1) that Bryson declared the circle to be greater than *all* inscribed, and less than *all* circumscribed, polygons, while he also says (2) that the assumed axiom is *true*, though not peculiar to geometry. This suggests a possible explanation of what otherwise seems to be an absurd argument. Bryson may have multiplied the number of the sides of both the inscribed and circumscribed regular polygons as Antiphon did with inscribed polygons; he may then have argued that, if we continue this process long enough, we shall have an inscribed and a circumscribed polygon differing so little in area that, if we can describe a polygon intermediate between them in area, the circle, which is also intermediate in area between the inscribed and circumscribed polygons, must be equal to the intermediate polygon.[1] If this is the right explanation, Bryson's name by no means deserves to be banished from histories of Greek mathematics; on the contrary, in so far as he suggested the necessity of considering circumscribed as well as inscribed polygons, he went a step further than Antiphon; and the importance of the idea is attested by the fact that, in the regular method of exhaustion as practised by Archimedes, use is made of both inscribed and circumscribed figures, and this *compression*, as it were, of a circumscribed and an inscribed figure into one so that they ultimately coincide with one another, and with the

proper method of finding the area of a circle, but that which has found the most favour is to take the geometric mean between the inscribed and circumscribed squares'. I am not aware that he quotes Bryson as the authority for this method, and it gives the inaccurate value $\pi = \sqrt{8}$ or 2·8284272 .... Isaac Argyrus (14th cent.) adds to his account of Bryson the following sentence: 'For the circumscribed square *seems* to exceed the circle by the same amount as the inscribed square is exceeded by the circle.'

[1] It is true that, according to Philoponus, Proclus had before him an explanation of this kind, but rejected it on the ground that it would mean that the circle must actually *be* the intermediate polygon and not only be equal to it, in which case Bryson's contention would be tantamount to Antiphon's, whereas according to Aristotle it was based on a quite different principle. But it is sufficient that the circle should be taken to be *equal* to any polygon that can be drawn intermediate between the two ultimate polygons, and this gets over Proclus's difficulty.

curvilinear figure to be measured, is particularly characteristic of Archimedes.

We come now to the real rectifications or quadratures of circles effected by means of higher curves, the construction of which is more 'mechanical' than that of the circle. Some of these curves were applied to solve more than one of the three classical problems, and it is not always easy to determine for which purpose they were originally destined by their inventors, because the accounts of the different authorities do not quite agree. Iamblichus, speaking of the quadrature of the circle, said that

'Archimedes effected it by means of the spiral-shaped curve, Nicomedes by means of the curve known by the special name *quadratrix* (τετραγωνίζουσα), Apollonius by means of a certain curve which he himself calls "sister of the cochloid" but which is the same as Nicomedes's curve, and finally Carpus by means of a certain curve which he simply calls (the curve arising) "from a double motion".'[1]

Pappus says that

'for the squaring of the circle Dinostratus, Nicomedes and certain other and later geometers used a certain curve which took its name from its property; for those geometers called it *quadratrix*.'[2]

Lastly, Proclus, speaking of the trisection of any angle, says that

'Nicomedes trisected any rectilineal angle by means of the conchoidal curves, the construction, order and properties of which he handed down, being himself the discoverer of their peculiar character. Others have done the same thing by means of the *quadratrices* of Hippias and Nicomedes. . . . Others again, starting from the spirals of Archimedes, divided any given rectilineal angle in any given ratio.'[3]

All these passages refer to the *quadratrix* invented by Hippias of Elis. The first two seem to imply that it was not used by Hippias himself for squaring the circle, but that it was Dinostratus (a brother of Menaechmus) and other later geometers who first applied it to that purpose; Iamblichus and Pappus do not even mention the name of Hippias. We might conclude that Hippias originally intended his curve to

[1] Iambl. ap. Simpl. *in Categ.*, p. 192. 19–24 K., 64 b 13–18 Br.
[2] Pappus, iv, pp. 250. 33–252. 3. [3] Proclus on Eucl. I, p. 272. 1–12.

be used for trisecting an angle. But this becomes more doubtful when the passages of Proclus are considered. Pappus's authority seems to be Sporus, who was only slightly older than Pappus himself (towards the end of the third century A.D.), and who was the author of a compilation called *Κηρία* containing, among other things, mathematical extracts on the quadrature of the circle and the duplication of the cube. Proclus's authority, on the other hand, is doubtless Geminus, who was much earlier (first century B.C.) Now not only does the above passage of Proclus make it possible that the name *quadratrix* may have been used by Hippias himself, but in another place Proclus (i.e. Geminus) says that different mathematicians have explained the properties of particular kinds of curves:

'thus Apollonius shows in the case of each of the conic curves what is its property, and similarly Nicomedes with the conchoids, *Hippias with the quadratrices*, and Perseus with the spiric curves.'[1]

This suggests that Geminus had before him a regular treatise by Hippias on the properties of the *quadratrix* (which may have disappeared by the time of Sporus), and that Nicomedes did not write any such general work on that curve; and, if this is so, it seems not impossible that Hippias himself discovered that it would serve to rectify, and therefore to square, the circle.

(α) *The Quadratrix of Hippias.*

The method of constructing the curve is described by Pappus.[2] Suppose that $ABCD$ is a square, and BED a quadrant of a circle with centre A.

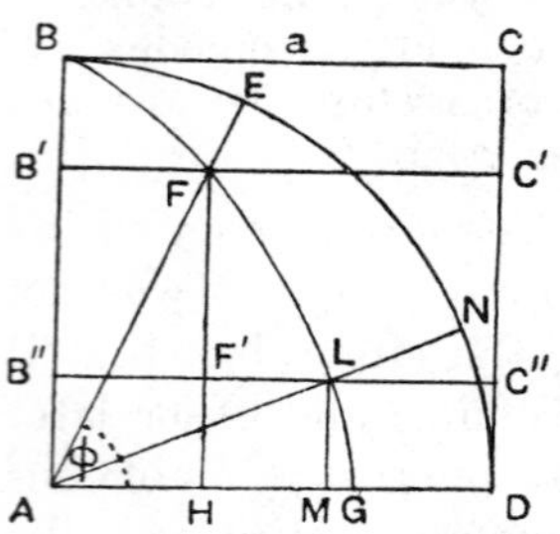

Suppose (1) that a radius of the circle moves uniformly about A from the position AB to the position AD, and (2) that *in the same time* the line BC moves uniformly, always parallel to itself and with its extremity B moving along BA, from the position BC to the position AD.

[1] Proclus on Eucl. I, p. 356. 6–12.

[2] Pappus, iv, pp. 252 sq.

Then, in their ultimate positions, the moving straight line and the moving radius will both coincide with AD; and at any previous instant during the motion the moving line and the moving radius will by their intersection determine a point, as F or L.

The locus of these points is the *quadratrix*.

The property of the curve is that

$$\angle BAD : \angle EAD = (\text{arc } BED) : (\text{arc } ED) = AB : FH.$$

In other words, if ϕ is the angle FAD made by any radius vector AF with AD, ρ the length of AF, and a the length of the side of the square,

$$\frac{\rho \sin \phi}{a} = \frac{\phi}{\frac{1}{2}\pi}.$$

Now clearly, when the curve is once constructed, it enables us not only to *trisect* the angle EAD but also to *divide it in any given ratio*.

For let FH be divided at F' in the given ratio. Draw $F'L$ parallel to AD to meet the curve in L: join AL, and produce it to meet the circle in N.

Then the angles EAN, NAD are in the ratio of FF' to $F'H$, as is easily proved.

Thus the quadratrix lends itself quite readily to the division of any angle in a given ratio.

The application of the *quadratrix* to the rectification of the circle is a more difficult matter, because it requires us to know the position of G, the point where the quadratrix intersects AD. This difficulty was fully appreciated in ancient times, as we shall see.

Meantime, assuming that the quadratrix intersects AD in G, we have to prove the proposition which gives the length of the arc of the quadrant BED and therefore of the circumference of the circle. This proposition is to the effect that

$$(\text{arc of quadrant } BED) : AB = AB : AG.$$

This is proved by *reductio ad absurdum*.

If the former ratio is not equal to $AB : AG$, it must be equal to $AB : AK$, where AK is either (1) greater or (2) less than AG.

(1) Let AK be greater than AG; and with A as centre

and AK as radius, draw the quadrant KFL cutting the quadratrix in F and AB in L.

Join AF, and produce it to meet the circumference BED in E; draw FH perpendicular to AD.

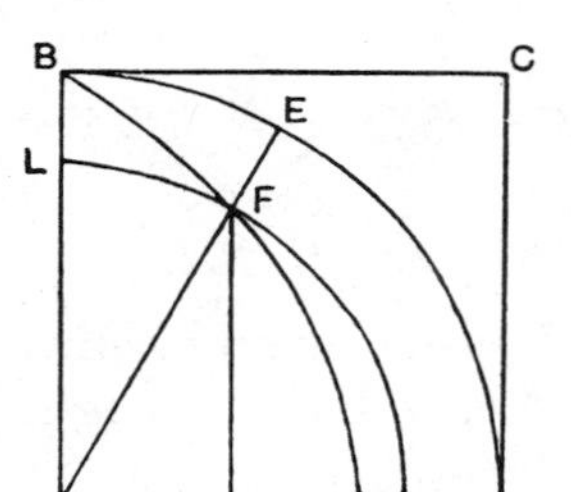

Now, by hypothesis,

$$(\text{arc } BED) : AB = AB : AK$$
$$= (\text{arc } BED) : (\text{arc } LFK);$$

therefore $AB = (\text{arc } LFK)$.

But, by the property of the *quadratrix*,

$$AB : FH = (\text{arc } BED) : (\text{arc } ED)$$
$$= (\text{arc } LFK) : (\text{arc } FK);$$

and it was proved that $AB = (\text{arc } LFK)$;

therefore $FH = (\text{arc } FK)$:

which is absurd. Therefore AK is not greater than AG.

(2) Let AK be less than AG.

With centre A and radius AK draw the quadrant KML.

Draw KF at right angles to AD meeting the quadratrix in F; join AF, and let it meet the quadrants in M, E respectively.

Then, as before, we prove that

$$AB = (\text{arc } LMK).$$

And, by the property of the *quadratrix*,

$$AB : FK = (\text{arc } BED) : (\text{arc } DE)$$
$$= (\text{arc } LMK) : (\text{arc } MK).$$

Therefore, since $AB = (\text{arc } LMK)$,

$$FK = (\text{arc } KM):$$

which is absurd. Therefore AK is not less than AG.

Since then AK is neither less nor greater than AG, it is equal to it, and

$$(\text{arc } BED) : AB = AB : AG.$$

[The above proof is presumably due to Dinostratus (if not to Hippias himself), and, as Dinostratus was a brother of Menaechmus, a pupil of Eudoxus, and therefore probably flourished about 350 B.C., that is to say, some time before Euclid, it is worth while to note certain propositions which are assumed as known. These are, in addition to the theorem of Eucl. VI. 33, the following: (1) the circumferences of circles are as their respective radii; (2) any arc of a circle is greater than the chord subtending it; (3) any arc of a circle less than a quadrant is less than the portion of the tangent at one extremity of the arc cut off by the radius passing through the other extremity. (2) and (3) are of course equivalent to the facts that, if α be the circular measure of an angle less than a right angle, $\sin\alpha < \alpha < \tan\alpha$.]

Even now we have only rectified the circle. To square it we have to use the proposition (1) in Archimedes's *Measurement of a Circle*, to the effect that the area of a circle is equal to that of a right-angled triangle in which the perpendicular is equal to the radius, and the base to the circumference, of the circle. This proposition is proved by the method of exhaustion and may have been known to Dinostratus, who was later than Eudoxus, if not to Hippias.

The criticisms of Sporus,[1] in which Pappus concurs, are worth quoting:

(1) 'The very thing for which the construction is thought to serve is actually assumed in the hypothesis. For how is it possible, with two points starting from B, to make one of them move along a straight line to A and the other along a circumference to D in an equal time, unless you first know the ratio of the straight line AB to the circumference BED? In fact this ratio must also be that of the speeds of motion. For, if you employ speeds not definitely adjusted (to this ratio), how can you make the motions end at the same moment, unless this should sometime happen by pure chance? Is not the thing thus shown to be absurd?

(2) 'Again, the extremity of the curve which they employ for squaring the circle, I mean the point in which the curve cuts the straight line AD, is not found at all. For if, in the figure, the straight lines CB, BA are made to end their motion together, they will then coincide with AD itself and will not cut one another any more. In fact they cease to intersect before they coincide with AD, and yet it was the intersection of these lines which was supposed to give the extremity of the

[1] Pappus, iv, pp. 252. 26–254. 22.

curve, where it met the straight line AD. Unless indeed any one should assert that the curve is conceived to be produced further, in the same way as we suppose straight lines to be produced, as far as AD. But this does not follow from the assumptions made; the point G can only be found by first assuming (as known) the ratio of the circumference to the straight line.'

The second of these objections is undoubtedly sound. The point G can in fact only be found by applying the method of exhaustion in the orthodox Greek manner; e.g. we may first bisect the angle of the quadrant, then the half towards AD, then the half of that and so on, drawing each time from the points F in which the bisectors cut the quadratrix perpendiculars FH on AD and describing circles with AF as radius cutting AD in K. Then, if we continue this process long enough, HK will get smaller and smaller and, as G lies between H and K, we can approximate to the position of G as nearly as we please. But this process is the equivalent of approximating to π, which is the very object of the whole construction.

As regards objection (1) Hultsch has argued that it is not valid because, with our modern facilities for making instruments of precision, there is no difficulty in making the two uniform motions take the same time. Thus an accurate clock will show the minute hand describing an exact quadrant in a definite time, and it is quite practicable now to contrive a uniform rectilinear motion taking exactly the same time. I suspect, however, that the rectilinear motion would be the result of converting some one or more circular motions into rectilinear motions; if so, they would involve the use of an approximate value of π, in which case the solution would depend on the assumption of the very thing to be found. I am inclined, therefore, to think that both Sporus's objections are valid.

(β) *The Spiral of Archimedes.*

We are assured that Archimedes actually used the spiral for squaring the circle. He does in fact show how to rectify a circle by means of a polar subtangent to the spiral. The spiral is thus generated: suppose that a straight line with one extremity fixed starts from a fixed position (the initial line) and revolves uniformly about the fixed extremity, while a point also moves uniformly along the moving straight line starting from the fixed extremity (the origin) at the commencement of the straight line's motion; the curve described is a spiral.

The polar equation of the curve is obviously $\rho = a\theta$.

Suppose that the tangent at any point P of the spiral is met at T by a straight line drawn from O, the origin or pole, perpendicular to the radius vector OP; then OT is the polar subtangent.

Now in the book *On Spirals* Archimedes proves generally the equivalent of the fact that, if ρ be the radius vector to the point P,

$$OT = \rho^2/a.$$

If P is on the nth turn of the spiral, the moving straight line will have moved through an angle $2(n-1)\pi + \theta$, say.

Hence $$\rho = a\{2(n-1)\pi + \theta\},$$

and $$OT = \rho^2/a = \rho\{2(n-1)\pi + \theta\}.$$

Archimedes's way of expressing this is to say (Prop. 20) that, if p be the circumference of the circle with radius $OP\,(=\rho)$, and if this circle cut the initial line in the point K,

$$OT = (n-1)\,p + \text{arc}\,KP \text{ measured 'forward' from } K \text{ to } P.$$

If P is the end of the nth turn, this reduces to

$$OT = n \text{ (circumf. of circle with radius } OP\text{)},$$

and, if P is the end of the first turn in particular,

$$OT = \text{(circumf. of circle with radius } OP\text{)}. \quad \text{(Prop. 19.)}$$

The spiral can thus be used for the rectification of any circle. And the quadrature follows directly from *Measurement of a Circle*, Prop. 1.

(γ) *Solutions by Apollonius and Carpus.*

Iamblichus says that Apollonius himself called the curve by means of which he squared the circle 'sister of the cochloid'. What this curve was is uncertain. As the passage goes on to say that it was really 'the same as the (curve) of Nicomedes', and the quadratrix has just been mentioned as the curve used by Nicomedes, some have supposed the 'sister of the cochloid' (or conchoid) to be the *quadratrix*, but this seems highly improbable. There is, however, another possibility. Apollonius is known to have written a regular treatise on the *Cochlias*, which was the cylindrical helix.[1] It is conceivable that he might call the *cochlias* the 'sister of the *cochloid*' on the ground of the similarity of the names, if not of the curves. And, as a matter of fact, the drawing of a tangent to the helix enables the circular section of the cylinder to be squared. For, if a plane be drawn at right angles to the axis of the cylinder through the initial position of the moving radius which describes the helix, and if we project on this plane the portion of the tangent at any point of the helix intercepted between the point and the plane, the projection is equal to an arc of the circular section of the cylinder subtended by an angle at the centre equal to the angle through which the plane through the axis and the moving radius has turned from its original position. And this squaring by means of what we may call the 'subtangent' is sufficiently parallel to the use by Archimedes of the polar subtangent to the spiral for the same purpose to make the hypothesis attractive.

Nothing whatever is known of Carpus's curve 'of double motion'. Tannery thought it was the cycloid; but there is no evidence for this.

(δ) *Approximations to the value of π.*

As we have seen, Archimedes, by inscribing and circumscribing regular polygons of 96 sides, and calculating their perimeters respectively, obtained the approximation $3\frac{1}{7} > \pi > 3\frac{10}{71}$ (*Measurement of a Circle*, Prop. 3). But we now learn[2] that, in a work on *Plinthides and Cylinders*, he made a nearer approximation still. Unfortunately the figures as they stand in the Greek text are incorrect, the lower limit being given as the ratio of $\overset{\kappa\alpha}{\mu},\alpha\omega o\epsilon$ to $\overset{\varsigma}{\mu},\zeta\upsilon\mu\alpha$, or 211875 : 67441 (= 3·141635), and the higher limit as the ratio of $\overset{\iota\theta}{\mu},\zeta\omega\pi\eta$ to $\overset{\varsigma}{\mu},\beta\tau\nu\alpha$ or 197888 : 62351 (= 3·17377), so that the lower limit

[1] Pappus, viii, p. 1110. 20; Proclus on Eucl. I, p. 105. 5.
[2] Heron, *Metrica*, i. 26, p. 66. 13–17.

as given is greater than the true value, and the higher limit is greater than the earlier upper limit $3\frac{1}{7}$. Slight corrections by Tannery ($\overset{\kappa\alpha}{\mu}{,}\alpha\omega o\beta$ for $\overset{\kappa\alpha}{\mu}{,}\alpha\omega o\epsilon$ and $\overset{\iota\theta}{\mu}{,}\epsilon\omega\pi\beta$ for $\overset{\iota\theta}{\mu}{,}\zeta\omega\pi\eta$) give better figures, namely

$$\frac{195882}{62351} > \pi > \frac{211872}{67441}$$

or
$$3\cdot1416016 > \pi > 3\cdot1415904....$$

Another suggestion[1] is to correct $\overset{\varsigma}{\mu}{,}\zeta\upsilon\mu\alpha$ into $\overset{\varsigma}{\mu}{,}\zeta\upsilon\mu\delta$ and $\overset{\iota\theta}{\mu}{,}\zeta\omega\pi\eta$ into $\overset{\iota\theta}{\mu}{,}\epsilon\omega\pi\eta$, giving

$$\frac{195888}{62351} > \pi > \frac{211875}{67444}$$

or
$$3\cdot141697... > \pi > 3\cdot141495....$$

If either suggestion represents the true reading, the mean between the two limits gives the same remarkably close approximation $3\cdot141596$.

Ptolemy[2] gives a value for the ratio of the circumference of a circle to its diameter expressed thus in sexagesimal fractions, $\gamma\ \eta\ \lambda$, i.e. $3 + \frac{8}{60} + \frac{30}{60^2}$ or $3\cdot1416$. He observes that this is almost exactly the mean between the Archimedean limits $3\frac{1}{7}$ and $3\frac{10}{71}$. It is, however, more exact than this mean, and Ptolemy no doubt obtained his value independently. He had the basis of the calculation ready to hand in his Table of Chords. This Table gives the lengths of the chords of a circle subtended by arcs of $\frac{1}{2}°$, $1°$, $1\frac{1}{2}°$, and so on by half degrees. The chords are expressed in terms of 120th parts of the length of the diameter. If one such part be denoted by 1^p, the chord subtended by an arc of $1°$ is given by the Table in terms of this unit and sexagesimal fractions of it thus, $1^p\, 2'\, 50''$. Since an angle of $1°$ at the centre subtends a side of the regular polygon of 360 sides inscribed in the circle, the perimeter of this polygon is 360 times $1^p\, 2'\, 50''$ or, since $1^p = 1/120$th of the diameter, the perimeter of the polygon expressed in terms of the diameter is 3 times $1\ 2'\, 50''$, that is $3\ 8'\, 30''$, which is Ptolemy's figure for π.

[1] J. L. Heiben in *Nordisk Tidsskrift for Filologi*, 3e Sér. xx. Fasc. 1–2.
[2] Ptolemy, *Syntaxis*, vi. 7, p. 513. 1–5, Heib.

There is evidence of a still closer calculation than Ptolemy's due to some Greek whose name we do not know. The Indian mathematician Aryabhaṭṭa (born A.D. 476) says in his *Lessons in Calculation*:

'To 100 add 4; multiply the sum by 8; add 62000 more and thus (we have), for a diameter of 2 myriads, the approximate length of the circumference of the circle';

that is, he gives $\frac{62832}{20000}$ or 3·1416 as the value of π. But the way in which he expresses it points indubitably to a Greek source, 'for the Greeks alone of all peoples made the myriad the unit of the second order' (Rodet).

This brings us to the notice at the end of Eutocius's commentary on the *Measurement of a Circle* of Archimedes, which records[1] that other mathematicians made similar approximations, though it does not give their results.

'It is to be observed that Apollonius of Perga solved the same problem in his Ὠκυτόκιον ("means of quick delivery"), using other numbers and making the approximation closer [than that of Archimedes]. While Apollonius's figures seem to be more accurate, they do not serve the purpose which Archimedes had in view; for, as we said, his object in this book was to find an approximate figure suitable for use in daily life. Hence we cannot regard as appropriate the censure of Sporus of Nicaea, who seems to charge Archimedes with having failed to determine with accuracy (the length of) the straight line which is equal to the circumference of the circle, to judge by the passage in his *Keria* where Sporus observes that his own teacher, meaning Philon of Gadara, reduced (the matter) to more exact numerical expression than Archimedes did, I mean in his $\frac{1}{7}$ and $\frac{10}{71}$; in fact people seem, one after the other, to have failed to appreciate Archimedes's object. They have also used multiplications and divisions of myriads, a method not easy to follow for any one who has not gone through a course of Magnus's *Logistica*.'

It is possible that, as Apollonius used myriads, 'second myriads', 'third myriads', &c., as orders of integral numbers, he may have worked with the fractions $\frac{1}{10000}, \frac{1}{10000^2}$, &c.;

[1] Archimedes, ed. Heib., vol. iii, pp. 258–9.

in any case Magnus (apparently later than Sporus, and therefore perhaps belonging to the fourth or fifth century A. D.) would seem to have written an exposition of such a method, which, as Eutocius indicates, must have been very much more troublesome than the method of sexagesimal fractions used by Ptolemy.

The Trisection of any Angle.

This problem presumably arose from attempts to continue the construction of regular polygons after that of the pentagon had been discovered. The trisection of an angle would be necessary in order to construct a regular polygon the sides of which are nine, or any multiple of nine, in number. A regular polygon of seven sides, on the other hand, would no doubt be constructed with the help of the first discovered method of dividing any angle in a given ratio, i.e. by means of the *quadratrix*. This method covered the case of trisection, but other more practicable ways of effecting this particular construction were in due time evolved.

We are told that the ancients attempted, and failed, to solve the problem by 'plane' methods, i.e. by means of the straight line and circle; they failed because the problem is not 'plane' but 'solid'. Moreover, they were not yet familiar with conic sections, and so were at a loss; afterwards, however, they succeeded in trisecting an angle by means of conic sections, a method to which they were led by the reduction of the problem to another, of the kind known as νεύσεις (*inclinationes*, or *vergings*).[1]

(α) *Reduction to a certain νεῦσις, solved by conics.*

The reduction is arrived at by the following analysis. It is only necessary to deal with the case where the given angle to be trisected is acute, since a right angle can be trisected by drawing an equilateral triangle.

Let ABC be the given angle, and let AC be drawn perpendicular to BC. Complete the parallelogram $ACBF$, and produce the side FA to E.

[1] Pappus, iv, p. 272. 7–14.

Suppose E to be such a point that, if BE be joined meeting AC in D, the intercept DE between AC and AE is equal to $2\,AB$.

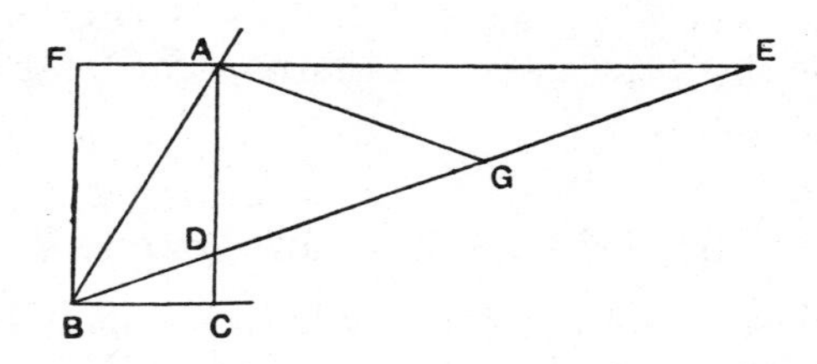

Bisect DE at G, and join AG.

Then $$DG = GE = AG = AB.$$

Therefore $$\angle ABG = \angle AGB = 2\angle AEG$$

$$= 2\angle DBC, \text{ since } FE, BC \text{ are parallel.}$$

Hence $$\angle DBC = \tfrac{1}{3}\angle ABC,$$

and the angle ABC is trisected by BE.

Thus the problem is reduced to *drawing BE from B to cut AC and AE in such a way that the intercept $DE = 2\,AB$.*

In the phraseology of the problems called νεύσεις the problem is to insert a straight line ED of given length $2\,AB$ between AE and AC in such a way that ED *verges* towards B.

Pappus shows how to solve this problem in a more general form. Given a parallelogram $ABCD$ (which need not be rectangular, as Pappus makes it), to draw AEF to meet CD and BC produced in points E and F such that EF has a given length.

Suppose the problem solved, EF being of the given length.

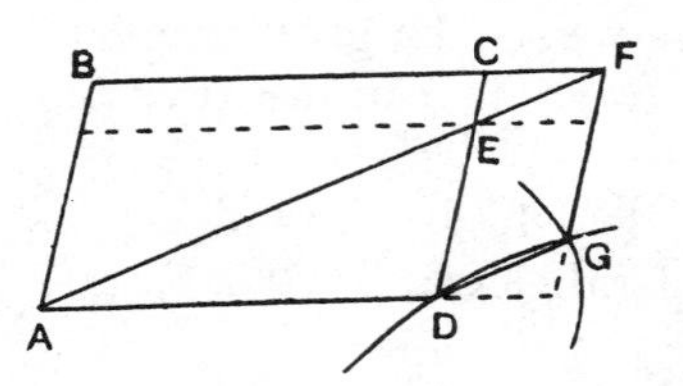

Complete the parallelogram $EDGF$.

Then, EF being given in length, DG is given in length.

Therefore G lies on a circle with centre D and radius equal to the given length.

Again, by the help of Eucl. I. 43 relating to the complements of the parallelograms about the diagonal of the complete parallelogram, we see that

$$BC \,.\, CD = BF \,.\, ED$$
$$= BF \,.\, FG.$$

Consequently G lies on a hyperbola with BF, BA as asymptotes and passing through D.

Thus, in order to effect the construction, we have only to draw this hyperbola as well as the circle with centre D and radius equal to the given length. Their intersection gives the point G, and E, F are then determined by drawing GF parallel to DC to meet BC produced in F and joining AF.

(β) *The νεῦσις equivalent to a cubic equation.*

It is easily seen that the solution of the νεῦσις is equivalent to the solution of a cubic equation. For in the first figure on p. 236, if FA be the axis of x, FB the axis of y, $FA = a$, $FB = b$, the solution of the problem by means of conics as Pappus gives it is the equivalent of finding a certain point as the intersection of the conics

$$xy = ab,$$
$$(x-a)^2 + (y-b)^2 = 4\,(a^2+b^2).$$

The second equation gives

$$(x+a)\,(x-3\,a) = (y+b)\,(3\,b-y).$$

From the first equation it is easily seen that

$$(x+a):(y+b) = a:y,$$

and that $$(x-3\,a)\,y = a(b-3\,y)\,;$$

therefore, eliminating x, we have

$$a^2(b-3\,y) = y^2(3\,b-y),$$

or $$y^3 - 3\,by^2 - 3\,a^2y + a^2b = 0.$$

Now suppose that $\angle ABC = \theta$, so that $\tan\theta = b/a$;

and suppose that $$t = \tan DBC,$$

so that $$y = at.$$

We have then

$$a^3t^3 - 3\,ba^2t^2 - 3\,a^3t + a^2b = 0,$$

or $$at^3 - 3bt^2 - 3at + b = 0,$$

whence $$b(1-3t^2) = a(3t-t^3),$$

or $$\tan\theta = \frac{b}{a} = \frac{3t-t^3}{1-3t^2},$$

so that, by the well-known trigonometrical formula,

$$t = \tan\tfrac{1}{3}\theta;$$

that is, BD trisects the angle ABC.

(γ) *The Conchoids of Nicomedes.*

Nicomedes invented a curve for the specific purpose of solving such νεύσεις as the above. His date can be fixed with sufficient accuracy by the facts (1) that he seems to have criticized unfavourably Eratosthenes's solution of the problem of the two mean proportionals or the duplication of the cube, and (2) that Apollonius called a certain curve the 'sister of the cochloid', evidently out of compliment to Nicomedes. Nicomedes must therefore have been about intermediate between Eratosthenes (a little younger than Archimedes, and therefore born about 280 B.C.) and Apollonius (born probably about 264 B.C.).

The curve is called by Pappus the *cochloid* (κοχλοειδὴς γραμμή), and this was evidently the original name for it; later, e.g. by Proclus, it was called the *conchoid* (κογχοειδὴς γραμμή). There were varieties of the cochloidal curves; Pappus speaks of the 'first', 'second', 'third' and 'fourth', observing that the 'first' was used for trisecting an angle and duplicating the cube, while the others were useful for other investigations.[1] It is the 'first' which concerns us here. Nicomedes constructed it by means of a mechanical device which may be described thus.[2] AB is a ruler with a slot in it parallel to its length, FE a second ruler fixed at right angles to the first, with a peg C fixed in it. A third ruler PC pointed at P has a slot in it parallel to its length which fits the peg C. D is a fixed peg on PC in a straight line with the slot, and D can move freely along the slot in AB. If then the ruler PC moves so that the peg D describes the

[1] Pappus, iv, p. 244. 18–20. [2] *Ib.*, pp. 242–4.

length of the slot in AB on each side of F, the extremity P of the ruler describes the curve which is called a conchoid or cochloid. Nicomedes called the straight line AB the *ruler*

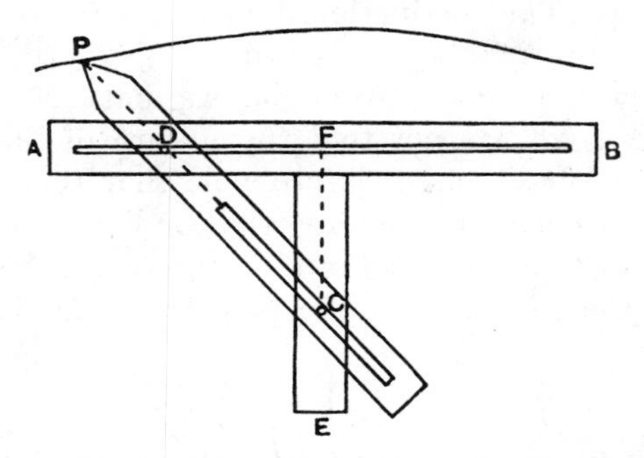

(κανών), the fixed point C the *pole* (πόλος), and the constant length PD the *distance* (διάστημα).

The fundamental property of the curve, which in polar coordinates would now be denoted by the equation

$$r = a + b \sec \theta,$$

is that, if any radius vector be drawn from C to the curve, as CP, the length intercepted on the radius vector between the curve and the straight line AB is constant. Thus any νεῦσις in which one of the two given lines (between which the straight line of given length is to be placed) is a straight line can be solved by means of the intersection of the other line with a certain conchoid having as its pole the fixed point to which the inserted straight line must *verge* (νεύειν). Pappus tells us that in practice the conchoid was not always actually drawn but that 'some', for greater convenience, moved a ruler about the fixed point until by trial the intercept was found to be equal to the given length.[1]

In the figure above (p. 236) showing the reduction of the trisection of an angle to a νεῦσις the conchoid to be used would have B for its *pole*, AC for the '*ruler*' or *base*, a length equal to $2AB$ for its *distance*; and E would be found as the intersection of the conchoid with FA produced.

Proclus says that Nicomedes gave the construction, the order, and the properties of the conchoidal lines[2]; but nothing

[1] Pappus, iv, p. 246. 15.

[2] Proclus on Eucl. I, p. 272. 3–7.

of his treatise has come down to us except the construction of the 'first' conchoid, its fundamental property, and the fact that the curve has the *ruler* or *base* as an asymptote in each direction. The distinction, however, drawn by Pappus between the 'first', 'second', 'third' and 'fourth' conchoids may well have been taken from the original treatise, directly or indirectly. We are not told the nature of the conchoids other than the 'first', but it is probable that they were three other curves produced by varying the conditions in the figure. Let a be the distance or fixed intercept between the curve and the base, b the distance of the pole from the base. Then

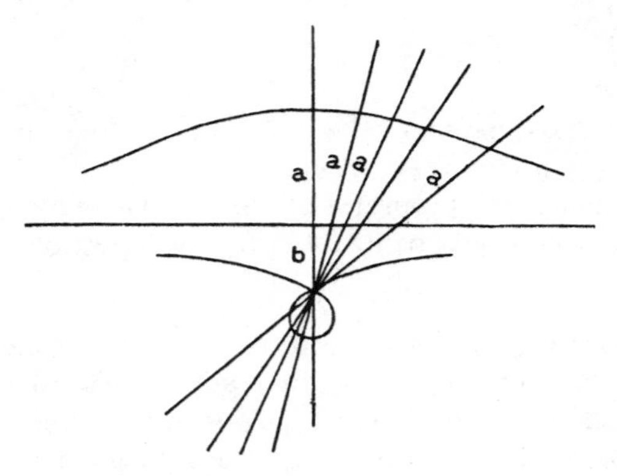

clearly, if along each radius vector drawn through the pole we measure a backwards from the base towards the pole, we get a conchoidal figure on the side of the base towards the pole. This curve takes three forms according as a is greater than, equal to, or less than b. Each of them has the base for asymptote, but in the first of the three cases the curve has a loop as shown in the figure, in the second case it has a cusp at the pole, in the third it has no double point. The most probable hypothesis seems to be that the other three cochloidal curves mentioned by Pappus are these three varieties.

(δ) *Another reduction to a* νεῦσις (*Archimedes*).

A proposition leading to the reduction of the trisection of an angle to another νεῦσις is included in the collection of Lemmas (*Liber Assumptorum*) which has come to us under the name of Archimedes through the Arabic. Though the Lemmas cannot have been written by Archimedes in their present form, because his name is quoted in them more than once, it is probable that some of them are of Archimedean origin, and especially is this the case with Prop. 8, since the *νεῦσις* suggested by it is of very much the same kind as those the solution of which is assumed in the treatise *On Spirals*, Props. 5–8. The proposition is as follows.

If AB be any chord of a circle with centre O, and AB be produced to C so that BC is equal to the radius, and if CO meet the circle in D, E, then the arc AE will be equal to three times the arc BD.

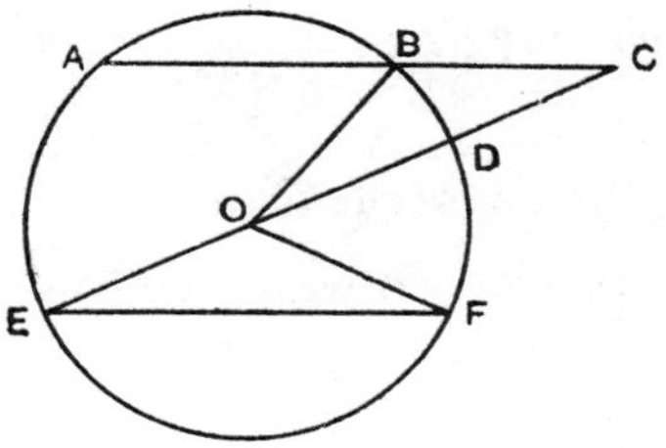

Draw the chord EF parallel to AB, and join OB, OF.

Since $BO = BC$,

$$\angle BOC = \angle BCO.$$

Now $\qquad \angle COF = 2 \angle OEF,$

$$= 2 \angle BCO, \text{ by parallels,}$$

$$= 2 \angle BOC.$$

Therefore $\qquad \angle BOF = 3 \angle BOD,$

and $\qquad (\text{arc } BF) = (\text{arc } AE) = 3\,(\text{arc } BD).$

By means of this proposition we can reduce the trisection of the arc AE to a *νεῦσις*. For, in order to find an arc which is one-third of the arc AE, we have only to draw through A a straight line ABC meeting the circle again in B and EO produced in C, and such that BC is equal to the radius of the circle.

(ε) *Direct solutions by means of conics.*

Pappus gives two solutions of the trisection problem in which conics are applied directly without any preliminary reduction of the problem to a *νεῦσις*.[1]

1. The analysis leading to the first method is as follows.

Let AC be a straight line, and B a point without it such that, if BA, BC be joined, the angle BCA is double of the angle BAC.

[1] Pappus, iv, pp. 282–4.

Draw BD perpendicular to AC, and cut off DE along DA equal to DC. Join BE.

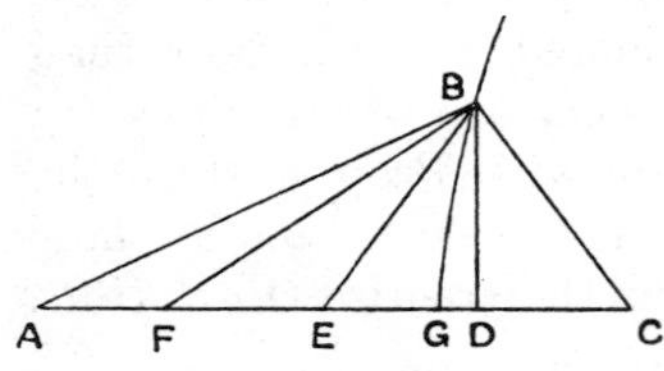

Then, since $BE = BC$,

$$\angle BEC = BCE.$$

But $\angle BEC = \angle BAE + \angle EBA$,

and, by hypothesis,

$$\angle BCA = 2\angle BAE.$$

Therefore $\qquad \angle BAE + \angle EBA = 2\angle BAE;$

therefore $\qquad \angle BAE = \angle ABE,$

or $\qquad AE = BE.$

Divide AC at G so that $AG = 2\,GC$, or $CG = \frac{1}{3}AC$.

Also let FE be made equal to ED, so that $CD = \frac{1}{3}CF$.

It follows that $GD = \frac{1}{3}(AC - CF) = \frac{1}{3}AF$.

Now
$$BD^2 = BE^2 - ED^2$$
$$= BE^2 - EF^2.$$

Also
$$DA\,.\,AF = AE^2 - EF^2 \qquad \text{(Eucl. II. 6)}$$
$$= BE^2 - EF^2.$$

Therefore
$$BD^2 = DA\,.\,AF$$
$$= 3\,AD\,.\,DG, \text{ from above,}$$

so that
$$BD^2 : AD\,.\,DG = 3 : 1$$
$$= 3\,AG^2 : AG^2.$$

Hence D lies on a hyperbola with AG as transverse axis and with conjugate axis equal to $\sqrt{3}\,.\,AG$.

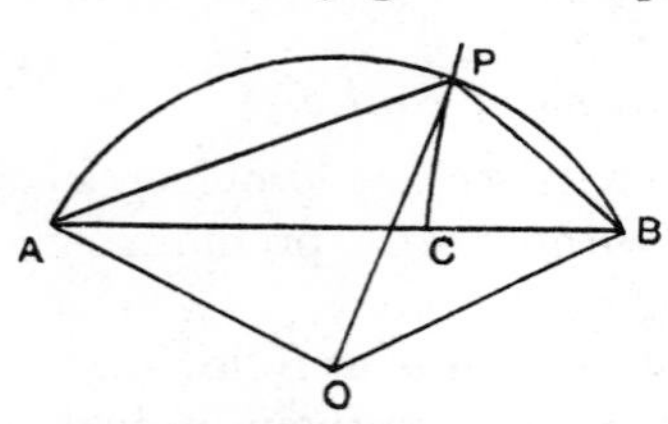

Now suppose we are required to trisect an arc AB of a circle with centre O.

Draw the chord AB, divide it at C so that $AC = 2\,CB$, and construct the hyperbola which has AC for transverse axis and a straight line equal to $\sqrt{3}\,.\,AC$ for conjugate axis.

Let the hyperbola meet the circular arc in P. Join PA, PO, PB.

Then, by the above proposition,

$$\angle PBA = 2\angle PAB.$$

Therefore their doubles are equal,

or $$\angle POA = 2\angle POB,$$

and OP accordingly trisects the arc APB and the angle AOB.

2. 'Some', says Pappus, set out another solution not involving recourse to a νεῦσις, as follows.

Let RPS be an arc of a circle which it is required to trisect.

Suppose it done, and let the arc SP be one-third of the arc SPR.

Join RP, SP.

Then the angle RSP is equal to twice the angle SRP.

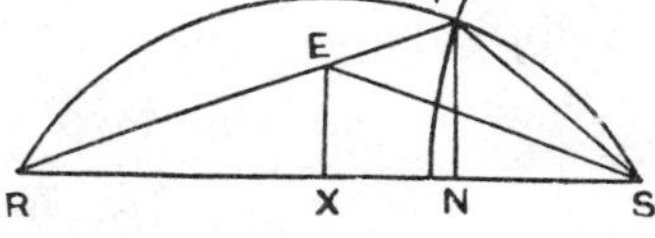

Let SE bisect the angle RSP, meeting RP in E, and draw EX, PN perpendicular to RS.

Then $\angle ERS = \angle ESR$, so that $RE = ES$.

Therefore $RX = XS$, and X is given.

Again $RS : SP = RE : EP = RX : XN$;

therefore $$RS : RX = SP : NX.$$

But $$RS = 2RX;$$

therefore $$SP = 2NX.$$

It follows that P lies on a hyperbola with S as focus and XE as directrix, and with eccentricity 2.

Hence, in order to trisect the arc, we have only to bisect RS at X, draw XE at right angles to RS, and then draw a hyperbola with S as focus, XE as directrix, and 2 as the eccentricity. The hyperbola is the same as that used in the first solution.

The passage of Pappus from which this solution is taken is remarkable as being one of three passages in Greek mathematical works still extant (two being in Pappus and one in a fragment of Anthemius on burning mirrors) which refer to the focus-and-directrix property of conics. The second passage in Pappus comes under the heading of Lemmas to the *Surface-Loci* of Euclid.[1] Pappus there gives a complete proof of the

[1] Pappus, vii, pp. 1004–1114.

theorem that, *if the distance of a point from a fixed point is in a given ratio to its distance from a fixed line, the locus of the point is a conic section which is an ellipse, a parabola, or a hyperbola according as the given ratio is less than, equal to, or greater than, unity.* The importance of these passages lies in the fact that the Lemma was required for the understanding of Euclid's treatise. We can hardly avoid the conclusion that the property was used by Euclid in his *Surface-Loci*, but was assumed as well known. It was, therefore, probably taken from some treatise current in Euclid's time, perhaps from Aristaeus's work on *Solid Loci*.

The Duplication of the Cube, or the problem of the two mean proportionals.

(α) *History of the problem.*

In his commentary on Archimedes, *On the Sphere and Cylinder*, II. 1, Eutocius has preserved for us a precious collection of solutions of this famous problem.[1] One of the solutions is that of Eratosthenes, a younger contemporary of Archimedes, and it is introduced by what purports to be a letter from Eratosthenes to Ptolemy. This was Ptolemy Euergetes, who at the beginning of his reign (245 B.C.) persuaded Eratosthenes to come from Athens to Alexandria to be tutor to his son (Philopator). The supposed letter gives the tradition regarding the origin of the problem and the history of its solution up to the time of Eratosthenes. Then, after some remarks on its usefulness for practical purposes, the author describes the construction by which Eratosthenes himself solved it, giving the proof of it at some length and adding directions for making the instrument by which the construction could be effected in practice. Next he says that the mechanical contrivance represented by Eratosthenes was, 'in the votive monument', actually of bronze, and was fastened on with lead close under the στεφάνη of the pillar. There was, further, on the pillar the proof in a condensed form, with one figure, and, at the end, an epigram. The supposed letter of Eratosthenes is a forgery, but the author rendered a real service

[1] Archimedes, ed. Heib., vol. iii, pp. 54. 26–106. 24.

by actually quoting the proof and the epigram, which are the genuine work of Eratosthenes.

Our document begins with the story that an ancient tragic poet had represented Minos as putting up a tomb to Glaucus but being dissatisfied with its being only 100 feet each way; Minos was then represented as saying that it must be made double the size, by increasing each of the dimensions in that ratio. Naturally the poet 'was thought to have made a mistake'. Von Wilamowitz has shown that the verses which Minos is made to say cannot have been from any play by Aeschylus, Sophocles, or Euripides. They are the work of some obscure poet, and the ignorance of mathematics shown by him is the only reason why they became notorious and so survived. The letter goes on to say that

'Geometers took up the question and sought to find out how one could double a given solid while keeping the same shape; the problem took the name of "the duplication of the cube" because they started from a cube and sought to double it. For a long time all their efforts were vain; then Hippocrates of Chios discovered for the first time that, if we can devise a way of finding two mean proportionals in continued proportion between two straight lines the greater of which is double of the less, the cube will be doubled; that is, one puzzle (ἀπόρημα) was turned by him into another not less difficult. After a time, so goes the story, certain Delians, who were commanded by the oracle to double a certain altar, fell into the same quandary as before.'

At this point the versions of the story diverge somewhat. The pseudo-Eratosthenes continues as follows:

'They therefore sent over to beg the geometers who were with Plato in the Academy to find them the solution. The latter applying themselves diligently to the problem of finding two mean proportionals between two given straight lines, Archytas of Taras is said to have found them by means of a half cylinder, and Eudoxus by means of the so-called curved lines; but, as it turned out, all their solutions were theoretical, and no one of them was able to give a practical construction for ordinary use, save to a certain small extent Menaechmus, and that with difficulty.'

Fortunately we have Eratosthenes's own version in a quotation by Theon of Smyrna:

'Eratosthenes in his work entitled *Platonicus* relates that, when the god proclaimed to the Delians by the oracle that, if they would get rid of a plague, they should construct an altar double of the existing one, their craftsmen fell into great perplexity in their efforts to discover how a solid could be made double of a (similar) solid; they therefore went to ask Plato about it, and he replied that the oracle meant, not that the god wanted an altar of double the size, but that he wished, in setting them the task, to shame the Greeks for their neglect of mathematics and their contempt for geometry.' [1]

Eratosthenes's version may well be true; and there is no doubt that the question was studied in the Academy, solutions being attributed to Eudoxus, Menaechmus, and even (though erroneously) to Plato himself. The description by the pseudo-Eratosthenes of the three solutions by Archytas, Eudoxus and Menaechmus is little more than a paraphrase of the lines about them in the genuine epigram of Eratosthenes,

'Do not seek to do the difficult business of the cylinders of Archytas, or to cut the cones in the triads of Menaechmus, or to draw such a curved form of lines as is described by the god-fearing Eudoxus.'

The different versions are reflected in Plutarch, who in one place gives Plato's answer to the Delians in almost the same words as Eratosthenes,[2] and in another place tells us that Plato referred the Delians to Eudoxus and Helicon of Cyzicus for a solution of the problem.[3]

After Hippocrates had discovered that the duplication of the cube was equivalent to finding two mean proportionals in continued proportion between two given straight lines, the problem seems to have been attacked in the latter form exclusively. The various solutions will now be reproduced in chronological order.

(β) *Archytas.*

The solution of Archytas is the most remarkable of all, especially when his date is considered (first half of fourth century B.C.), because it is not a construction in a plane but a bold

[1] Theon of Smyrna, p. 2. 3-12.
[2] Plutarch, *De E apud Delphos*, c. 6, 386 E.
[3] *De genio Socratis*, c. 7, 579 C, D.

construction in three dimensions, determining a certain point as the intersection of three surfaces of revolution, (1) a right cone, (2) a cylinder, (3) a *tore* or anchor-ring with inner diameter *nil*. The intersection of the two latter surfaces gives (says Archytas) a certain curve (which is in fact a curve

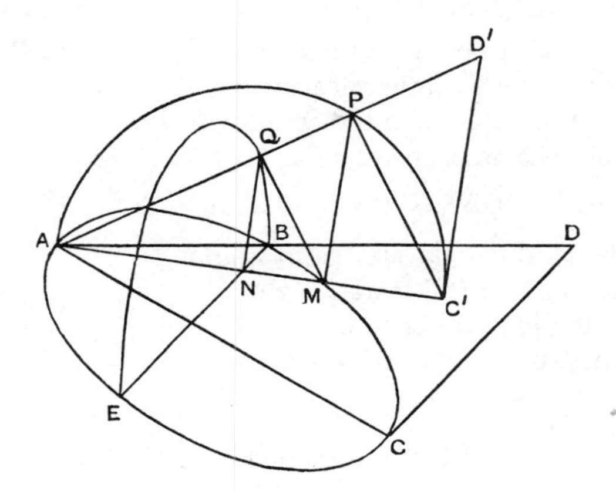

of double curvature), and the point required is found as the point in which the cone meets this curve.

Suppose that AC, AB are the two straight lines between which two mean proportionals are to be found, and let AC be made the diameter of a circle and AB a chord in it.

Draw a semicircle with AC as diameter, but in a plane at right angles to the plane of the circle ABC, and imagine this semicircle to revolve about a straight line through A perpendicular to the plane of ABC (thus describing half a *tore* with inner diameter *nil*).

Next draw a right half-cylinder on the semicircle ABC as base; this will cut the surface of the half-*tore* in a certain curve.

Lastly let CD, the tangent to the circle ABC at the point C, meet AB produced in D; and suppose the triangle ADC to revolve about AC as axis. This will generate the surface of a right circular cone; the point B will describe a semicircle BQE at right angles to the plane of ABC and having its diameter BE at right angles to AC; and the surface of the cone will meet in some point P the curve which is the intersection of the half-cylinder and the half-*tore*.

Let APC' be the corresponding position of the revolving semicircle, and let AC' meet the circumference ABC in M.

Drawing PM perpendicular to the plane of ABC, we see that it must meet the circumference of the circle ABC because P is on the cylinder which stands on ABC as base.

Let AP meet the circumference of the semicircle BQE in Q, and let AC' meet its diameter in N. Join PC', QM, QN.

Then, since both semicircles are perpendicular to the plane ABC, so is their line of intersection QN [Eucl. XI. 19].

Therefore QN is perpendicular to BE.

Therefore $$QN^2 = BN \,.\, NE = AN \,.\, NM, \quad \text{[Eucl. III. 35]}$$

so that the angle AQM is a right angle.

But the angle APC' is also right;
therefore MQ is parallel to $C'P$.

It follows, by similar triangles, that

$$C'A : AP = AP : AM = AM : AQ;$$

that is, $$AC : AP = AP : AM = AM : AB,$$

and AB, AM, AP, AC are in continued proportion, so that AM, AP are the two mean proportionals required.

In the language of analytical geometry, if AC is the axis of x, a line through A perpendicular to AC in the plane of ABC the axis of y, and a line through A parallel to PM the axis of z, then P is determined as the intersection of the surfaces

$$(1) \qquad x^2 + y^2 + z^2 = \frac{a^2}{b^2} x^2, \qquad \text{(the cone)}$$

$$(2) \qquad x^2 + y^2 = ax, \qquad \text{(the cylinder)}$$

$$(3) \qquad x^2 + y^2 + z^2 = a \sqrt{(x^2 + y^2)}, \qquad \text{(the } tore\text{)}$$

where $$AC = a, \quad AB = b.$$

From the first two equations we obtain

$$x^2 + y^2 + z^2 = (x^2 + y^2)^2 / b^2,$$

and from this and (3) we have

$$\frac{a}{\sqrt{(x^2 + y^2 + z^2)}} = \frac{\sqrt{(x^2 + y^2 + z^2)}}{\sqrt{(x^2 + y^2)}} = \frac{\sqrt{(x^2 + y^2)}}{b},$$

or $$AC : AP = AP : AM = AM : AB.$$

Compounding the ratios, we have

$$AC : AB = (AM : AB)^3;$$

therefore the cube of side AM is to the cube of side AB as AC is to AB.

In the particular case where $AC = 2AB$, $AM^3 = 2AB^3$, and the cube is doubled.

(γ) *Eudoxus.*

Eutocius had evidently seen some document purporting to give Eudoxus's solution, but it is clear that it must have been an erroneous version. The epigram of Eratosthenes says that Eudoxus solved the problem by means of lines of a 'curved or bent form' (καμπύλον εἶδος ἐν γραμμαῖς). According to Eutocius, while Eudoxus said in his preface that he had discovered a solution by means of 'curved lines', yet, when he came to the proof, he made no use of such lines, and further he committed an obvious error in that he treated a certain discrete proportion as if it were continuous.[1] It may be that, while Eudoxus made use of what was really a curvilinear locus, he did not actually draw the whole curve but only indicated a point or two upon it sufficient for his purpose. This may explain the first part of Eutocius's remark, but in any case we cannot believe the second part; Eudoxus was too accomplished a mathematician to make any confusion between a discrete and a continuous proportion. Presumably the mistake which Eutocius found was made by some one who wrongly transcribed the original; but it cannot be too much regretted, because it caused Eutocius to omit the solution altogether from his account.

Tannery[2] made an ingenious suggestion to the effect that Eudoxus's construction was really adapted from that of Archytas by what is practically projection on the plane of the circle ABC in Archytas's construction. It is not difficult to represent the projection on that plane of the curve of intersection between the cone and the *tore*, and, when this curve is drawn in the plane ABC, its intersection with the circle ABC itself gives the point M in Archytas's figure.

[1] Archimedes, ed. Heib., vol. iii, p. 56. 4–8.
[2] Tannery, *Mémoires scientifiques*, vol. i, pp. 53–61.

The projection on the plane ABC of the intersection between the cone and the *tore* is seen, by means of their equations (1) and (3) above, to be

$$x^2 = \frac{b^2}{a}\sqrt{(x^2+y^2)},$$

or, in polar coordinates referred to A as origin and AC as axis,

$$\rho = \frac{b^2}{a\cos^2\theta}.$$

It is easy to find any number of points on the curve. Take the circle ABC, and let AC the diameter and AB a chord

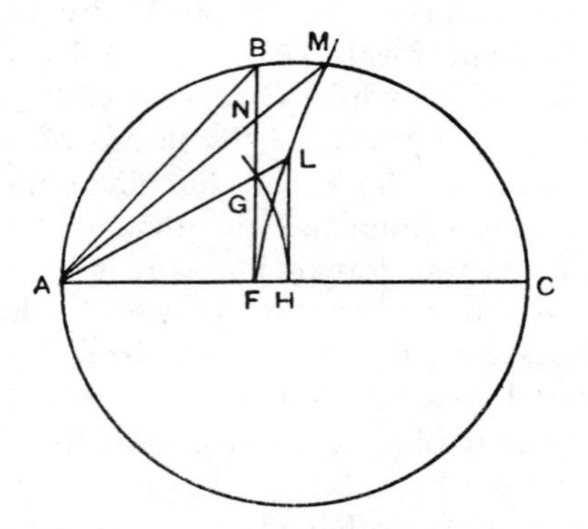

be the two given straight lines between which two mean proportionals have to be found.

With the above notation

$$AC = a, \quad AB = b;$$

and, if BF be drawn perpendicular to AC,

$$AB^2 = AF \,.\, AC,$$

or

$$AF = b^2/a.$$

Take any point G on BF and join AG.

Then, if $\angle GAF = \theta, \quad AG = AF\sec\theta.$

With A as centre and AG as radius draw a circle meeting AC in H, and draw HL at right angles to AC, meeting AG produced in L.

Then $\quad AL = AH \sec\theta = AG \sec\theta = AF \sec^2\theta.$

That is, if $\rho = AL, \quad \rho = \dfrac{b^2}{a} \sec^2\theta,$

and L is a point on the curve.

Similarly any number of other points on the curve may be found. If the curve meets the circle ABC in M, the length AM is the same as that of AM in the figure of Archytas's solution.

And AM is the first of the two mean proportionals between AB and AC. The second ($= AP$ in the figure of Archytas's solution) is easily found from the relation $AM^2 = AB \,.\, AP$, and the problem is solved.

It must be admitted that Tannery's suggestion as to Eudoxus's method is attractive; but of course it is only a conjecture. To my mind the objection to it is that it is too close an adaptation of Archytas's ideas. Eudoxus was, it is true, a pupil of Archytas, and there is a good deal of similarity of character between Archytas's construction of the curve of double curvature and Eudoxus's construction of the spherical lemniscate by means of revolving concentric spheres; but Eudoxus was, I think, too original a mathematician to content himself with a mere adaptation of Archytas's method of solution.

(δ) *Menaechmus.*

Two solutions by Menaechmus of the problem of finding two mean proportionals are described by Eutocius; both find a certain point as the intersection between two conics, in the one case two parabolas, in the other a parabola and a rectangular hyperbola. The solutions are referred to in Eratosthenes's epigram: 'do not', says Eratosthenes, 'cut the cone in the triads of Menaechmus.' From the solutions coupled with this remark it is inferred that Menaechmus was the discoverer of the conic sections.

Menaechmus, brother of Dinostratus, who used the *quadratrix* to square the circle, was a pupil of Eudoxus and flourished about the middle of the fourth century B.C. The most attractive form of the story about the geometer and the king who wanted a short cut to geometry is told of Menaechmus and

Alexander: 'O king,' said Menaechmus, 'for travelling over the country there are royal roads and roads for common citizens, but in geometry there is one road for all.'[1] A similar story is indeed told of Euclid and Ptolemy; but there would be a temptation to transfer such a story at a later date to the more famous mathematician. Menaechmus was evidently a considerable mathematician; he is associated by Proclus with Amyclas of Heraclea, a friend of Plato, and with Dinostratus as having 'made the whole of geometry more perfect'.[2] Beyond, however, the fact that the discovery of the conic sections is attributed to him, we have very few notices relating to his work. He is mentioned along with Aristotle and Callippus as a supporter of the theory of concentric spheres invented by Eudoxus, but as postulating a larger number of spheres.[3] We gather from Proclus that he wrote on the technology of mathematics; he discussed for instance the difference between the broader meaning of the word *element* (in which any proposition leading to another may be said to be an element of it) and the stricter meaning of something simple and fundamental standing to consequences drawn from it in the relation of a *principle*, which is capable of being universally applied and enters into the proof of all manner of propositions.[4] Again, he did not agree in the distinction between theorems and problems, but would have it that they were all *problems*, though directed to two different objects[5]; he also discussed the important question of the convertibility of theorems and the conditions necessary to it.[6]

If x, y are two mean proportionals between straight lines a, b,

that is, if $$a : x = x : y = y : b,$$

then clearly $$x^2 = ay,\ y^2 = bx, \text{ and } xy = ab.$$

It is easy for us to recognize here the Cartesian equations of two parabolas referred to a diameter and the tangent at its extremity, and of a hyperbola referred to its asymptotes. But Menaechmus appears to have had not only to recognize,

[1] Stobaeus, *Eclogae*, ii. 31, 115 (vol. ii, p. 228. 30, Wachsmuth).
[2] Proclus on Eucl. I, p. 67. 9.
[3] Theon of Smyrna, pp. 201. 22–202. 2.
[4] Proclus on Eucl. I, pp. 72. 23–73. 14.
[5] *Ib.*, p. 78. 8–13.
[6] *Ib.*, p. 254. 4–5.

but to discover, the existence of curves having the properties corresponding to the Cartesian equations. He discovered them in plane sections of right circular cones, and it would doubtless be the properties of the *principal* ordinates in relation to the abscissae on the axes which he would arrive at first. Though only the parabola and the hyperbola are wanted for the particular problem, he would certainly not fail to find the ellipse and its property as well. But in the case of the hyperbola he needed the property of the curve with reference to the *asymptotes,* represented by the equation $xy = ab$; he must therefore have discovered the existence of the asymptotes, and must have proved the property, at all events for the rectangular hyperbola. The original method of discovery of the conics will occupy us later. In the meantime it is obvious that the use of any two of the curves $x^2 = ay$, $y^2 = bx$, $xy = ab$ gives the solution of our problem, and it was in fact the intersection of the second and third which Menaechmus used in his first solution, while for his second solution he used the first two. Eutocius gives the analysis and synthesis of each solution in full. I shall reproduce them as shortly as possible, only suppressing the use of four separate lines representing the two given straight lines and the two required means in the figure of the first solution.

First solution.

Suppose that AO, OB are two given straight lines of which $AO > OB$, and let them form a right angle at O.

Suppose the problem solved, and let the two mean proportionals be OM measured along BO produced and ON measured along AO produced. Complete the rectangle $OMPN$.

Then, since $AO:OM = OM:ON = ON:OB$,

we have (1) $$OB.OM = ON^2 = PM^2,$$

so that P lies on a parabola which has O for vertex, OM for axis, and OB for *latus rectum*;

and (2) $$AO.OB = OM.ON = PN.PM,$$

so that P lies on a hyperbola with O as centre and OM, ON as asymptotes.

Accordingly, to find the point P, we have to construct (1) a parabola with O as vertex, OM as axis, and *latus rectum* equal to OB,

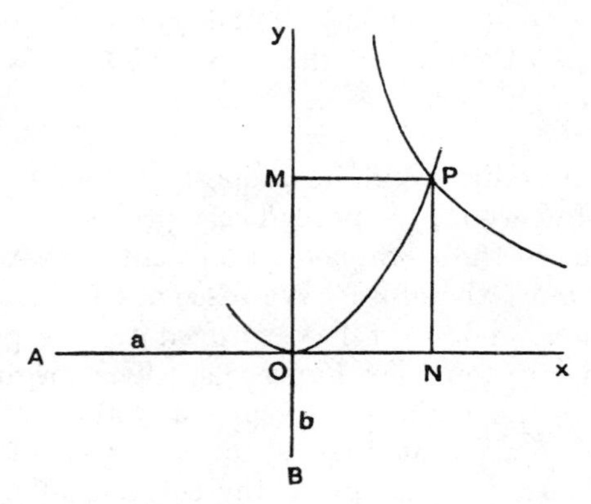

(2) a hyperbola with asymptotes OM, ON and such that the rectangle contained by straight lines PM, PN drawn from any point P on the curve parallel to one asymptote and meeting the other is equal to the rectangle $AO \,.\, OB$.

The intersection of the parabola and hyperbola gives the point P which solves the problem, for

$$AO : PN = PN : PM = PM : OB.$$

Second solution.

Supposing the problem solved, as in the first case, we have, since $$AO : OM = OM : ON = ON : OB,$$

(1) the relation $OB \,.\, OM = ON^2 = PM^2$,

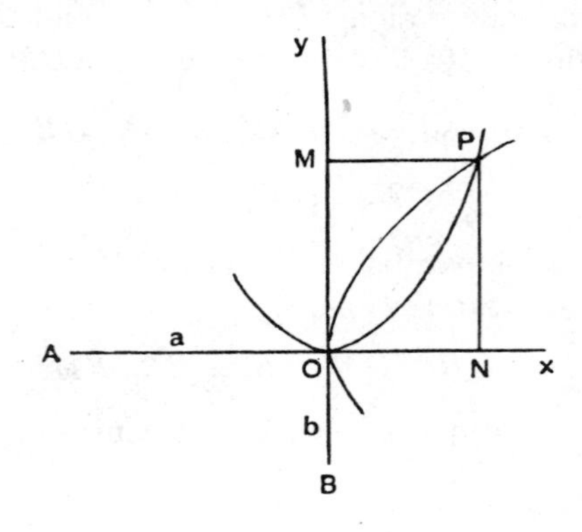

so that P lies on a parabola which has O for vertex, OM for axis, and OB for *latus rectum*,

(2) the similar relation $AO \,.\, ON = OM^2 = PN^2$,

so that P lies on a parabola which has O for vertex, ON for axis, and OA for *latus rectum*.

In order therefore to find P, we have only to construct the two parabolas with OM, ON for axes and OB, OA for *latera recta* respectively; the intersection of the two parabolas gives a point P such that

$$AO : PN = PN : PM = PM : OB,$$

and the problem is solved.

(We shall see later on that Menaechmus did not use the names *parabola* and *hyperbola* to describe the curves, those names being due to Apollonius.)

(ϵ) *The solution attributed to Plato.*

This is the first in Eutocius's arrangement of the various solutions reproduced by him. But there is almost conclusive reason for thinking that it is wrongly attributed to Plato. No one but Eutocius mentions it, and there is no reference to it in Eratosthenes's epigram, whereas, if a solution by Plato had then been known, it could hardly fail to have been mentioned along with those of Archytas, Menaechmus, and Eudoxus. Again, Plutarch says that Plato told the Delians that the problem of the two mean proportionals was no easy one, but that Eudoxus or Helicon of Cyzicus would solve it for them; he did not apparently propose to attack it himself. And, lastly, the solution attributed to him is mechanical, whereas we are twice told that Plato objected to mechanical solutions as destroying the good of geometry.[1] Attempts have been made to reconcile the contrary traditions. It is argued that, while Plato objected to mechanical solutions on principle, he wished to show how easy it was to discover such solutions and put forward that attributed to him as an illustration of the fact. I prefer to treat the silence of Eratosthenes as conclusive on the point, and to suppose that the solution was invented in the Academy by some one contemporary with or later than Menaechmus.

[1] Plutarch, *Quaest. Conviv.* 8. 2. 1, p. 718 E, F; *Vita Marcelli*, c. 14. 5.

For, if we look at the figure of Menaechmus's second solution, we shall see that the given straight lines and the two means between them are shown in cyclic order (clockwise) as straight lines radiating from O and separated by right angles. This is exactly the arrangement of the lines in 'Plato's' solution. Hence it seems probable that some one who had Menaechmus's second solution before him wished to show how the same representation of the four straight lines could be got by a mechanical construction as an alternative to the use of conics.

Drawing the two given straight lines with the means, that is to say, OA, OM, ON, OB, in cyclic clockwise order, as in Menaechmus's second solution, we have

$$AO : OM = OM : ON = ON : OB,$$

and it is clear that, if AM, MN, NB are joined, the angles AMN, MNB are both right angles. The problem then is, given OA, OB at right angles to one another, to contrive the rest of the figure so that the angles at M, N are right.

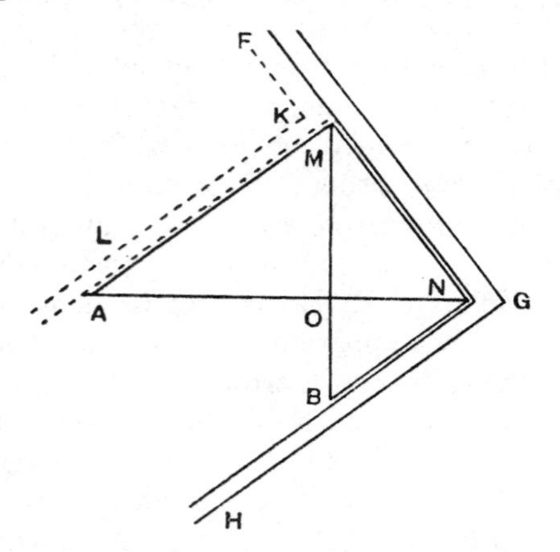

The instrument used is somewhat like that which a shoemaker uses to measure the length of the foot. FGH is a rigid right angle made, say, of wood. KL is a strut which, fastened, say, to a stick KF which slides along GF, can move while remaining always parallel to GH or at right angles to GF.

Now place the rigid right angle FGH so that the leg GH passes through B, and turn it until the angle G lies on AO

produced. Then slide the movable strut KL, which remains always parallel to GH, until its edge (towards GH) passes through A. If now the inner angular point between the strut KL and the leg FG does not lie on BO produced, the machine has to be turned again and the strut moved until the said point does lie on BO produced, as M, care being taken that during the whole of the motion the inner edges of KL and HG pass through A, B respectively and the inner angular point at G moves along AO produced.

That it is possible for the machine to take up the desired position is clear from the figure of Menaechmus, in which MO, NO are the means between AO and BO and the angles AMN, MNB are right angles, although to get it into the required position is perhaps not quite easy.

The matter may be looked at analytically thus. Let us take any other position of the machine in which the strut and the leg GH pass through A, B respectively, while G lies on AO produced, but P, the angular point between the strut KL and

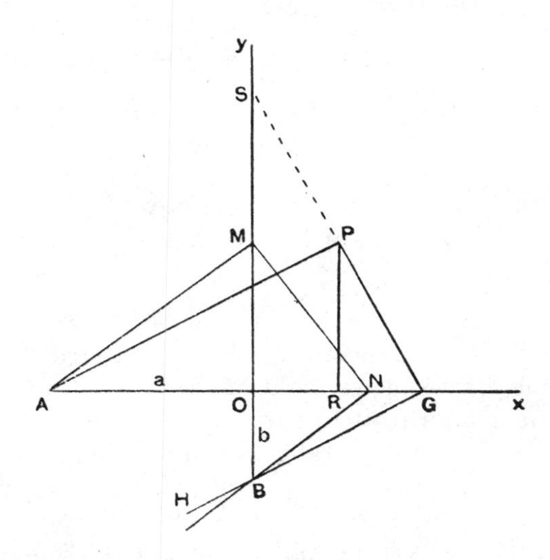

the leg FG, does not lie on OM produced. Take ON, OM as the axes of x, y respectively. Draw PR perpendicular to OG, and produce GP to meet OM produced in S.

Let $$AO = a, \quad BO = b, \quad OG = r.$$

Then $$AR \,.\, RG = PR^2,$$

or $$(a+x)(r-x) = y^2. \qquad (1)$$

Also, by similar triangles,

$$PR : RG = SO : OG$$
$$= OG : OB;$$

or $$\frac{y}{r-x} = \frac{r}{b}. \qquad (2)$$

From the equation (1) we obtain

$$r = \frac{x^2+y^2+ax}{a+x},$$

and, by multiplying (1) and (2), we have

$$by(a+x) = ry^2,$$

whence, substituting the value of r, we obtain, as the locus of P, a curve of the third degree,

$$b(a+x)^2 = y(x^2+y^2+ax).$$

The intersection (M) of this curve with the axis of y gives

$$OM^3 = a^2b.$$

As a theoretical solution, therefore, 'Plato's' solution is more difficult than that of Menaechmus.

(ζ) *Eratosthenes.*

This is also a mechanical solution effected by means of three plane figures (equal right-angled triangles or rectangles) which can move parallel to one another and to their original positions between two parallel rulers forming a sort of frame and fitted with grooves so arranged that the figures can move over one another. Pappus's account makes the figures triangles,[1] Eutocius has parallelograms with diagonals drawn; triangles seem preferable. I shall use the lettering of Eutocius for the second figure so far as it goes, but I shall use triangles instead of rectangles.

[1] Pappus, iii, pp. 56–8.

Suppose the frame bounded by the parallels AX, EY. The

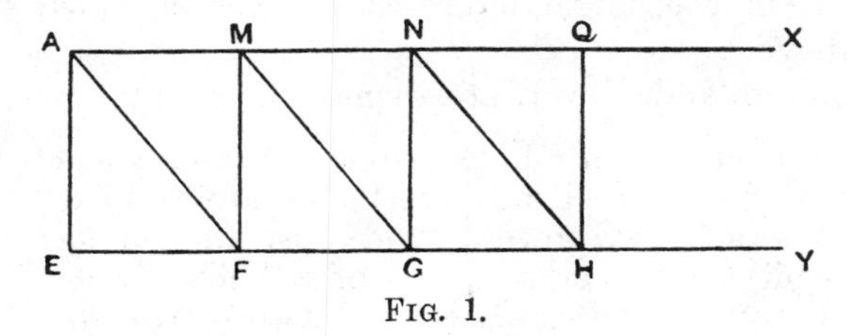

Fig. 1.

initial position of the triangles is that shown in the first figure, where the triangles are AMF, MNG, NQH.

In the second figure the straight lines AE, DH which are

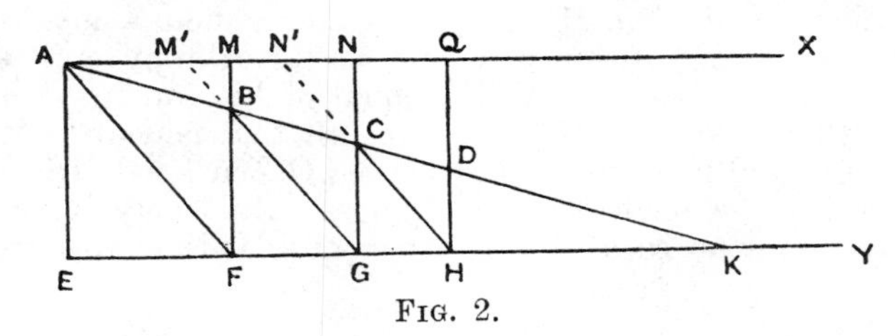

Fig. 2.

parallel to one another are those between which two mean proportionals have to be found.

In the second figure the triangles (except AMF, which remains fixed) are moved parallel to their original positions towards AMF so that they overlap (as AMF, $M'NG$, $N'QH$), NQH taking the position $N'QH$ in which QH passes through D, and MNG a position $M'NG$ such that the points B, C where MF, $M'G$ and NG, $N'H$ respectively intersect are in a straight line with A, D.

Let AD, EH meet in K.

Then $$EK : KF = AK : KB$$
$$= FK : KG,$$

and $EK : KF = AE : BF$, while $FK : KG = BF : CG$;

therefore $$AE : BF = BF : CG.$$

Similarly $$BF : CG = CG : DH,$$

so that AE, BF, CG, DH are in continued proportion, and BF, CG are the required mean proportionals.

This is substantially the short proof given in Eratosthenes's inscription on the column; the construction was left to be inferred from the single figure which corresponded to the second above.

The epigram added by Eratosthenes was as follows:

'If, good friend, thou mindest to obtain from a small (cube) a cube double of it, and duly to change any solid figure into another, this is in thy power; thou canst find the measure of a fold, a pit, or the broad basin of a hollow well, by this method, that is, if thou (thus) catch between two rulers (two) means with their extreme ends converging.[1] Do not thou seek to do the difficult business of Archytas's cylinders, or to cut the cone in the triads of Menaechmus, or to compass such a curved form of lines as is described by the god-fearing Eudoxus. Nay thou couldst, on these tablets, easily find a myriad of means, beginning from a small base. Happy art thou, Ptolemy, in that, as a father the equal of his son in youthful vigour, thou hast thyself given him all that is dear to Muses and Kings, and may he in the future,[2] O Zeus, god of heaven, also receive the sceptre at thy hands. Thus may it be, and let any one who sees this offering say "This is the gift of Eratosthenes of Cyrene".'

(η) *Nicomedes.*

The solution by Nicomedes was contained in his book on conchoids, and, according to Eutocius, he was inordinately proud of it, claiming for it much superiority over the method of Eratosthenes, which he derided as being impracticable as well as ungeometrical.

Nicomedes reduced the problem to a νεῦσις which he solved by means of the conchoid. Both Pappus and Eutocius explain the method (the former twice over[3]) with little variation.

Let AB, BC be the two straight lines between which two means are to be found. Complete the parallelogram $ABCL$.

Bisect AB, BC in D and E.

Join LD, and produce it to meet CB produced in G.

Draw EF at right angles to BC and of such length that $CF = AD$.

Join GF, and draw CH parallel to it.

[1] Lit. 'converging with their extreme ends' (τέρμασιν ἄκροις συνδρομάδας).

[2] Reading with v. Wilamowitz ὃ δ' ἐς ὕστερον.

[3] Pappus, iii, pp. 58. 23–62. 13; iv, pp. 246. 20–250. 25.

Then from the point F draw FHK cutting CH and EC produced in H and K in such a way that the intercept $HK = CF = AD$.

(This is done by means of a conchoid constructed with F as pole, CH as 'ruler', and 'distance' equal to AD or CF. This

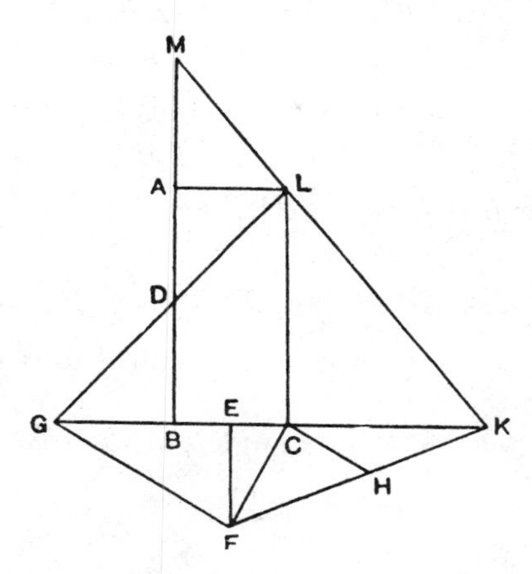

conchoid meets EC produced in a point K. We then join FK and, by the property of the conchoid, HK = the 'distance'.)

Join KL, and produce it to meet BA produced in M.

Then shall CK, MA be the required mean proportionals.

For, since BC is bisected at E and produced to K,

$$BK \,.\, KC + CE^2 = EK^2.$$

Add EF^2 to each;

therefore $$BK \,.\, KC + CF^2 = KF^2. \quad (1)$$

Now, by parallels, $MA : AB = ML : LK$

$$= BC : CK.$$

But $AB = 2\,AD$, and $BC = \frac{1}{2}GC$;

therefore $$MA : AD = GC : CK$$

$$= FH : HK,$$

and, *componendo*, $$MD : DA = FK : HK.$$

But, by construction, $DA = HK$;

therefore $MD = FK$, and $MD^2 = FK^2$.

Now $$MD^2 = BM \,.\, MA + DA^2,$$
while, by (1), $$FK^2 = BK \,.\, KC + CF^2;$$
therefore $$BM \,.\, MA + DA^2 = BK \,.\, KC + CF^2.$$

But $DA = CF$; therefore $BM \,.\, MA = BK \,.\, KC$.

Therefore $$CK : MA = BM : BK$$
$$= LC : CK;$$
while, at the same time, $BM : BK = MA : AL$.

Therefore $$LC : CK = CK : MA = MA : AL,$$
or $$AB : CK = CK : MA = MA : BC.$$

(θ) *Apollonius, Heron, Philon of Byzantium.*

I give these solutions together because they really amount to the same thing.[1]

Let AB, AC, placed at right angles, be the two given straight

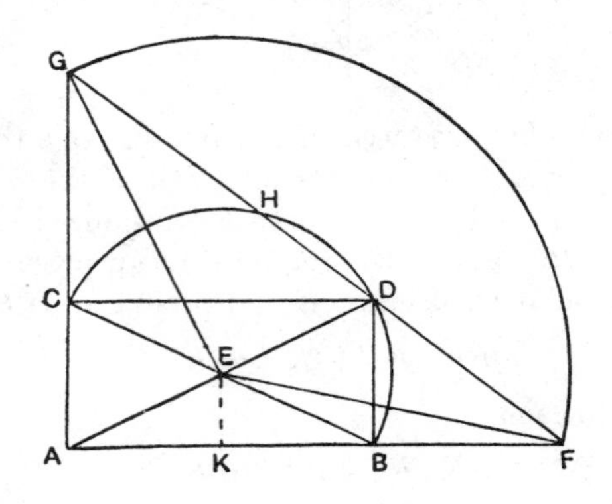

lines. Complete the rectangle $ABDC$, and let E be the point at which the diagonals bisect one another.

Then a circle with centre E and radius EB will circumscribe the rectangle $ABDC$.

Now (Apollonius) draw with centre E a circle cutting AB, AC produced in F, G but such that F, D, G are in one straight line.

Or (Heron) place a ruler so that its edge passes through D,

[1] Heron's solution is given in his *Mechanics* (i. 11) and *Belopoeica*, and is reproduced by Pappus (iii, pp. 62–4) as well as by Eutocius (loc. cit.).

and move it about D until the edge intersects AB, AC produced in points (F, G) which are equidistant from E.

Or (Philon) place a ruler so that it passes through D and turn it round D until it cuts AB, AC produced and the circle about $ABDC$ in points F, G, H such that the intercepts FD, HG are equal.

Clearly all three constructions give the same points F, G. For in Philon's construction, since $FD = HG$, the perpendicular from E on DH, which bisects DH, must also bisect FG, so that $EF = EG$.

We have first to prove that $AF . FB = AG . GC$.

(a) With Apollonius's and Heron's constructions we have, if K be the middle point of AB,

$$AF . FB + BK^2 = FK^2.$$

Add KE^2 to both sides;

therefore $$AF . FB + BE^2 = EF^2.$$

Similarly $$AG . GC + CE^2 = EG^2.$$

But $BE = CE$, and $EF = EG$;

therefore $$AF . FB = AG . GC.$$

(b) With Philon's construction, since $GH = FD$,

$$HF . FD = DG . GH.$$

But, since the circle $BDHC$ passes through A,

$$HF . FD = AF . FB, \text{ and } DG . GH = AG . GC;$$

therefore $$AF . FB = AG . GC.$$

Therefore $$FA : AG = CG : FB.$$

But, by similar triangles,

$$FA : AG = DC : CG, \text{ and also } = FB : BD;$$

therefore $$DC : CG = CG : FB = FB : BD,$$

or $$AB : CG = CG : FB = FB : AC.$$

The connexion between this solution and that of Menaechmus can be seen thus. We saw that, if $a : x = x : y = y : b$,

$$x^2 = ay, \quad y^2 = bx, \quad xy = ab,$$

which equations represent, in Cartesian coordinates, two parabolas and a hyperbola. Menaechmus in effect solved the problem of the two mean proportionals by means of the points of intersection of any two of these conics.

But, if we add the first two equations, we have

$$x^2+y^2-bx-ay = 0,$$

which is a circle passing through the points common to the two parabolas $x^2 = ay$, $y^2 = bx$.

Therefore we can equally obtain a solution by means of the intersections of the circle $x^2+y^2-bx-ay = 0$ and the rectangular hyperbola $xy = ab$.

This is in effect what Philon does, for, if AF, AG are the coordinate axes, the circle $x^2+y^2-bx-ay = 0$ is the circle $BDHC$, and $xy = ab$ is the rectangular hyperbola with AF, AG as asymptotes and passing through D, which hyperbola intersects the circle again in H, a point such that $FD = HG$.

(ι) *Diocles and the cissoid.*

We gather from allusions to the cissoid in Proclus's commentary on Eucl. I that the curve which Geminus called by that name was none other than the curve invented by Diocles and used by him for doubling the cube or finding two mean proportionals. Hence Diocles must have preceded Geminus (fl. 70 B.C.). Again, we conclude from the two fragments preserved by Eutocius of a work by him, περὶ πυρείων, *On burning-mirrors*, that he was later than Archimedes and Apollonius. He may therefore have flourished towards the end of the second century or at the beginning of the first century B.C. Of the two fragments given by Eutocius one contains a solution by means of conics of the problem of dividing a sphere by a plane in such a way that the volumes of the resulting segments shall be in a given ratio—a problem equivalent to the solution of a certain cubic equation—while the other gives the solution of the problem of the two mean proportionals by means of the cissoid.

Suppose that AB, DC are diameters of a circle at right angles to one another. Let E, F be points on the quadrants BD, BC respectively such that the arcs BE, BF are equal.

Draw EG, FH perpendicular to DC. Join CE, and let P be the point in which CE, FH intersect.

The cissoid is the locus of all the points P corresponding to different positions of E on the quadrant BD and of F at an equal distance from B on the quadrant BC.

If P is any point found by the above construction, it is

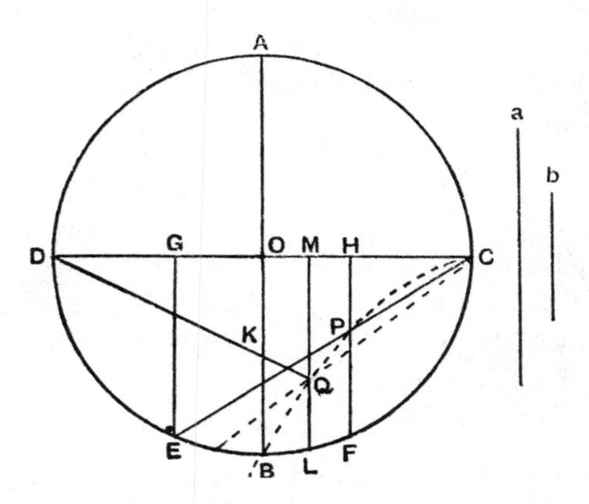

required to prove that FH, HC are two mean proportionals in continued proportion between DH and HP, or that

$$DH : HF = HF : HC = HC : HP.$$

Now it is clear from the construction that $EG = FH$, $DG = HC$, so that $CG : GE \equiv DH : HF$.

And, since FH is a mean proportional between DH, HC,

$$DH : HF = HF : CH.$$

But, by similar triangles,

$$CG : GE = CH : HP.$$

It follows that

$$DH : HF = HF : CH = CH : HP,$$

or FH, HC are the two mean proportionals between DH, HP.

[Since $DH . HP = HF . CH$, we have, if a is the radius of the circle and if $OH = x$, $HP = y$, or (in other words) if we use OC, OB as axes of coordinates,

$$(a+x)y = \sqrt{(a^2-x^2)} . (a-x)$$

or

$$y^2(a+x) = (a-x)^3,$$

which is the Cartesian equation of the curve. It has a cusp at C, and the tangent to the circle at D is an asymptote to it.]

Suppose now that the cissoid has been drawn as shown by the dotted line in the figure, and that we are required to find two mean proportionals between two straight lines a, b.

Take the point K on OB such that $DO:OK = a:b$.

Join DK, and produce it to meet the cissoid in Q.

Through Q draw the ordinate LM perpendicular to DC.

Then, by the property of the cissoid, LM, MC are the two mean proportionals between DM, MQ. And

$$DM:MQ = DO:OK = a:b.$$

In order, then, to obtain the two mean proportionals between a and b, we have only to take straight lines which bear respectively the same ratio to DM, LM, MC, MQ as a bears to DM. The extremes are then a, b, and the two mean proportionals are found.

(κ) *Sporus and Pappus.*

The solutions of Sporus and Pappus are really the same as that of Diocles, the only difference being that, instead of using the cissoid, they use a ruler which they turn about a certain point until certain intercepts which it cuts off between two pairs of lines are equal.

In order to show the identity of the solutions, I shall draw Sporus's figure with the same lettering as above for corresponding points, and I shall add dotted lines to show the additional auxiliary lines used by Pappus.[1] (Compared with my figure, Sporus's is the other way up, and so is Pappus's where it occurs in his own *Synagoge*, though not in Eutocius.)

Sporus was known to Pappus, as we have gathered from Pappus's reference to his criticisms on the *quadratrix*, and it is not unlikely that Sporus was either Pappus's master or a fellow-student of his. But when Pappus gives (though in better form, if we may judge by Eutocius's reproduction of Sporus) the same solution as that of Sporus, and calls it a solution καθ' ἡμᾶς, he clearly means 'according to my method', not '*our* method', and it appears therefore that he claimed the credit of it for himself.

Sporus makes DO, OK (at right angles to one another) the actual given straight lines; Pappus, like Diocles, only takes

[1] Pappus, iii, pp. 64–8; viii, pp. 1070–2.

them in the same proportion as the given straight lines. Otherwise the construction is the same.

A circle being drawn with centre O and radius DO, we join DK and produce it to meet the circle in I.

Now conceive a ruler to pass through C and to be turned about C until it cuts DI, OB and the circumference of the

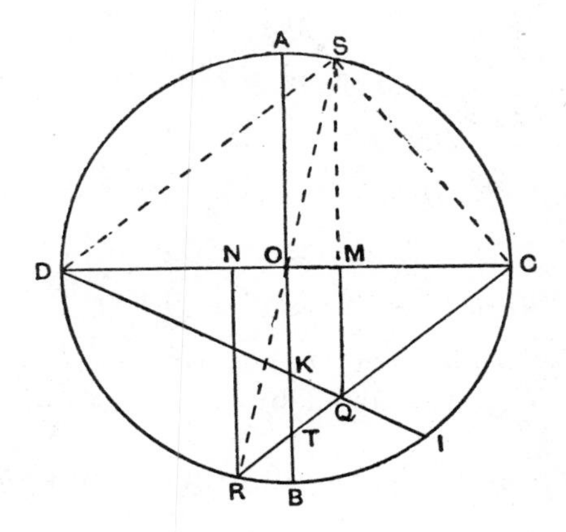

circle in points Q, T, R such that $QT = TR$. Draw QM, RN perpendicular to DC.

Then, since $QT = TR$, $MO = ON$, and MQ, NR are equidistant from OB. Therefore in reality Q lies on the cissoid of Diocles, and, as in the first part of Diocles's proof, we prove (since RN is equal to the ordinate through Q, the foot of which is M) that

$$DM : RN = RN : MC = MC : MQ,$$

and we have the two means between DM, MQ, so that we can easily construct the two means between DO, OK.

But Sporus actually proves that the first of the two means between DO and OK is OT. This is obvious from the above relations, because

$$RN : OT = CN : CO = DM : DO = MQ : OK.$$

Sporus has an *ab initio* proof of the fact, but it is rather confused, and Pappus's proof is better worth giving, especially as it includes the actual duplication of the cube.

It is required to prove that $DO : OK = DO^3 : OT^3$.

Join RO, and produce it to meet the circle at S. Join DS, SC.

Then, since $RO = OS$ and $RT = TQ$, SQ is parallel to AB and meets OC in M.

Now

$$DM : MC = SM^2 : MC^2 = CM^2 : MQ^2 \text{ (since } \angle RCS \text{ is right).}$$

Multiply by the ratio $CM : MQ$;

therefore $(DM : MC) . (CM : MQ) = (CM^2 : MQ^2) . (CM : MQ)$

or $$DM : MQ = CM^3 : MQ^3.$$

But $$DM : MQ = DO : OK,$$

and $$CM : MQ = CO : OT.$$

Therefore $DO : OK = CO^3 : OT^3 = DO^3 : OT^3$.

Therefore OT is the first of the two mean proportionals to DO, OK; the second is found by taking a third proportional to DO, OT.

And a cube has been increased in any given ratio.

(λ) *Approximation to a solution by plane methods only.*

There remains the procedure described by Pappus and criticized by him at length at the beginning of Book III of his *Collection*.[1] It was suggested by some one 'who was thought to be a great geometer', but whose name is not given. Pappus maintains that the author did not understand what he was about, 'for he claimed that he was in possession of a method of finding two mean proportionals between two straight lines by means of plane considerations only'; he gave his construction to Pappus to examine and pronounce upon, while Hierius the philosopher and other friends of his supported his request for Pappus's opinion. The construction is as follows.

Let the given straight lines be AB, AD placed at right angles to one another, AB being the greater.

Draw BC parallel to AD and equal to AB. Join CD meeting BA produced in E. Produce BC to L, and draw EL' through E parallel to BL. Along CL cut off lengths CF, FG, GK, KL,

[1] Pappus, iii, pp. 30–48.

each of which is equal to BC. Draw CC', FF', GG', KK', LL' parallel to BA.

On LL', KK' take LM, KR equal to BA, and bisect LM in N.

Take P, Q on LL' such that $L'L$, $L'N$, $L'P$, $L'Q$ are in continued proportion; join QR, RL, and through N draw NS parallel to QR meeting RL in S.

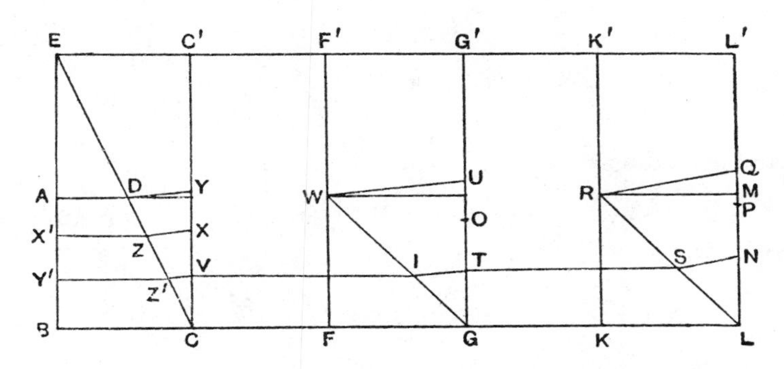

Draw ST parallel to BL meeting GG' in T.

To $G'G$, $G'T$ take continued proportionals $G'O$, $G'U$, as before. Take W on FF' such that $FW = BA$, join UW, WG, and through T draw TI parallel to UW meeting WG in I.

Through I draw IV parallel to BC meeting CC' in V.

Take continued proportionals $C'C$, $C'V$, $C'X$, $C'Y$, and draw XZ, VZ' parallel to YD meeting EC in Z, Z'. Lastly draw ZX', $Z'Y'$ parallel to BC.

Then, says the author, it is required to prove that ZX', $Z'Y'$ are two mean proportionals in continued proportion between AD, BC.

Now, as Pappus noticed, the supposed conclusion is clearly not true unless DY is parallel to BC, which in general it is not. But what Pappus failed to observe is that, if the operation of taking the continued proportionals as described is repeated, not three times, but an infinite number of times, the length of the line $C'Y$ tends continually towards equality with EA. Although, therefore, by continuing the construction we can never exactly determine the required means, the method gives an endless series of approximations tending towards the true lengths of the means.

Let $LL' = BE = a$, $AB = b$, $L'N = \alpha$ (for there is no necessity to take N at the middle point of LM).

Then $$L'Q = \alpha^3/a^2,$$

therefore $$LQ = (a^3 - \alpha^3)/a^2.$$

And $$\frac{TG}{RK} = \frac{SL}{RL} = \frac{NL}{QL} = \frac{(a-\alpha)a^2}{a^3-\alpha^3};$$

therefore $$TG = \frac{(a-\alpha)a^2b}{a^3-\alpha^3},$$

and accordingly $$G'T = a - \frac{(a-\alpha)a^2b}{a^3-\alpha^3}.$$

Now let α_n be the length corresponding to $G'T$ after n operations; then it is clear that

$$a - \alpha_{n+1} = \frac{(a-\alpha_n)a^2b}{a^3-\alpha_n{}^3}.$$

α_n must approach some finite limit when $n = \infty$. Taking ξ as this limit, we have

$$a - \xi = \frac{(a-\xi)a^2b}{a^3-\xi^3},$$

and, $\xi = a$ not being a root of this equation, we get at once

$$\xi^3 = a^3 - a^2b = a^2(a-b).$$

Therefore, ultimately $C'V$ is one of the mean proportionals between EA and EB, whence $Y'Z'$ will be one of the mean proportionals between AD, BC, that is, between AD and AB.

The above was pointed out for the first time by R. Pendlebury,[1] and I have followed his way of stating the matter.

[1] *Messenger of Mathematics*, ser. 2, vol. ii (1873), pp. 166–8.

VIII

ZENO OF ELEA

We have already seen how the consideration of the subject of infinitesimals was forced upon the Greek mathematicians so soon as they came to close grips with the problem of the quadrature of the circle. Antiphon the Sophist was the first to indicate the correct road upon which the solution was to be found, though he expressed his idea in a crude form which was bound to provoke immediate and strong criticism from logical minds. Antiphon had inscribed a series of successive regular polygons in a circle, each of which had double as many sides as the preceding, and he asserted that, by continuing this process, we should at length exhaust the circle: 'he thought that in this way the area of the circle would sometime be used up and a polygon would be inscribed in the circle the sides of which on account of their smallness would coincide with the circumference.'[1] Aristotle roundly said that this was a fallacy which it was not even necessary for a geometer to trouble to refute, since an expert in any science is not called upon to refute *all* fallacies, but only those which are false deductions from the admitted principles of the science; if the fallacy is based on anything which is in contradiction to any of those principles, it may at once be ignored.[2] Evidently therefore, in Aristotle's view, Antiphon's argument violated some 'geometrical principle', whether this was the truth that a straight line, however short, can never coincide with an arc of a circle, or the principle assumed by geometers that geometrical magnitudes can be divided *ad infinitum*.

But Aristotle is only a representative of the criticisms directed against the ideas implied in Antiphon's argument; those ideas had already, as early as the time of Antiphon

[1] Simpl. *in Arist. Phys.*, p. 55. 6 Diels.
[2] Arist. *Phys.* i. 2, 185 a 14–17.

himself (a contemporary of Socrates), been subjected to a destructive criticism expressed with unsurpassable piquancy and force. No wonder that the subsequent course of Greek geometry was profoundly affected by the arguments of Zeno on motion. Aristotle indeed called them 'fallacies', without being able to refute them. The mathematicians, however, knew better, and, realizing that Zeno's arguments were fatal to infinitesimals, they saw that they could only avoid the difficulties connected with them by once for all banishing the idea of the infinite, even the potentially infinite, altogether from their science; thenceforth, therefore, they made no use of magnitudes increasing or diminishing *ad infinitum*, but contented themselves with finite magnitudes that can be made as great or as small *as we please*.[1] If they used infinitesimals at all, it was only as a tentative means of *discovering* propositions; they *proved* them afterwards by rigorous geometrical methods. An illustration of this is furnished by the *Method* of Archimedes. In that treatise Archimedes finds (*a*) the areas of curves, and (*b*) the volumes of solids, by treating them respectively as the sums of an infinite number (*a*) of parallel *lines*, i.e. infinitely narrow strips, and (*b*) of parallel *planes*, i.e. infinitely thin laminae; but he plainly declares that this method is only useful for discovering results and does not furnish a proof of them, but that to establish them scientifically a geometrical proof by the method of exhaustion, with its double *reductio ad absurdum*, is still necessary.

Notwithstanding that the criticisms of Zeno had so important an influence upon the lines of development of Greek geometry, it does not appear that Zeno himself was really a mathematician or even a physicist. Plato mentions a work of his (τὰ τοῦ Ζήνωνος γράμματα, or τὸ σύγγραμμα) in terms which imply that it was his only known work.[2] Simplicius too knows only one work of his, and this the same as that mentioned by Plato[3]; when Suidas mentions four, a *Commentary on* or *Exposition of Empedocles*, *Controversies*, *Against the philosophers* and *On Nature*, it may be that the last three titles are only different designations for the one work, while the book on Empedocles may have been wrongly attributed

[1] Cf. Arist. *Phys.* iii. 7, 207 b 31. [2] Plato, *Parmenides*, 127 c sq.
[3] Simpl. *in Phys.*, pp. 139. 5, 140. 27 Diels.

to Zeno.[1] Plato puts into the mouth of Zeno himself an explanation of the character and object of his book.[2] It was a youthful effort, and it was stolen by some one, so that the author had no opportunity of considering whether to publish it or not. Its object was to defend the system of Parmenides by attacking the common conceptions of things. Parmenides held that only the One exists; whereupon common sense pointed out that many contradictions and absurdities will follow if this be admitted. Zeno replied that, if the popular view that Many exist be accepted, still more absurd results will follow. The work was divided into several parts (λόγοι according to Plato) and each of these again into sections ('hypotheses' in Plato, 'contentions', ἐπιχειρήματα, in Simplicius): each of the latter (which according to Proclus numbered forty in all[3]) seems to have taken one of the assumptions made on the ordinary view of life and to have shown that it leads to an absurdity. It is doubtless on account of this systematic use of indirect proof by the *reductio ad absurdum* of particular hypotheses that Zeno is said to have been called by Aristotle the discoverer of Dialectic[4]; Plato, too, says of him that he understood how to make one and the same thing appear like and unlike, one and many, at rest and in motion.[5]

Zeno's arguments about motion.

It does not appear that the full significance and value of Zeno's paradoxes have ever been realized until these latter days. The most modern view of them shall be expressed in the writer's own words:

'In this capricious world nothing is more capricious than posthumous fame. One of the most notable victims of posterity's lack of judgement is the Eleatic Zeno. Having invented four arguments all immeasurably subtle and profound, the grossness of subsequent philosophers pronounced him to be a mere ingenious juggler, and his arguments to be

[1] Zeller, i⁵, p. 587 note.
[2] Plato, *Parmenides* 128 C–E.
[3] Proclus *in Parm.*, p. 694. 23 seq.
[4] Diog. L. viii. 57, ix. 25; Sext. Emp. *Math.* vii. 6.
[5] Plato, *Phaedrus* 261 D.

one and all sophisms. After two thousand years of continual refutation, these sophisms were reinstated, and made the foundation of a mathematical renaissance, by a German professor who probably never dreamed of any connexion between himself and Zeno. Weierstrass, by strictly banishing all infinitesimals, has at last shown that we live in an unchanging world, and that the arrow, at every moment of its flight, is truly at rest. The only point where Zeno probably erred was in inferring (if he did infer) that, because there is no change, the world must be in the same state at one time as at another. This consequence by no means follows, and in this point the German professor is more constructive than the ingenious Greek. Weierstrass, being able to embody his opinions in mathematics, where familiarity with truth eliminates the vulgar prejudices of common sense, has been able to give to his propositions the respectable air of platitudes; and if the result is less delightful to the lover of reason than Zeno's bold defiance, it is at any rate more calculated to appease the mass of academic mankind.'[1]

Thus, while in the past the arguments of Zeno have been treated with more or less disrespect as mere sophisms, we have now come to the other extreme. It appears to be implied that Zeno anticipated Weierstrass. This, I think, a calmer judgement must pronounce to be incredible. If the arguments of Zeno are found to be 'immeasurably subtle and profound' because they contain ideas which Weierstrass used to create a great mathematical theory, it does not follow that for Zeno they meant at all the same thing as for Weierstrass. On the contrary, it is probable that Zeno happened upon these ideas without realizing any of the significance which Weierstrass was destined to give them; nor shall we give Zeno any less credit on this account.

It is time to come to the arguments themselves. It is the four arguments on the subject of motion which are most important from the point of view of the mathematician; but they have points of contact with the arguments which Zeno used to prove the non-existence of Many, in refutation of those who attacked Parmenides's doctrine of the One. According to Simplicius, he showed that, if Many exist, they must

[1] Bertrand Russell, *The Principles of Mathematics*, vol. i, 1903, pp. 347, 348.

be both great and small, so great on the one hand as to be infinite in size and so small on the other as to have no size.[1] To prove the latter of these contentions, Zeno relied on the infinite divisibility of bodies as evident; assuming this, he easily proved that division will continually give smaller and smaller parts, there will be no limit to the diminution, and, if there is a final element, it must be absolutely *nothing*. Consequently to add any number of these *nil*-elements to anything will not increase its size, nor will the subtraction of them diminish it; and of course to add them to one another, even in infinite number, will give *nothing* as the total. (The second horn of the dilemma, not apparently stated by Zeno in this form, would be this. A critic might argue that infinite division would only lead to parts having *some* size, so that the last element would itself have some size; to this the answer would be that, as there would, by hypothesis, be an infinite number of such parts, the original magnitude which was divided would be infinite in size.) The connexion between the arguments against the Many and those against motion lies in the fact that the former rest on the assumption of the divisibility of matter *ad infinitum*, and that this is the hypothesis assumed in the first two arguments against motion. We shall see that, while the first two arguments proceed on this hypothesis, the last two appear to proceed on the opposite hypothesis that space and time are not infinitely divisible, but that they are composed of *indivisible* elements; so that the four arguments form a complete dilemma.

The four arguments against motion shall be stated in the words of Aristotle.

I. The *Dichotomy*.

'There is no motion because that which is moved must arrive at the middle (of its course) before it arrives at the end.'[2] (And of course it must traverse the half of the half before it reaches the middle, and so on *ad infinitum*.)

II. The *Achilles*.

'This asserts that the slower when running will never be

[1] Simpl. *in Phys.*, p. 139. 5, Diels.
[2] Aristotle, *Phys.* vi. 9, 239 b 11.

overtaken by the quicker; for that which is pursuing must first reach the point from which that which is fleeing started, so that the slower must necessarily always be some distance ahead.'[1]

III. The *Arrow*.

'If, says Zeno, everything is either at rest or moving when it occupies a space equal (to itself), while the object moved is always in the instant (ἔστι δ' ἀεὶ τὸ φερόμενον ἐν τῷ νῦν, in the *now*), the moving arrow is unmoved.'[2]

I agree in Brochard's interpretation of this passage,[3] from which Zeller[4] would banish ἢ κινεῖται, 'or is moved'. The argument is this. It is strictly impossible that the arrow can move in the *instant*, supposed indivisible, for, if it changed its position, the instant would be at once divided. Now the moving object is, in the instant, either at rest or in motion; but, as it is not in motion, it is at rest, and as, by hypothesis, time is composed of nothing but instants, the moving object is always at rest. This interpretation has the advantage of agreeing with that of Simplicius,[5] which seems preferable to that of Themistius[6] on which Zeller relies.

IV. The *Stadium*. I translate the first two sentences of Aristotle's account[7]:

'The fourth is the argument concerning the two rows of bodies each composed of an equal number of bodies of equal size, which pass one another on a race-course as they proceed with equal velocity in opposite directions, one row starting from the end of the course and the other from the middle. This, he thinks, involves the conclusion that half a given time is equal to its double. The fallacy of the reasoning lies in the assumption that an equal magnitude occupies an equal time in passing with equal velocity a magnitude that is in motion and a magnitude that is at rest, an assumption which is false.'

Then follows a description of the process by means of

[1] Aristotle, *Phys.* vi. 9, 239 b 14. [2] *Ib.* 239 b 5–7.
[3] V. Brochard, *Études de Philosophie ancienne et de Philosophie moderne*, Paris 1912, p. 6.
[4] Zeller, i⁵, p. 599. [5] Simpl. *in Phys.*, pp. 1011–12, Diels.
[6] Them. (*ad loc.*, p. 392 Sp., p. 199 Sch.)
[7] *Phys.* vi, 9, 239 b 33–240 a 18.

letters A, B, C the *exact* interpretation of which is a matter of some doubt [1]; the essence of it, however, is clear. The first diagram below shows the original positions of the rows of

A_1 A_2 A_3 A_4 A_5 A_6 A_7 A_8

B_8 B_7 B_6 B_5 B_4 B_3 B_2 B_1 ⟶

⟵ C_1 C_2 C_3 C_4 C_5 C_6 C_7 C_8

bodies (say eight in number). The A's represent a row which is stationary, the B's and C's are rows which move with equal velocity alongside the A's and one another, in the directions shown by the arrows. Then clearly there will be (1) a moment

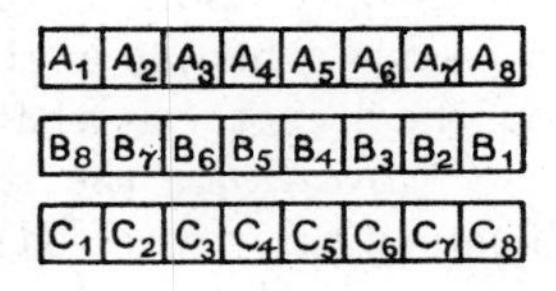

when the B's and C's will be exactly under the respective A's, as in the second diagram, and after that (2) a moment when the B's and C's will have exactly reversed their positions relatively to the A's, as in the third figure.

A_1 A_2 A_3 A_4 A_5 A_6 A_7 A_8

B_8 B_7 B_6 B_5 B_4 B_3 B_2 B_1

C_1 C_2 C_3 C_4 C_5 C_6 C_7 C_8

The observation has been made [2] that the four arguments form a system curiously symmetrical. The first and fourth consider the continuous and movement within given limits, the second and third the continuous and movement over

[1] The interpretation of the passage 240 a 4–18 is elaborately discussed by R. K. Gaye in the *Journal of Philology*, xxxi, 1910, pp. 95–116. It is a question whether in the above quotation Aristotle means that Zeno argued that half the given time would be equal to double the half, i.e. the whole time simply, or to double the whole, i.e. *four* times the half. Gaye contends (unconvincingly, I think) for the latter.

[2] Brochard, *loc. cit.*, pp. 4, 5.

lengths which are indeterminate. In the first and third there is only one moving object, and it is shown that it cannot even begin to move. The second and fourth, comparing the motions of two objects, make the absurdity of the hypothesis even more palpable, so to speak, for they prove that the movement, even if it has once begun, cannot continue, and that relative motion is no less impossible than absolute motion. The first two establish the impossibility of movement by the nature of space, supposed continuous, without any implication that time is otherwise than continuous in the same way as space; in the last two it is the nature of time (considered as made up of indivisible elements or instants) which serves to prove the impossibility of movement, and without any implication that space is not likewise made up of indivisible elements or points. The second argument is only another form of the first, and the fourth rests on the same principle as the third. Lastly, the first pair proceed on the hypothesis that continuous magnitudes are divisible *ad infinitum*; the second pair give the other horn of the dilemma, being directed against the assumption that continuous magnitudes are made up of *indivisible* elements, an assumption which would scarcely suggest itself to the imagination until the difficulties connected with the other were fully realized. Thus the logical order of the arguments corresponds exactly to the historical order in which Aristotle has handed them down and which was certainly the order adopted by Zeno.

Whether or not the paradoxes had for Zeno the profound meaning now claimed for them, it is clear that they have been very generally misunderstood, with the result that the criticisms directed against them have been wide of the mark. Aristotle, it is true, saw that the first two arguments, the *Dichotomy* and the *Achilles*, come to the same thing, the latter differing from the former only in the fact that the ratio of each space traversed by Achilles to the preceding space is not that of 1 : 2 but a ratio of 1 : n, where n may be any number, however large; but, he says, both proofs rest on the fact that a certain moving object 'cannot reach the end of the course if the magnitude is divided in a certain way'.[1] But another passage shows that he mistook the character of the argument

[1] Arist. *Phys.* vi. 9, 239 b 18–24.

in the *Dichotomy*. He observes that time is divisible in exactly the same way as a length; if therefore a length is infinitely divisible, so is the corresponding time; he adds '*this is why* (διό) Zeno's argument falsely assumes that it is not possible to traverse or touch each of an infinite number of points in a finite time',[1] thereby implying that Zeno did not regard time as divisible *ad infinitum* like space. Similarly, when Leibniz declares that a space divisible *ad infinitum* is traversed in a time divisible *ad infinitum*, he, like Aristotle, is entirely beside the question. Zeno was perfectly aware that, in respect of divisibility, time and space have the same property, and that they are alike, always, and concomitantly, divisible *ad infinitum*. The question is how, in the one as in the other, this series of divisions, by definition inexhaustible, can be exhausted; and it must be exhausted if motion is to be possible. It is not an answer to say that the two series are exhausted simultaneously.

The usual mode of refutation given by mathematicians from Descartes to Tannery, correct in a sense, has an analogous defect. To show that the sum of the infinite series $1 + \frac{1}{2} + \frac{1}{4} + \ldots$ is equal to 2, or to calculate (in the *Achilles*) the exact moment when Achilles will overtake the tortoise, is to answer the question *when*? whereas the question actually asked is *how*? On the hypothesis of divisibility *ad infinitum* you will, in the *Dichotomy*, never reach the limit, and, in the *Achilles*, the distance separating Achilles from the tortoise, though it continually decreases, will never vanish. And if you introduce the limit, or, with a numerical calculation, the discontinuous, Zeno is quite aware that his arguments are no longer valid. We are then in presence of another hypothesis as to the composition of the continuum; and this hypothesis is dealt with in the third and fourth arguments.[2]

It appears then that the first and second arguments, in their full significance, were not really met before G. Cantor formulated his new theory of continuity and infinity. On this I can only refer to Chapters xlii and xliii of Mr. Bertrand Russell's *Principles of Mathematics*, vol. i. Zeno's argument in the *Dichotomy* is that, whatever motion we assume to have taken place, this presupposes another motion; this in turn

[1] *Ib.* vi. 2, 233 a 16–23.

[2] Brochard, *loc. cit.*, p. 9.

another, and so on *ad infinitum*. Hence there is an endless regress in the mere idea of any assigned motion. Zeno's argument has then to be met by proving that the 'infinite regress' in this case is 'harmless'.

As regards the *Achilles*, Mr. G. H. Hardy remarks that 'the kernel of it lies in the perfectly valid proof which it affords that the tortoise passes through as many points as Achilles, a view which embodies an accepted doctrine of modern mathematics'.[1]

The argument in the *Arrow* is based on the assumption that time is made up of *indivisible* elements or instants. Aristotle meets it by denying the assumption. 'For time is not made up of indivisible instants (*nows*), any more than any other magnitude is made up of indivisible elements.' '(Zeno's result) follows through assuming that time is made up of (indivisible) instants (*nows*); if this is not admitted, his conclusion does not follow.'[2] On the other hand, the modern view is that Zeno's contention is *true*: 'If' (said Zeno) 'everything is at rest or in motion when it occupies a space equal to itself, and if what moves is always in the instant, it follows that the moving arrow is unmoved.' Mr. Russell[3] holds that this is 'a very plain statement of an elementary fact';

'it is a very important and very widely applicable platitude, namely "Every possible value of a variable is a constant". If x be a variable which can take all values from 0 to 1, all the values it can take are definite numbers such as $\frac{1}{2}$ or $\frac{1}{3}$, which are all absolute constants ... Though a variable is always connected with some class, it is not the class, nor a particular member of the class, nor yet the whole class, but *any* member of the class.' The usual x in algebra 'denotes the disjunction formed by the various members' ... 'The values of x are then the terms of the disjunction; and each of these is a constant. This simple logical fact seems to constitute the essence of Zeno's contention that the arrow is always at rest.' 'But Zeno's argument contains an element which is specially applicable to continua. In the case of motion it denies that there is such a thing as a *state* of motion. In the general case of a continuous variable, it may be taken as denying actual infinitesimals. For infinitesimals are an

[1] *Encyclopaedia Britannica*, art. Zeno.
[2] Arist. *Phys.* vi. 9, 239 b 8, 31.
[3] Russell, *Principles of Mathematics*, i, pp. 350, 351.

attempt to extend to the *values* of a variable the variability which belongs to it alone. When once it is firmly realized that all the values of a variable are constants, it becomes easy to see, by taking *any* two such values, that their difference is always finite, and hence that there are no infinitesimal differences. If x be a variable which may take all real values from 0 to 1, then, taking any two of these values, we see that their difference is finite, although x is a continuous variable. It is true the difference might have been less than the one we chose; but if it had been, it would still have been finite. The lower limit to possible differences is zero, but all possible differences are finite; and in this there is no shadow of contradiction. This static theory of the variable is due to the mathematicians, and its absence in Zeno's day led him to suppose that continuous change was impossible without a state of change, which involves infinitesimals and the contradiction of a body's being where it is not.'

In his later chapter on Motion Mr. Russell concludes as follows:[1]

'It is to be observed that, in consequence of the denial of the infinitesimal and in consequence of the allied purely technical view of the derivative of a function, we must entirely reject the notion of a *state* of motion. Motion consists *merely* in the occupation of different places at different times, subject to continuity as explained in Part V. There is no transition from place to place, no consecutive moment or consecutive position, no such thing as velocity except in the sense of a real number which is the limit of a certain set of quotients. The rejection of velocity and acceleration as physical facts (i. e. as properties belonging *at each instant* to a moving point, and not merely real numbers expressing limits of certain ratios) involves, as we shall see, some difficulties in the statement of the laws of motion; but the reform introduced by Weierstrass in the infinitesimal calculus has rendered this rejection imperative.'

We come lastly to the fourth argument (the *Stadium*). Aristotle's representation of it is obscure through its extreme brevity of expression, and the matter is further perplexed by an uncertainty of reading. But the meaning intended to be conveyed is fairly clear. The eight *A*'s, *B*'s and *C*'s being

[1] *Op. cit.*, p. 473.

initially in the position shown in Figure 1, suppose, e.g., that the *B*'s move to the right and the *C*'s to the left with equal

A_1 A_2 A_3 A_4 A_5 A_6 A_7 A_8

B_8 B_7 B_6 B_5 B_4 B_3 B_2 B_1 →

← C_1 C_2 C_3 C_4 C_5 C_6 C_7 C_8

velocity until the rows are vertically under one another as in Figure 2. Then C_1 has passed alongside all the eight *B*'s (and B_1

A_1 A_2 A_3 A_4 A_5 A_6 A_7 A_8

B_8 B_7 B_6 B_5 B_4 B_3 B_2 B_1

C_1 C_2 C_3 C_4 C_5 C_6 C_7 C_8

alongside all the eight *C*'s), while B_1 has passed alongside only half the *A*'s (and similarly for C_1). But (Aristotle makes Zeno say) *C_1 is the same time in passing each of the B's as it is in passing each of the A's.* It follows that the time occupied by C_1 in passing all the *A*'s is the same as the time occupied by C_1 in passing half the *A*'s, or a given time is equal to its half. Aristotle's criticism on this is practically that Zeno did not understand the difference between absolute and relative motion. This is, however, incredible, and another explanation must be found. The real explanation seems to be that given by

A_1 A_2 A_3 A_4 A_5 A_6 A_7 A_8

B_8 B_7 B_6 B_5 B_4 B_3 B_2 B_1

C_1 C_2 C_3 C_4 C_5 C_6 C_7 C_8

Brochard, Noël and Russell. Zeno's object is to prove that time is not made up of indivisible elements or instants. Suppose the *B*'s have moved one place to the right and the *C*'s one place to the left, so that B_1, which was under A_4, is now under A_5, and C_1, which was under A_5, is now under A_4. We must suppose that B_1 and C_1 are absolute indivisible elements of space, and that they move to their new positions in an instant, the absolute indivisible element of time; this is Zeno's hypothesis. But, in order that B_1, C_1 may have taken up their new positions, there must have been a moment at which they crossed or B_1 was vertically over C_1. Yet the motion has, by hypothesis, taken place in an indivisible instant. Therefore, either they have *not* crossed (in which case there is no movement), or in the particular indivisible instant two positions have been occupied by the two moving objects, that is to say, the instant is no longer indivisible. And, if the instant is divided into two equal parts, this, on the hypothesis of indivisibles, is equivalent to saying that an instant is double of itself.

Two remarks may be added. Though the first two arguments are directed against those who assert the divisibility *ad infinitum* of magnitudes and times, there is no sufficient justification for Tannery's contention that they were specially directed against a view, assumed by him to be Pythagorean, that bodies, surfaces and lines are made up of *mathematical* points. There is indeed no evidence that the Pythagoreans held this view at all; it does not follow from their definition of a point as a 'unit having position' (μονὰς θέσιν ἔχουσα); and, as we have seen, Aristotle says that the Pythagoreans maintained that units and numbers have magnitude.[1]

It would appear that, after more than 2,300 years, controversy on Zeno's arguments is yet by no means at an end. But the subject cannot here be pursued further.[2]

[1] Arist. *Metaph.* M. 6, 1080 b 19, 32.

[2] It is a pleasure to be able to refer the reader to a most valuable and comprehensive series of papers by Professor Florian Cajori, under the title 'The History of Zeno's arguments on Motion', published in the *American Mathematical Monthly* of 1915, and also available in a reprint. This work carries the history of the various views and criticisms of Zeno's arguments down to 1914. I may also refer to the portions of Bertrand Russell's work, *Our Knowledge of the External World as a Field for Scientific Method in Philosophy*, 1914, which deal with Zeno, and to Philip E. B. Jourdain's article, 'The Flying Arrow; an Anachronism', in *Mind*, January 1916, pp. 42–55.

IX

PLATO

It is in the Seventh Book of the *Republic* that we find the most general statement of the attitude of Plato towards mathematics. Plato regarded mathematics in its four branches, arithmetic, geometry, stereometry and astronomy, as the first essential in the training of philosophers and of those who should rule his ideal State; 'let no one destitute of geometry enter my doors', said the inscription over the door of his school. There could be no better evidence of the supreme importance which he attached to the mathematical sciences.

What Plato emphasizes throughout when speaking of mathematics is its value for the training of the mind; its practical utility is of no account in comparison. Thus arithmetic must be pursued for the sake of knowledge, not for any practical ends such as its use in trade[1]; the real science of arithmetic has nothing to do with actions, its object is knowledge.[2] A very little geometry and arithmetical calculation suffices for the commander of an army; it is the higher and more advanced portions which tend to lift the mind on high and to enable it ultimately to see the final aim of philosophy, the idea of the Good[3]; the value of the two sciences consists in the fact that they draw the soul towards truth and create the philosophic attitude of mind, lifting on high the things which our ordinary habit would keep down.[4]

The extent to which Plato insisted on the purely theoretical character of the mathematical sciences is illustrated by his peculiar views about the two subjects which the ordinary person would regard as having, at least, an important practical side, namely astronomy and music. According to Plato, true astronomy is not concerned with the movements of the visible

[1] *Rep.* vii. 525 C, D.
[2] *Politicus* 258 D.
[3] *Rep.* 526 D, E.
[4] *Ib.* 527 B.

heavenly bodies. The arrangement of the stars in the heaven and their apparent movements are indeed wonderful and beautiful, but the observation of and the accounting for them falls far short of true astronomy. Before we can attain to this we must get beyond mere observational astronomy, 'we must leave the heavens alone'. The true science of astronomy is in fact a kind of ideal kinematics, dealing with the laws of motion of true stars in a sort of mathematical heaven of which the visible heaven is an imperfect expression in time and space. The visible heavenly bodies and their apparent motions we are to regard merely as illustrations, comparable to the diagrams which the geometer draws to illustrate the true straight lines, circles, &c., about which his science reasons; they are to be used as 'problems' only, with the object of ultimately getting rid of the apparent irregularities and arriving at 'the true motions with which essential speed and essential slowness move in relation to one another in the true numbers and the true forms, and carry their contents with them' (to use Burnet's translation of τὰ ἐνόντα).[1] 'Numbers' in this passage correspond to the periods of the apparent motions; the 'true forms' are the true orbits contrasted with the apparent. It is right to add that according to one view (that of Burnet) Plato means, not that true astronomy deals with an 'ideal heaven' different from the apparent, but that it deals with the true motions of the visible bodies as distinct from their apparent motions. This would no doubt agree with Plato's attitude in the *Laws*, and at the time when he set to his pupils as a problem for solution the question by what combinations of uniform circular revolutions the apparent movements of the heavenly bodies can be accounted for. But, except on the assumption that an ideal heaven is meant, it is difficult to see what Plato can mean by the contrast which he draws between the visible broideries of heaven (the visible stars and their arrangement), which are indeed beautiful, and the true broideries which they only imitate and which are infinitely more beautiful and marvellous.

This was not a view of astronomy that would appeal to the ordinary person. Plato himself admits the difficulty.

[1] *Rep.* vii. 529 c–530 c.

When Socrates's interlocutor speaks of the use of astronomy for distinguishing months and seasons, for agriculture and navigation, and even for military purposes, Socrates rallies him on his anxiety that his curriculum should not consist of subjects which the mass of people would regard as useless: 'it is by no means an easy thing, nay it is difficult, to believe that in studying these subjects a certain organ in the mind of every one is purified and rekindled which is destroyed and blinded by other pursuits, an organ which is more worthy of preservation than ten thousand eyes; for by it alone is truth discerned.'[1]

As with astronomy, so with harmonics.[2] The true science of harmonics differs from that science as commonly understood. Even the Pythagoreans, who discovered the correspondence of certain intervals to certain numerical ratios, still made their theory take too much account of audible sounds. The true science of harmonics should be altogether independent of observation and experiment. Plato agreed with the Pythagoreans as to the nature of sound. Sound is due to concussion of air, and when there is rapid motion in the air the tone is high-pitched, when the motion is slow the tone is low; when the speeds are in certain arithmetical proportions, consonances or harmonies result. But audible movements produced, say, by different lengths of strings are only useful as illustrations; they are imperfect representations of those mathematical movements which produce mathematical consonances, and it is these true consonances which the true *ἁρμονικός* should study.

We get on to easier ground when Plato discusses geometry. The importance of geometry lies, not in its practical use, but in the fact that it is a study of objects eternal and unchangeable, and tends to lift the soul towards truth. The essence of geometry is therefore directly opposed even to the language which, for want of better terms, geometers are obliged to use; thus they speak of 'squaring', 'applying (a rectangle)', 'adding', &c., as if the object were to *do* something, whereas the true purpose of geometry is knowledge.[3] Geometry is concerned, not with material things, but with mathematical

[1] *Rep.* 527 D, E. [2] *Ib.* 531 A–C.
[3] *Ib.* vii. 526 D–527 B.

points, lines, triangles, squares, &c., as objects of pure thought. If we use a diagram in geometry, it is only as an illustration; the triangle which we draw is an imperfect representation of the real triangle of which we think. *Constructions*, then, or the *processes* of squaring, adding, and so on, are not of the essence of geometry, but are actually antagonistic to it. With these views before us, we can without hesitation accept as well founded the story of Plutarch that Plato blamed Eudoxus, Archytas and Menaechmus for trying to reduce the duplication of the cube to mechanical constructions by means of instruments, on the ground that 'the good of geometry is thereby lost and destroyed, as it is brought back to things of sense instead of being directed upward and grasping at eternal and incorporeal images'.[1] It follows almost inevitably that we must reject the tradition attributing to Plato himself the elegant mechanical solution of the problem of the two mean proportionals which we have given in the chapter on Special Problems (pp. 256–7). Indeed, as we said, it is certain on other grounds that the so-called Platonic solution was later than that of Eratosthenes; otherwise Eratosthenes would hardly have failed to mention it in his epigram, along with the solutions by Archytas and Menaechmus. Tannery, indeed, regards Plutarch's story as an invention based on nothing more than the general character of Plato's philosophy, since it took no account of the real nature of the solutions of Archytas and Menaechmus; these solutions are in fact purely theoretical and would have been difficult or impossible to carry out in practice, and there is no reason to doubt that the solution by Eudoxus was of a similar kind.[2] This is true, but it is evident that it was the practical difficulty quite as much as the theoretical elegance of the constructions which impressed the Greeks. Thus the author of the letter, wrongly attributed to Eratosthenes, which gives the history of the problem, says that the earlier solvers had all solved the problem in a theoretical manner but had not been able to reduce their solutions to practice, except to a certain small extent Menaechmus, and that with difficulty; and the epigram of Eratosthenes himself says, 'do not attempt the impracticable

[1] Plutarch, *Quaest. Conviv.* viii. 2. 1, p. 718 F.
[2] Tannery, *La géométrie grecque*, pp. 79, 80.

business of the cylinders of Archytas or the cutting of the cone in the three curves of Menaechmus'. It would therefore be quite possible for Plato to regard Archytas and Menaechmus as having given constructions that were ultra-mechanical, since they were more mechanical than the ordinary constructions by means of the straight line and circle; and even the latter, which alone are required for the processes of 'squaring', 'applying (a rectangle)' and 'adding', are according to Plato no part of theoretic geometry. This banning even of simple constructions from true geometry seems, incidentally, to make it impossible to accept the conjecture of Hankel that we owe to Plato the limitation, so important in its effect on the later development of geometry, of the instruments allowable in constructions to the ruler and compasses.[1] Indeed, there are signs that the limitation began before Plato's time (e.g. this may be the explanation of the two constructions attributed to Oenopides), although no doubt Plato's influence would help to keep the restriction in force; for other instruments, and the use of curves of higher order than circles in constructions, were expressly barred in any case where the ruler and compasses could be made to serve (cf. Pappus's animadversion on a solution of a 'plane' problem by means of conics in Apollonius's *Conics*, Book V).

Contributions to the philosophy of mathematics.

We find in Plato's dialogues what appears to be the first serious attempt at a philosophy of mathematics. Aristotle says that between sensible objects and the ideas Plato placed 'things mathematical' (τὰ μαθηματικά), which differed from sensibles in being eternal and unmoved, but differed again from the ideas in that there can be many mathematical objects of the same kind, while the idea is one only; e.g. the idea of triangle is one, but there may be any number of mathematical triangles as of visible triangles, namely the perfect triangles of which the visible triangles are imperfect copies. A passage in one of the *Letters* (No. 7, to the friends of Dion) is interesting in this connexion.[2] Speaking of a circle by way of example, Plato says there is (1) some-

[1] Hankel, *op. cit.*, p. 156.

[2] Plato, *Letters*, 342 B, C, 343 A, B.

thing called a circle and known by that name; next there is (2) its definition as that in which the distances from its extremities in all directions to the centre are always equal, for this may be said to be the definition of that to which the names 'round' and 'circle' are applied; again (3) we have the circle which is drawn or turned: this circle is perishable and perishes; not so, however, with (4) αὐτὸς ὁ κύκλος, the essential circle, or the idea of circle: it is by reference to this that the other circles exist, and it is different from each of them. The same distinction applies to anything else, e.g. the straight, colour, the good, the beautiful, or any natural or artificial object, fire, water, &c. Dealing separately with the four things above distinguished, Plato observes that there is nothing essential in (1) the name: it is merely conventional; there is nothing to prevent our assigning the name 'straight' to what we now call 'round' and vice versa; nor is there any real definiteness about (2) the definition, seeing that it too is made up of parts of speech, nouns and verbs. The circle (3), the particular circle drawn or turned, is not free from admixture of other things: it is even full of what is opposite to the true nature of a circle, for it will anywhere touch a straight line', the meaning of which is presumably that we cannot in practice draw a circle and a tangent with only *one* point common (although a mathematical circle and a mathematical straight line touching it meet in one point only). It will be observed that in the above classification there is no place given to the many particular mathematical circles which correspond to those which we draw, and are intermediate between these imperfect circles and the idea of circle which is one only.

(a) *The hypotheses of mathematics.*

The *hypotheses* of mathematics are discussed by Plato in the *Republic*.

'I think you know that those who occupy themselves with geometries and calculations and the like take for granted the odd and the even, figures, three kinds of angles, and other things cognate to these in each subject; assuming these things as known, they take them as hypotheses and thenceforward they do not feel called upon to give any explanation with regard to them either to themselves or any one else, but treat them as manifest to every one; basing themselves on these hypotheses, they proceed at once to go through the rest of the argument till they arrive, with general assent, at the particular conclusion to which their inquiry was directed. Further you know that they make use of visible figures and argue about them, but in doing so they are not thinking of these figures but of the things which they represent; thus it is the absolute square and the absolute diameter which is the object of their argument, not the diameter which they draw; and similarly, in other cases, the things which they actually model or draw, and which may also have their images in shadows or in water, are themselves in turn used as images, the object of the inquirer being to see their absolute counterparts which cannot be seen otherwise than by thought.' [1]

(β) *The two intellectual methods.*

Plato distinguishes two processes: both begin from hypotheses. The one method cannot get above these hypotheses, but, treating them as if they were first principles, builds upon them and, with the aid of diagrams or images, arrives at conclusions: this is the method of geometry and mathematics in general. The other method treats the hypotheses as being really hypotheses and nothing more, but uses them as stepping-stones for mounting higher and higher until the principle of all things is reached, a principle about which there is nothing hypothetical; when this is reached, it is possible to descend again, by steps each connected with the preceding step, to the conclusion, a process which has no need of any sensible images but deals in ideas only and ends in them [2]; this method, which rises above and puts an end to hypotheses, and reaches the first principle in this way, is the dialectical method. For want of this, geometry and the other sciences which in some sort lay hold of truth are comparable to one dreaming about truth, nor can they have a waking sight of it so long as they treat their hypotheses as immovable truths, and are unable to give any account or explanation of them.[3]

[1] *Republic*, vi. 510 C–E. [2] *Ib.* vi. 510 B 511 A–C.
[3] *Ib.* vii. 533 B–E.

With the above quotations we should read a passage of Proclus.

'Nevertheless certain methods have been handed down. The finest is the method which by means of *analysis* carries the thing sought up to an acknowledged principle; a method which Plato, as they say, communicated to Leodamas, and by which the latter too is said to have discovered many things in geometry. The second is the method of *division*, which divides into its parts the genus proposed for consideration, and gives a starting-point for the demonstration by means of the elimination of the other elements in the construction of what is proposed, which method also Plato extolled as being of assistance to all sciences.'[1]

The first part of this passage, with a like dictum in Diogenes Laertius that Plato 'explained to Leodamas of Thasos the method of inquiry by analysis',[2] has commonly been understood as attributing to Plato the *invention* of the method of mathematical analysis. But, analysis being according to the ancient view nothing more than a series of successive reductions of a theorem or problem till it is finally reduced to a theorem or problem already known, it is difficult to see in what Plato's supposed discovery could have consisted; for analysis in this sense must have been frequently used in earlier investigations. Not only did Hippocrates of Chios reduce the problem of duplicating the cube to that of finding two mean proportionals, but it is clear that the method of analysis in the sense of reduction must have been in use by the Pythagoreans. On the other hand, Proclus's language suggests that what he had in mind was the philosophical method described in the passage of the *Republic*, which of course does not refer to mathematical analysis at all; it may therefore well be that the idea that Plato discovered the method of analysis is due to a misapprehension. But analysis and synthesis following each other are related in the same way as the upward and downward progressions in the dialectician's intellectual method. It has been suggested, therefore, that Plato's achievement was to observe the importance from the point of view of logical rigour, of the confirmatory synthesis following analysis. The method of *division*

[1] Proclus, *Comm. on Eucl.* I, pp. 211. 18–212. 1.
[2] Diog. L. iii. 24, p. 74, Cobet.

mentioned by Proclus is the method of successive bipartitions of genera into species such as we find in the *Sophist* and the *Politicus*, and has little to say to geometry; but the mention of it side by side with analysis itself suggests that Proclus confused the latter with the philosophical method referred to.

(γ) *Definitions.*

Among the fundamentals of mathematics Plato paid a good deal of attention to definitions. In some cases his definitions connect themselves with Pythagorean tradition; in others he seems to have struck out a new line for himself. The division of numbers into odd and even is one of the most common of his illustrations; number, he says, is divided equally, i. e. there are as many odd numbers as even, and this is the true division of number; to divide number (e. g.) into myriads and what are not myriads is not a proper division.[1] An even number is defined as a number divisible into two equal parts[2]; in another place it is explained as that which is not scalene but isosceles[3]: a curious and apparently unique application of these terms to number, and in any case a defective statement unless the term 'scalene' is restricted to the case in which one part of the number is odd and the other even; for of course an even number can be divided into two unequal odd numbers or two unequal even numbers (except 2 in the first case and 2 and 4 in the second). The further distinction between even-times-even, odd-times-even, even-times-odd and odd-times-odd occurs in Plato[4]: but, as thrice two is called odd-times-even and twice three is even-times-odd, the number in both cases being the same, it is clear that, like Euclid, Plato regarded even-times-odd and odd-times-even as convertible terms, and did not restrict their meaning in the way that Nicomachus and the neo-Pythagoreans did.

Coming to geometry we find an interesting view of the term 'figure'. What is it, asks Socrates, that is true of the round, the straight, and the other things that you call figures, and is the same for all? As a suggestion for a definition of 'figure', Socrates says, 'let us regard as *figure* that which alone of existing things is associated with colour'. Meno

[1] *Politicus*, 262 D, E.
[2] *Laws*, 895 E.
[3] *Euthyphro*, 12 D.
[4] *Parmenides*, 143 E–144 A.

asks what is to be done if the interlocutor says he does not know what colour is; what alternative definition is there? Socrates replies that it will be admitted that in geometry there are such things as what we call a surface or a solid, and so on; from these examples we may learn what we mean by figure; figure is that in which a solid ends, or figure is the limit (or extremity, πέρας) of a solid.[1] Apart from 'figure' as form or shape, e.g. the round or straight, this passage makes 'figure' practically equivalent to surface, and we are reminded of the Pythagorean term for surface, χροιά, colour or skin, which Aristotle similarly explains as χρῶμα, colour, something inseparable from πέρας, extremity.[2] In Euclid of course ὅρος, limit or boundary, is defined as the extremity (πέρας) of a thing, while 'figure' is that which is contained by one or more boundaries.

There is reason to believe, though we are not specifically told, that the definition of a line as 'breadthless length' originated in the Platonic School, and Plato himself gives a definition of a straight line as 'that of which the middle covers the ends'[3] (i.e. to an eye placed at either end and looking along the straight line); this seems to me to be the origin of the Euclidean definition 'a line which lies evenly with the points on it', which, I think, can only be an attempt to express the sense of Plato's definition in terms to which a geometer could not take exception as travelling outside the subject matter of geometry, i.e. in terms excluding any appeal to vision. A *point* had been defined by the Pythagoreans as a 'monad having position'; Plato apparently objected to this definition and substituted no other; for, according to Aristotle, he regarded the genus of points as being a 'geometrical fiction', calling a point the beginning of a line, and often using the term 'indivisible lines' in the same sense.[4] Aristotle points out that even indivisible lines must have extremities, and therefore they do not help, while the definition of a point as 'the extremity of a line' is unscientific.[5]

The 'round' (στρογγύλον) or the circle is of course defined as 'that in which the furthest points (τὰ ἔσχατα) in all

[1] *Meno*, 75 A–76 A.

[2] Arist. *De sensu*, 439 a 31, &c.

[3] *Parmenides*, 137 E.

[4] Arist. *Metaph.* A. 9, 992 a 20.

[5] Arist. *Topics*, vi. 4, 141 b 21.

directions are at the same distance from the middle (centre)'.[1] The 'sphere' is similarly defined as 'that which has the distances from its centre to its terminations or ends in every direction equal', or simply as that which is 'equal (equidistant) from the centre in all directions'.[2]

The *Parmenides* contains certain phrases corresponding to what we find in Euclid's preliminary matter. Thus Plato speaks of something which is 'a part' but not 'parts' of the One,[3] reminding us of Euclid's distinction between a fraction which is 'a part', i. e. an aliquot part or submultiple, and one which is 'parts', i. e. some number more than one of such parts, e. g. $\frac{3}{7}$. If equals be added to unequals, the sums differ by the same amount as the original unequals did:[4] an axiom in a rather more complete form than that subsequently interpolated in Euclid.

Summary of the mathematics in Plato.

The actual arithmetical and geometrical propositions referred to or presupposed in Plato's writings are not such as to suggest that he was in advance of his time in mathematics; his knowledge does not appear to have been more than up to date. In the following paragraphs I have attempted to give a summary, as complete as possible, of the mathematics contained in the dialogues.

A proposition in proportion is quoted in the *Parmenides*,[5] namely that, if $a > b$, then $(a+c):(b+c) < a:b$.

In the *Laws* a certain number, 5,040, is selected as a most convenient number of citizens to form a state; its advantages are that it is the product of 12, 21 and 20, that a twelfth part of it is again divisible by 12, and that it has as many as 59 different divisors in all, including all the natural numbers from 1 to 12 with the exception of 11, while it is nearly divisible by 11 (5038 being a multiple of 11).[6]

(α) *Regular and semi-regular solids.*

The 'so-called Platonic figures', by which are meant the five regular solids, are of course not Plato's discovery, for they had been partly investigated by the Pythagoreans, and very

[1] *Parmenides*, 137 E.
[2] *Timaeus*, 33 B, 34 B.
[3] *Parmenides*, 153 D.
[4] *Ib.* 154 B.
[5] *Ib.* 154 D.
[6] *Laws*, 537 E–538 A.

fully by Theaetetus; they were evidently only called Platonic because of the use made of them in the *Timaeus*, where the particles of the four elements are given the shapes of the first four of the solids, the pyramid or tetrahedron being appropriated to fire, the octahedron to air, the icosahedron to water, and the cube to earth, while the Creator used the fifth solid, the dodecahedron, for the universe itself.[1]

According to Heron, however, Archimedes, who discovered thirteen semi-regular solids inscribable in a sphere, said that

'Plato also knew one of them, the figure with fourteen faces, of which there are two sorts, one made up of eight triangles and six squares, of earth and air, and already known to some of the ancients, the other again made up of eight squares and six triangles, which seems to be more difficult.'[2]

The first of these is easily obtained; if we take each square face of a cube and make in it a smaller square by joining the middle points of each pair of consecutive sides, we get six squares (one in each face); taking the three out of the twenty-four sides of these squares which are about any one angular point of the cube, we have an equilateral triangle; there are eight of these equilateral triangles, and if we cut off from the corners of the cube the pyramids on these triangles as bases, we have a semi-regular polyhedron inscribable in a sphere and having as faces eight equilateral triangles and six squares. The description of the second semi-regular figure with fourteen faces is wrong: there are only two more such figures, (1) the figure obtained by cutting off from the corners of the cube smaller pyramids on equilateral triangular bases such that regular *octagons*, and not squares, are left in the six square faces, the figure, that is, contained by eight triangles and six octagons, and (2) the figure obtained by cutting off from the corners of an *octahedron* equal pyramids with square bases such as to leave eight regular hexagons in the eight faces, that is, the figure contained by six squares and eight hexagons.

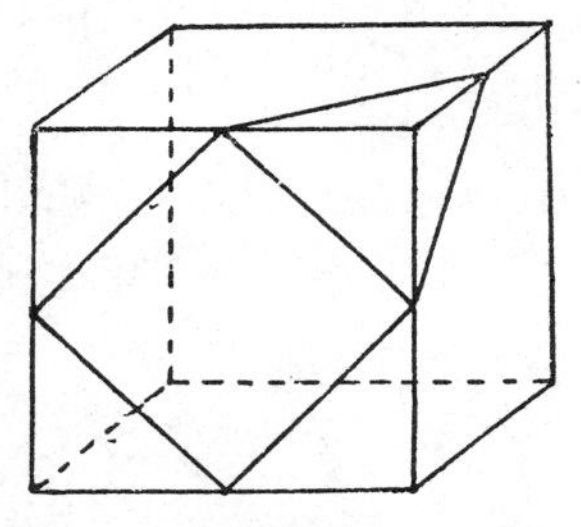

[1] *Timaeus*, 55 D–56 B, 55 C.
[2] Heron, *Definitions*, 104, p. 66, Heib.

(β) *The construction of the regular solids.*

Plato, of course, constructs the regular solids by simply putting together the plane faces. These faces are, he observes, made up of triangles; and all triangles are decomposable into two right-angled triangles. Right-angled triangles are either (1) isosceles or (2) not isosceles, having the two acute angles unequal. Of the latter class, which is unlimited in number, one triangle is the most beautiful, that in which the square on the perpendicular is triple of the square on the base (i.e. the triangle which is the half of an equilateral triangle obtained by drawing a perpendicular from a vertex on the opposite side). (Plato is here Pythagorizing.[1]) One of the regular solids, the cube, has its faces (squares) made up of the first kind of right-angled triangle, the isosceles, four of them being put together to form the square; three others with equilateral triangles for faces, the tetrahedron, octahedron and icosahedron, depend upon the other species of right-angled triangle only, each face being made up of six (not two) of those right-angled triangles, as shown in the figure; the fifth solid, the dodecahedron, with twelve regular pentagons for faces, is merely alluded to, not described, in the passage before us, and Plato is aware that its faces cannot be constructed out of the two elementary right-angled triangles on which the four other solids depend. That an attempt was made to divide the pentagon into a number of triangular elements is clear from three passages, two in Plutarch[2] and one in Alcinous.[3] Plutarch says that each of the twelve faces of a dodecahedron is made up of thirty elementary scalene triangles which are different from the elementary triangle of the solids with triangular faces. Alcinous speaks of the 360 elements which are produced when each pentagon is divided into five isosceles triangles and each of the

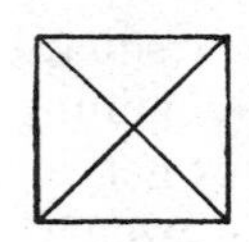

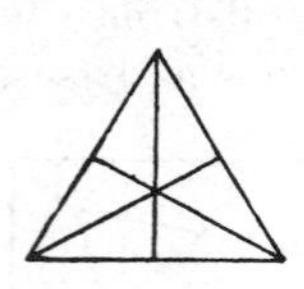

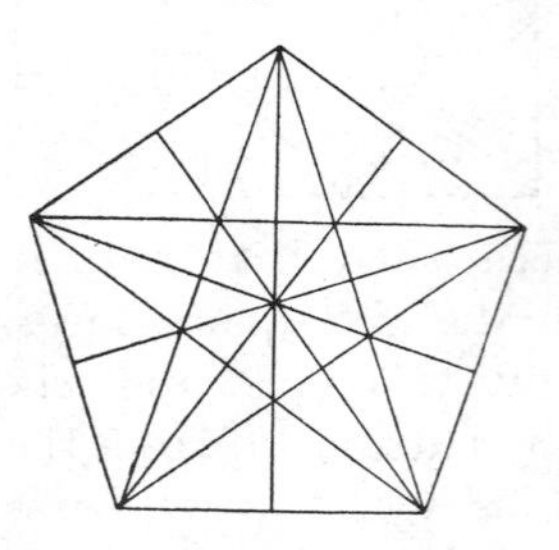

[1] Cf. Speusippus in *Theol. Ar.*, p. 61, Ast.
[2] Plutarch, *Quaest. Plat.* 5. 1, 1003 D; *De defectu Oraculorum*, c. 33, 428 A.
[3] Alcinous, *De Doctrina Platonis*, c. 11.

latter into six scalene triangles. If we draw lines in a pentagon as shown in the accompanying figure, we obtain such a set of triangles in a way which also shows the Pythagorean pentagram (cf. p. 161, above).

(γ) *Geometric means between two square numbers or two cubes.*

In the *Timaeus* Plato, speaking of numbers 'whether solid or square' with a (geometric) mean or means between them, observes that between *planes* one mean suffices, but to connect two *solids* two means are necessary.[1] By *planes* and *solids* Plato probably meant *square* and *cube numbers* respectively, so that the theorems quoted are probably those of Eucl. VIII. 11, 12, to the effect that between two square numbers there is one mean proportional number, and between two cube numbers two mean proportional numbers. Nicomachus quotes these very propositions as constituting 'a certain Platonic theorem'.[2] Here, too, it may be that the theorem is called 'Platonic' for the sole reason that it is quoted by Plato in the *Timaeus*; it may well be older, for the idea of two mean proportionals between two straight lines had already appeared in Hippocrates's reduction of the problem of doubling the cube. Plato's allusion does not appear to be to the duplication of the cube in this passage any more than in the expression κύβων αὔξη, 'cubic increase', in the *Republic*,[3] which appears to be nothing but the addition of the third dimension to a square, making a cube (cf. τρίτη αὔξη, 'third increase',[4] meaning a cube number as compared with δύναμις, a square number, terms which are applied, e.g. to the numbers 729 and 81 respectively).

(δ) *The two geometrical passages in the* MENO.

We come now to the two geometrical passages in the *Meno*. In the first[5] Socrates is trying to show that teaching is only reawaking in the mind of the learner the memory of something. He illustrates by putting to the slave a carefully prepared series of questions, each requiring little more than

[1] *Timaeus*, 31 C–32 B.
[2] Nicom. ii. 24. 6.
[3] *Republic*, 528 B.
[4] *Ib.* 587 D.
[5] *Meno*, 82 B–85 B.

'yes' or 'no' for an answer, but leading up to the geometrical construction of $\sqrt{2}$. Starting with a straight line *AB* 2 feet long, Socrates describes a square *ABCD* upon it and easily shows that the area is 4 square feet. Producing the sides *AB*, *AD* to *G*, *K* so that *BG*, *DK* are equal to *AB*, *AD*, and completing the figure, we have a square of side 4 feet, and this square is equal to four times the original square and therefore has an area of 16 square feet. Now, says Socrates, a square 8 feet in area must have its side greater than 2 and less than 4 feet. The slave suggests that it is 3 feet in length. By taking *N* the middle point of *DK* (so that *AN* is 3 feet) and completing the square on *AN*, Socrates easily shows that the square on *AN* is not 8 but 9 square feet in area. If *L*, *M* be the middle points of *GH*, *HK* and *CL*, *CM* be joined, we have four squares in the figure, one of which is *ABCD*, while each of the others is equal to it. If now we draw the diagonals *BL*, *LM*, *MD*, *DB* of the four squares, each diagonal bisects its square, and the four make a square *BLMD*, the area of which is half that of the square *AGHK*, and is therefore 8 square feet; *BL* is a side of this square. Socrates concludes with the words:

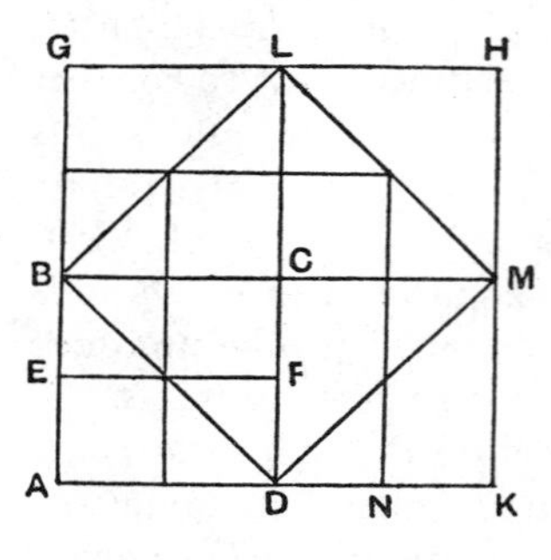

'The Sophists call this straight line (*BD*) the *diameter* (diagonal); this being its name, it follows that the square which is double (of the original square) has to be described on the diameter.'

The other geometrical passage in the *Meno* is much more difficult,[1] and it has gathered round it a literature almost comparable in extent to the volumes that have been written to explain the Geometrical Number of the *Republic*. C. Blass, writing in 1861, knew thirty different interpretations; and since then many more have appeared. Of recent years Benecke's interpretation[2] seems to have enjoyed the most

[1] *Meno*, 86 E–87 C.

[2] Dr. Adolph Benecke, *Ueber die geometrische Hypothesis in Platon's Menon* (Elbing, 1867). See also below, pp. 302-3.

acceptance; nevertheless, I think that it is not the right one, but that the essentials of the correct interpretation were given by S. H. Butcher[1] (who, however, seems to have been completely anticipated by E. F. August, the editor of Euclid, in 1829). It is necessary to begin with a literal translation of the passage. Socrates is explaining a procedure 'by way of hypothesis', a procedure which, he observes, is illustrated by the practice of geometers

'when they are asked, for example, as regards a given area, whether it is possible for this area to be inscribed in the form of a triangle in a given circle. The answer might be, "I do not yet know whether this area is such as can be so inscribed, but I think I can suggest a hypothesis which will be useful for the purpose; I mean the following. If the given area is such as, when one has applied it (as a rectangle) to the given straight line in the circle [τὴν δοθεῖσαν αὐτοῦ γραμμήν, the given straight line *in it*, cannot, I think, mean anything but the *diameter* of the circle[2]], it is deficient by a figure (rectangle) similar to the very figure which is applied, then one alternative seems to me to result, while again another results if it is impossible for what I said to be done with it. Accordingly, by using a hypothesis, I am ready to tell you what results with regard to the inscribing of the figure in the circle, namely, whether the problem is possible or impossible."'

Let AEB be a circle on AB as diameter, and let AC be the tangent at A. Take E any point on the circle and draw ED perpendicular to AB. Complete the rectangles $ACED$, $EDBF$.

Then it is clear that the rectangle $CEDA$ is 'applied' to the diameter AB, and also that it 'falls short' by a figure, the rectangle $EDBF$, similar to the 'applied' rectangle, for

$$AD : DE = ED : DB.$$

Also, if ED be produced to meet the circle again in G, AEG is an isosceles triangle bisected by the diameter AB, and therefore equal in area to the rectangle $ACED$.

If then the latter rectangle, 'applied' to AB in the manner

[1] *Journal of Philology*, vol. xvii, pp. 219-25; cf. E. S. Thompson's edition of the *Meno*.

[2] The obvious 'line' of a circle is its diameter, just as, in the first geometrical passage about the squares, the γραμμή, the 'line', of a square is its *side*.

described, is equal to the given area, that area is inscribed in the form of a triangle in the given circle.[1]

In order, therefore, to inscribe in the circle an isosceles triangle equal to the given area (X), we have to find a point E on the circle such that, if ED be drawn perpendicular to AB,

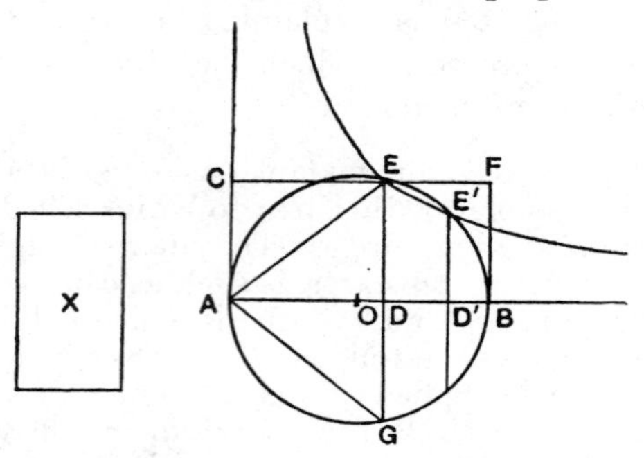

the rectangle $AD \cdot DE$ is equal to the given area X ('applying' to AB a rectangle equal to X and falling short by a figure similar to the 'applied' figure is only another way of expressing it). Evidently E lies on the rectangular hyperbola

[1] Butcher, after giving the essentials of the interpretation of the passage quite correctly, finds a difficulty. 'If', he says, 'the condition' (as interpreted by him) 'holds good, the given χωρίον can be inscribed in a circle. But the converse proposition is not true. The χωρίον can still be inscribed, as required, even if the condition laid down is not fulfilled; the true and necessary condition being that the given area is not greater than that of the equilateral triangle, i.e. the *maximum* triangle, which can be inscribed in the given circle.' The difficulty arises in this way. Assuming (quite fairly) that the given area is given in the form of a rectangle (for any given rectilineal figure can be transformed into a rectangle of equal area), Butcher seems to suppose that it is identically the given rectangle that is applied to AB. But this is not necessary. The terminology of mathematics was not quite fixed in Plato's time, and he allows himself some latitude of expression, so that we need not be surprised to find him using the phrase 'to apply the area (χωρίον) to a given straight line' as short for 'to apply to a given straight line a *rectangle equal* (but not similar) to the given area' (cf. Pappus vi, p. 544. 8–10 μὴ πᾶν τὸ δοθὲν παρὰ τὴν δοθεῖσαν παραβάλλεσθαι ἐλλεῖπον τετραγώνῳ, 'that it is not every given (area) that can be applied (in the form of a rectangle) falling short by a square figure'). If we interpret the expression in this way, the converse *is* true; if we cannot apply, in the way described, a rectangle *equal* to the given rectangle, it is because the given rectangle is greater than the equilateral, i.e. the maximum, triangle that can be inscribed in the circle, and the problem is therefore impossible of solution. (It was not till long after the above was written that my attention was drawn to the article on the same subject in the *Journal of Philology*, xxviii, 1903, pp. 222–40, by Professor Cook Wilson. I am gratified to find that my interpretation of the passage agrees with his.)

the equation of which referred to AB, AC as axes of x, y is $xy = b^2$, where b^2 is equal to the given area. For a real solution it is necessary that b^2 should be not greater than the equilateral triangle inscribed in the circle, i.e. not greater than $3\sqrt{3}\,.\,a^2/4$, where a is the radius of the circle. If b^2 is equal to this area, there is only one solution (the hyperbola in that case touching the circle); if b^2 is less than this area, there are two solutions corresponding to two points E, E' in which the hyperbola cuts the circle. If $AD = x$, we have $OD = x-a$, $DE = \sqrt{(2ax-x^2)}$, and the problem is the equivalent of solving the equation

$$x\sqrt{(2ax-x^2)} = b^2,$$

or
$$x^2(2ax-x^2) = b^4.$$

This is an equation of the fourth degree which can be solved by means of conics, but not by means of the straight line and circle. The solution is given by the points of intersection of the hyperbola $xy = b^2$ and the circle $y^2 = 2ax-x^2$ or $x^2+y^2 = 2ax$. In this respect therefore the problem is like that of finding the two mean proportionals, which was likewise solved, though not till later, by means of conics (Menaechmus). I am tempted to believe that we have here an allusion to another actual problem, requiring more than the straight line and circle for its solution, which had exercised the minds of geometers by the time of Plato, the problem, namely, of inscribing in a circle a triangle equal to a given area, a problem which was still awaiting a solution, although it had been reduced to the problem of

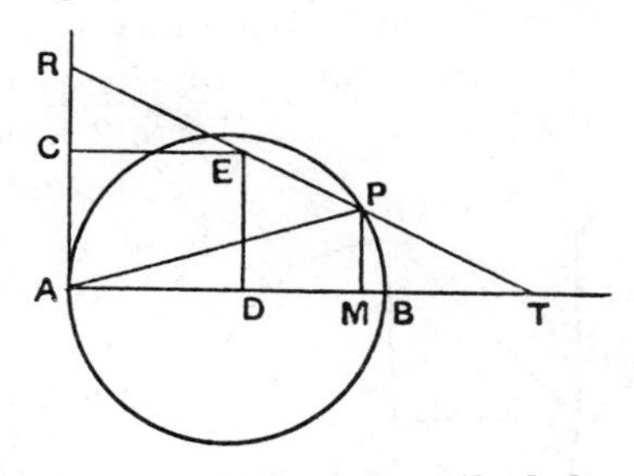

applying a rectangle satisfying the condition described by Plato, just as the duplication of the cube had been reduced to the problem of finding two mean proportionals. Our problem can, like the latter problem, easily be solved by the 'mechanical' use of a ruler. Suppose that the given rectangle is placed so that the side AD lies along the diameter AB of the circle. Let E be the angle of the rectangle $ADEC$ opposite to A. Place a ruler so that it passes through E and turn it about E until it passes through a point P of the circle such that, if EP meets AB and AC produced in T, R, PT shall be equal to ER. Then, since $RE = PT$, $AD = MT$, where M is the foot of the ordinate PM.

Therefore $DT = AM$, and

$$AM : AD = DT : MT$$
$$= ED : PM,$$

whence $$PM . MA = ED . DA,$$

and APM is the half of the required (isosceles) triangle.

Benecke criticizes at length the similar interpretation of the passage given by E. F. August. So far, however, as his objections relate to the translation of particular words in the Greek text, they are, in my opinion, not well founded.[1] For the rest, Benecke holds that, in view of the difficulty of the problem which emerges, Plato is unlikely to have introduced it in such an abrupt and casual way into the conversation between Socrates and Meno. But the problem is only one of the same nature as that of the finding of two mean proportionals which was already a famous problem, and, as regards the form of the allusion, it is to be noted that Plato was fond of dark hints in things mathematical.

If the above interpretation is too difficult (which I, for one, do not admit), Benecke's is certainly too easy. He connects his interpretation of the passage with the earlier passage about the square of side 2 feet; according to him the problem is, can an isosceles *right-angled* triangle equal to the said square be inscribed in the given circle? This is of course only possible if the radius of the circle is 2 feet in length. If AB, DE be two diameters at right angles, the inscribed triangle is ADE; the square $ACDO$ formed by the radii AO, OD and the tangents at D, A is then the 'applied' rectangle, and

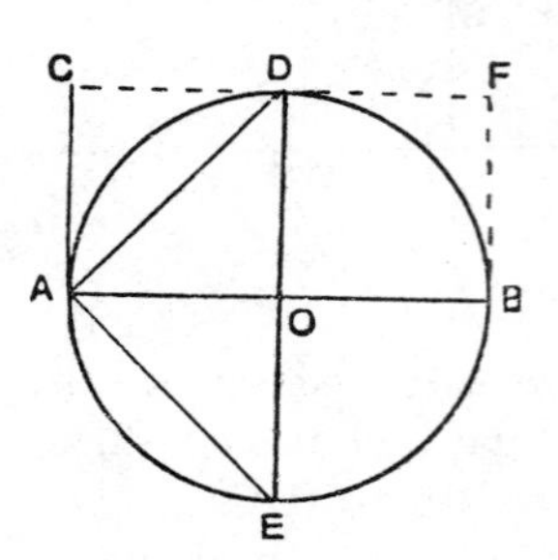

the rectangle by which it falls short is also a square and equal

[1] The main point of Benecke's criticisms under this head has reference to τοιούτῳ χωρίῳ οἷον in the phrase ἐλλείπειν τοιούτῳ χωρίῳ οἷον ἂν αὐτὸ τὸ παρατ.ταμένον ᾖ. He will have it that τοιούτῳ οἷον cannot mean 'similar to',

to the other square. If this were the correct interpretation, Plato is using much too general language about the applied rectangle and that by which it is deficient; it would be extraordinary that he should express the condition in this elaborate way when he need only have said that the radius of the circle must be equal to the side of the square and therefore 2 feet in length. The explanation seems to me incredible. The criterion sought by Socrates is evidently intended to be a real *διορισμός*, or determination of the conditions or limits of the possibility of a solution of the problem whether in its original form or in the form to which it is reduced; but it is no real *διορισμός* to say what is equivalent to saying that the problem is possible of solution if the circle is of a particular size, but impossible if the circle is greater or less than that size.

The passage incidentally shows that the idea of a formal *διορισμός* defining the limits of possibility of solution was familiar even before Plato's time, and therefore that Proclus must be in error when he says that Leon, the pupil of Neoclides, '*invented διορισμοί* (determining) when the problem which is the subject of investigation is possible and when impossible',[1] although Leon may have been the first to introduce the term or to recognize formally the essential part played by *διορισμοί* in geometry.

(ε) *Plato and the doubling of the cube.*

The story of Plato's relation to the problem of doubling the cube has already been told (pp. 245–6, 255). Although the solution attributed to him is not his, it may have been with this problem in view that he complained that the study of solid geometry had been unduly neglected up to his time.[2]

and he maintains that, if Plato had meant it in this sense, he should have added that the 'defect', although 'similar', is not similarly situated. I see no force in this argument in view of the want of fixity in mathematical terminology in Plato's time, and of his own habit of varying his phrases for literary effect. Benecke makes the words mean 'of the same *kind*', e. g. a square with a square or a rectangle with a rectangle. But this would have no point unless the figures are *squares*, which begs the whole question.

[1] Proclus on Eucl. I, p. 66. 20–2.

[2] *Republic*, vii. 528 A–C.

(ζ) *Solution of $x^2+y^2=z^2$ in integers.*

We have already seen (p. 81) that Plato is credited with a rule (complementary to the similar rule attributed to Pythagoras) for finding a whole series of square numbers the sum of which is also a square; the formula is

$$(2n)^2+(n^2-1)^2=(n^2+1)^2.$$

(η) *Incommensurables.*

On the subject of incommensurables or irrationals we have first the passage of the *Theaetetus* recordin that Theodorus proved the incommensurability of $\sqrt{3}$, $\sqrt{5}\ldots\sqrt{17}$, after which Theaetetus generalized the theory of such 'roots'. This passage has already been fully discussed (pp. 203–9). The subject of incommensurables comes up again in the *Laws*, where Plato inveighs against the ignorance prevailing among the Greeks of his time of the fact that lengths, breadths and depths may be incommensurable as well as commensurable with one another, and appears to imply that he himself had not learnt the fact till late (ἀκούσας ὀψέ ποτε), so that he was ashamed for himself as well as for his countrymen in general.[1] But the irrationals known to Plato included more than mere 'surds' or the sides of non-squares; in one place he says that, just as an even number may be the sum of either two odd or two even numbers, the sum of two irrationals may be either rational or irrational.[2] An obvious illustration of the former case is afforded by a rational straight line divided 'in extreme and mean ratio'. Euclid (XIII. 6) proves that each of the segments is a particular kind of irrational straight line called by him in Book X an *apotome*; and to suppose that the irrationality of the two segments was already known to Plato is natural enough if we are correct in supposing that 'the theorems which' (in the words of Proclus) 'Plato originated regarding *the section*'[3] were theorems about what came to be called the 'golden section', namely the division of a straight line in extreme and mean ratio as in Eucl. II. 11 and VI. 30. The appearance of the latter problem in Book II, the content of which is probably all Pythagorean, suggests that the incommensurability of the segments with

[1] *Laws*, 819 D–820 C. [2] *Hippias Maior*, 303 B, C.
[3] Proclus on Eucl. I, p. 67. 6.

the whole line was discovered before Plato's time, if not as early as the irrationality of $\sqrt{2}$.

(θ) *The Geometrical Number.*

This is not the place to discuss at length the famous passage about the Geometrical Number in the *Republic*.[1] Nor is its mathematical content of importance; the whole thing is mystic rather than mathematical, and is expressed in rhapsodical language, veiling by fanciful phraseology a few simple mathematical conceptions. The numbers mentioned are supposed to be two. Hultsch and Adam arrive at the same two numbers, though by different routes. The first of these numbers is 216, which according to Adam is the sum of three cubes $3^3 + 4^3 + 5^3$; $2^3 . 3^3$ is the form in which Hultsch obtains it.[2]

[1] *Republic*, viii. 546 B–D. The number of interpretations of this passage is legion. For an exhaustive discussion of the language as well as for one of the best interpretations that has been put forward, see Dr. Adam's edition of the *Republic*, vol. ii, pp. 204–8, 264–312.

[2] The Greek is *ἐν ᾧ πρώτῳ αὐξήσεις δυνάμεναί τε καὶ δυναστευόμεναι, τρεῖς ἀποστάσεις, τέτταρας δὲ ὅρους λαβοῦσαι ὁμοιούντων τε καὶ ἀνομοιούντων καὶ αὐξόντων καὶ φθινόντων, πάντα προσήγορα καὶ ῥητὰ πρὸς ἄλληλα ἀπέφηναν*, which Adam translates by 'the first number in which root and square increases, comprehending three distances and four limits, of elements that make like and unlike and wax and wane, render all things conversable and rational with one another'. *αὐξήσεις* are clearly multiplications. *δυνάμεναί τε καὶ δυναστευόμεναι* are explained in this way. A straight line is said *δύνασθαι* ('to be capable of') an area, e. g. a rectangle, when the square on it is equal to the rectangle; hence *δυναμένη* should mean a side of a square. *δυναστευομένη* represents a sort of passive of *δυναμένη*, meaning that of which the *δυναμένη* is 'capable'; hence Adam takes it here to be the square of which the *δυναμένη* is the side, and the whole expression to mean the product of a square and its side, i. e. simply the cube of the side. The cubes 3^3, 4^3, 5^3 are supposed to be meant because the words in the description of the second number 'of which the ratio in its lowest terms 4:3 when joined to 5' clearly refer to the right-angled triangle 3, 4, 5, and because at least three authors, Plutarch (*De Is. et Os.* 373 F), Proclus (on Eucl. I, p. 428. 1) and Aristides Quintilianus (*De mus.*, p. 152 Meibom. = p. 90 Jahn) say that Plato used the Pythagorean or 'cosmic' triangle in his Number. The 'three distances' are regarded as 'dimensions', and the 'three distances and four limits' are held to confirm the interpretation 'cube', because a solid (parallelepiped) was said to have 'three dimensions and four limits' (*Theol. Ar.*, p. 16 Ast, and Iambl. *in Nicom.*, p. 93. 10), the limits being bounding points as A, B, C, D in the accompanying figure. 'Making like and unlike' is supposed to refer to the square and oblong forms in which the second

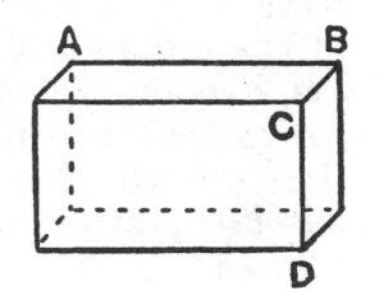

The second number is described thus:

'The ratio 4 : 3 in its lowest terms ('the base', πυθμήν, of the ratio ἐπίτριτος) joined or wedded to 5 yields two harmonies when thrice increased (τρὶς αὐξηθείς), the one equal an equal number of times, so many times 100, the other of equal length one way, but oblong, consisting on the one hand of 100 squares of rational diameters of 5 diminished by one each or, if of

number is stated.

Another view of the whole passage has recently appeared (A. G. Laird, *Plato's Geometrical Number and the comment of Proclus*, Madison, Wisconsin, 1918). Like all other solutions, it is open to criticism in some details, but it is attractive in so far as it makes greater use of Proclus (*in Platonis remp.*, vol. ii, p. 36 seq. Kroll) and especially of the passage (p. 40) in which he illustrates the formation of the 'harmonies' by means of geometrical figures. According to Mr. Laird there are not *two* separate numbers, and the description from which Hultsch and Adam derive the number 216 is not a description of a number but a statement of a general method of formation of 'harmonies', which is then applied to the triangle 3, 4, 5 as a particular case, in order to produce the one Geometrical Number. The basis of the whole thing is the use of figures like that of Eucl. VI. 8 (a right-angled triangle divided by a perpendicular from the right angle on the opposite side into two right-angled triangles similar to one another and to the original triangle). Let ABC be a right-angled triangle in which the sides CB, BA containing the right angle are rational numbers a, b respectively. Draw AF at right angles to AC meeting CB produced in F. Then the figure AFC is that of Eucl. VI. 8, and of course $AB^2 = CB.BF$. Complete the rectangle $ABFL$, and produce FL, CA to meet at K. Then, by similar triangles, CB, BA, FB $(= AL)$ and KL are four straight lines in continued proportion, and their lengths are a, b, b^2/a, b^3/a^2 respectively. Multiplying throughout by a^2 in order to get rid of fractions, we may take the lengths to be a^3, a^2b, ab^2, b^3 respectively. Now, on Mr. Laird's view, αὐξήσεις δυνάμεναι are *squares*, as AB^2, and αὐξήσεις δυναστευόμεναι *rectangles*, as *FB*, *BC*, *to which the squares are equal*. 'Making like and unlike' refers to the equal factors of a^3, b^3 and the unequal factors of a^2b, ab^2; the terms a^3, a^2b, ab^2, b^3 are four *terms* (ὅροι) of a continued proportion with three *intervals* (ἀποστάσεις), and of course are all 'conversable and rational with one another'. (Incidentally, out of such terms we can even obtain the number 216, for if we put $a = 2$, $b = 3$, we have 8, 12, 18, 27, and the product of the extremes $8 . 27 =$ the product of the means $12 . 18 = 216$). Applying the method to the triangle 3, 4, 5 (as Proclus does) we have the terms 27, 36, 48, 64, and the first three numbers, multiplied respectively by 100, give the elements of the Geometrical Number $3600^2 = 2700 . 4800$. On this interpretation τρὶς αὐξηθείς simply means raised to the third dimension or 'made solid' (as Aristotle says, *Politics* Θ (E). 12, 1316 a 8), the factors being of course $3 . 3 . 3 = 27$, $3 . 3 . 4 = 36$, and $3 . 4 . 4 = 48$; and 'the ratio 4 : 3 joined to 5' does not mean either the product or the sum of 3, 4, 5, but simply the triangle 3, 4, 5.

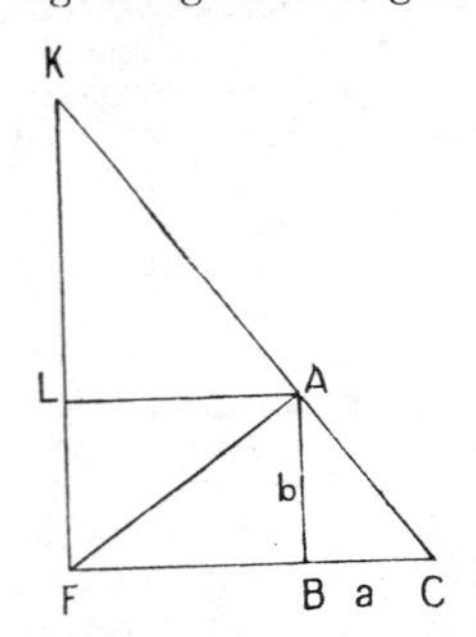

irrational diameters, by two, and on the other hand of 100 cubes of 3.'

The ratio 4 : 3 must be taken in the sense of 'the numbers 4 and 3', and Adam takes 'joined with 5' to mean that 4, 3 and 5 are multiplied together, making 60; 60 'thrice increased' he interprets as '60 thrice multiplied by 60', that is to say, $60 \times 60 \times 60 \times 60$ or 3600^2; 'so many times 100' must then be the 'equal' side of this, or 36 times 100; this 3600^2, or 12960000, is one of the 'harmonies'. The other is the same number expressed as the product of two unequal factors, an 'oblong' number; the first factor is 100 times a number which can be described either as 1 less than the square of the 'rational diameter of 5', or as 2 less than the square of the 'irrational diameter' of 5, where the irrational diameter of 5 is the diameter of a square of side 5, i.e. $\sqrt{50}$, and the rational diameter is the nearest whole number to this, namely 7, so that the number which is multiplied by 100 is $49-1$, or $50-2$, i.e. 48, and the first factor is therefore 4800; the second factor is 100 cubes of 3, or 2700; and of course $4800 \times 2700 = 3600^2$ or 12960000. Hultsch obtains the side, 3600, of the first 'harmony' in another way; he takes 4 and 3 joined to 5 to be the *sum* of 4, 3 and 5, i.e. 12, and τρὶς αὐξηθείς, 'thrice increased', to mean that the 12 is 'multiplied by three'[1] making 36; 'so many times 100' is then 36 times 100, or 3600.

But the main interest of the passage from the historical

[1] Adam maintains that τρὶς αὐξηθείς cannot mean 'multiplied by 3'. He observes (p. 278, note) that the Greek for 'multiplied by 3', if we use αὐξάνω, would be τριάδι αὐξηθείς, this being the construction used by Nicomachus (ii. 15. 2 ἵνα ὁ θ τρὶς γ ὢν πάλιν τριάδι ἐπ' ἄλλο διάστημα αὐξηθῇ καὶ γένηται ὁ κζ) and in *Theol. Ar.* (p. 39, Ast ἑξάδι αὐξηθείς). Nevertheless I think that τρὶς αὐξηθείς would not be an unnatural expression for a mathematician to use for 'multiplied by 3', let alone Plato in a passage like this. It is to be noted that πολλαπλασιάζω and πολλαπλάσιος are likewise commonly used with the dative of the multiplier; yet ἰσάκις πολλαπλάσιος is the regular expression for 'equimultiple'. And αὐξάνω is actually found with τοσαυτάκις: see Pappus ii, p. 28. 15, 22, where τοσαυτάκις αὐξήσομεν means 'we have to multiply by such a power' of 10000 or of 10 (although it is true that the chapter in which the expression occurs may be a late addition to Pappus's original text). On the whole, I prefer Hultsch's interpretation to Adam's. τρὶς αὐξηθείς can hardly mean that 60 is raised to the *fourth* power, 60^4; and if it did, 'so many times 100', immediately following the expression for 3600^2, would be pointless and awkward. On the other hand, 'so many times 100' following the expression for 36 would naturally indicate 3600.

point of view lies in the terms 'rational' and 'irrational diameter of 5'. A fair approximation to $\sqrt{2}$ was obtained by selecting a square number such that, if 2 be multiplied by it, the product is nearly a square; 25 is such a square number, since 25 times 2, or 50, only differs by 1 from 7^2; consequently $\frac{7}{5}$ is an approximation to $\sqrt{2}$. It may have been arrived at in the tentative way here indicated; we cannot doubt that it was current in Plato's time; nay, we know that the general solution of the equations

$$x^2 - 2y^2 = \pm 1$$

by means of successive 'side-' and 'diameter-' numbers was Pythagorean, and Plato was therefore, here as in so many other places, 'Pythagorizing'.

The diameter is again mentioned in the *Politicus*, where Plato speaks of 'the diameter which is in square (δυνάμει) two feet', meaning the diagonal of the square with side 1 foot, and again of the diameter of the square on this diameter, i.e. the diagonal of a square 2 square feet in area, in other words, the side of a square 4 square feet in area, or a straight line 2 feet in length.[1]

Enough has been said to show that Plato was abreast of the mathematics of his day, and we can understand the remark of Proclus on the influence which he exerted upon students and workers in that field:

'he caused mathematics in general and geometry in particular to make a very great advance by reason of his enthusiasm for them, which of course is obvious from the way in which he filled his books with mathematical illustrations and everywhere tries to kindle admiration for these subjects in those who make a pursuit of philosophy.'[2]

Mathematical 'arts'.

Besides the purely theoretical subjects, Plato recognizes the practical or applied mathematical 'arts'; along with arithmetic, he mentions the art of measurement (for purposes of trade or craftsmanship) and that of weighing[3]; in the former connexion he speaks of the instruments of the craftsman, the circle-drawer (τόρνος), the compasses (διαβήτης), the rule

[1] *Politicus*, 266 B.

[2] Proclus on Eucl. I, p. 66. 8–14.

[3] *Philebus*, 55 E–56 E.

(στάθμη) and 'a certain elaborate προσαγώγιον' (? approximator). The art of weighing, he says,[1] 'is concerned with the heavier and lighter weight', as 'logistic' deals with odd and even in their relation to one another, and geometry with magnitudes greater and less or equal; in the *Protagoras* he speaks of the man skilled in weighing

'who puts together first the pleasant, and second the painful things, and adjusts the near and the far on the balance'[2];

the principle of the lever was therefore known to Plato, who was doubtless acquainted with the work of Archytas, the reputed founder of the science of mechanics.[3]

(α) *Optics.*

In the physical portion of the *Timaeus* Plato gives his explanation of the working of the sense organs. The account of the process of vision and the relation of vision to the light of day is interesting,[4] and at the end of it is a reference to the properties of mirrors, which is perhaps the first indication of a science of optics. When, says Plato, we see a thing in a mirror, the fire belonging to the face combines about the bright surface of the mirror with the fire in the visual current; the right portion of the face appears as the left in the image seen, and vice versa, because it is the mutually opposite parts of the visual current and of the object seen which come into contact, contrary to the usual mode of impact. (That is, if you imagine your reflection in the mirror to be another person looking at you, *his* left eye is the image of your right, and the left side of *his* left eye is the image of the right side of your right.) But, on the other hand, the right side really becomes the right side and the left the left when the light in combination with that with which it combines is transferred from one side to the other; this happens when the smooth part of the mirror is higher at the sides than in the middle (i. e. the mirror is a hollow cylindrical mirror held with its axis vertical), and so diverts the right portion of the visual current to the left and vice versa. And if you turn the mirror so that its axis is horizontal, everything appears upside down.

[1] *Charmides*, 166 B.

[2] *Protagoras*, 356 B.

[3] Diog. L. viii. 83.

[4] *Timaeus*, 45 B–46 C.

(β) *Music.*

In music Plato had the advantage of the researches of Archytas and the Pythagorean school into the numerical relations of tones. In the *Timaeus* we find an elaborate filling up of intervals by the interposition of arithmetic and harmonic means[1]; Plato is also clear that higher and lower pitch are due to the more or less rapid motion of the air.[2] In like manner the different notes in the 'harmony of the spheres', poetically turned into Sirens sitting on each of the eight whorls of the Spindle and each uttering a single sound, a single musical note, correspond to the different speeds of the eight circles, that of the fixed stars and those of the sun, the moon, and the five planets respectively.[3]

(γ) *Astronomy.*

This brings us to Plato's astronomy. His views are stated in their most complete and final form in the *Timaeus*, though account has to be taken of other dialogues, the *Phaedo*, the *Republic*, and the *Laws*. He based himself upon the early Pythagorean system (that of Pythagoras, as distinct from that of his successors, who were the first to abandon the geocentric system and made the earth, with the sun, the moon and the other planets, revolve in circles about the 'central fire'); while of course he would take account of the results of the more and more exact observations made up to his own time. According to Plato, the universe has the most perfect of all shapes, that of a sphere. In the centre of this sphere rests the earth, immovable and kept there by the equilibrium of symmetry as it were ('for a thing in equilibrium in the middle of any uniform substance will not have cause to incline more or less in any direction'[4]). The axis of the sphere of the universe passes through the centre of the earth, which is also spherical, and the sphere revolves uniformly about the axis in the direction from east to west. The fixed stars are therefore carried round in small circles of the sphere. The sun, the moon and the five planets are also carried round in the motion of the outer sphere, but they have independent circular movements of their own in addition.

[1] *Timaeus*, 35 C–36 B.
[2] *Ib.* 67 B.
[3] *Republic*, 617 B.
[4] *Phaedo*, 109 A.

These latter movements take place in a plane which cuts at an angle the equator of the heavenly sphere; the several orbits are parts of what Plato calls the 'circle of the Other', as distinguished from the 'circle of the Same', which is the daily revolution of the heavenly sphere as a whole and which, carrying the circle of the Other and the seven movements therein along with it, has the mastery over them. The result of the combination of the two movements in the case of any one planet is to twist its actual path in space into a spiral[1]; the spiral is of course included between two planes parallel to that of the equator at a distance equal to the maximum deviation of the planet in its course from the equator on either side. The speeds with which the sun, the moon and the five planets describe their own orbits (independently of the daily rotation) are in the following order; the moon is the quickest; the sun is the next quickest and Venus and Mercury travel in company with it, each of the three taking about a year to describe its orbit; the next in speed is Mars, the next Jupiter, and the last and slowest is Saturn; the speeds are of course angular speeds, not linear. The order of distances from the earth is, beginning with the nearest, as follows: moon, sun, Venus, Mercury, Mars, Jupiter, Saturn. In the *Republic* all these heavenly bodies describe their own orbits in a sense opposite to that of the daily rotation, i. e. in the direction from west to east; this is what we should expect; but in the *Timaeus* we are distinctly told, in one place, that the seven circles move 'in opposite senses to one another',[2] and, in another place, that Venus and Mercury have 'the contrary tendency' to the sun.[3] This peculiar phrase has not been satisfactorily interpreted. The two statements taken together in their literal sense appear to imply that Plato actually regarded Venus and Mercury as describing their orbits the contrary way to the sun, incredible as this may appear (for on this hypothesis the angles of divergence between the two planets and the sun would be capable of any value up to 180°, whereas observation shows that they are never far from the sun). Proclus and others refer to attempts to explain the passages by means of the theory of epicycles; Chalcidius in particular indicates that the sun's motion on its

[1] *Timaeus*, 38 E–39 B. [2] *Ib.* 36 D. [3] *Ib.* 38 D.

epicycle (which is from east to west) is in the contrary sense to the motion of Venus and Mercury on their epicycles respectively (which is from west to east)[1]; and this would be a satisfactory explanation if Plato could be supposed to have been acquainted with the theory of epicycles. But the probabilities are entirely against the latter supposition. All, therefore, that can be said seems to be this. Heraclides of Pontus, Plato's famous pupil, is known on clear evidence to have discovered that Venus and Mercury revolve round the sun like satellites. He may have come to the same conclusion about the superior planets, but this is not certain; and in any case he must have made the discovery with reference to Mercury and Venus first. Heraclides's discovery meant that Venus and Mercury, while accompanying the sun in its annual motion, described what are really epicycles about it. Now discoveries of this sort are not made without some preliminary seeking, and it may have been some vague inkling of the truth that prompted the remark of Plato, whatever the precise meaning of the words.

The differences between the angular speeds of the planets account for the overtakings of one planet by another, and the combination of their independent motions with that of the daily rotation causes one planet to *appear* to be overtaking another when it is really being overtaken by it and vice versa.[2] The sun, moon and planets are instruments for measuring time.[3] Even the earth is an instrument for making night and day by virtue of its *not* rotating about its axis, while the rotation of the fixed stars carrying the sun with it is completed once in twenty-four hours; a month has passed when the moon after completing her own orbit overtakes the sun (the 'month' being therefore the *synodic* month), and a year when the sun has completed its own circle. According to Plato the time of revolution of the other planets (except Venus and Mercury, which have the same speed as the sun) had not been exactly calculated; nevertheless the Perfect Year is completed 'when the relative speeds of all the eight revolutions [the seven independent revolutions and the daily rotation] accomplish their course together and reach their

[1] Chalcidius on *Timaeus*, cc. 81, 109, 112. [2] *Timaeus*, 39 A.
[3] *Ib.* 41 E, 42 D.

starting-point'.[1] There was apparently a tradition that the Great Year of Plato was 36000 years: this corresponds to the minimum estimate of the precession of the equinoxes quoted by Ptolemy from Hipparchus's treatise on the length of the year, namely at least one-hundredth of a degree in a year, or 1° in 100 years,[2] that is to say, 360° in 36000 years. The period is connected by Adam with the Geometrical Number 12960000 because this number of days, at the rate of 360 days in the year, makes 36000 years. The coincidence may, it is true, have struck Ptolemy and made him describe the Great Year arrived at on the basis of 1° per 100 years as the 'Platonic' year; but there is nothing to show that Plato himself calculated a Great Year with reference to precession: on the contrary, precession was first discovered by Hipparchus.

As regards the distances of the sun, moon and planets Plato has nothing more definite than that the seven circles are 'in the proportion of the double intervals, three of each'[3]: the reference is to the Pythagorean τετρακτύς represented in the annexed figure, the numbers after 1 being on the one side successive powers of 2, and on the other side successive powers of 3. This gives 1, 2, 3, 4, 8, 9, 27 in ascending order. What precise estimate of relative distances Plato based upon these figures is uncertain. It is generally supposed (1) that the radii of the successive orbits are in the ratio of the numbers; but (2) Chalcidius considered that 2, 3, 4 ... are the successive differences between these radii,[4] so that the radii themselves are in the ratios of 1, $1+2 = 3$, $1+2+3 = 6$, &c.; and again (3), according to Macrobius,[5] the Platonists held that the successive radii are as 1, $1 \cdot 2 = 2$, $1 \cdot 2 \cdot 3 = 6$, $6 \cdot 4 = 24$, $24 \cdot 9 = 216$, $216 \cdot 8 = 1728$ and $1728 \cdot 27 = 46656$. In any case the figures have no basis in observation.

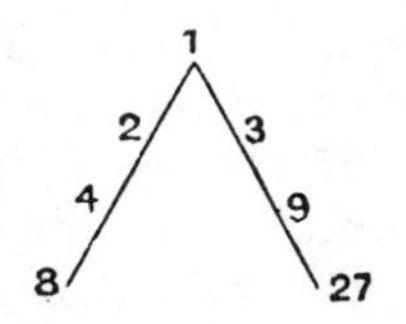

We have said that Plato made the earth occupy the centre of the universe and gave it no movement of any kind. Other

[1] *Timaeus*, 39 B–D.
[2] Ptolemy, *Syntaxis*, vii. 2, vol. ii, p. 15. 9–17, Heib.
[3] *Timaeus*, 36 D.
[4] Chalcidius on *Timaeus*, c. 96, p. 167, Wrobel
[5] Macrobius, *In somn. Scip.* ii. 3. 14.

views, however, have been attributed to Plato by later writers. In the *Timaeus* Plato had used of the earth the expression which has usually been translated 'our nurse, globed (ἰλλομένην) round the axis stretched from pole to pole through the universe'.[1] It is well known that Aristotle refers to the passage in these terms:

> 'Some say that the earth, actually lying at the centre (καὶ κειμένην ἐπὶ τοῦ κέντρου), is yet wound *and moves* (ἴλλεσθαι καὶ κινεῖσθαι) about the axis stretched through the universe from pole to pole.'[2]

This naturally implies that Aristotle attributed to Plato the view that the earth rotates about its axis. Such a view is, however, entirely inconsistent with the whole system described in the *Timaeus* (and also in the *Laws*, which Plato did not live to finish), where it is the sphere of the fixed stars which by its revolution about the earth in 24 hours makes night and day; moreover, there is no reason to doubt the evidence that it was Heraclides of Pontus who was the first to affirm the rotation of the earth about its own axis in 24 hours. The natural inference seems to be that Aristotle either misunderstood or misrepresented Plato, the ambiguity of the word ἰλλομένην being the contributing cause or the pretext as the case may be. There are, however, those who maintain that Aristotle *must* have known what Plato meant and was incapable of misrepresenting him on a subject like this. Among these is Professor Burnet,[3] who, being satisfied that Aristotle understood ἰλλομένην to mean motion of some sort, and on the strength of a new reading which he has adopted from two MSS. of the first class, has essayed a new interpretation of Plato's phrase. The new reading differs from the former texts in having the article τὴν after ἰλλομένην, which makes the phrase run thus, γῆν δὲ τροφὸν μὲν ἡμετέραν, ἰλλομένην δὲ τὴν περὶ τὸν διὰ παντὸς πόλον τεταμένον. Burnet, holding that we can only supply with τὴν some word like ὁδόν, understands περίοδον or περιφοράν, and translates 'earth our nurse going to and fro on its path round the axis which stretches right through the universe'.

[1] *Timaeus*, 40 B.

[2] Arist. *De caelo*, ii. 13, 293 b 20; cf. ii. 14, 296 a 25.

[3] *Greek Philosophy*, Part I, Thales to Plato, pp. 347-8.

In confirmation of this Burnet cites the 'unimpeachable testimony' of Theophrastus, who said that

'Plato in his old age repented of having given the earth the central place in the universe, to which it had no right'[1];

and he concludes that, according to Plato in the *Timaeus*, the earth is not the centre of the universe. But the sentences in which Aristotle paraphrases the ἰλλομένην in the *Timaeus* by the words ἴλλεσθαι καὶ κινεῖσθαι both make it clear that the persons who held the view in question also declared that the earth *lies* or *is placed at the centre* (κειμένην ἐπὶ τοῦ κέντρου), or 'placed the earth at the centre' (ἐπὶ τοῦ μέσου θέντες). Burnet's explanation is therefore in contradiction to part of Aristotle's statement, if not to the rest; so that he does not appear to have brought the question much nearer to a solution. Perhaps some one will suggest that the rotation or oscillation about the axis of the universe is *small*, so small as to be fairly consistent with the statement that the earth remains at the centre. Better, I think, admit that, on our present information, the puzzle is insoluble.

The dictum of Theophrastus that Plato in his old age repented of having placed the earth in the centre is inconsistent with the theory of the *Timaeus*, as we have said. Boeckh explained it as a misapprehension. There appear to have been among Plato's immediate successors some who altered Plato's system in a Pythagorean sense and who may be alluded to in another passage of the *De caelo*[2]; Boeckh suggested, therefore, that the views of these Pythagorizing Platonists may have been put down to Plato himself. But the tendency now seems to be to accept the testimony of Theophrastus literally. Heiberg does so, and so does Burnet, who thinks it probable that Theophrastus heard the statement which he attributes to Plato from Plato himself. But I would point out that, if the *Timaeus*, as Burnet contends, contained Plato's explicit recantation of his former view that the earth was at the centre, there was no need to supplement it by an oral communication to Theophrastus. In any case the question has no particular importance in comparison with the developments which have next to be described.

[1] Plutarch, *Quaest. Plat.* 8. 1, 1006 C; cf. *Life of Numa*, c. 11.
[2] Arist. *De caelo*, ii. 13, 293 a 27–b 1.

X

FROM PLATO TO EUCLID

WHATEVER original work Plato himself did in mathematics (and it may not have been much), there is no doubt that his enthusiasm for the subject in all branches and the pre-eminent place which he gave it in his system had enormous influence upon its development in his lifetime and the period following. In astronomy we are told that Plato set it as a problem to all earnest students to find 'what are the uniform and ordered movements by the assumption of which the apparent movements of the planets can be accounted for'; our authority for this is Sosigenes, who had it from Eudemus.[1] One answer to this, representing an advance second to none in the history of astronomy, was given by Heraclides of Pontus, one of Plato's pupils (*circa* 388–310 B.C.); the other, which was by Eudoxus and on purely mathematical lines, constitutes one of the most remarkable achievements in pure geometry that the whole of the history of mathematics can show. Both were philosophers of extraordinary range. Heraclides wrote works of the highest class both in matter and style: the catalogue of them covers subjects ethical, grammatical, musical and poetical, rhetorical, historical; and there were geometrical and dialectical treatises as well. Similarly Eudoxus, celebrated as philosopher, geometer, astronomer, geographer, physician and legislator, commanded and enriched almost the whole field of learning.

Heraclides of Pontus: astronomical discoveries.

Heraclides held that the apparent daily revolution of the heavenly bodies round the earth was accounted for, not by

[1] Simpl. on *De caelo*, ii. 12 (292 b 10), p. 488. 20–34, Heib.

the circular motion of the stars round the earth, but by the rotation of the earth about its own axis; several passages attest this, e.g.

'Heraclides of Pontus supposed that the earth is in the centre and rotates (lit. 'moves in a circle') while the heaven is at rest, and he thought by this supposition to save the phenomena.'[1]

True, Heraclides may not have been alone in holding this view, for we are told that Ecphantus of Syracuse, a Pythagorean, also asserted that 'the earth, being in the centre of the universe, moves about its own centre in an eastward direction'[2]; when Cicero[3] says the same thing of Hicetas, also of Syracuse, this is probably due to a confusion. But there is no doubt of the originality of the other capital discovery made by Heraclides, namely that Venus and Mercury revolve, like satellites, round the sun as centre. If, as Schiaparelli argued, Heraclides also came to the same conclusion about Mars, Jupiter and Saturn, he anticipated the hypothesis of Tycho Brahe (or rather improved on it), but the evidence is insufficient to establish this, and I think the probabilities are against it; there is some reason for thinking that it was Apollonius of Perga who thus completed what Heraclides had begun and put forward the full Tychonic hypothesis.[4] But there is nothing to detract from the merit of Heraclides in having pointed the way to it.

Eudoxus's theory of concentric spheres is even more remarkable as a mathematical achievement; it is worthy of the man who invented the great theory of proportion set out in Euclid, Book V, and the powerful *method of exhaustion* which not only enabled the areas of circles and the volumes of pyramids, cones, spheres, &c., to be obtained, but is at the root of all Archimedes's further developments in the mensuration of plane and solid figures. But, before we come to Eudoxus, there are certain other names to be mentioned.

[1] Simpl. on *De caelo*, p. 519. 9–11, Heib.; cf. pp. 441. 31–445. 5, pp. 541. 27–542. 2; Proclus *in Tim.* 281 E.

[2] Hippolytus, *Refut.* i. 15 (*Vors.* i³, p. 340. 31), cf. Aëtius, iii. 13. 3 (*Vors.* i³, p. 341. 8–10).

[3] Cic. *Acad. Pr.* ii. 39, 123.

[4] *Aristarchus of Samos, the ancient Copernicus*, ch. xviii.

Theory of numbers (Speusippus, Xenocrates).

To begin with arithmetic or the theory of numbers. SPEUSIPPUS, nephew of Plato, who succeeded him as head of the school, is said to have made a particular study of Pythagorean doctrines, especially of the works of Philolaus, and to have written a small treatise *On the Pythagorean Numbers* of which a fragment, mentioned above (pp. 72, 75, 76) is preserved in the *Theologumena Arithmetices*.[1] To judge by the fragment, the work was not one of importance. The arithmetic in it was evidently of the geometrical type (polygonal numbers, for example, being represented by dots making up the particular figures). The portion of the book dealing with 'the five figures (the regular solids) which are assigned to the cosmic elements, their particularity and their community with one another', can hardly have gone beyond the putting together of the figures by faces, as we find it in the *Timaeus*. To Plato's distinction of the fundamental triangles, the equilateral, the isosceles right-angled, and the half of an equilateral triangle cut off by a perpendicular from a vertex on the opposite side, he adds a distinction ('passablement futile', as is the whole fragment in Tannery's opinion) of four pyramids (1) the regular pyramid, with an equilateral triangle for base and all the edges equal, (2) the pyramid on a square base, and (evidently) having its four edges terminating at the corners of the base equal, (3) the pyramid which is the half of the preceding one obtained by drawing a plane through the vertex so as to cut the base perpendicularly in a diagonal of the base, (4) a pyramid constructed on the half of an equilateral triangle as base; the object was, by calling these pyramids a monad, a dyad, a triad and a tetrad respectively, to make up the number 10, the special properties and virtues of which as set forth by the Pythagoreans were the subject of the second half of the work. Proclus quotes a few opinions of Speusippus; e.g., in the matter of theorems and problems, he differed from Menaechmus, since he regarded both alike as being more properly *theorems*, while Menaechmus would call both alike *problems*.[2]

[1] *Theol. Ar.*, Ast, p. 61.
[2] Proclus on Eucl. I, pp. 77. 16; 78. 14.

Xenocrates of Chalcedon (396–314 B.C.), who succeeded Speusippus as head of the school, having been elected by a majority of only a few votes over Heraclides, is also said to have written a book *On Numbers* and a *Theory of Numbers*, besides books on geometry.[1] These books have not survived, but we learn that Xenocrates upheld the Platonic tradition in requiring of those who would enter the school a knowledge of music, geometry and astronomy; to one who was not proficient in these things he said 'Go thy way, for thou hast not the means of getting a grip of philosophy'. Plutarch says that he put at 1,002,000,000,000 the number of syllables which could be formed out of the letters of the alphabet.[2] If the story is true, it represents the first attempt on record to solve a difficult problem in permutations and combinations. Xenocrates was a supporter of 'indivisible lines' (and magnitudes) by which he thought to get over the paradoxical arguments of Zeno.[3]

The Elements. Proclus's summary (*continued*).

In geometry we have more names mentioned in the summary of Proclus.[4]

'Younger than Leodamas were Neoclides and his pupil Leon, who added many things to what was known before their time, so that Leon was actually able to make a collection of the elements more carefully designed in respect both of the number of propositions proved and of their utility, besides which he invented *diorismi* (the object of which is to determine) when the problem under investigation is possible of solution and when impossible.'

Of Neoclides and Leon we know nothing more than what is here stated; but the definite recognition of the διορισμός, that is, of the necessity of finding, as a preliminary to the solution of a problem, the conditions for the possibility of a solution, represents an advance in the philosophy and technology of mathematics. Not that the thing itself had not been met with before: there is, as we have seen, a

[1] Diog. L. iv. 13, 14.
[2] Plutarch, *Quaest. Conviv.* viii. 9. 13, 733 A.
[3] Simpl. *in Phys.*, p. 138. 3, &c.
[4] Proclus on Eucl. I, p. 66. 18–67. 1.

διορισμός indicated in the famous geometrical passage of the *Meno*[1]; no doubt, too, the geometrical solution by the Pythagoreans of the quadratic equation would incidentally make clear to them the limits of possibility corresponding to the διορισμός in the solution of the most general form of quadratic in Eucl. VI. 27–9, where, in the case of the 'deficient' parallelogram (Prop. 28), the enunciation states that 'the given rectilineal figure must not be greater than the parallelogram described on half of the straight line and similar to the defect'. Again, the condition of the possibility of constructing a triangle out of three given straight lines (Eucl. I. 22), namely that any two of them must be together greater than the third, must have been perfectly familiar long before Leon or Plato.

Proclus continues:[2]

'Eudoxus of Cnidos, a little younger than Leon, who had been associated with the school of Plato, was the first to increase the number of the so-called general theorems; he also added three other proportions to the three already known, and multiplied the theorems which originated with Plato about the section, applying to them the method of analysis. Amyclas [more correctly Amyntas] of Heraclea, one of the friends of Plato, Menaechmus, a pupil of Eudoxus who had also studied with Plato, and Dinostratus, his brother, made the whole of geometry still more perfect. Theudius of Magnesia had the reputation of excelling in mathematics as well as in the other branches of philosophy; for he put together the elements admirably and made many partial (or limited) theorems more general. Again, Athenaeus of Cyzicus, who lived about the same time, became famous in other branches of mathematics and most of all in geometry. These men consorted together in the Academy and conducted their investigations in common. Hermotimus of Colophon carried further the investigations already opened up by Eudoxus and Theaetetus, discovered many propositions of the Elements and compiled some portion of the theory of Loci. Philippus of Medma, who was a pupil of Plato and took up mathematics at his instance, not only carried out his investigations in accordance with Plato's instructions but also set himself to do whatever in his view contributed to the philosophy of Plato.'

[1] Plato, *Meno*, 87 A.

[2] Proclus on Eucl. I., p. 67. 2–68. 4.

It will be well to dispose of the smaller names in this list before taking up Eudoxus, the principal subject of this chapter. The name of Amyclas should apparently be Amyntas,[1] although Diogenes Laertius mentions Amyclos of Heraclea in Pontus as a pupil of Plato[2] and has elsewhere an improbable story of one Amyclas, a Pythagorean, who with Clinias is supposed to have dissuaded Plato from burning the works of Democritus in view of the fact that there were many other copies in circulation.[3] Nothing more is known of Amyntas, Theudius, Athenaeus and Hermotimus than what is stated in the above passage of Proclus. It is probable, however, that the propositions, &c., in elementary geometry which are quoted by Aristotle were taken from the Elements of Theudius, which would no doubt be the text-book of the time just preceding Euclid. Of Menaechmus and Dinostratus we have already learnt that the former discovered conic sections, and used them for finding two mean proportionals, and that the latter applied the quadratrix to the squaring of the circle. Philippus of Medma (vulg. Mende) is doubtless the same person as Philippus of Opus, who is said to have revised and published the *Laws* of Plato which had been left unfinished, and to have been the author of the *Epinomis*. He wrote upon astronomy chiefly; the astronomy in the *Epinomis* follows that of the *Laws* and the *Timaeus*; but Suidas records the titles of other works by him as follows: *On the distance of the sun and moon, On the eclipse of the moon, On the size of the sun, the moon and the earth, On the planets.* A passage of Aëtius[4] and another of Plutarch[5] alluding to his *proofs* about the shape of the moon may indicate that Philippus was the first to establish the complete theory of the phases of the moon. In mathematics, according to the same notice by Suidas, he wrote *Arithmetica, Means, On polygonal numbers, Cyclica, Optics, Enoptrica* (On mirrors); but nothing is known of the contents of these works.

[1] See *Ind. Hercul.*, ed. B cheler, *Ind. Schol. Gryphisw.*, 1869/70, col. 6 in.

[2] Diog. L. iii. 46.

[3] *Ib.* ix. 40.

[4] *Dox. Gr.*, p. 360.

[5] *Non posse suaviter vivi secundum Epicurum*, c. 11, 1093 E.

According to Apollodorus, EUDOXUS flourished in Ol. 103 = 368–365 B.C., from which we infer that he was born about 408 B.C., and (since he lived 53 years) died about 355 B.C. In his 23rd year he went to Athens with the physician Theomedon, and there for two months he attended lectures on philosophy and oratory, and in particular the lectures of Plato; so poor was he that he took up his abode at the Piraeus and trudged to Athens and back on foot each day. It would appear that his journey to Italy and Sicily to study geometry with Archytas, and medicine with Philistion, must have been earlier than the first visit to Athens, for from Athens he returned to Cnidos, after which he went to Egypt with a letter of introduction to King Nectanebus, given him by Agesilaus; the date of this journey was probably 381–380 B.C. or a little later, and he stayed in Egypt sixteen months. After that he went to Cyzicus, where he collected round him a large school which he took with him to Athens in 368 B.C. or a little later. There is apparently no foundation for the story mentioned by Diogenes Laertius that he took up a hostile attitude to Plato,[1] nor on the other side for the statements that he went with Plato to Egypt and spent thirteen years in the company of the Egyptian priests, or that he visited Plato when Plato was with the younger Dionysius on his third visit to Sicily in 361 B.C. Returning later to his native place, Eudoxus was by a popular vote entrusted with legislative office.

When in Egypt Eudoxus assimilated the astronomical knowledge of the priests of Heliopolis and himself made observations. The observatory between Heliopolis and Cercesura used by him was still pointed out in Augustus's time; he also had one built at Cnidos, and from there he observed the star Canopus which was not then visible in higher latitudes. It was doubtless to record the observations thus made that he wrote the two books attributed to him by Hipparchus, the *Mirror* and the *Phaenomena*[2]; it seems, however, unlikely that there could have been two independent works dealing with the same subject, and the latter, from which

[1] Diog. L. viii. 87.
[2] Hipparchus, *in Arati et Eudoxi phaenomena commentarii*, i. 2. 2, p. 8. 15–20 Manitius.

the poem of Aratus was drawn, so far as verses 19–732 are concerned, may have been a revision of the former work and even, perhaps, posthumous.

But it is the theoretical side of Eudoxus's astronomy rather than the observational that has importance for us; and, indeed, no more ingenious and attractive hypothesis than that of Eudoxus's system of concentric spheres has ever been put forward to account for the apparent motions of the sun, moon and planets. It was the first attempt at a purely mathematical theory of astronomy, and, with the great and immortal contributions which he made to geometry, puts him in the very first rank of mathematicians of all time. He was a *man of science* if there ever was one. No occult or superstitious lore appealed to him; Cicero says that Eudoxus, 'in astrologia iudicio doctissimorum hominum facile princeps', expressed the opinion and left it on record that no sort of credence should be given to the Chaldaeans in their predictions and their foretelling of the life of individuals from the day of their birth.[1] Nor would he indulge in vain physical speculations on things which were inaccessible to observation and experience in his time; thus, instead of guessing at the nature of the sun, he said that he would gladly be burnt up like Phaethon if at that price he could get to the sun and so ascertain its form, size, and nature.[2] Another story (this time presumably apocryphal) is to the effect that he grew old at the top of a very high mountain in the attempt to discover the movements of the stars and the heavens.[3]

In our account of his work we will begin with the sentence about him in Proclus's summary. First, he is said to have increased 'the number of the *so-called general* theorems'. 'So-called general theorems' is an odd phrase; it occurred to me whether this could mean theorems which were true of everything falling under the conception of magnitude, as are the definitions and theorems forming part of Eudoxus's own theory of proportion, which applies to numbers, geometrical magnitudes of all sorts, times, &c. A number of propositions

[1] Cic., *De div.* ii. 42.
[2] Plutarch, *Non posse suaviter vivi secundum Epicurum*, c. 11, 1094 B.
[3] Petronius Arbiter, *Satyricon*, 88.

at the beginning of Euclid's Book X similarly refer to magnitudes in general, and the proposition X. 1 or its equivalent was actually used by Eudoxus in his *method of exhaustion*, as it is by Euclid in his application of the same method to the theorem (among others) of XII. 2 that circles are to one another as the squares on their diameters.

The three 'proportions' or means added to the three previously known (the arithmetic, geometric and harmonic) have already been mentioned (p. 86), and, as they are alternatively attributed to others, they need not detain us here.

Thirdly, we are told that Eudoxus 'extended' or 'increased the number of the (propositions) about *the section* (τὰ περὶ τὴν τομήν) which originated with Plato, applying to them the method of analysis'. What is *the section*? The suggestion which has been received with most favour is that of Bretschneider,[1] who pointed out that up to Plato's time there was only one 'section' that had any real significance in geometry, namely the section of a straight line in extreme and mean ratio which is obtained in Eucl. II. 11 and is used again in Eucl. IV. 10–14 for the construction of a pentagon. These theorems were, as we have seen, pretty certainly Pythagorean, like the whole of the substance of Euclid, Book II. Plato may therefore, says Bretschneider, have directed attention afresh to this subject and investigated the metrical relations between the segments of a straight line so cut, while Eudoxus may have continued the investigation where Plato left off. Now the passage of Proclus says that, in extending the theorems about 'the section', Eudoxus applied the method of analysis; and we actually find in Eucl. XIII. 1–5 five propositions about straight lines cut in extreme and mean ratio followed, in the MSS., by definitions of analysis and synthesis, and alternative proofs of the same propositions in the form of analysis followed by synthesis. Here, then, Bretschneider thought he had found a fragment of some actual work by Eudoxus corresponding to Proclus's description. But it is certain that the definitions and the alternative proofs were interpolated by some scholiast, and, judging by the figures (which are merely straight lines) and by comparison

[1] Bretschneider, *Die Geometrie und die Geometer vor Eukleides*, pp. 167–9.

with the remarks on analysis and synthesis quoted from Heron by An-Nairīzī at the beginning of his commentary on Eucl. Book II, it seems most likely that the interpolated definitions and proofs were taken from Heron. Bretschneider's argument based on Eucl. XIII. 1–5 accordingly breaks down, and all that can be said further is that, if Eudoxus investigated the relation between the segments of the straight line, he would find in it a case of incommensurability which would further enforce the necessity for a theory of proportion which should be applicable to incommensurable as well as to commensurable magnitudes. Proclus actually observes that 'theorems about sections like those in Euclid's Second Book are common to both [arithmetic and geometry] *except that in which the straight line is cut in extreme and mean ratio*'[1] (cf. Eucl. XIII. 6 for the actual proof of the irrationality in this case). Opinion, however, has not even in recent years been unanimous in favour of Bretschneider's interpretation; Tannery[2] in particular preferred the old view, which prevailed before Bretschneider, that 'section' meant section *of solids*, e.g. by planes, a line of investigation which would naturally precede the discovery of conics; he pointed out that the use of the singular, τὴν τομήν, which might no doubt be taken as 'section' in the abstract, is no real objection, that there is no other passage which speaks of a certain section *par excellence*, and that Proclus in the words just quoted expresses himself quite differently, speaking of 'sections' of which the particular section in extreme and mean ratio is only one. Presumably the question will never be more definitely settled unless by the discovery of new documents.

(α) *Theory of proportion.*

The anonymous author of a scholium to Euclid's Book V, who is perhaps Proclus, tells us that 'some say' that this Book, containing the general theory of proportion which is equally applicable to geometry, arithmetic, music and all mathematical science, 'is the discovery of Eudoxus, the teacher of Plato'.[3] There is no reason to doubt the truth of this

[1] Proclus on Eucl. I, p. 60. 16–19.
[2] Tannery, *La géométrie grecque*, p. 76.
[3] Euclid, ed. Heib., vol. v, p. 280.

statement. The new theory appears to have been already familiar to Aristotle. Moreover, the fundamental principles show clear points of contact with those used in the *method of exhaustion*, also due to Eudoxus. I refer to the definition (Eucl. V, Def. 4) of magnitudes having a ratio to one another, which are said to be 'such as are capable, when (sufficiently) multiplied, of exceeding one another'; compare with this Archimedes's 'lemma' by means of which he says that the theorems about the volume of a pyramid and about circles being to one another as the squares on their diameters were proved, namely that 'of unequal lines, unequal surfaces, or unequal solids, the greater exceeds the less by such a magnitude as is capable, if added (continually) to itself, of exceeding any magnitude of those which are comparable to one another', i.e. of magnitudes of the same kind as the original magnitudes.

The essence of the new theory was that it was applicable to incommensurable as well as commensurable quantities; and its importance cannot be overrated, for it enabled geometry to go forward again, after it had received the blow which paralysed it for the time. This was the discovery of the irrational, at a time when geometry still depended on the Pythagorean theory of proportion, that is, the numerical theory which was of course applicable only to commensurables. The discovery of incommensurables must have caused what Tannery described as 'un véritable scandale logique' in geometry, inasmuch as it made inconclusive all the proofs which had depended on the old theory of proportion. One effect would naturally be to make geometers avoid the use of proportions as much as possible; they would have to use other methods wherever they could. Euclid's Books I–IV no doubt largely represent the result of the consequent remodelling of fundamental propositions; and the ingenuity of the substitutes devised is nowhere better illustrated than in I. 44, 45, where the equality of the complements about the diagonal of a parallelogram is used (instead of the construction, as in Book VI, of a fourth proportional) for the purpose of applying to a given straight line a parallelogram in a given angle and equal to a given triangle or rectilineal area.

The greatness of the new theory itself needs no further

argument when it is remembered that the definition of equal ratios in Eucl. V, Def. 5 corresponds exactly to the modern theory of irrationals due to Dedekind, and that it is word for word the same as Weierstrass's definition of equal numbers.

(β) *The method of exhaustion.*

In the preface to Book I of his treatise *On the Sphere and Cylinder* Archimedes attributes to Eudoxus the proof of the theorems that the volume of a pyramid is one-third of the volume of the prism which has the same base and equal height, and that the volume of a cone is one-third of the cylinder with the same base and height. In the *Method* he says that these facts were discovered, though not proved (i. e. in Archimedes's sense of the word), by Democritus, who accordingly deserved a great part of the credit for the theorems, but that Eudoxus was the first to supply the scientific proof. In the preface to the *Quadrature of the Parabola* Archimedes gives further details. He says that for the proof of the theorem that the area of a segment of a parabola cut off by a chord is $\frac{4}{3}$rds of the triangle on the same base and of equal height with the segment he himself used the 'lemma' quoted above (now known as the Axiom of Archimedes), and he goes on:

'The earlier geometers have also used this lemma; for it is by the use of this lemma that they have proved the propositions (1) that circles are to one another in the duplicate ratio of their diameters, (2) that spheres are to one another in the triplicate ratio of their diameters, and further (3) that every pyramid is one third part of the prism which has the same base with the pyramid and equal height; also (4) that every cone is one third part of the cylinder having the same base with the cone and equal height they proved by assuming a certain lemma similar to that aforesaid.'

As, according to the other passage, it was Eudoxus who first proved the last two of these theorems, it is a safe inference that he used for this purpose the 'lemma' in question or its equivalent. But was he the first to use the lemma? This has been questioned on the ground that one of the theorems mentioned as having been proved by 'the earlier geometers' in this way is the theorem that circles are to one another as the squares on their diameters, which proposition, as we are told on the authority of Eudemus, was proved (δεῖξαι) by Hippocrates of Chios. This suggested to Hankel that the lemma in question must have been formulated by Hippocrates and used in his proof.[1] But seeing that, according to Archimedes, 'the earlier geometers' proved by means of the same lemma *both* Hippocrates's proposition, (1) above, and the theorem (3) about the volume of a pyramid, while the first proof of the latter was certainly given by Eudoxus, it is simplest to suppose that it was Eudoxus who first formulated the 'lemma' and used it to prove both propositions, and that Hippocrates's 'proof' did not amount to a rigorous demonstration such as would have satisfied Eudoxus or Archimedes. Hippocrates may, for instance, have proceeded on the lines of Antiphon's 'quadrature', gradually exhausting the circles and *taking the limit,* without clinching the proof by the formal *reductio ad absurdum* used in the method of exhaustion as practised later. Without therefore detracting from the merit of Hippocrates, whose argument may have contained the germ of the method of exhaustion, we do not seem to have any sufficient reason to doubt that it was Eudoxus who established this method as part of the regular machinery of geometry.

The 'lemma' itself, we may observe, is not found in Euclid in precisely the form that Archimedes gives it, though it is equivalent to Eucl. V, Def. 4 (Magnitudes are said to have a ratio to one another which are capable, when multiplied, of exceeding one another). When Euclid comes to prove the propositions about the content of circles, pyramids and cones (XII. 2, 4–7 Por., and 10), he does not use the actual lemma of Archimedes, but another which forms Prop. 1 of Book X, to the effect that, if there are two unequal magnitudes and from the greater there be subtracted more than its half (or the half itself), from the remainder more than its half (or the half), and if this be done continually, there will be left some magnitude which will be less than the lesser of the given magnitudes. This last lemma is frequently used by Archimedes himself (notably in the second proof of the proposition about the area

[1] Hankel, *Zur Geschichte der Mathematik in Alterthum und Mittelalter,* p. 122.

of a parabolic segment), and it may be the 'lemma similar to the aforesaid' which he says was used in the case of the cone. But the existence of the two lemmas constitutes no real difficulty, because Archimedes's lemma (under the form of Eucl. V, Def. 4) is in effect used by Euclid to prove X. 1.

We are not told whether Eudoxus proved the theorem that spheres are to one another in the triplicate ratio of their diameters. As the proof of this in Eucl. XII. 16–18 is likewise based on X. 1 (which is used in XII. 16), it is probable enough that this proposition, mentioned along with the others by Archimedes, was also first proved by Eudoxus.

Eudoxus, as we have seen, is said to have solved the problem of the two mean proportionals by means of 'curved lines'. This solution has been dealt with above (pp. 249–51).

We pass on to the

(γ) *Theory of concentric spheres.*

This was the first attempt to account by purely geometrical hypotheses for the apparent irregularities of the motions of the planets; it included similar explanations of the apparently simpler movements of the sun and moon. The ancient evidence of the details of the system of concentric spheres (which Eudoxus set out in a book entitled *On speeds*, *Περὶ ταχῶν*, now lost) is contained in two passages. The first is in Aristotle's *Metaphysics*, where a short notice is given of the numbers and relative positions of the spheres postulated by Eudoxus for the sun, moon and planets respectively, the additions which Callippus thought it necessary to make to the numbers of those spheres, and lastly the modification of the system which Aristotle himself considers necessary 'if the phenomena are to be produced by all the spheres acting in combination'.[1] A more elaborate and detailed account of the system is contained in Simplicius's commentary on the *De caelo* of Aristotle[2]; Simplicius quotes largely from Sosigenes the Peripatetic (second century A. D.), observing that Sosigenes drew from Eudemus, who dealt with the subject in the second book of his *History of Astronomy*. Ideler was

[1] Aristotle, *Metaph.* Λ. 8. 1073 b 17–1074 a 14.

[2] Simpl. on *De caelo*, p. 488. 18–24, pp. 493. 4–506. 18 Heib.; p. 498 a 45–b 3, pp. 498 b 27–503 a 33.

the first to appreciate the elegance of the theory and to attempt to explain its working (1828, 1830); E. F. Apelt, too, gave a fairly full exposition of it in a paper of 1849. But it was reserved for Schiaparelli to work out a complete restoration of the theory and to investigate in detail the extent to which it could be made to account for the phenomena; his paper has become a classic,[1] and all accounts must necessarily follow his.

I shall here only describe the system so far as to show its mathematical interest. I have given fuller details elsewhere.[2] Eudoxus adopted the view which prevailed from the earliest times to the time of Kepler, that circular motion was sufficient to account for the movements of all the heavenly bodies. With Eudoxus this circular motion took the form of the revolution of different spheres, each of which moves about a diameter as axis. All the spheres were concentric, the common centre being the centre of the earth; hence the name of 'homocentric' spheres used in later times to describe the system. The spheres were of different sizes, one inside the other. Each planet was fixed at a point in the equator of the sphere which carried it, the sphere revolving at uniform speed about the diameter joining the corresponding poles; that is, the planet revolved uniformly in a great circle of the sphere perpendicular to the axis of rotation. But one such circular motion was not enough; in order to explain the changes in the apparent speed of the planets' motion, their stations and retrogradations, Eudoxus had to assume a number of such circular motions working on each planet and producing by their combination that single apparently irregular motion which observation shows us. He accordingly held that the poles of the sphere carrying the planet are not fixed, but themselves move on a greater sphere concentric with the carrying sphere and moving about two different poles with uniform speed. The poles of the second sphere were similarly placed on a third sphere concentric with and larger than the first and second, and moving about separate poles

[1] Schiaparelli, *Le sfere omocentriche di Eudosso, di Callippo e di Aristotele*, Milano 1875; Germ. trans. by W. Horn in *Abh. zur Gesch. d. Math.*, i. Heft, 1877, pp. 101–98.

[2] *Aristarchus of Samos, the ancient Copernicus*, pp. 193–224.

of its own with a speed peculiar to itself. For the planets yet a fourth sphere was required, similarly related to the others; for the sun and moon Eudoxus found that, by a suitable choice of the positions of the poles and of speeds of rotation, he could make three spheres suffice. Aristotle and Simplicius describe the spheres in the reverse order, the sphere carrying the planet being the last; this makes the description easier, because we begin with the sphere representing the daily rotaton of the heavens. The spheres which move each planet Eudoxus made quite separate from those which move the others; but one sphere sufficed to produce the daily rotation of the heavens. The hypothesis was purely mathematical; Eudoxus did not trouble himself about the material of the spheres or their mechanical connexion.

The moon has a motion produced by three spheres; the first or outermost moves in the same sense as the fixed stars from east to west in 24 hours; the second moves about an axis perpendicular to the plane of the zodiac circle or the ecliptic, and in the sense of the daily rotation, i. e. from east to west; the third again moves about an axis inclined to the axis of the second at an angle equal to the highest latitude attained by the moon, and from west to east; the moon is fixed on the equator of this third sphere. The speed of the revolution of the second sphere was very slow (a revolution was completed in a period of 223 lunations); the third sphere produced the revolution of the moon from west to east in the draconitic or nodal month (of 27 days, 5 hours, 5 minutes, 36 seconds) round a circle inclined to the ecliptic at an angle equal to the greatest latitude of the moon.[1] The moon described the latter circle, while the circle itself was carried round by the second sphere in a retrograde sense along the ecliptic in a period of 223 lunations; and both the inner spheres were bodily carried round by the first sphere in 24 hours in the sense of the daily rotation. The three spheres thus produced the motion of the moon in an orbit inclined to the ecliptic, and the retrogression of the nodes, completed in a period of about $18\frac{1}{2}$ years.

[1] Simplicius (and presumably Aristotle also) confused the motions of the second and third spheres. The above account represents what Eudoxus evidently intended.

The system of three spheres for the sun was similar, except that the orbit was less inclined to the ecliptic than that of the moon, and the second sphere moved from west to east instead of from east to west, so that the nodes moved slowly forward in the direct order of the signs instead of backward.

But the case to which the greatest mathematical interest attaches is that of the planets, the motion of which is produced by sets of four spheres for each. Of each set the first and outermost produced the daily rotation in 24 hours; the second, the motion round the zodiac in periods which in the case of superior planets are equal to the sidereal periods of revolution, and for Mercury and Venus (on a geocentric system) one year. The third sphere had its poles fixed at two opposite points on the zodiac circle, the poles being carried round in the motion of the second sphere; the revolution of the third sphere about its poles was again uniform and was completed in the synodic period of the planet or the time which elapsed between two successive oppositions or conjunctions with the sun. The poles of the third sphere were the same for Mercury and Venus but different for all the other planets. On the surface of the third sphere the poles of the fourth sphere were fixed, the axis of the latter being inclined to that of the former at an angle which was constant for each planet but different for the different planets. The rotation of the fourth sphere about its axis took place in the same time as the rotation of the third about its axis but in the opposite sense. On the equator of the fourth sphere the planet was fixed. Consider now the actual path of a planet subject to the rotations of the third and fourth spheres only, leaving out of account for the moment the first two spheres the motion of which produces the daily rotation and the motion along the zodiac respectively. The problem is the following. A sphere rotates uniformly about the fixed diameter AB. P, P' are two opposite poles on this sphere, and a second sphere concentric with the first rotates uniformly about the diameter PP' in the same time as the former sphere rotates about AB, but in the opposite direction. M is a point on the second sphere equidistant from P, P', i.e. a point on the equator of the second sphere. Required to find the path of the point M. This is not difficult nowadays for any one familiar with spherical trigonometry and analytical geometry; but Schiaparelli showed, by means of a series of seven propositions or problems involving only elementary geometry, that it was well within the powers of such a geometer as Eudoxus. The path of M in space turns out in fact to be a curve like a lemniscate or figure-of-eight described on the surface of a sphere, namely the fixed sphere about AB as diameter. This

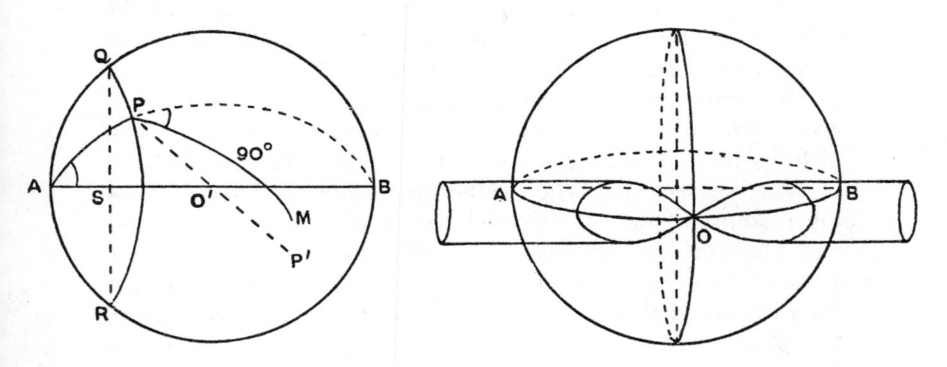

'spherical lemniscate' is roughly shown in the second figure above. The curve is actually the intersection of the sphere with a certain cylinder touching it internally at the double point O, namely a cylinder with diameter equal to AS the *sagitta* (shown in the other figure) of the diameter of the small circle on which P revolves. But the curve is also the intersection of *either* the sphere *or* the cylinder with a certain cone with vertex O, axis parallel to the axis of the cylinder (i. e. touching the circle AOB at O) and vertical angle equal to the 'inclination' (the angle $AO'P$ in the first figure). That this represents the actual result obtained by Eudoxus himself is conclusively proved by the facts that Eudoxus called the curve described by the planet about the zodiac circle the *hippopede* or *horse-fetter*, and that the same term *hippopede* is used by Proclus to describe the plane curve of similar shape formed by a plane section of an anchor-ring or *tore* touching the tore internally and parallel to its axis.[1]

So far account has only been taken of the motion due to the combination of the rotations of the third and fourth

[1] Proclus on Eucl. I, p. 112. 5.

spheres. But A, B, the poles of the third sphere, are carried round the zodiac or ecliptic by the motion of the second sphere in a time equal to the 'zodiacal' period of the planet. Now the axis of symmetry of the 'spherical lemniscate' (the arc of the great circle bisecting it longitudinally) always lies on the ecliptic. We may therefore substitute for the third and fourth spheres the 'lemniscate' moving bodily round the ecliptic. The combination of the two motions (that of the 'lemniscate' and that of the planet on it) gives the motion of the planet through the constellations. The motion of the planet round the curve is an oscillatory motion, now forward in acceleration of the motion round the ecliptic due to the motion of the second sphere, now backward in retardation of the same motion; the period of the oscillation is the period of the synodic revolution, and the acceleration and retardation occupy half the period respectively. When the retardation in the sense of longitude due to the backward oscillation is greater than the speed of the forward motion of the lemniscate itself, the planet will for a time have a retrograde motion, at the beginning and end of which it will appear stationary for a little while, when the two opposite motions balance each other.

It will be admitted that to produce the retrogradations in this theoretical way by superimposed axial rotations of spheres was a remarkable stroke of genius. It was no slight geometrical achievement, for those days, to demonstrate the *effect* of the hypotheses; but this is nothing in comparison with the speculative power which enabled the man to invent the hypothesis which would produce the effect. It was, of course, a much greater achievement than that of Eudoxus's teacher Archytas in finding the two mean proportionals by means of the intersection of three surfaces in space, a *tore* with internal diameter *nil*, a cylinder and a cone; the problem solved by Eudoxus was much more difficult, and yet there is the curious resemblance between the two solutions that Eudoxus's *hippopede* is actually the section of a sphere with a cylinder touching it internally and also with a certain cone; the two cases together show the freedom with which master and pupil were accustomed to work with figures in three dimensions, and in particular with surfaces of revolution, their intersections, &c.

Callippus (about 370–300 B.C.) tried to make the system of concentric spheres suit the phenomena more exactly by adding other spheres; he left the number of the spheres at four in the case of Jupiter and Saturn, but added one each to the other planets and two each in the case of the sun and moon (making five in all). This would substitute for the hippopede a still more complicated elongated figure, and the matter is not one to be followed out here. Aristotle modified the system in a mechanical sense by introducing between each planet and the one below it reacting spheres one less in number than those acting on the former planet, and with motions equal and opposite to each of them, except the outermost, respectively; by neutralizing the motions of all except the outermost sphere acting on any planet he wished to enable that outermost to be the outermost acting on the planet below, so that the spheres became one connected system, each being in actual contact with the one below and acting on it, whereas with Eudoxus and Callippus the spheres acting on each planet formed a separate set independent of the others. Aristotle's modification was not an improvement, and has no mathematical interest.

The works of ARISTOTLE are of the greatest importance to the history of mathematics and particularly of the Elements. His date (384–322/1) comes just before that of Euclid, so that from the differences between his statement of things corresponding to what we find in Euclid and Euclid's own we can draw a fair inference as to the innovations which were due to Euclid himself. Aristotle was no doubt a competent mathematician, though he does not seem to have specialized in mathematics, and fortunately for us he was fond of mathematical illustrations. His allusions to particular definitions, propositions, &c., in geometry are in such a form as to suggest that his pupils must have had at hand some text-book where they could find the things he mentions. The particular text-book then in use would presumably be that which was the immediate predecessor of Euclid's, namely the Elements of Theudius; for Theudius is the latest of pre-Euclidean geometers whom the summary of Proclus mentions as a compiler of Elements.[1]

[1] Proclus on Eucl. I, p. 67. 12–16.

The mathematics in Aristotle comes under the following heads.

(a) *First principles.*

On no part of the subject does Aristotle throw more light than on the first principles as then accepted. The most important passages dealing with this subject are in the *Posterior Analytics*.[1] While he speaks generally of 'demonstrative sciences', his illustrations are mainly mathematical, doubtless because they were readiest to his hand. He gives the clearest distinctions between axioms (which are common to all sciences), definitions, hypotheses and postulates (which are different for different sciences since they relate to the subject-matter of the particular science). If we exclude from Euclid's axioms (1) the assumption that two straight lines cannot enclose a space, which is interpolated, and (2) the so-called 'Parallel-Axiom' which is the 5th Postulate, Aristotle's explanation of these terms fits the classification of Euclid quite well. Aristotle calls the axioms by various terms, '*common* (things)', 'common axioms', 'common opinions', and this seems to be the origin of 'common notions' (*κοιναὶ ἔννοιαι*), the term by which they are described in the text of Euclid; the particular axiom which Aristotle is most fond of quoting is No. 3, stating that, if equals be subtracted from equals, the remainders are equal. Aristotle does not give any instance of a geometrical postulate. From this we may fairly make the important inference that Euclid's Postulates are all his own, the momentous Postulate 5 as well as Nos. 1, 2, 3 relating to constructions of lines and circles, and No. 4 that all right angles are equal. These postulates as well as those which Archimedes lays down at the beginning of his book *On Plane Equilibriums* (e.g. that 'equal weights balance at equal lengths, but equal weights at unequal lengths do not balance but incline in the direction of the weight which is at the greater length') correspond exactly enough to Aristotle's idea of a postulate. This is something which, e.g., the geometer assumes (for reasons known to himself) without demonstration (though properly a subject for demonstration)

[1] *Anal. Post.* i. 6. 74 b 5, i. 10. 76 a 31–77 a 4.

and without any assent on the part of the learner, or even against his opinion rather than otherwise. As regards definitions, Aristotle is clear that they do not assert existence or non-existence; they only require to be understood. The only exception he makes is in the case of the *unit* or *monad* and *magnitude*, the existence of which has to be assumed, while the existence of everything else has to be proved; the things actually necessary to be assumed in geometry are points and lines only; everything constructed out of them, e.g. triangles, squares, tangents, and their properties, e.g. incommensurability, has to be *proved* to exist. This again agrees substantially with Euclid's procedure. Actual construction is with him the proof of existence. If triangles other than the equilateral triangle constructed in I. 1 are assumed in I. 4–21, it is only provisionally, pending the construction of a triangle out of three straight lines in I. 22; the drawing and producing of straight lines and the describing of circles is postulated (Postulates 1–3). Another interesting statement on the philosophical side of geometry has reference to the geometer's hypotheses. It is untrue, says Aristotle, to assert that a geometer's hypotheses are false because he assumes that a line which he has drawn is a foot long when it is not, or straight when it is not straight. The geometer bases no conclusion on the particular line being that which he has assumed it to be; he argues about what it *represents*, the figure itself being a mere illustration.[1]

Coming now to the first definitions of Euclid, Book I, we find that Aristotle has the equivalents of Defs. 1–3 and 5, 6. But for a straight line he gives Plato's definition only: whence we may fairly conclude that Euclid's definition was his own, as also was his definition of a plane which he adapted from that of a straight line. Some terms seem to have been defined in Aristotle's time which Euclid leaves undefined, e.g. *κεκλάσθαι*, 'to be inflected', *νεύειν*, to 'verge'.[2] Aristotle seems to have known Eudoxus's new theory of proportion, and he uses to a considerable extent the usual

[1] Arist. *Anal. Post.* i. 10. 76 b 39–77 a 2; cf. *Anal. Prior.* i. 41. 49 b 34 sq.; *Metaph.* N. 2. 1089 a 20–5.
[2] *Anal. Post.* i. 10. 76 b 9.

terminology of proportions; he defines similar figures as Euclid does.

(β) *Indications of proofs differing from Euclid's.*

Coming to theorems, we find in Aristotle indications of proofs differing entirely from those of Euclid. The most remarkable case is that of the theorem of I. 5. For the purpose of illustrating the statement that in any syllogism one of the propositions must be affirmative and universal he gives a proof of the proposition as follows.[1]

'For let A, B be drawn [i. e. joined] to the centre.

'If then we assumed (1) that the angle AC [i. e. $A+C$] is equal to the angle BD [i. e. $B+D$] without asserting generally that *the angles of semicircles are equal,* and again (2) that the angle C is equal to the angle D without making the further assumption that *the two angles of all segments are equal*, and if we then inferred, lastly, that since the whole angles are equal, and equal angles are subtracted from them, the angles which remain, namely E, F, are equal, without assuming generally that, if equals be subtracted from equals, the remainders are equal, we should commit a *petitio principii.*'

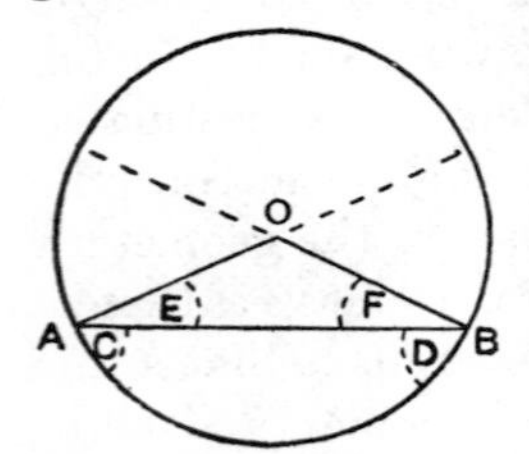

There are obvious peculiarities of notation in this extract; the angles are indicated by single letters, and sums of two angles by two letters in juxtaposition (cf. DE for $D+E$ in the proof cited from Archytas above, p. 215). The angles A, B are the angles at A, B of the *isosceles triangle* OAB, the same angles as are afterwards spoken of as E, F. But the differences of substance between this and Euclid's proof are much more striking. First, it is clear that 'mixed' angles ('angles' formed by straight lines with circular arcs) played a much larger part in earlier text-books than they do in Euclid, where indeed they only appear once or twice as a survival. Secondly, it is remarkable that the equality of the two 'angles' of a semicircle and of the two 'angles' of any segment is assumed as a means of proving a proposition so

[1] *Anal. Prior.* i. 24. 41 b 13-22.

elementary as I. 5, although one would say that the assumptions are no more obvious than the proposition to be proved; indeed some kind of proof, e.g. by superposition, would doubtless be considered necessary to justify the assumptions. It is a natural inference that Euclid's proof of I. 5 was his own, and it would appear that his innovations as regards order of propositions and methods of proof began at the very threshold of the subject.

There are two passages[1] in Aristotle bearing on the theory of parallels which seem to show that the theorems of Eucl. I. 27, 28 are pre-Euclidean; but another passage[2] appears to indicate that there was some vicious circle in the theory of parallels then current, for Aristotle alludes to a *petitio principii* committed by 'those who think that they draw parallels' (or 'establish the theory of parallels', τὰς παραλλήλους γράφειν), and, as I have tried to show elsewhere,[3] a note of Philoponus makes it possible that Aristotle is criticizing a *direction*-theory of parallels such as has been adopted so often in modern text-books. It would seem, therefore, to have been Euclid who first got rid of the *petitio principii* in earlier text-books by formulating the famous Postulate 5 and basing I. 29 upon it.

A difference of method is again indicated in regard to the theorem of Eucl. III. 31 that the angle in a semicircle is right. Two passages of Aristotle taken together[4] show that before Euclid the proposition was proved by means of the radius drawn to the middle point of the arc of the semicircle. Joining the extremity of this radius to the extremities of the diameter respectively, we have two isosceles right-angled triangles, and the two angles, one in each triangle, which are at the middle point of the arc, being both of them halves of right angles, make the angle in the semicircle *at that point* a right angle. The proof of the theorem must have been completed by means of the theorem

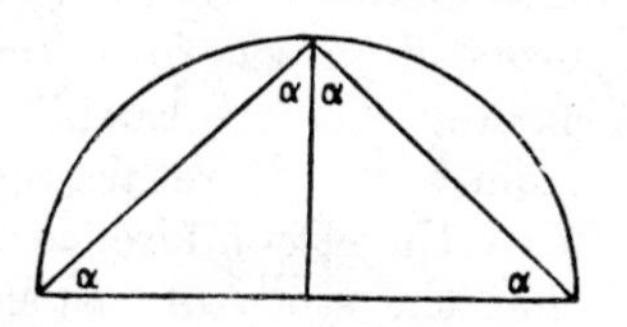

[1] *Anal. Post.* i. 5. 74 a 13–16; *Anal. Prior.* ii. 17. 66 a 11–15.

[2] *Anal. Prior.* ii. 16. 65 a 4.

[3] See *The Thirteen Books of Euclid's Elements*, vol. i, pp. 191–2 (cf. pp. 308–9).

[4] *Anal. Post.* ii. 11. 94 a 28; *Metaph.* Θ. 9. 1051 a 26.

of III. 21 that angles in the same segment are equal, a proposition which Euclid's more general proof does not need to use.

These instances are sufficient to show that Euclid was far from taking four complete Books out of an earlier text-book without change; his changes began at the very beginning, and there are probably few, if any, groups of propositions in which he did not introduce some improvements of arrangement or method.

It is unnecessary to go into further detail regarding Euclidean theorems found in Aristotle except to note the interesting fact that Aristotle already has the principle of the method of exhaustion used by Eudoxus: 'If I continually add to a finite magnitude, I shall exceed every assigned ('defined', ὡρισμένου) magnitude, and similarly, if I subtract, I shall fall short (of any assigned magnitude).'[1]

(γ) *Propositions not found in Euclid.*

Some propositions found in Aristotle but not in Euclid should be mentioned. (1) The exterior angles of any polygon are together equal to four right angles[2]; although omitted in Euclid and supplied by Proclus, this is evidently a Pythagorean proposition. (2) The locus of a point such that its distances from two given points are in a given ratio (not being a ratio of equality) is a circle[3]; this is a proposition quoted by Eutocius from Apollonius's *Plane Loci*, but the proof given by Aristotle differs very little from that of Apollonius as reproduced by Eutocius, which shows that the proposition was fully known and a standard proof of it was in existence before Euclid's time. (3) Of all closed lines starting from a point, returning to it again, and including a given area, the circumference of a circle is the shortest[4]; this shows that the study of isoperimetry (comparison of the perimeters of different figures having the same area) began long before the date of Zenodorus's treatise quoted by Pappus and Theon of Alexandria. (4) Only two solids can fill up space, namely the pyramid and the cube[5]; this is the complement of the Pythagorean statement that the only three figures which can

[1] Arist. *Phys.* viii. 10. 266 b 2.
[2] *Anal. Post.* i. 24. 85 b 38; ii. 17. 99 a 19.
[3] *Meteorologica*, iii. 5. 376 a 3 sq.
[4] *De caelo*, ii. 4. 287 a 27.
[5] *Ib.* iii. 8. 306 b 7.

by being put together fill up space in a plane are the equilateral triangle, the square and the regular hexagon.

(δ) *Curves and solids known to Aristotle.*

There is little beyond elementary plane geometry in Aristotle. He has the distinction between straight and 'curved' lines (καμπύλαι γραμμαί), but the only curve mentioned specifically, besides circles, seems to be the spiral[1]; this term may have no more than the vague sense which it has in the expression 'the spirals of the heaven'[2]; if it really means the cylindrical helix, Aristotle does not seem to have realized its property, for he includes it among things which are not such that 'any part will coincide with any other part', whereas Apollonius later proved that the cylindrical helix has precisely this property.

In solid geometry he distinguishes clearly the three dimensions belonging to 'body', and, in addition to parallelepipedal solids, such as cubes, he is familiar with spheres, cones and cylinders. A sphere he defines as the figure which has all its radii ('lines from the centre') equal,[3] from which we may infer that Euclid's definition of it as the solid generated by the revolution of a semicircle about its diameter is his own (Eucl. XI, Def. 14). Referring to a cone, he says[4] 'the straight lines thrown out from K in the form of a cone make GK *as a sort of axis* (ὥσπερ ἄξονα)', showing that the use of the word 'axis' was not yet quite technical; of conic sections he does not seem to have had any knowledge, although he must have been contemporary with Menaechmus. When he alludes to 'two cubes being a cube' he is not speaking, as one might suppose, of the duplication of the cube, for he is saying that no science is concerned to prove anything outside its own subject-matter; thus geometry is not required to prove 'that two cubes are a cube'[5]; hence the sense of this expression must be not geometrical but arithmetical, meaning that the product of two cube numbers is also a cube number. In the Aristotelian *Problems* there is a question which, although not mathematical in intention, is perhaps the first suggestion of

[1] *Phys.* v. 4. 228 b 24.
[2] *Metaph.* B. 2. 998 a 5.
[3] *Phys.* ii. 4. 287 a 19.
[4] *Meteorologica*, iii. 5. 375 b 21.
[5] *Anal. Post.* i. 7. 75 b 12.

a certain class of investigation. If a book in the form of a cylindrical roll is cut by a plane and then unrolled, why is it that the cut edge appears as a straight line if the section is parallel to the base (i.e. is a right section), but as a crooked line if the section is obliquely inclined (to the axis).[1] The *Problems* are not by Aristotle; but, whether this one goes back to Aristotle or not, it is unlikely that he would think of investigating the form of the curve mathematically.

(ε) *The continuous and the infinite.*

Much light was thrown by Aristotle on certain general conceptions entering into mathematics such as the 'continuous' and the 'infinite'. The continuous, he held, could not be made up of indivisible parts; the continuous is that in which the boundary or limit between two consecutive parts, where they touch, is one and the same, and which, as the name itself implies, is *kept together*, which is not possible if the extremities are two and not one.[2] The 'infinite' or 'unlimited' only exists potentially, not in actuality. The infinite is so in virtue of its endlessly changing into something else, like day or the Olympic games, and is manifested in different forms, e.g. in time, in Man, and in the division of magnitudes. For, in general, the infinite consists in something new being continually taken, that something being itself always finite but always different. There is this distinction between the forms above mentioned that, whereas in the case of magnitudes what is once taken remains, in the case of time and Man it passes or is destroyed, but the succession is unbroken. The case of addition is in a sense the same as that of division; in the finite magnitude the former takes place in the converse way to the latter; for, as we see the finite magnitude divided *ad infinitum*, so we shall find that addition gives a sum tending to a definite limit. Thus, in the case of a finite magnitude, you may take a definite fraction of it and add to it continually in the same ratio; if now the successive added terms do not include one and the same magnitude, whatever it is [i.e. if the successive terms diminish in geometrical progression], you will not come to the end of the finite magnitude, but, if the ratio is increased so that each term

[1] *Probl.* xvi. 6. 914 a 25. [2] *Phys.* v. 3. 227 a 11; vii. 1. 231 a 24.

does include one and the same magnitude, whatever it is, you will come to the end of the finite magnitude, for every finite magnitude is exhausted by continually taking from it any definite fraction whatever. In no other sense does the infinite exist but only in the sense just mentioned, that is, potentially and by way of diminution.[1] And in this sense you may have potentially infinite addition, the process being, as we say, in a manner the same as with division *ad infinitum*; for in the case of addition you will always be able to find something outside the total for the time being, but the total will never exceed every definite (or assigned) magnitude in the way that, in the direction of division, the result will pass every definite magnitude, that is, by becoming smaller than it. The infinite therefore cannot exist, even potentially, in the sense of exceeding every finite magnitude as the result of successive addition. It follows that the correct view of the infinite is the opposite of that commonly held; it is not that which has nothing outside it, but that which always has something outside it.[2] Aristotle is aware that it is essentially of physical magnitudes that he is speaking: it is, he says, perhaps a more general inquiry that would be necessary to determine whether the infinite is possible in mathematics and in the domain of thought and of things which have no magnitude.[3]

'But', he says, 'my argument does not anyhow rob mathematicians of their study, although it denies the existence of the infinite in the sense of actual existence as something increased to such an extent that it cannot be gone through (ἀδιεξίτητον); for, as it is, they do not even need the infinite or use it, but only require that the finite (straight line) shall be as long *as they please*. . . . Hence it will make no difference to them for the purpose of demonstration.'[4]

The above disquisition about the infinite should, I think, be interesting to mathematicians for the distinct expression of Aristotle's view that the existence of an infinite series the terms of which are *magnitudes* is impossible unless it is convergent and (with reference to Riemann's developments) that it does not matter to geometry if the straight line is not infinite in length provided that it is as long as we please.

[1] *Phys.* iii. 6. 206 a 15–b 13.
[2] *Ib.* iii. 6. 206 b 16–207 a 1.
[3] *Ib.* iii. 5. 204 a 34.
[4] *Ib.* iii. 7. 207 b 27.

Aristotle's denial of even the potential existence of a sum of magnitudes which shall exceed every definite magnitude was, as he himself implies, inconsistent with the lemma or assumption used by Eudoxus in his method of exhaustion. We can, therefore, well understand why, a century later, Archimedes felt it necessary to justify his own use of the lemma:

'the earlier geometers too have used this lemma: for it is by its help that they have proved that circles have to one another the duplicate ratio of their diameters, that spheres have to one another the triplicate ratio of their diameters, and so on. And, in the result, each of the said theorems has been accepted no less than those proved without the aid of this lemma.' [1]

(ζ) *Mechanics.*

An account of the mathematics in Aristotle would be incomplete without a reference to his ideas in mechanics, where he laid down principles which, even though partly erroneous, held their ground till the time of Benedetti (1530–90) and Galilei (1564–1642). The *Mechanica* included in the Aristotelian writings is not indeed Aristotle's own work, but it is very close in date, as we may conclude from its terminology; this shows more general agreement with the terminology of Euclid than is found in Aristotle's own writings, but certain divergences from Euclid's terms are common to the latter and to the *Mechanica*; the conclusion from which is that the *Mechanica* was written before Euclid had made the terminology of mathematics more uniform and convenient, or, in the alternative, that it was composed after Euclid's time by persons who, though they had partly assimilated Euclid's terminology, were close enough to Aristotle's date to be still influenced by his usage. But the Aristotelian origin of many of the ideas in the *Mechanica* is proved by their occurrence in Aristotle's genuine writings. Take, for example, the principle of the lever. In the *Mechanica* we are told that,

'as the weight moved is to the moving weight, so is the length (or distance) to the length inversely. In fact the moving weight will more easily move (the system) the farther it is away from the fulcrum. The reason is that aforesaid,

[1] Archimedes, *Quadrature of a Parabola*, Preface.

namely that the line which is farther from the centre describes the greater circle, so that, if the power applied is the same, that which moves (the system) will change its position the more, the farther it is away from the fulcrum.'[1]

The idea then is that the greater power exerted by the weight at the greater distance corresponds to its greater velocity. Compare with this the passage in the *De caelo* where Aristotle is speaking of the speeds of the circles of the stars:

'it is not at all strange, nay it is inevitable, that the speeds of circles should be in the proportion of their sizes.'[2] . . . 'Since in two concentric circles the segment (sector) of the outer cut off between two radii common to both circles is greater than that cut off on the inner, it is reasonable that the greater circle should be carried round in the same time.'[3]

Compare again the passage of the *Mechanica*:

'what happens with the balance is reduced to (the case of the) circle, the case of the lever to that of the balance, and practically everything concerning mechanical movements to the case of the lever. Further it is the fact that, given a radius of a circle, no two points of it move at the same speed (as the radius itself revolves), but the point more distant from the centre always moves more quickly, and this is the reason of many remarkable facts about the movements of circles which will appear in the sequel.'[4]

The axiom which is regarded as containing the germ of the principle of virtual velocities is enunciated, in slightly different forms, in the *De caelo* and the *Physics*:

'A smaller and lighter weight will be given more movement if the force acting on it is the same. . . . The speed of the lesser body will be to that of the greater as the greater body is to the lesser.'[5]

'If A be the movent, B the thing moved, C the length through which it is moved, D the time taken, then

A will move $\frac{1}{2}B$ over the distance $2\,C$ in the time D,
and A ,, $\frac{1}{2}B$,, ,, C ,, ,, $\frac{1}{2}D$;
thus proportion is maintained.'[6]

[1] *Mechanica*, 3. 850 b 1.
[2] *De caelo*, ii. 8. 289 b 15.
[3] *Ib.* 290 a 2.
[4] *Mechanica*, 848 a 11.
[5] *De caelo*, iii. 2. 301 b 4, 11.
[6] *Phys.* vii. 5. 249 b 30–250 a 4.

Again, says Aristotle,

	A will move	B over the distance	$\frac{1}{2}C$	in the time	$\frac{1}{2}D$,
and	$\frac{1}{2}A$,,	$\frac{1}{2}B$ a distance	C	,, ,,	D;[1]

and so on.

Lastly, we have in the *Mechanica* the parallelogram of velocities:

'When a body is moved in a certain ratio (i.e. has two linear movements in a constant ratio to one another), the body must move in a straight line, and this straight line is the diameter of the figure (parallelogram) formed from the straight lines which have the given ratio.'[2]

The author goes on to say[3] that, if the ratio of the two movements does not remain the same from one instant to the next, the motion will not be in a straight line but in a curve. He instances a circle in a vertical plane with a point moving along it downwards from the topmost point; the point has two simultaneous movements; one is in a vertical line, the other displaces this vertical line parallel to itself away from the position in which it passes through the centre till it reaches the position of a tangent to the circle; if during this time the ratio of the two movements were constant, say one of equality, the point would not move along the circumference at all but along the diagonal of a rectangle.

The parallelogram of *forces* is easily deduced from the parallelogram of velocities combined with Aristotle's axiom that the force which moves a given weight is directed along the line of the weight's motion and is proportional to the distance described by the weight in a given time.

Nor should we omit to mention the Aristotelian tract *On indivisible lines*. We have seen (p. 293) that, according to Aristotle, Plato objected to the genus 'point' as a geometrical fiction, calling a point the beginning of a line, and often positing 'indivisible lines' in the same sense.[4] The idea of indivisible lines appears to have been only vaguely conceived by Plato, but it took shape in his school, and with Xenocrates

[1] *Phys.* vii. 5. 250 a 4-7.
[2] *Mechanica*, 2. 848 b 10.
[3] *Ib.* 848 b 26 sq.
[4] *Metaph.* A. 9. 992 a 20.

became a definite doctrine. There is plenty of evidence for this[1]; Proclus, for instance, tells us of 'a discourse or argument by Xenocrates introducing indivisible lines'.[2] The tract *On indivisible lines* was no doubt intended as a counterblast to Xenocrates. It can hardly have been written by Aristotle himself; it contains, for instance, some expressions without parallel in Aristotle. But it is certainly the work of some one belonging to the school; and we can imagine that, having on some occasion to mention 'indivisible lines', Aristotle may well have set to some pupil, as an exercise, the task of refuting Xenocrates. According to Simplicius and Philoponus, the tract was attributed by some to Theophrastus[3]; and this seems the most likely supposition, especially as Diogenes Laertius mentions, in a list of works by Theophrastus, '*On indivisible lines*, one Book'. The text is in many places corrupt, so that it is often difficult or impossible to restore the argument. In reading the book we feel that the writer is for the most part chopping logic rather than contributing seriously to the philosophy of mathematics. The interest of the work to the historian of mathematics is of the slightest. It does indeed cite the equivalent of certain definitions and propositions in Euclid, especially Book X (on irrationals), and in particular it mentions the irrationals called 'binomial' or 'apotome', though, as far as irrationals are concerned, the writer may have drawn on Theaetetus rather than Euclid. The mathematical phraseology is in many places similar to that of Euclid, but the writer shows a tendency to hark back to older and less fixed terminology such as is usual in Aristotle. The tract begins with a section stating the arguments for indivisible lines, which we may take to represent Xenocrates's own arguments. The next section purports to refute these arguments one by one, after which other considerations are urged against indivisible lines. It is sought to show that the hypothesis of indivisible lines is not reconcilable with the principles assumed, or the conclusions proved, in mathematics; next, it is argued that, if a line is made up of indivisible lines (whether an odd or even number of such lines), or if the indivisible line has any point in it, or points

[1] Cf. Zeller, ii. 1^4, p. 1017.

[2] Proclus on Eucl. I, p. 279. 5.

[3] See Zeller, ii. 2^3, p. 90, note.

terminating it, the indivisible line must be divisible; and, lastly, various arguments are put forward to show that a line can no more be made up of points than of indivisible lines, with more about the relation of points to lines, &c.[1]

Sphaeric.

AUTOLYCUS of Pitane was the teacher of Arcesilaus (about 315–241/40 B.C.), also of Pitane, the founder of the so-called Middle Academy. He may be taken to have flourished about 310 B.C. or a little earlier, so that he was an elder contemporary of Euclid. We hear of him in connexion with Eudoxus's theory of concentric spheres, to which he adhered. The great difficulty in the way of this theory was early seen, namely the impossibility of reconciling the assumption of the invariability of the distance of each planet with the observed differences in the brightness, especially of Mars and Venus, at different times, and the apparent differences in the relative sizes of the sun and moon. We are told that no one before Autolycus had even attempted to deal with this difficulty 'by means of hypotheses', i. e. (presumably) in a theoretical manner, and even he was not successful, as clearly appeared from his controversy with Aristotherus[2] (who was the teacher of Aratus); this implies that Autolycus's argument was in a written treatise.

Two works by Autolycus have come down to us. They both deal with the geometry of the sphere in its application to astronomy. The definite place which they held among Greek astronomical text-books is attested by the fact that, as we gather from Pappus, one of them, the treatise *On the moving Sphere*, was included in the list of works forming the 'Little Astronomy', as it was called afterwards, to distinguish it from the 'Great Collection' (μεγάλη σύνταξις) of Ptolemy; and we may doubtless assume that the other work *On Risings and Settings* was similarly included.

[1] A revised text of the work is included in Aristotle, *De plantis*, edited by O. Apelt, who also gave a German translation of it in *Beiträge zur Geschichte der griechischen Philosophie* (1891), pp. 271–86. A translation by H. H. Joachim has since appeared (1908) in the series of Oxford Translations of Aristotle's works.

[2] Simplicius on *De caelo*, p. 504. 22–5 Heib.

Both works have been well edited by Hultsch with Latin translation.[1] They are of great interest for several reasons. First, Autolycus is the earliest Greek mathematician from whom original treatises have come down to us entire, the next being Euclid, Aristarchus and Archimedes. That he wrote earlier than Euclid is clear from the fact that Euclid, in his similar work, the *Phaenomena*, makes use of propositions appearing in Autolycus, though, as usual in such cases, giving no indication of their source. The form of Autolycus's propositions is exactly the same as that with which we are familiar in Euclid; we have first the enunciation of the proposition in general terms, then the particular enunciation with reference to a figure with letters marking the various points in it, then the demonstration, and lastly, in some cases but not in all, the conclusion in terms similar to those of the enunciation. This shows that Greek geometrical propositions had already taken the form which we recognize as classical, and that Euclid did not invent this form or introduce any material changes.

A lost text-book on Sphaeric.

More important still is the fact that Autolycus, as well as Euclid, makes use of a number of propositions relating to the sphere without giving any proof of them or quoting any authority. This indicates that there was already in existence in his time a text-book of the elementary geometry of the sphere, the propositions of which were generally known to mathematicians. As many of these propositions are proved in the *Sphaerica* of Theodosius, a work compiled two or three centuries later, we may assume that the lost text-book proceeded on much the same lines as that of Theodosius, with much the same order of propositions. Like Theodosius's *Sphaerica* it treated of the stationary sphere, its sections (great and small circles) and their properties. The geometry of the sphere at rest is of course prior to the consideration of the sphere in motion, i. e. the sphere rotating about its axis, which is the subject of Autolycus's works. Who was the author of the lost pre-Euclidean text-book it is impossible to say;

[1] *Autolyci De sphaera quae movetur liber, De ortibus et occasibus libri duo* edidit F. Hultsch (Teubner 1885).

Tannery thought that we could hardly help attributing it to Eudoxus. The suggestion is natural, seeing that Eudoxus showed, in his theory of concentric spheres, an extraordinary mastery of the geometry of the sphere; on the other hand, as Loria observes, it is, speaking generally, dangerous to assume that a work of an unknown author appearing in a certain country at a certain time must have been written by a particular man of science simply because he is the only man of the time of whom we can certainly say that he was capable of writing it.[1] The works of Autolycus also serve to confirm the pre-Euclidean origin of a number of propositions in the *Elements*. Hultsch[2] examined this question in detail in a paper of 1886. There are (1) the propositions presupposed in one or other of Autolycus's theorems. We have also to take account of (2) the propositions which would be required to establish the propositions in sphaeric assumed by Autolycus as known. The best clue to the propositions under (2) is the actual course of the proofs of the corresponding propositions in the *Sphaerica* of Theodosius; for Theodosius was only a compiler, and we may with great probability assume that, where Theodosius uses propositions from Euclid's *Elements*, propositions corresponding to them were used to prove the analogous propositions in the fourth-century *Sphaeric*. The propositions which, following this criterion, we may suppose to have been directly used for this purpose are, roughly, those represented by Eucl. I. 4, 8, 17, 19, 26, 29, 47; III. 1–3, 7, 10, 16 Cor., 26, 28, 29; IV. 6; XI. 3, 4, 10, 11, 12, 14, 16, 19, and the interpolated 38. It is, naturally, the subject-matter of Books I, III, and XI that is drawn upon, but, of course, the propositions mentioned by no means exhaust the number of pre-Euclidean propositions even in those Books. When, however, Hultsch increased the list of propositions by adding the whole chain of propositions (including Postulate 5) leading up to them in Euclid's arrangement, he took an unsafe course, because it is clear that many of Euclid's proofs were on different lines from those used by his predecessors.

[1] Loria, *Le scienze esatte nell' antica Grecia*, 1914, p. 496–7.

[2] *Berichte der Kgl. Sächs. Gesellschaft der Wissenschaften zu Leipzig*, Phil.-hist. Classe, 1886, pp. 128–55.

The work *On the moving Sphere* assumes abstractly a sphere moving about the axis stretching from pole to pole, and different series of circular sections, the first series being great circles passing through the poles, the second small circles (as well as the equator) which are sections of the sphere by planes at right angles to the axis and are called the 'parallel circles', while the third kind are great circles inclined obliquely to the axis of the sphere; the motion of points on these circles is then considered in relation to the section by a fixed plane through the centre of the sphere. It is easy to recognize in the oblique great circle in the sphere the ecliptic or zodiac circle, and in the section made by the fixed plane the horizon, which is described as the circle in the sphere 'which defines (ὁρίζων) the visible and the invisible portions of the sphere'. To give an idea of the content of the work, I will quote a few enunciations from Autolycus and along with two of them, for the sake of comparison with Euclid, the corresponding enunciations from the *Phaenomena*.

Autolycus.	Euclid.
1. If a sphere revolve uniformly about its own axis, all the points on the surface of the sphere which are not on the axis will describe parallel circles which have the same poles as the sphere and are also at right angles to the axis.	
7. If the circle in the sphere defining the visible and the invisible portions of the sphere be obliquely inclined to the axis, the circles which are at right angles to the axis and cut the defining circle [horizon] always make both their risings and settings at the same points of the defining circle [horizon] and further will also be similarly inclined to that circle.	3. The circles which are at right angles to the axis and cut the horizon make both their risings and settings at the same points of the horizon.

Autolycus.	Euclid.
9. If in a sphere a great circle which is obliquely inclined to the axis define the visible and the invisible portions of the sphere, then, of the points which rise at the same time, those towards the visible pole set later and, of those which set at the same time, those towards the visible pole rise earlier.	
11. If in a sphere a great circle which is obliquely inclined to the axis define the visible and the invisible portions of the sphere, and any other oblique great circle touch greater (parallel) circles than those which the defining circle (horizon) touches, the said other oblique circle makes its risings and settings over the whole extent of the circumference (arc) of the defining circle included between the parallel circles which it touches.	7. That the circle of the zodiac rises and sets over the whole extent of the horizon between the tropics is manifest, forasmuch as it touches circles greater than those which the horizon touches.

It will be noticed that Autolycus's propositions are more abstract in so far as the 'other oblique circle' in Autolycus is any other oblique circle, whereas in Euclid it definitely becomes the zodiac circle. In Euclid 'the great circle defining the visible and the invisible portions of the sphere' is already shortened into the technical term 'horizon' (ὁρίζων), which is defined as if for the first time: 'Let the name *horizon* be given to the plane through us (as observers) passing through the universe and separating off the hemisphere which is visible above the earth.'

The book *On Risings and Settings* is of astronomical interest only, and belongs to the region of *Phaenomena* as understood by Eudoxus and Aratus, that is, observational astronomy. It begins with definitions distinguishing between 'true' and 'apparent' morning- and evening-risings and settings of fixed stars. The 'true' morning-rising (setting) is when the star rises (sets) at the moment of the sun's rising; the 'true' morning-rising (setting) is, therefore invisible to us, and so is the 'true' evening-rising (setting) which takes place at the moment when the sun is setting. The 'apparent' morning-rising (setting) takes place when the star is first seen rising (setting) before the sun rises, and the 'apparent' evening-rising (setting) when the star is last seen rising (setting) after the sun has set. The following are the enunciations of a few of the propositions in the treatise.

I. 1. In the case of each of the fixed stars the apparent morning-risings and settings are later than the true, and the apparent evening-risings and settings are earlier than the true.

I. 2. Each of the fixed stars is seen rising each night from the (time of its) apparent morning-rising to the time of its apparent evening-rising but at no other period, and the time during which the star is seen rising is less than half a year.

I. 5. In the case of those of the fixed stars which are on the zodiac circle, the interval from the time of their apparent evening-rising to the time of their apparent evening-setting is half a year, in the case of those north of the zodiac circle more than half a year, and in the case of those south of the zodiac circle less than half a year.

II. 1. The twelfth part of the zodiac circle in which the sun is, is neither seen rising nor setting, but is hidden; and similarly the twelfth part which is opposite to it is neither seen setting nor rising but is visible above the earth the whole of the nights.

II. 4. Of the fixed stars those which are cut off by the zodiac circle in the northerly or the southerly direction will reach their evening-setting at an interval of five months from their morning-rising.

II. 9. Of the stars which are carried on the same (parallel-) circle those which are cut off by the zodiac circle in the northerly direction will be hidden a shorter time than those on the southern side of the zodiac.

XI

EUCLID

Date and traditions.

WE have very few particulars of the lives of the great mathematicians of Greece. Even Euclid is no exception. Practically all that is known about him is contained in a few sentences of Proclus's summary:

'Not much younger than these (sc. Hermotimus of Colophon and Philippus of Mende or Medma) is Euclid, who put together the Elements, collecting many of Eudoxus's theorems, perfecting many of Theaetetus's, and also bringing to irrefragable demonstration the things which were only somewhat loosely proved by his predecessors. This man lived in the time of the first Ptolemy. For Archimedes, who came immediately after the first (Ptolemy), makes mention of Euclid; and further they say that Ptolemy once asked him if there was in geometry any shorter way than that of the Elements, and he replied that there was no royal road to geometry. He is then younger than the pupils of Plato, but older than Eratosthenes and Archimedes, the latter having been contemporaries, as Eratosthenes somewhere says.'[1]

This passage shows that even Proclus had no direct knowledge of Euclid's birthplace, or of the dates of his birth and death; he can only infer generally at what period he flourished. All that is certain is that Euclid was later than the first pupils of Plato and earlier than Archimedes. As Plato died in 347 B.C. and Archimedes lived from 287 to 212 B.C., Euclid must have flourished about 300 B.C., a date which agrees well with the statement that he lived under the first Ptolemy, who reigned from 306 to 283 B.C.

[1] Proclus on Eucl. I, p. 68. 6–20.

More particulars are, it is true, furnished by Arabian authors. We are told that

'Euclid, son of Naucrates, and grandson of Zenarchus [the *Fihrist* has 'son of Naucrates, the son of Berenice (?)'], called the author of geometry, a philosopher of somewhat ancient date, a Greek by nationality, domiciled at Damascus, born at Tyre, most learned in the science of geometry, published a most excellent and most useful work entitled the foundation or elements of geometry, a subject in which no more general treatise existed before among the Greeks: nay, there was no one even of later date who did not walk in his footsteps and frankly profess his doctrine. Hence also Greek, Roman, and Arabian geometers not a few, who undertook the task of illustrating this work, published commentaries, scholia, and notes upon it, and made an abridgement of the work itself. For this reason the Greek philosophers used to post up on the doors of their schools the well-known notice, "Let no one come to our school, who has not first learnt the elements of Euclid".'[1]

This shows the usual tendency of the Arabs to romance. They were in the habit of recording the names of grandfathers, while the Greeks were not; Damascus and Tyre were no doubt brought in to gratify the desire which the Arabians always showed to connect famous Greeks in some way or other with the east (thus they described Pythagoras as a pupil of the wise Salomo, and Hipparchus as 'the Chaldaean'). We recognize the inscription over the doors of the schools of the Greek philosophers as a variation of Plato's μηδεὶς ἀγεωμέτρητος εἰσίτω; the philosopher has become Greek philosophers in general, the school their schools, while geometry has become the *Elements* of Euclid. The Arabs even explained that the name of Euclid, which they pronounced variously as *Uclides* or *Icludes*, was compounded of *Ucli*, a key, and *Dis*, a measure, or, as some say, geometry, so that Uclides is equivalent to the *key of geometry*!

In the Middle Ages most translators and editors spoke of Euclid as Euclid *of Megara*, confusing our Euclid with Euclid the philosopher, and the contemporary of Plato, who lived about 400 B.C. The first trace of the confusion appears in Valerius

[1] Casiri, *Bibliotheca Arabico-Hispana Escurialensis*, i, p. 339 (Casiri's source is the *Ta'rīkh al-Ḥukamā* of al-Qifṭī (d. 1248).

Maximus (in the time of Tiberius) who says[1] that Plato, on being appealed to for a solution of the problem of doubling the cube, sent the inquirers to 'Euclid the geometer'. The mistake was seen by one Constantinus Lascaris (d. about 1493), and the first translator to point it out clearly was Commandinus (in his translation of Euclid published in 1572).

Euclid may have been a Platonist, as Proclus says, though this is not certain. In any case, he probably received his mathematical training in Athens from the pupils of Plato; most of the geometers who could have taught him were of that school. But he himself taught and founded a school at Alexandria, as we learn from Pappus's statement that Apollonius 'spent a very long time with the pupils of Euclid at Alexandria'.[2] Here again come in our picturesque Arabians,[3] who made out that the *Elements* were originally written by a man whose name was Apollonius, a carpenter, who wrote the work in fifteen books or sections (this idea seems to be based on some misunderstanding of Hypsicles's preface to the so-called Book XIV of Euclid), and that, as some of the work was lost in course of time and the rest disarranged, one of the kings at Alexandria who desired to study geometry and to master this treatise in particular first questioned about it certain learned men who visited him, and then sent for Euclid, who was at that time famous as a geometer, and asked him to revise and complete the work and reduce it to order, upon which Euclid rewrote the work in thirteen books, thereafter known by his name.

On the character of Euclid Pappus has a remark which, however, was probably influenced by his obvious animus against Apollonius, whose preface to the *Conics* seemed to him to give too little credit to Euclid for his earlier work in the same subject. Pappus contrasts Euclid's attitude to his predecessors. Euclid, he says, was no such boaster or controversialist: thus he regarded Aristaeus as deserving credit for the discoveries he had made in conics, and made no attempt to anticipate him or to construct afresh the same system, such was his scrupulous fairness and his exemplary kindliness to all who

[1] viii. 12, ext. 1.

[2] Pappus, vii, p. 678. 10–12.

[3] The authorities are al-Kindī, *De instituto libri Euclidis* and a commentary by Qāḍīzāde on the *Ashkal at-ta'sīs* of Ashraf Shamsaddīn as-Samarqandī (quoted by Casiri and Ḥājī Khalfa).

could advance mathematical science to however small an extent.[1] Although, as I have indicated, Pappus's motive was rather to represent Apollonius in a relatively unfavourable light than to state a historical fact about Euclid, the statement accords well with what we should gather from Euclid's own works. These show no sign of any claim to be original; in the *Elements*, for instance, although it is clear that he made great changes, altering the arrangement of whole Books, redistributing propositions between them, and inventing new proofs where the new order made the earlier proofs inapplicable, it is safe to say that he made no more alterations than his own acumen and the latest special investigations (such as Eudoxus's theory of proportion) showed to be imperative in order to make the exposition of the whole subject more scientific than the earlier efforts of writers of elements. His respect for tradition is seen in his retention of some things which were out of date and useless, e.g. certain definitions never afterwards used, the solitary references to the angle of a semicircle or the angle of a segment, and the like; he wrote no sort of preface to his work (would that he had!) such as those in which Archimedes and Apollonius introduced their treatises and distinguished what they claimed as new in them from what was already known: he plunges at once into his subject, '*A point is that which has no part*'!

And what a teacher he must have been! One story enables us to picture him in that capacity. According to Stobaeus,

'some one who had begun to read geometry with Euclid, when he had learnt the first theorem, asked Euclid, "what shall I get by learning these things?" Euclid called his slave and said, "Give him threepence, since he must make gain out of what he learns".'[2]

Ancient commentaries, criticisms, and references.

Euclid has, of course, always been known almost exclusively as the author of the *Elements*. From Archimedes onwards the Greeks commonly spoke of him as ὁ στοιχειώτης, the writer of the *Elements*, instead of using his name. This wonderful book, with all its imperfections, which indeed are slight enough when account is taken of the date at which

[1] Pappus, vii, pp. 676. 25–678. 6. [2] Stobaeus, *Floril.* iv. p. 205.

it appeared, is and will doubtless remain the greatest mathematical text-book of all time. Scarcely any other book except the Bible can have circulated more widely the world over, or been more edited and studied. Even in Greek times the most accomplished mathematicians occupied themselves with it; Heron, Pappus, Porphyry, Proclus and Simplicius wrote commentaries; Theon of Alexandria re-edited it, altering the language here and there, mostly with a view to greater clearness and consistency, and interpolating intermediate steps, alternative proofs, separate 'cases', porisms (corollaries) and lemmas (the most important addition being the second part of VI. 33 relating to *sectors*). Even the great Apollonius was moved by Euclid's work to discuss the first principles of geometry; his treatise on the subject was in fact a criticism of Euclid, and none too successful at that; some alternative definitions given by him have point, but his alternative solutions of some of the easy problems in Book I do not constitute any improvement, and his attempt to prove the axioms (if one may judge by the case quoted by Proclus, that of Axiom 1) was thoroughly misconceived.

Apart from systematic commentaries on the whole work or substantial parts of it, there were already in ancient times discussions and controversies on special subjects dealt with by Euclid, and particularly his theory of parallels. The fifth Postulate was a great stumbling-block. We know from Aristotle that up to his time the theory of parallels had not been put on a scientific basis[1]: there was apparently some *petitio principii* lurking in it. It seems therefore clear that Euclid was the first to apply the bold remedy of laying down the indispensable principle of the theory in the form of an indemonstrable Postulate. But geometers were not satisfied with this solution. Posidonius and Geminus tried to get over the difficulty by substituting an *equidistance* theory of parallels. Ptolemy actually tried to prove Euclid's postulate, as also did Proclus, and (according to Simplicius) one Diodorus, as well as 'Aganis'; the attempt of Ptolemy is given by Proclus along with his own, while that of 'Aganis' is reproduced from Simplicius by the Arabian commentator an-Nairīzī.

[1] *Anal. Prior*. ii. 16. 65 a 4.

Other very early criticisms there were, directed against the very first steps in Euclid's work. Thus Zeno of Sidon, an Epicurean, attacked the proposition I. 1 on the ground that it is not conclusive unless it be first assumed that neither two straight lines nor two circumferences can have a common segment; and this was so far regarded as a serious criticism that Posidonius wrote a whole book to controvert Zeno.[1] Again, there is the criticism of the Epicureans that I. 20, proving that any two sides in a triangle are together greater than the third, is evident even to an ass and requires no proof. I mention these isolated criticisms to show that the *Elements*, although they superseded all other Elements and never in ancient times had any rival, were not even at the first accepted without question.

The first Latin author to mention Euclid is Cicero; but it is not likely that the *Elements* had then been translated into Latin. Theoretical geometry did not appeal to the Romans, who only cared for so much of it as was useful for measurements and calculations. Philosophers studied Euclid, but probably in the original Greek; Martianus Capella speaks of the effect of the mention of the proposition 'how to construct an equilateral triangle on a given straight line' among a company of philosophers, who, recognizing the first proposition of the *Elements*, straightway break out into encomiums on Euclid.[2] Beyond a fragment in a Verona palimpsest of a free rendering or rearrangement of some propositions from Books XII and XIII dating apparently from the fourth century, we have no trace of any Latin version before Boëtius (born about A. D. 480), to whom Magnus Aurelius Cassiodorus and Theodoric attribute a translation of Euclid. The so-called geometry of Boëtius which has come down to us is by no means a translation of Euclid; but even the redaction of this in two Books which was edited by Friedlein is not genuine, having apparently been put together in the eleventh century from various sources; it contains the definitions of Book I, the Postulates (five in number), the Axioms (three only), then some definitions from Eucl. II, III, IV, followed by the *enunciations* only (without proofs) of Eucl. I, ten propositions

[1] Proclus on Eucl. I, p. 200. 2.

[2] Mart. Capella, vi. 724.

of Book II, and a few of Books III and IV, and lastly a passage indicating that the editor will now give something of his own, which turns out to be a literal translation of the proofs of Eucl. I. 1–3. This proves that the Pseudo-Boëtius had a Latin translation of Euclid from which he extracted these proofs; moreover, the text of the definitions from Book I shows traces of perfectly correct readings which are not found even in the Greek manuscripts of the tenth century, but which appear in Proclus and other ancient sources. Fragments of such a Latin translation are also found in the *Gromatici veteres*.[1]

The text of the Elements.

All our Greek texts of the *Elements* up to a century ago depended upon manuscripts containing Theon's recension of the work; these manuscripts purport, in their titles, to be either 'from the edition of Theon' (ἐκ τῆς Θέωνος ἐκδόσεως) or 'from the lectures of Theon' (ἀπὸ συνουσιῶν τοῦ Θέωνος). Sir Henry Savile in his *Praelectiones* had drawn attention to the passage in Theon's Commentary on Ptolemy[2] quoting the second part of VI. 33 about sectors as having been proved by *himself* in his edition of the *Elements*; but it was not till Peyrard discovered in the Vatican the great MS. gr. 190, containing neither the words from the titles of the other manuscripts quoted above nor the addition to VI. 33, that scholars could get back from Theon's text to what thus represents, on the face of it, a more ancient edition than Theon's. It is also clear that the copyist of P (as the manuscript is called after Peyrard), or rather of its archetype, had before him the two recensions and systematically gave the preference to the earlier one; for at XIII. 6 in P the first hand has a marginal note, 'This theorem is not given in most copies of the *new edition*, but is found in those of the old'. The *editio princeps* (Basel, 1533) edited by Simon Grynaeus was based on two manuscripts (Venetus Marcianus 301 and Paris. gr. 2343) of the sixteenth century, which are among the worst. The Basel edition was again the foundation of the text of Gregory (Oxford, 1703), who only consulted the

[1] Ed. Lachmann, pp. 377 sqq.

[2] I, p. 201, ed. Halma.

manuscripts bequeathed by Savile to the University in places where the Basel text differed from the Latin version of Commandinus which he followed in the main. It was a pity that even Peyrard in his edition (1814–18) only corrected the Basel text by means of P, instead of rejecting it altogether and starting afresh; but he adopted many of the readings of P and gave a conspectus of them in an appendix. E. F. August's edition (1826–9) followed P more closely, and he consulted the Viennese MS. gr. 103 also; but it was left for Heiberg to bring out a new and definitive Greek text (1883–8) based on P and the best of the Theonine manuscripts, and taking account of external sources such as Heron and Proclus. Except in a few passages, Proclus's manuscript does not seem to have been of the best, but authors earlier than Theon, e.g. Heron, generally agree with our best manuscripts. Heiberg concludes that the *Elements* were most spoiled by interpolations about the third century, since Sextus Empiricus had a correct text, while Iamblicus had an interpolated one.

The differences between the inferior Theonine manuscripts and the best sources are perhaps best illustrated by the arrangement of postulates and axioms in Book I. Our ordinary editions based on Simson have three postulates and twelve axioms. Of these twelve axioms the eleventh (stating that all right angles are equal) is, in the genuine text, the fourth Postulate, and the twelfth Axiom (the Parallel-Postulate) is the fifth Postulate; the Postulates were thus originally five in number. Of the ten remaining Axioms or Common Notions Heron only recognized the first three, and Proclus only these and two others (that things which coincide are equal, and that the whole is greater than the part); it is fairly certain, therefore, that the rest are interpolated, including the assumption that two straight lines cannot enclose a space (Euclid himself regarded this last fact as involved in Postulate 1, which implies that a straight line joining one point to another is *unique*).

Latin and Arabic translations.

The first Latin translations which we possess in a complete form were made not from the Greek but from the Arabic. It was as early as the eighth century that the *Elements* found their way to Arabia. The Caliph al-Manṣūr (754–75), as the result of a mission to the Byzantine Emperor, obtained a copy of Euclid among other Greek books, and the Caliph al-Ma'mūn (813–33) similarly obtained manuscripts of Euclid, among others, from the Byzantines. Al-Hajjāj b. Yūsuf b. Maṭar made two versions of the *Elements*, the first in the reign of Hārūn ar-Rashīd (786–809), the second for al-Ma'mūn; six Books of the second of these versions survive in a Leyden manuscript (Cod. Leidensis 399. 1) which is being edited along with an-Nairīzī's commentary by Besthorn and Heiberg[1]; this edition was abridged, with corrections and explanations, but without change of substance, from the earlier version, which appears to be lost. The work was next translated by Abū Ya'qūb Isḥāq b. Ḥunain b. Isḥāq al-'Ibādī (died 910), evidently direct from the Greek; this translation seems itself to have perished, but we have it as revised by Thābit b. Qurra (died 901) in two manuscripts (No. 279 of the year 1238 and No. 280 written in 1260–1) in the Bodleian Library; Books I–XIII in these manuscripts are in the Isḥāq-Thābit version, while the non-Euclidean Books XIV, XV are in the translation of Qusṭā b. Lūqā al-Ba'labakkī (died about 912). Isḥāq's version seems to be a model of good translation; the technical terms are simply and consistently rendered, the definitions and enunciations differ only in isolated cases from the Greek, and the translator's object seems to have been only to get rid of difficulties and unevennesses in the Greek text while at the same time giving a faithful reproduction of it. The third Arabic version still accessible to us is that of Naṣīraddīn aṭ-Ṭūsī (born in 1201 at Ṭūs in Khurāsān); this, however, is not a translation of Euclid but a rewritten version based upon the older Arabic translations. On the whole, it appears probable that the Arabic tradition (in spite of its omission of lemmas and porisms, and, except in a very few cases, of the interpolated alternative proofs) is not to be preferred to that of the Greek manuscripts, but must be regarded as inferior in authority.

The known Latin translations begin with that of Athelhard, an Englishman, of Bath; the date of it is about 1120. That

[1] Parts I, i. 1893, I, ii. 1897, II, i. 1900, II, ii. 1905, III, i. 1910 (Copenhagen).

it was made from the Arabic is clear from the occurrence of Arabic words in it; but Athelhard must also have had before him a translation of (at least) the enunciations of Euclid based ultimately upon the Greek text, a translation going back to the old Latin version which was the common source of the passage in the *Gromatici* and 'Boëtius'. But it would appear that even before Athelhard's time some sort of translation, or at least fragments of one, were available even in England if one may judge by the Old English verses:

> 'The clerk Euclide on this wyse hit fonde
> Thys craft of gemetry yn Egypte londe
> Yn Egypte he tawghte hyt ful wyde,
> In dyvers londe on every syde.
> Mony erys afterwarde y understonde
> Yer that the craft com ynto thys londe.
> Thys craft com into England, as y yow say,
> Yn tyme of good Kyng Adelstone's day',

which would put the introduction of Euclid into England as far back as A.D. 924–40.

Next, Gherard of Cremona (1114–87) is said to have translated the '15 Books of Euclid' from the Arabic as he undoubtedly translated an-Nairīzī's commentary on Books I–X; this translation of the *Elements* was till recently supposed to have been lost, but in 1904 A. A. Björnbo discovered in manuscripts at Paris, Boulogne-sur-Mer and Bruges the whole, and at Rome Books X–XV, of a translation which he gives good ground for identifying with Gherard's. This translation has certain Greek words such as *rombus, romboides*, where Athelhard keeps the Arabic terms; it was thus clearly independent of Athelhard's, though Gherard appears to have had before him, in addition, an old translation of Euclid from the Greek which Athelhard also used. Gherard's translation is much clearer than Athelhard's; it is neither abbreviated nor 'edited' in the same way as Athelhard's, but it is a word for word translation of an Arabic manuscript containing a revised and critical edition of Thābit's version.

A third translation from the Arabic was that of Johannes Campanus, which came some 150 years after that of Athelhard. That Campanus's translation was not independent of Athelhard's is proved by the fact that, in all manuscripts and editions, the definitions, postulates and axioms, and the 364 enunciations are word for word identical in Athelhard and Campanus. The exact relation between the two seems even yet not to have been sufficiently elucidated. Campanus may have used Athelhard's translation and only developed the proofs by means of another redaction of the Arabian Euclid. Campanus's translation is the clearer and more complete, following the Greek text more closely but still at some distance; the arrangement of the two is different; in Athelhard the proofs regularly precede the enunciations, while Campanus follows the usual order. How far the differences in the proofs and the additions in each are due to the translators themselves or go back to Arabic originals is a moot question; but it seems most probable that Campanus stood to Athelhard somewhat in the relation of a commentator, altering and improving his translation by means of other Arabic originals.

The first printed editions.

Campanus's translation had the luck to be the first to be put into print. It was published at Venice by Erhard Ratdolt in 1482. This beautiful and very rare book was not only the first printed edition of Euclid, but also the first printed mathematical book of any importance. It has margins of $2\frac{1}{2}$ inches and in them are placed the figures of the propositions. Ratdolt says in his dedication that, at that time, although books by ancient and modern authors were being printed every day in Venice, little or nothing mathematical had appeared; this fact he puts down to the difficulty involved by the figures, which no one had up to that time succeeded in printing; he adds that after much labour he had discovered a method by which figures could be produced as easily as letters. Experts do not seem even yet to be agreed as to the actual way in which the figures were made, whether they were woodcuts or whether they were made by putting together lines and circular arcs as letters are put together to make words. How eagerly the opportunity of spreading geometrical knowledge was seized upon is proved by the number of editions which followed in the next few years. Even the year 1482 saw two forms of the book, though they only differ in the first sheet. Another edition came out at Ulm in 1486, and another at Vicenza in 1491.

In 1501 G. Valla gave in his encyclopaedic work *De expetendis et fugiendis rebus* a number of propositions with proofs and scholia translated from a Greek manuscript which was once in his possession; but Bartolomeo Zamberti (Zambertus) was the first to bring out a translation from the Greek text of the whole of the *Elements*, which appeared at Venice in 1505. The most important Latin translation is, however, that of Commandinus (1509–75), who not only followed the Greek text more closely than his predecessors, but added to his translation some ancient scholia as well as good notes of his own; this translation, which appeared in 1572, was the foundation of most translations up to the time of Peyrard, including that of Simson, and therefore of all those editions, numerous in England, which gave Euclid 'chiefly after the text of Dr. Simson'.

The study of Euclid in the Middle Ages.

A word or two about the general position of geometry in education during the Middle Ages will not be out of place in a book for English readers, in view of the unique place which Euclid has till recently held as a text-book in this country. From the seventh to the tenth century the study of geometry languished: 'We find in the whole literature of that time hardly the slightest sign that any one had gone farther in this department of the Quadrivium than the definitions of a triangle, a square, a circle, or of a pyramid or cone, as Martianus Capella and Isidorus (Hispalensis, died as Bishop of Seville in 636) left them.'[1] (Isidorus had disposed of the four subjects of Arithmetic, Geometry, Music and Astronomy in *four pages* of his encyclopaedic work *Origines* or *Etymologiae*). In the tenth century appeared a 'reparator studiorum' in the person of the great Gerbert, who was born at Aurillac, in Auvergne, in the first half of the tenth century, and after a very varied life ultimately (in 999) became Pope Sylvester II; he died in 1003. About 967 he went on

[1] Hankel, *op. cit.*, pp. 311–12.

a journey to Spain, where he studied mathematics. In 970 he went to Rome with Bishop Hatto of Vich (in the province of Barcelona), and was there introduced by Pope John XIII to the German king Otto I. To Otto, who wished to find him a post as a teacher, he could say that 'he knew enough of mathematics for this, but wished to improve his knowledge of logic'. With Otto's consent he went to Reims, where he became Scholasticus or teacher at the Cathedral School, remaining there for about ten years, 972 to 982. As the result of a mathematico-philosophic argument in public at Ravenna in 980, he was appointed by Otto II to the famous monastery at Bobbio in Lombardy, which, fortunately for him, was rich in valuable manuscripts of all sorts. Here he found the famous 'Codex Arcerianus' containing fragments of the works of the *Gromatici*, Frontinus, Hyginus, Balbus, Nipsus, Epaphroditus and Vitruvius Rufus. Although these fragments are not in themselves of great merit, there are things in them which show that the authors drew upon Heron of Alexandria, and Gerbert made the most of them. They formed the basis of his own 'Geometry', which may have been written between the years 981 and 983. In writing this book Gerbert evidently had before him Boëtius's *Arithmetic*, and in the course of it he mentions Pythagoras, Plato's *Timaeus*, with Chalcidius's commentary thereon, and Eratosthenes. The geometry in the book is mostly practical; the theoretical part is confined to necessary preliminary matter, definitions, &c., and a few proofs; the fact that the sum of the angles of a triangle is equal to two right angles is proved in Euclid's manner. A great part is taken up with the solution of triangles, and with heights and distances. The Archimedean value of π $(\frac{22}{7})$ is used in stating the area of a circle; the surface of a sphere is given as $\frac{11}{21}D^3$. The plan of the book is quite different from that of Euclid, showing that Gerbert could neither have had Euclid's *Elements* before him, nor, probably, Boëtius's *Geometry*, if that work in its genuine form was a version of Euclid. When in a letter written probably from Bobbio in 983 to Adalbero, Archbishop of Reims, he speaks of his expectation of finding 'eight volumes of Boëtius on astronomy, also the most famous of figures (presumably propositions) in geometry and other things not less admirable', it is not clear that he actually found these things, and it is still less certain that the geometrical matter referred to was Boëtius's *Geometry*.

From Gerbert's time, again, no further progress was made until translations from the Arabic began with Athelhard and the rest. Gherard of Cremona (died 1187), who translated the *Elements* and an-Nairīzī's commentary thereon, is credited with a whole series of translations from the Arabic of Greek authors; they included the *Data* of Euclid, the *Sphaerica* of Theodosius, the *Sphaerica* of Menelaus, the *Syntaxis* of Ptolemy; besides which he translated Arabian geometrical works such as the *Liber trium fratrum*, and also the algebra of Muḥammad b. Mūsā. One of the first results of the interest thus aroused in Greek and Arabian mathematics was seen in the very remarkable works of Leonardo of Pisa (Fibonacci). Leonardo first published in 1202, and then brought out later (1228) an improved edition of, his *Liber abaci* in which he gave the whole of arithmetic and algebra as known to the Arabs, but in a free and independent style of his own; in like manner in his *Practica geometriae* of 1220 he collected (1) all that the *Elements* of Euclid and Archimedes's books on the *Measurement of a Circle* and *On the Sphere and Cylinder* had taught him about the measurement of plane figures bounded by straight lines, solid figures bounded by planes, the circle and the sphere respectively, (2) divisions of figures in different proportions, wherein he based himself on Euclid's book *On the divisions of figures*, but carried the subject further, (3) some trigonometry, which he got from Ptolemy and Arabic sources (he uses the terms *sinus rectus* and *sinus versus*); in the treatment of these varied subjects he showed the same mastery and, in places, distinct originality. We should have expected a great general advance in the next centuries after such a beginning, but, as Hankel says, when we look at the work of Luca Paciuolo nearly three centuries later, we find that the talent which Leonardo had left to the Latin world had lain hidden in a napkin and earned no interest. As regards the place of geometry in education during this period we have the evidence of Roger Bacon (1214–94), though he, it is true, seems to have taken an exaggerated view of the incompetence of the mathematicians and teachers of his time; the philosophers of his day, he says, despised geometry, languages, &c., declaring that they were useless; people in general, not finding utility in any science such as geometry, at once recoiled, unless they were boys forced to it by the rod, from the idea of studying it, so that they would hardly learn as much as three or four propositions; the fifth proposition of Euclid was called *Elefuga* or *fuga miserorum.*[1]

As regards Euclid at the Universities, it may be noted that the study of geometry seems to have been neglected at the University of Paris. At the reformation of the University in 1336 it was only provided that no one should take a Licentiate who had not attended lectures on some mathematical books; the same requirement reappears in 1452 and 1600. From the preface to a commentary on Euclid which appeared in 1536 we learn that a candidate for the degree of M.A. had to take a solemn oath that he had attended lectures on the first six Books; but it is doubtful whether for the examinations more than Book I was necessary, seeing that the proposition I. 47 was known as *Magister matheseos.* At the University of Prague (founded in 1348) mathematics were more regarded. Candidates for the Baccalaureate had to attend lectures on the *Tractatus de Sphaera materiali*, a treatise on the fundamental ideas of spherical astronomy, mathematical geography and the ordinary astronomical phenomena, but without the help of mathematical propositions, written by Johannes de Sacrobosco (i. e. of Holywood, in Yorkshire) in 1250, a book which was read at all Universities for four centuries and many times commented upon; for the Master's degree lectures on the first six Books of Euclid were compulsory. Euclid was lectured upon at the Universities of Vienna (founded 1365), Heidelberg (1386), Cologne (1388); at Heidelberg an oath was required from the candidate for the Licentiate corresponding to M.A. that he had attended lectures on some whole books and not merely parts of several books (not necessarily, it appears, of Euclid); at Vienna, the first five Books of Euclid were required; at Cologne, no mathematics were required for the Baccalaureate, but the candidate for M.A. must have attended

[1] Roger Bacon, *Opus Tertium*, cc. iv, vi.

lectures on the *Sphaera mundi*, planetary theory, three Books of Euclid, optics and arithmetic. At Leipzig (founded 1409), as at Vienna and Prague, there were lectures on Euclid for some time at all events, though Hankel says that he found no mention of Euclid in a list of lectures given in the consecutive years 1437–8, and Regiomontanus, when he went to Leipzig, found no fellow-students in geometry. At Oxford, in the middle of the fifteenth century, the first two Books of Euclid were read, and doubtless the Cambridge course was similar.

The first English editions.

After the issue of the first printed editions of Euclid, beginning with the translation of Campano, published by Ratdolt, and of the *editio princeps* of the Greek text (1533), the study of Euclid received a great impetus, as is shown by the number of separate editions and commentaries which appeared in the sixteenth century. The first complete English translation by Sir Henry Billingsley (1570) was a monumental work of 928 pages of folio size, with a preface by John Dee, and notes extracted from all the most important commentaries from Proclus down to Dee himself, a magnificent tribute to the immortal Euclid. About the same time Sir Henry Savile began to give *unpaid* lectures on the Greek geometers; those on Euclid do not indeed extend beyond I. 8, but they are valuable because they deal with the difficulties connected with the preliminary matter, the definitions, &c., and the tacit assumptions contained in the first propositions. But it was in the period from about 1660 to 1730, during which Wallis and Halley were Professors at Oxford, and Barrow and Newton at Cambridge, that the study of Greek mathematics was at its height in England. As regards Euclid in particular Barrow's influence was doubtless very great. His Latin version (*Euclidis Elementorum Libri XV breviter demonstrati*) came out in 1655, and there were several more editions of the same published up to 1732; his first English edition appeared in 1660, and was followed by others in 1705, 1722, 1732, 1751. This brings us to Simson's edition, first published both in Latin and English in 1756. It is presumably from this time onwards that Euclid acquired the unique status as a text-book which it maintained till recently. I cannot help thinking that it was Barrow's influence which contributed most powerfully to this. We are told that Newton, when he first bought a Euclid in 1662 or 1663, thought it 'a trifling book', as the propositions seemed to him obvious; afterwards, however, on Barrow's advice, he studied the *Elements* carefully and derived, as he himself stated, much benefit therefrom.

Technical terms connected with the classical form of a proposition.

As the classical form of a proposition in geometry is that which we find in Euclid, though it did not originate with him, it is desirable, before we proceed to an analysis of the *Elements*, to give some account of the technical terms used by the Greeks in connexion with such propositions and their proofs. We will take first the terms employed to describe the formal divisions of a proposition.

(α) *Terms for the formal divisions of a proposition.*

In its completest form a proposition contained six parts, (1) the πρότασις, or *enunciation* in general terms, (2) the ἔκθεσις, or *setting-out*, which states the particular *data*, e.g. a given straight line *AB*, two given triangles *ABC*, *DEF*, and the like, generally shown in a figure and constituting that upon which the proposition is to operate, (3) the διορισμός, *definition* or *specification*, which means the restatement of what it is required to do or to prove in terms of the particular data, the object being to fix our ideas, (4) the κατασκευή, the *construction* or *machinery* used, which includes any additions to the original figure by way of construction that are necessary to enable the proof to proceed, (5) the ἀπόδειξις, or the *proof* itself, and (6) the συμπέρασμα, or *conclusion*, which reverts to the enunciation, and states what has been proved or done; the conclusion can, of course, be stated in as general terms as the enunciation, since it does not depend on the particular figure drawn; that figure is only an illustration, a type of the *class* of figure, and it is legitimate therefore, in stating the conclusion, to pass from the particular to the general.

In particular cases some of these formal divisions may be absent, but three are always found, the *enunciation*, *proof* and *conclusion*. Thus in many propositions no construction is needed, the given figure itself sufficing for the proof; again, in IV. 10 (to construct an isosceles triangle with each of the base angles double of the vertical angle) we may, in a sense, say with Proclus [1] that there is neither *setting-out* nor *definition*, for there is nothing *given* in the enunciation, and we set out, not a given straight line, but any straight line AB, while the proposition does not state (what might be said by way of *definition*) that the required triangle is to have AB for one of its equal sides.

(β) *The διορισμός or statement of conditions of possibility.*

Sometimes to the statement of a problem there has to be added a διορισμός in the more important and familiar sense of a criterion of the conditions of possibility or, in its most complete form, a criterion as to 'whether what is sought is impossible or possible and how far it is practicable and in how many ways'.[2] Both kinds of διορισμός begin with the words δεῖ δή, which should be translated, in the case of the *definition*, 'thus it is required (to prove or do so and so)' and, in the case of the criterion of possibility, 'thus it is necessary that . . .' (not '*but* it is necessary . . .'). Cf. I. 22, 'Out of three straight lines which are equal to three given straight lines to construct a triangle: thus it is necessary that two of the straight lines taken together in any manner should be greater than the remaining straight line'.

(γ) *Analysis, synthesis, reduction, reductio ad absurdum.*

The *Elements* is a synthetic treatise in that it goes directly forward the whole way, always proceeding from the known to the unknown, from the simple and particular to the more complex and general; hence *analysis*, which reduces the unknown or the more complex to the known, has no place in the exposition, though it would play an important part in the discovery of the proofs. A full account of the Greek *analysis* and *synthesis* will come more conveniently elsewhere.

[1] Proclus on Eucl. I, p. 203. 23 sq. [2] *Ib.*, p. 202. 3.

In the meantime we may observe that, where a proposition is worked out by analysis followed by synthesis, the analysis comes between the *definition* and the *construction* of the proposition; and it should not be forgotten that *reductio ad absurdum* (called in Greek ἡ εἰς τὸ ἀδύνατον ἀπαγωγή, 'reduction to the impossible', or ἡ διὰ τοῦ ἀδυνάτου δεῖξις or ἀπόδειξις, 'proof *per impossibile*'), a method of proof common in Euclid as elsewhere, is a variety of analysis. For analysis begins with *reduction* (ἀπαγωγή) of the original proposition, which we hypothetically assume to be true, to something simpler which we can recognize as being either true or false; the case where it leads to a conclusion known to be false is the *reductio ad absurdum.*

(δ) *Case, objection, porism, lemma.*

Other terms connected with propositions are the following. A proposition may have several *cases* according to the different arrangements of points, lines, &c., in the figure that may result from variations in the positions of the elements given; the word for *case* is πτῶσις. The practice of the great geometers was, as a rule, to give only one case, leaving the others for commentators or pupils to supply for themselves. But they were fully alive to the existence of such other cases; sometimes, if we may believe Proclus, they would even give a proposition solely with a view to its use for the purpose of proving a case of a later proposition which is actually omitted. Thus, according to Proclus,[1] the second part of I. 5 (about the angles beyond the base) was intended to enable the reader to meet an *objection* (ἔνστασις) that might be raised to I. 7 as given by Euclid on the ground that it was incomplete, since it took no account of what was given by Proclus himself, and is now generally given in our text-books, as the second case.

What we call a *corollary* was for the Greeks a *porism* (πόρισμα), i. e. something provided or ready-made, by which was meant some result incidentally revealed in the course of the demonstration of the main proposition under discussion, a sort of incidental gain' arising out of the demonstration,

[1] Proclus on Eucl. I, pp. 248. 8–11; 263. 4–8.

as Proclus says.[1] The name *porism* was also applied to a special kind of substantive proposition, as in Euclid's separate work in three Books entitled *Porisms* (see below, pp. 431–8).

The word *lemma* (λῆμμα) simply means something *assumed*. Archimedes uses it of what is now known as the Axiom of Archimedes, the principle assumed by Eudoxus and others in the method of exhaustion; but it is more commonly used of a subsidiary proposition requiring proof, which, however, it is convenient to assume in the place where it is wanted in order that the argument may not be interrupted or unduly lengthened. Such a lemma might be proved in advance, but the proof was often postponed till the end, the assumption being marked as something to be afterwards proved by some such words as ὡς ἑξῆς δειχθήσεται, 'as will be proved in due course'.

Analysis of the *Elements*.

Book I of the *Elements* necessarily begins with the essential preliminary matter classified under the headings *Definitions* (ὅροι), *Postulates* (αἰτήματα) and *Common Notions* (κοιναὶ ἔννοιαι). In calling the axioms *Common Notions* Euclid followed the lead of Aristotle, who uses as alternatives for 'axioms' the terms 'common (things)', 'common opinions'.

Many of the *Definitions* are open to criticism on one ground or another. Two of them at least seem to be original, namely, the definitions of a straight line (4) and of a plane surface (7); unsatisfactory as these are, they seem to be capable of a simple explanation. The definition of a straight line is apparently an attempt to express, without any appeal to sight, the sense of Plato's definition 'that of which the middle covers the ends' (*sc.* to an eye placed at one end and looking along it); and the definition of a plane surface is an adaptation of the same definition. But most of the definitions were probably adopted from earlier text-books; some appear to be inserted merely out of respect for tradition, e.g. the definitions of *oblong*, *rhombus*, *rhomboid*, which are never used in the *Elements*. The definitions of various figures assume the existence of the thing defined, e.g. the square, and the

[1] *Ib.*, p. 212. 16.

different kinds of triangle under their twofold classification (*a*) with reference to their sides (equilateral, isosceles and scalene), and (*b*) with reference to their angles (right-angled, obtuse-angled and acute-angled); such definitions are provisional pending the proof of existence by means of actual construction. A *parallelogram* is not defined; its existence is first proved in I. 33, and in the next proposition it is called a 'parallelogrammic area', meaning an area contained by parallel lines, in preparation for the use of the simple word 'parallelogram' from I. 35 onwards. The definition of a diameter of a circle (17) includes a theorem; for Euclid adds that 'such a straight line also bisects the circle', which is one of the theorems attributed to Thales; but this addition was really necessary in view of the next definition (18), for, without this explanation, Euclid would not have been justified in describing a *semi*-circle as a portion of a circle cut off by a diameter.

More important by far are the five Postulates, for it is in them that Euclid lays down the real principles of Euclidean geometry; and nothing shows more clearly his determination to reduce his original assumptions to the very minimum. The first three Postulates are commonly regarded as the postulates of *construction*, since they assert the possibility (1) of drawing the straight line joining two points, (2) of producing a straight line in either direction, and (3) of describing a circle with a given centre and 'distance'. But they imply much more than this. In Postulates 1 and 3 Euclid postulates the existence of straight lines and circles, and implicitly answers the objections of those who might say that, as a matter of fact, the straight lines and circles which we can draw are not mathematical straight lines and circles; Euclid may be supposed to assert that we can nevertheless assume our straight lines and circles to be such for the purpose of our proofs, since they are only illustrations enabling us to *imagine* the real things which they imperfectly represent. But, again, Postulates 1 and 2 further imply that the straight line drawn in the first case and the produced portion of the straight line in the second case are *unique*; in other words, Postulate 1 implies that two straight lines cannot enclose a space, and so renders unnecessary the 'axiom' to that effect interpolated in Proposition 4, while Postulate 2 similarly implies the theorem that two straight lines cannot have a common segment, which Simson gave as a corollary to I. 11.

At first sight the Postulates 4 (that all right angles are equal) and 5 (the Parallel-Postulate) might seem to be of an altogether different character, since they are rather of the nature of theorems unproved. But Postulate 5 is easily seen to be connected with constructions, because so many constructions depend on the existence and use of points in which straight lines intersect; it is therefore absolutely necessary to lay down some criterion by which we can judge whether two straight lines in a figure will or will not meet if produced. Postulate 5 serves this purpose as well as that of providing a basis for the theory of parallel lines. Strictly speaking, Euclid ought to have gone further and given criteria for judging whether other pairs of lines, e.g. a straight line and a circle, or two circles, in a particular figure will or will not intersect one another. But this would have necessitated a considerable series of propositions, which it would have been difficult to frame at so early a stage, and Euclid preferred to assume such intersections provisionally in certain cases, e.g. in I. 1.

Postulate 4 is often classed as a theorem. But it had in any case to be placed before Postulate 5 for the simple reason that Postulate 5 would be no criterion at all unless right angles were determinate magnitudes; Postulate 4 then declares them to be such. But this is not all. If Postulate 4 were to be proved as a theorem, it could only be proved by applying one pair of 'adjacent' right angles to another pair. This method would not be valid unless on the assumption of the *invariability of figures*, which would therefore have to be asserted as an antecedent postulate. Euclid preferred to assert as a postulate, directly, the fact that all right angles are equal; hence his postulate may be taken as equivalent to the principle of the *invariability of figures*, or, what is the same thing, the *homogeneity of space*.

For reasons which I have given above (pp. 339, 358), I think that the great Postulate 5 is due to Euclid himself; and it seems probable that Postulate 4 is also his, if not Postulates 1–3 as well.

Of the *Common Notions* there is good reason to believe that only five (at the most) are genuine, the first three and two others, namely 'Things which coincide when applied to one another are equal to one another' (4), and 'The whole is greater than the part' (5). The objection to (4) is that it is incontestably geometrical, and therefore, on Aristotle's principles, should not be classed as an 'axiom'; it is a more or less sufficient definition of geometrical equality, but not a real axiom. Euclid evidently disliked the method of superposition for proving equality, no doubt because it assumes the possibility of motion *without deformation.* But he could not dispense with it altogether. Thus in I. 4 he practically had to choose between using the method and assuming the whole proposition as a postulate. But he does not there quote *Common Notion* 4; he says 'the base *BC* will coincide with the base *EF* and will be equal to it'. Similarly in I. 6 he does not quote *Common Notion* 5, but says 'the triangle *DBC* will be equal to the triangle *ACB*, the less to the greater, which is absurd'. It seems probable, therefore, that even these two *Common Notions*, though apparently recognized by Proclus, were generalizations from particular inferences found in Euclid and were inserted after his time.

The propositions of Book I fall into three distinct groups. The first group consists of Propositions 1–26, dealing mainly with triangles (without the use of parallels) but also with perpendiculars (11, 12), two intersecting straight lines (15), and one straight line standing on another but not cutting it, and making 'adjacent' or supplementary angles (13, 14). Proposition 1 gives the construction of an equilateral triangle on a given straight line as base; this is placed here not so much on its own account as because it is at once required for constructions (in 2, 9, 10, 11). The construction in 2 is a direct continuation of the minimum constructions assumed in Postulates 1–3, and enables us (as the Postulates do not) to transfer a given length of straight line from one place to another; it leads in 3 to the operation so often required of cutting off from one given straight line a length equal to another. 9 and 10 are the problems of bisecting a given angle and a given straight line respectively, and 11 shows how to erect a perpendicular to a given straight line from a given point on it. Construction as a means of proving existence is in evidence in the Book, not only in 1 (the equilateral triangle) but in 11, 12 (perpendiculars erected and let fall), and in 22 (construction of a triangle in the general case where the lengths of the sides are given); 23 constructs, by means of 22, an angle equal to a given rectilineal angle. The propositions about triangles include the congruence-theorems (4, 8, 26)—omitting the 'ambiguous case' which is only taken into account in the analogous proposition (7) of Book VI—and the theorems (allied to 4) about two triangles in which two sides of the one are respectively equal to two sides of the other, but of the included angles (24) or of the bases (25) one is greater than the other, and it is proved that the triangle in which the included angle is greater has the greater base and vice versa. Proposition 7, used to prove Proposition 8, is also important as being the Book I equivalent of III. 10 (that two circles cannot intersect in more points than two). Then we have theorems about single triangles in 5, 6 (isosceles triangles have the angles opposite to the equal sides equal—Thales's theorem—and the converse), the important propositions 16 (the exterior angle of a triangle is greater than either of the interior and opposite angles) and its derivative 17 (any two angles of a triangle are together less than two right angles), 18, 19 (greater angle subtended by greater side and vice versa), 20 (any two sides together greater than the third). This last furnishes the necessary *διορισμός*, or criterion of possibility, of the problem in 22 of constructing a triangle out of three straight lines of given length, which problem had therefore to come after and not before 20. 21 (proving that the two sides of a triangle other than the base are together greater, but include a lesser angle, than the two sides of any other triangle on the same base but with vertex within the original triangle) is useful for the proof of the proposition (not stated in Euclid) that of all straight lines drawn from an external point to a given straight line the perpendicular is the shortest, and the nearer to the perpendicular is less than the more remote.

The second group (27–32) includes the theory of parallels (27–31, ending with the construction through a given point of a parallel to a given straight line); and then, in 32, Euclid proves that the sum of the three angles of a triangle is equal to two right angles by means of a parallel to one side drawn from the opposite vertex (cf. the slightly different Pythagorean proof, p. 143).

The third group of propositions (33–48) deals generally with parallelograms, triangles and squares with reference to their areas. 33, 34 amount to the proof of the existence and the property of a parallelogram, and then we are introduced to a new conception, that of *equivalent* figures, or figures equal in area though not equal in the sense of congruent: parallelograms on the same base or on equal bases and between the same parallels are equal in area (35, 36); the same is true of triangles (37, 38), and a parallelogram on the same (or an equal) base with a triangle and between the same parallels is double of the triangle (41). 39 and the interpolated 40 are partial converses of 37 and 38. The theorem 41 enables us 'to construct in a given rectilineal angle a parallelogram equal to a given triangle' (42). Propositions 44, 45 are of the greatest importance, being the first cases of the Pythagorean method of 'application of areas', 'to apply to a given straight line, in a given rectilineal angle, a parallelogram equal to a given triangle (or rectilineal figure)'. The construction in 44 is remarkably ingenious, being based on that of 42 combined with the proposition (43) proving that the 'complements of the parallelograms about the diameter' in any parallelogram are equal. We are thus enabled to transform a parallelogram of any shape into another with the same angle and of equal area but with one side of any given length, say a *unit* length; this is the geometrical equivalent of the algebraic operation of dividing the product of two quantities by a third. Proposition 46 constructs a square on any given straight line as side, and is followed by the great Pythagorean theorem of the square on the hypotenuse of a right-angled triangle (47) and its converse (48). The remarkably clever proof of 47 by means of the well-known 'windmill' figure and the application to it of I. 41 combined with I. 4 seems to be due to Euclid himself; it is really equivalent to a proof by the methods of Book VI (Propositions 8, 17), and Euclid's achievement was that of avoiding the use of proportions and making the proof dependent upon Book I only.

I make no apology for having dealt at some length with Book I and, in particular, with the preliminary matter, in view of the unique position and authority of the *Elements* as an exposition of the fundamental principles of Greek geometry, and the necessity for the historian of mathematics of a clear understanding of their nature and full import. It will now be possible to deal more summarily with the other Books.

Book II is a continuation of the third section of Book I, relating to the transformation of areas, but is specialized in that it deals, not with parallelograms in general, but with *rectangles* and squares, and makes great use of the figure called the *gnomon*. The *rectangle* is introduced (Def. 1) as a 'rectangular parallelogram', which is said to be 'contained by the two straight lines containing the right angle'. The *gnomon* is defined (Def. 2) with reference to any parallelogram, but the only gnomon actually used is naturally that which belongs to a square. The whole Book constitutes an essential part of the *geometrical algebra* which really, in Greek geometry, took the place of our algebra. The first ten propositions give the equivalent of the following algebraical identities.

1. $a(b+c+d+\ldots) = ab+ac+ad+\ldots,$

2. $(a+b)a+(a+b)b = (a+b)^2,$

3. $(a+b)a = ab+a^2,$

4. $(a+b)^2 = a^2+b^2+2ab,$

5. $ab+\{\frac{1}{2}(a+b)-b\}^2 = \{\frac{1}{2}(a+b)\}^2,$

   or $(\alpha+\beta)(\alpha-\beta)+\beta^2 = \alpha^2,$

6. $(2a+b)b+a^2 = (a+b)^2,$

   or $(\alpha+\beta)(\beta-\alpha)+\alpha^2 = \beta^2,$

7. $(a+b)^2+a^2 = 2(a+b)a+b^2,$

   or $\alpha^2+\beta^2 = 2\alpha\beta+(\alpha-\beta)^2,$

8. $4(a+b)a+b^2 = \{(a+b)+a\}^2,$

   or $4\alpha\beta+(\alpha-\beta)^2 = (\alpha+\beta)^2,$

$$9. \quad a^2+b^2 = 2[\{\tfrac{1}{2}(a+b)\}^2+\{\tfrac{1}{2}(a+b)-b\}^2],$$

$$\text{or} \quad (\alpha+\beta)^2+(\alpha-\beta)^2 = 2(\alpha^2+\beta^2),$$

$$10. \quad (2a+b)^2+b^2 = 2\{a^2+(a+b)^2\},$$

$$\text{or} \quad (\alpha+\beta)^2+(\beta-\alpha)^2 = 2(\alpha^2+\beta^2).$$

As we have seen (pp. 151–3), Propositions 5 and 6 enable us to solve the quadratic equations

$$(1) \quad ax-x^2 = b^2 \quad \text{or} \quad \left.\begin{aligned} x+y &= a \\ xy &= b^2 \end{aligned}\right\},$$

and

$$(2) \quad ax+x^2 = b^2 \quad \text{or} \quad \left.\begin{aligned} y-x &= a \\ xy &= b^2 \end{aligned}\right\}.$$

The procedure is *geometrical* throughout; the areas in all the Propositions 1–8 are actually shown in the figures. Propositions 9 and 10 were really intended to solve a problem in *numbers*, that of finding any number of successive pairs of integral numbers ('side-' and 'diameter-' numbers) satisfying the equations

$$2x^2-y^2 = \pm 1$$

(see p. 93, above).

Of the remaining propositions, II. 11 and II. 14 give the geometrical equivalent of solving the quadratic equations

$$x^2+ax = a^2$$

and

$$x^2 = ab,$$

while the intervening propositions 12 and 13 prove, for any triangle with sides a, b, c, the equivalent of the formula

$$a^2 = b^2+c^2-2bc\cos A.$$

It is worth noting that, while I. 47 and its converse conclude Book I as if that Book was designed to lead up to the great proposition of Pythagoras, the last propositions but one of Book II give the generalization of the same proposition with *any* triangle substituted for a right-angled triangle.

The subject of Book III is the geometry of the circle, including the relations between circles cutting or touching each other. It begins with some definitions, which are generally of the same sort as those of Book I. Definition 1, stating that *equal circles* are those which have their diameters or their radii equal, might alternatively be regarded as a postulate or a theorem; if stated as a theorem, it could only be proved by superposition and the congruence-axiom. It is curious that the Greeks had no single word for *radius*, which was with them 'the (straight line) from the centre', ἡ ἐκ τοῦ κέντρου. A tangent to a circle is defined (Def. 2) as a straight line which meets the circle but, if produced, does not cut it; this is provisional pending the proof in III. 16 that such lines exist. The definitions (4, 5) of straight lines (in a circle), i. e. chords, equally distant or more or less distant from the centre (the test being the length of the perpendicular from the centre on the chord) might have referred, more generally, to the distance of any straight line from any point. The definition (7) of the 'angle *of* a segment' (the 'mixed' angle made by the circumference with the base at either end) is a survival from earlier text-books (cf. Props. 16, 31). The definitions of the 'angle *in* a segment' (8) and of 'similar segments' (11) assume (provisionally pending III. 21) that the angle in a segment is one and the same at whatever point of the circumference it is formed. A *sector* (τομεύς, explained by a scholiast as σκυτοτομικὸς τομεύς, a shoemaker's knife) is defined (10), but there is nothing about 'similar sectors' and no statement that similar segments belong to similar sectors.

Of the propositions of Book III we may distinguish certain groups. Central properties account for four propositions, namely 1 (to find the centre of a circle), 3 (any straight line through the centre which bisects any chord not passing through the centre cuts it at right angles, and vice versa), 4 (two chords not passing through the centre cannot bisect one another) and 9 (the centre is the only point from which more than two equal straight lines can be drawn to the circumference). Besides 3, which shows that any diameter bisects the whole series of chords at right angles to it, three other propositions throw light on the *form* of the circumference of a circle, 2 (showing that it is everywhere concave towards the centre), 7 and 8 (dealing with the varying lengths of straight lines drawn from any point, internal or external, to the concave or convex circumference, as the case may be, and proving that they are of maximum or minimum length when they pass through the centre, and that they diminish or increase as they diverge more and more from the maximum or minimum straight lines on either side, while the lengths of any two which are equally inclined to them, one on each side, are equal).

Two circles which cut or touch one another are dealt with in 5, 6 (the two circles cannot have the same centre), 10, 13 (they cannot cut in more points than two, or touch at more points than one), 11 and the interpolated 12 (when they touch, the line of centres passes through the point of contact).

14, 15 deal with chords (which are equal if equally distant from the centre and vice versa, while chords more distant from the centre are less, and chords less distant greater, and vice versa).

16–19 are concerned with tangent properties including the drawing of a tangent (17); it is in 16 that we have the survival of the 'angle *of* a semicircle', which is proved greater than any acute rectilineal angle, while the 'remaining' angle (the 'angle', afterwards called κερατοειδής, or 'hornlike', between the curve and the tangent at the point of contact) is less than any rectilineal angle. These 'mixed' angles, occurring in 16 and 31, appear no more in serious Greek geometry, though controversy about their nature went on in the works of commentators down to Clavius, Peletarius (Pelétier), Vieta, Galilei and Wallis.

We now come to propositions about segments. 20 proves that the angle at the centre is double of the angle at the circumference, and 21 that the angles in the same segment are all equal, which leads to the property of the quadrilateral in a circle (22). After propositions (23, 24) on 'similar segments', it is proved that in equal circles equal arcs subtend and are subtended by equal angles at the centre or circumference, and equal arcs subtend and are subtended by equal chords (26–9). 30 is the problem of bisecting a given arc, and 31 proves that the angle in a segment is right, acute or obtuse according as the segment is a semicircle, greater than a semicircle or less than a semicircle. 32 proves that the angle made by a tangent with a chord through the point of contact is equal to the angle in the alternate segment;

33, 34 are problems of constructing or cutting off a segment containing a given angle, and 25 constructs the complete circle when a segment of it is given.

The Book ends with three important propositions. Given a circle and any point O, internal (35) or external (36), then, if any straight line through O meets the circle in P, Q, the rectangle $PO \cdot OQ$ is constant and, in the case where O is an external point, is equal to the square on the tangent from O to the circle. Proposition 37 is the converse of 36.

Book IV, consisting entirely of problems, again deals with circles, but in relation to rectilineal figures inscribed or circumscribed to them. After definitions of these terms, Euclid shows, in the preliminary Proposition 1, how to fit into a circle a chord of given length, being less than the diameter. The remaining problems are problems of inscribing or circumscribing rectilineal figures. The case of the triangle comes first, and we learn how to inscribe in or circumscribe about a circle a triangle equiangular with a given triangle (2, 3) and to inscribe a circle in or circumscribe a circle about a given triangle (4, 5). 6–9 are the same problems for a square, 11–14 for a regular pentagon, and 15 (with porism) for a regular hexagon. The porism to 15 also states that the side of the inscribed regular hexagon is manifestly equal to the radius of the circle. 16 shows how to inscribe in a circle a regular polygon with fifteen angles, a problem suggested by astronomy, since the obliquity of the ecliptic was taken to be about 24°, or one-fifteenth of 360°. IV. 10 is the important proposition, required for the construction of a regular pentagon, 'to construct an isosceles triangle such that each of the base angles is double of the vertical angle', which is effected by dividing one of the equal sides in extreme and mean ratio (II. 11) and fitting into the circle with this side as radius a chord equal to the greater segment; the proof of the construction depends on III. 32 and 37.

We are not surprised to learn from a scholiast that the whole Book is 'the discovery of the Pythagoreans'.[1] The same scholium says that 'it is proved in this Book that the perimeter of a circle is not triple of its diameter, as many

[1] Euclid, ed. Heib., vol. v, pp. 272–3.

suppose, but greater than that (the reference is clearly to IV. 15 Por.), and likewise that neither is the circle three-fourths of the triangle circumscribed about it'. Were these fallacies perhaps exposed in the lost *Pseudaria* of Euclid?

Book V is devoted to the new theory of proportion, applicable to incommensurable as well as commensurable magnitudes, and to magnitudes of every kind (straight lines, areas, volumes, numbers, times, &c.), which was due to Eudoxus. Greek mathematics can boast no finer discovery than this theory, which first put on a sound footing so much of geometry as depended on the use of proportions. How far Eudoxus himself worked out his theory in detail is unknown; the scholiast who attributes the discovery of it to him says that 'it is recognized by all' that Book V is, as regards its arrangement and sequence in the *Elements,* due to Euclid himself.[1] The ordering of the propositions and the development of the proofs are indeed masterly and worthy of Euclid; as Barrow said, 'There is nothing in the whole body of the elements of a more subtile invention, nothing more solidly established, and more accurately handled, than the doctrine of proportionals'. It is a pity that, notwithstanding the pre-eminent place which Euclid has occupied in English mathematical teaching, Book V itself is little known in detail; if it were, there would, I think, be less tendency to seek for substitutes; indeed, after reading some of the substitutes, it is with relief that one turns to the original. For this reason, I shall make my account of Book V somewhat full, with the object of indicating not only the whole content but also the course of the proofs.

Of the Definitions the following are those which need separate mention. The definition (3) of *ratio* as 'a sort of relation (ποιὰ σχέσις) in respect of size (πηλικότης) between two magnitudes of the same kind' is as vague and of as little practical use as that of a straight line; it was probably inserted for completeness' sake, and in order merely to aid the conception of a ratio. Definition 4 ('Magnitudes are said to have a ratio to one another which are capable, when multiplied, of exceeding one another') is important not only because

[1] Euclid, ed. Heib., vol. v, p. 282.

it shows that the magnitudes must be of the same kind, but because, while it includes incommensurable as well as commensurable magnitudes, it excludes the relation of a finite magnitude to a magnitude of the same kind which is either infinitely great or infinitely small; it is also practically equivalent to the principle which underlies the method of exhaustion now known as the Axiom of Archimedes. Most important of all is the fundamental definition (5) of magnitudes which are in the same ratio: 'Magnitudes are said to be in the same ratio, the first to the second and the third to the fourth, when, if any equimultiples whatever be taken of the first and third, and any equimultiples whatever of the second and fourth, the former equimultiples alike exceed, are alike equal to, or alike fall short of, the latter equimultiples taken in corresponding order.' Perhaps the greatest tribute to this marvellous definition is its adoption by Weierstrass as a definition of equal numbers. For a most attractive explanation of its exact significance and its absolute sufficiency the reader should turn to De Morgan's articles on Ratio and Proportion in the *Penny Cyclopaedia*.[1] The definition (7) of *greater ratio* is an addendum to Definition 5: 'When, of the equimultiples, the multiple of the first exceeds the multiple of the second, but the multiple of the third does not exceed the multiple of the fourth, then the first is said to have a *greater ratio* to the second than the third has to the fourth'; this (possibly for brevity's sake) states only one criterion, the other possible criterion being that, while the multiple of the first is *equal* to that of the second, the multiple of the third is *less* than that of the fourth. A proportion may consist of three or four terms (Defs. 8, 9, 10); 'corresponding' or 'homologous' terms are antecedents in relation to antecedents and consequents in relation to consequents (11). Euclid proceeds to define the various transformations of ratios. *Alternation* (ἐναλλάξ, *alternando*) means taking the alternate terms in the proportion $a:b = c:d$, i.e. transforming it into $a:c = b:d$ (12). *Inversion* (ἀνάπαλιν, inversely) means turning the ratio $a:b$ into $b:a$ (13). *Composition* of a ratio, σύνθεσις λόγου (*componendo* is in Greek συνθέντι, 'to one who has compounded

[1] Vol. xix (1841). I have largely reproduced the articles in *The Thirteen Books of Euclid's Elements*, vol. ii, pp. 116–24.

or added', i.e. if one compounds or adds) is the turning of $a:b$ into $(a+b):b$ (14). *Separation*, διαίρεσις (διελόντι = *separando*) turns $a:b$ into $(a-b):b$ (15). *Conversion*, ἀναστροφή (ἀναστρέψαντι = *convertendo*) turns $a:b$ into $a:a-b$ (16). Lastly, *ex aequali* (sc. *distantia*), δι' ἴσου, and *ex aequali in disturbed proportion* (ἐν τεταραγμένῃ ἀναλογίᾳ) are defined (17, 18). If $a:b = A:B$, $b:c = B:C \ldots k:l = K:L$, then the inference *ex aequali* is that $a:l = A:L$ (proved in V. 22). If again $a:b = B:C$ and $b:c = A:B$, the inference *ex aequali in disturbed proportion* is $a:c = A:C$ (proved in V. 23).

In reproducing the content of the Book I shall express magnitudes in general (which Euclid represents by straight lines) by the letters $a, b, c \ldots$ and I shall use the letters $m, n, p \ldots$ to express integral numbers: thus ma, mb are equimultiples of a, b.

The first six propositions are simple theorems in concrete arithmetic, and they are practically all proved by separating into their units the multiples used.

$$\begin{cases} 1. & ma+mb+mc+\ldots = m(a+b+c+\ldots). \\ 5. & ma-mb = m(a-b). \end{cases}$$

5 is proved by means of 1. As a matter of fact, Euclid assumes the construction of a straight line equal to $1/m$th of $ma-mb$. This is an anticipation of VI. 9, but can be avoided; for we can draw a straight line equal to $m(a-b)$; then, by 1, $m(a-b)+mb = ma$, or $ma-mb = m(a-b)$.

$$\begin{cases} 2. & ma+na+pa+\ldots = (m+n+p+\ldots)a. \\ 6. & ma-na = (m-n)a. \end{cases}$$

Euclid actually expresses 2 and 6 by saying that $ma \pm na$ is the same multiple of a that $mb \pm nb$ is of b. By separation of m, n into units he in fact shows (in 2) that

$$ma+na = (m+n)a, \text{ and } mb+nb = (m+n)b.$$

6 is proved by means of 2, as 5 by means of 1.

3. If $m.na$, $m.nb$ are equimultiples of na, nb, which are themselves equimultiples of a, b, then $m.na$, $m.nb$ are also equimultiples of a, b.

By separating m, n into their units Euclid practically proves that $m.na = mn.a$ and $m.nb = mn.b$.

4. If $a : b = c : d$, then $ma : nb = mc : nd$.

Take any equimultiples $p.ma$, $p.mc$ of ma, mc, and any equimultiples $q.nb$, $q.nd$ of nb, nd. Then, by 3, these equimultiples are also equimultiples of a, c and b, d respectively, so that by Def. 5, since $a : b = c : d$,

$$p.ma > = < q.nb \text{ according as } p.mc > = < q.nd,$$

whence, again by Def. 5, since p, q are any integers,

$$ma : nb = mc : nd.$$

7, 9. If $a = b$, then $\left.\begin{array}{l} a : c = b : c \\ \text{and } c : a = c : b \end{array}\right\}$; and conversely.

8, 10. If $a > b$, then $\left.\begin{array}{l} a : c > b : c \\ \text{and } c : b > c : a \end{array}\right\}$; and conversely.

7 is proved by means of Def. 5. Take ma, mb equimultiples of a, b, and nc a multiple of c. Then, since $a = b$,

$$ma > = < nc \text{ according as } mb > = < nc,$$

and $$nc > = < ma \text{ according as } nc > = < mb,$$

whence the results follow.

8 is divided into two cases according to which of the two magnitudes $a - b$, b is the less. Take m such that

$$m(a-b) > c \text{ or } mb > c$$

in the two cases respectively. Next let nc be the first multiple of c which is greater than mb or $m(a-b)$ respectively, so that

$$nc > \begin{array}{c} mb \\ \text{or } m(a-b) \end{array} \geqq (n-1)c.$$

Then, (i) since $m(a-b) > c$, we have, by addition, $ma > nc$.

(ii) since $mb > c$, we have similarly $ma > nc$.

In either case $mb < nc$, since in case (ii) $m(a-b) > mb$.
Thus in either case, by the definition (7) of greater ratio,

$$a : c > b : c,$$

and $$c : b > c : a.$$

The converses 9, 10 are proved from 7, 8 by *reductio ad absurdum*.

11. If $$a:b = c:d,$$
and $$c:d = e:f,$$
then $$a:b = e:f.$$

Proved by taking any equimultiples of a, c, e and any other equimultiples of b, d, f, and using Def. 5.

12. If $$a:b = c:d = e:f = \dots$$
then $$a:b = (a+c+e+\dots):(b+d+f+\dots).$$

Proved by means of V. 1 and Def. 5, after taking equimultiples of a, c, $e\dots$ and other equimultiples of b, d, $f\dots$.

13. If $$a:b = c:d,$$
and $$c:d > e:f,$$
then $$a:b > e:f.$$

Equimultiples mc, me of c, e are taken and equimultiples nd, nf of d, f such that, while $mc > nd$, me is not greater than nf (Def. 7). Then the same equimultiples ma, mc of a, c and the same equimultiples nb, nd of b, d are taken, and Defs. 5 and 7 are used in succession.

14. If $a:b = c:d$, then, according as $a >=< c$, $b >=< d$.

The first case only is proved; the others are dismissed with 'Similarly'.

If $$a > c,\quad a:b > c:b. \qquad (8)$$

But $a:b = c:d$, whence (13) $c:d > c:b$, and therefore (10) $b > d$.

15. $a:b = ma:mb$.

Dividing the multiples into their units, we have m equal ratios $a:b$; the result follows by 12.

Propositions 16–19 prove certain cases of the transformation of proportions in the sense of Defs. 12–16. The case of *inverting* the ratios is omitted, probably as being obvious. For, if $a:b = c:d$, the application of Def. 5 proves simultaneously that $b:a = d:c$.

16. If $$a:b = c:d,$$
then, *alternando*, $$a:c = b:d.$$

Since $$a:b = ma:mb, \text{ and } c:d = nc:nd, \qquad (15)$$

we have $ma : mb = nc : nd,$ **(11)**

whence (14), according as $ma > = < nc$, $mb > = < nd$;

therefore (Def. 5) $a : c = b : d.$

17. If $a : b = c : d,$

then, *separando*, $(a-b) : b = (c-d) : d.$

Take ma, mb, mc, md equimultiples of all four magnitudes, and nb, nd other equimultiples of b, d. It follows (2) that $(m+n)b$, $(m+n)d$ are also equimultiples of b, d.

Therefore, since $a : b = c : d,$

$ma > = < (m+n)b$ according as $mc > = < (m+n)d$. (Def. 5)

Subtracting mb from both sides of the former relation and md from both sides of the latter, we have (5)

$$m(a-b) > = < nb \text{ according as } m(c-d) > = < nd.$$

Therefore (Def. 5) $a-b : b = c-d : d.$

(I have here abbreviated Euclid a little, without altering the substance.)

18. If $a : b = c : d,$

then, *componendo*, $(a+b) : b = (c+d) : d.$

Proved by *reductio ad absurdum*. Euclid assumes that $a+b : b = (c+d) : (d \pm x)$, if that is possible. (This implies that to any three given magnitudes, two of which at least are of the same kind, there exists a fourth proportional, an assumption which is not strictly legitimate until the fact has been proved by construction.)

Therefore, *separando* (17), $a : b = (c \mp x) : (d \pm x)$,

whence (11), $(c \mp x) : (d \pm x) = c : d$, which relations are impossible, by 14.

19. If $a : b = c : d,$

then $(a-c) : (b-d) = a : b.$

Alternately (16),

$$a : c = b : d, \text{ whence } (a-c) : c = (b-d) : d \ (17).$$

Alternately again, $(a-c) : (b-d) = c : d$ (16);

whence (11) $(a-c) : (b-d) = a : b.$

The transformation *convertendo* is only given in an interpolated Porism to 19. But it is easily obtained by using 17 (*separando*) combined with *alternando* (16). Euclid himself proves it in X. 14 by using successively *separando* (17), *inversion* and *ex aequali* (22).

The *composition* of ratios *ex aequali* and *ex aequali in disturbed proportion* is dealt with in 22, 23, each of which depends on a preliminary proposition.

20. If $$a : b = d : e,$$
and $$b : c = e : f,$$
then, *ex aequali*, according as $a > = < c$, $d > = < f$.

For, according as $a > = < c$, $a : b > = < c : b$ (7, 8),

and therefore, by means of the above relations and 13, 11,
$$d : e > = < f : e,$$
and therefore again (9, 10)
$$d > = < f.$$

21. If $$a : b = e : f,$$
and $$b : c = d : e,$$
then, *ex aequali in disturbed proportion*,

according as $a > = < c$, $d > = < f$.

For, according as $a > = < c$, $a : b > = < c : b$ (7, 8),

or $e : f > = < e : d$ (13, 11),

and therefore $d > = < f$ (9, 10).

22. If $$a : b = d : e,$$
and $$b : c = e : f,$$
then, *ex aequali*, $a : c = d : f$.

Take equimultiples ma, md; nb, ne; pc, pf, and it follows

$$\left.\begin{aligned} \text{that} \quad ma : nb &= md : ne, \\ \text{and} \quad nb : pc &= ne : pf \end{aligned}\right\} \quad (4)$$

Therefore (20), according as $ma > = < pc$, $md > = < pf$,

whence (Def. 5) $a : c = d : f$.

23. If $$a : b = e : f,$$
and $$b : c = d : e,$$
then, *ex aequali in disturbed proportion*, $a : c = d : f$.

Equimultiples ma, mb, md and nc, ne, nf are taken, and it is proved, by means of 11, 15, 16, that
$$ma : mb = ne : nf,$$
and $$mb : nc = md : ne,$$
whence (21) $ma > = < nc$ according as $md > = < nf$,
and (Def. 5) $$a : c = d : f.$$

24. If $$a : c = d : f,$$
and also $$b : c = e : f,$$
then $$(a+b) : c = (d+e) : f.$$

Invert the second proportion to $c : b = f : e$, and compound the first proportion with this (22);

therefore $$a : b = d : e.$$

Componendo, $(a+b) : b = (d+e) : e$, which compounded (22) with the second proportion gives $(a+b) : c = (d+e) : f$.

25. If $a : b = c : d$, and of the four terms a is the greatest (so that d is also the least), $a+d > b+c$.

Since $$a : b = c : d,$$
$$a-c : b-d = a : b; \qquad (19)$$
and, since $a > b$, $$(a-c) > (b-d). \qquad (16, 14)$$

Add $c+d$ to each;

therefore $$a+d > b+c.$$

Such slight defects as are found in the text of this great Book as it has reached us, like other slight imperfections of form in the *Elements*, point to the probability that the work never received its final touches from Euclid's hand; but they can all be corrected without much difficulty, as Simson showed in his excellent edition.

Book VI contains the application to plane geometry of the general theory of proportion established in Book V. It begins with definitions of 'similar rectilineal figures' and of what is meant by cutting a straight line 'in extreme and mean ratio'. The first and last propositions are analogous; 1 proves that triangles and parallelograms of the same height are to one another as their bases, and 33 that in equal circles angles at the centre or circumference are as the arcs on which they stand; both use the method of equimultiples and apply V, Def. 5 as the test of proportion. Equally fundamental are 2 (that two sides of a triangle cut by any parallel to the third side are divided proportionally, and the converse), and 3 (that the internal bisector of an angle of a triangle cuts the opposite side into parts which have the same ratio as the sides containing the angle, and the converse); 2 depends directly on 1 and 3 on 2. Then come the alternative conditions for the similarity of two triangles: equality of all the angles respectively (4), proportionality of pairs of sides in order (5), equality of one angle in each with proportionality of sides containing the equal angles (6), and the 'ambiguous case' (7), in which one angle is equal to one angle and the sides about other angles are proportional. After the important proposition (8) that the perpendicular from the right angle in a right-angled triangle to the opposite side divides the triangle into two triangles similar to the original triangle and to one another, we pass to the proportional division of straight lines (9, 10) and the problems of finding a third proportional to two straight lines (11), a fourth proportional to three (12), and a mean proportional to two straight lines (13, the Book VI version of II. 14). In 14, 15 Euclid proves the reciprocal proportionality of the sides about the equal angles in parallelograms or triangles of equal area which have one angle equal to one angle and the converse; by placing the equal angles vertically opposite to one another so that the sides about them lie along two straight lines, and completing the figure, Euclid is able to apply VI. 1. From 14 are directly deduced 16, 17 (that, if four or three straight lines be proportionals, the rectangle contained by the extremes is equal to the rectangle contained by the two means or the square on the one mean, and the converse). 18–22 deal with similar rectilineal figures; 19 (with Porism) and 20 are specially important, proving that similar triangles, and similar polygons generally, are to one another in the duplicate ratio of corresponding sides, and that, if three straight lines are proportional, then, as the first is to the third, so is the figure described on the first to the similar figure similarly described on the second. The fundamental case of the two similar triangles is prettily proved thus. The triangles being ABC, DEF, in which B, E are equal angles and BC, EF corresponding sides, find a third proportional to BC, EF and measure it off along BC as BG; join AG. Then the triangles ABG, DEF have their sides about the equal angles B, E reciprocally proportional and are therefore equal (VI. 15); the rest follows from VI. 1 and the definition of duplicate ratio (V, Def. 9).

Proposition 23 (equiangular parallelograms have to one another the ratio compounded of the ratios of their sides) is important in itself, and also because it introduces us to the practical use of the method of compounding, i.e. multiplying, ratios which is of such extraordinarily wide application in Greek geometry. Euclid has never defined 'compound ratio' or the 'compounding' of ratios; but the meaning of the terms

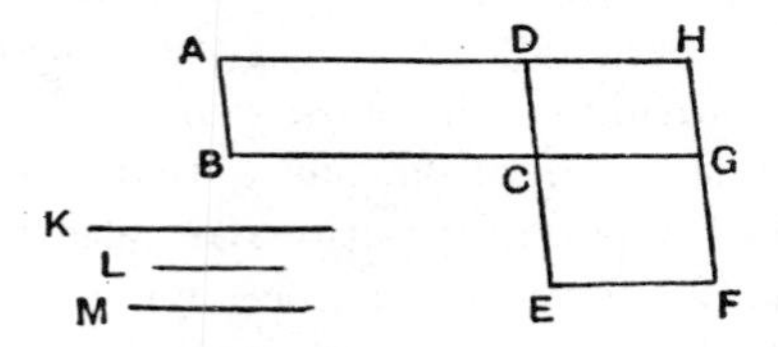

and the way to compound ratios are made clear in this proposition. The equiangular parallelograms are placed so that two equal angles as BCD, GCE are vertically opposite at C. Complete the parallelogram $DCGH$. Take any straight line K, and (12) find another, L, such that

$$BC : CG = K : L,$$

and again another straight line M, such that

$$DC : CE = L : M.$$

Now the ratio compounded of $K : L$ and $L : M$ is $K : M$; therefore $K : M$ is the 'ratio compounded of the ratios of the sides'.

And
$$(ABCD) : (DCGH) = BC : CG, \quad (1)$$
$$= K : L;$$
$$(DCGH) : (CEFG) = DC : CE \quad (1)$$
$$= L : M,$$

Therefore, *ex aequali* (V. 22),

$$(ABCD) : (CEFG) = K : M.$$

The important Proposition 25 (to construct a rectilineal figure similar to one, and equal to another, given rectilineal figure) is one of the famous problems alternatively associated with the story of Pythagoras's sacrifice[1]; it is doubtless Pythagorean. The given figure (P, say) to which the required figure is to be similar is transformed (I. 44) into a parallelogram on the same base BC. Then the other figure (Q, say) to which the required figure is to be *equal* is (I. 45) transformed into a parallelogram on the base CF (in a straight line with BC) and of equal height with the other parallelogram. Then $(P) : (Q) = BC : CF$ (1). It is then only necessary to take a straight line GH a mean proportional between BC and CF, and to describe on GH as base a rectilineal figure similar to P which has BC as base (VI. 18). The proof of the correctness of the construction follows from VI. 19 Por.

In 27, 28, 29 we reach the final problems in the Pythagorean *application of areas*, which are the geometrical equivalent of the algebraical solution of the most general form of quadratic equation where that equation has a real and positive root. Detailed notice of these propositions is necessary because of their exceptional historic importance, which arises from the fact that the method of these propositions was constantly used

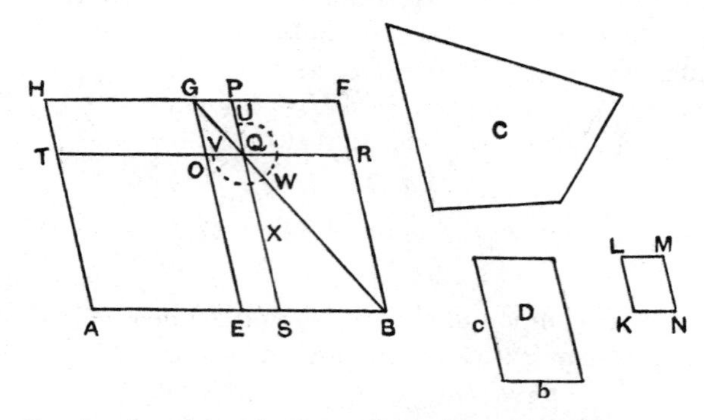

by the Greeks in the solution of problems. They constitute, for example, the foundation of Book X of the *Elements* and of

[1] Plutarch, *Non posse suaviter vivi secundum Epicurum*, c. 11.

the whole treatment of conic sections by Apollonius. The problems themselves are enunciated in 28, 29: 'To a given straight line to apply a parallelogram equal to a given rectilineal figure and *deficient* (or *exceeding*) by a parallelogrammic figure similar to a given parallelogram'; and **27** supplies the διορισμός, or determination of the condition of possibility, which is necessary in the case of *deficiency* (28): 'The given rectilineal figure must (in that case) not be greater than the parallelogram described on the half of the straight line and similar to the defect.' We will take the problem of 28 for examination.

We are already familiar with the notion of applying a parallelogram to a straight line AB so that it *falls short* or *exceeds* by a certain other parallelogram. Suppose that D is the given parallelogram to which the *defect* in this case has to be similar. Bisect AB in E, and on the half EB describe the parallelogram $GEBF$ similar and similarly situated to D. Draw the diagonal GB and complete the parallelogram $HABF$. Now, if we draw through any point T on HA a straight line TR parallel to AB meeting the diagonal GB in Q, and then draw PQS parallel to TA, the parallelogram $TASQ$ is a parallelogram applied to AB but falling short by a parallelogram similar and similarly situated to D, since the deficient parallelogram is $QSBR$ which is similar to EF (**24**). (In the same way, if T had been on HA *produced* and TR had met GB *produced* in R, we should have had a parallelogram applied to AB but *exceeding* by a parallelogram similar and similarly situated to D.)

Now consider the parallelogram AQ falling short by SR similar and similarly situated to D. Since $(AO) = (ER)$, and $(OS) = (QF)$, it follows that the parallelogram AQ is equal to the gnomon UWV, and the problem is therefore that of constructing the gnomon UWV such that its area is equal to that of the given rectilineal figure C. The gnomon obviously cannot be greater than the parallelogram EF, and hence the given rectilineal figure C must not be greater than that parallelogram. This is the διορισμός proved in **27**.

Since the gnomon is equal to C, it follows that the parallelogram $GOQP$ which with it makes up the parallelogram EF is equal to the difference between (EF) and C. Therefore, in order to construct the required gnomon, we have only to draw in the angle FGE the parallelogram $GOQP$ equal to $(EF)-C$ and similar and similarly situated to D. This is what Euclid in fact does; he constructs the parallelogram $LKNM$ equal to $(EF)-C$ and similar and similarly situated to D (by means of 25), and then draws $GOQP$ equal to it. The problem is thus solved, $TASQ$ being the required parallelogram.

To show the correspondence to the solution of a quadratic equation, let $AB = a$, $QS = x$, and let $b:c$ be the ratio of the sides of D; therefore $SB = \frac{b}{c}x$. Then, if m is a certain constant (in fact the sine of an angle of one of the parallelograms), $(AQ) = m\,(ax - \frac{b}{c}x^2)$, so that the equation solved is

$$m\left(ax - \frac{b}{c}x^2\right) = C.$$

The algebraical solution is $x = \frac{c}{b}\cdot\frac{a}{2} \pm \sqrt{\left\{\frac{c}{b}\left(\frac{c}{b}\cdot\frac{a^2}{4} - \frac{C}{m}\right)\right\}}$.

Euclid gives only one solution (that corresponding to the *negative* sign), but he was of course aware that there are two, and how he could exhibit the second in the figure.

For a real solution we must have C not greater than $m\frac{c}{b}\cdot\frac{a^2}{4}$, which is the area of EF. This corresponds to Proposition 27.

We observe that what Euclid in fact does is to find the parallelogram $GOQP$ which is of given shape (namely such that its area $m\,.\,GO\,.\,OQ = m\,.\,GO^2\frac{b}{c}$) and is equal to $(EF)-C$; that is, he finds GO such that $GO^2 = \frac{c}{b}\left(\frac{c}{b}\cdot\frac{a^2}{4} - \frac{C}{m}\right)$. In other words, he finds the straight line equal to $\sqrt{\left\{\frac{c}{b}\left(\frac{c}{b}\cdot\frac{a^2}{4} - \frac{C}{m}\right)\right\}}$; and x is thus known, since $x = GE - GO = \frac{c}{b}\cdot\frac{a}{2} - GO$. Euclid's procedure, therefore, corresponds closely to the algebraic solution.

The solution of 29 is exactly similar, *mutatis mutandis*. A solution is always possible, so that no διορισμός is required.

VI. 31 gives the extension of the Pythagorean proposition I. 47 showing that for squares in the latter proposition we may substitute similar plane figures of any shape whatever. 30 uses 29 to divide a straight line in extreme and mean ratio (the same problem as II. 11).

Except in the respect that it is based on the new theory of proportion, Book VI does not appear to contain any matter that was not known before Euclid's time. Nor is the generalization of I. 47 in VI. 31, for which Proclus professes such admiration, original on Euclid's part, for, as we have already seen (p. 191), Hippocrates of Chios assumes its truth for semicircles described on the three sides of a right-angled triangle.

We pass to the arithmetical Books, VII, VIII, IX. Book VII begins with a set of definitions applicable in all the three Books. They include definitions of a *unit*, a *number*, and the following varieties of numbers, *even*, *odd*, *even-times-even*, *even-times-odd*, *odd-times-odd*, *prime*, *prime to one another*, *composite*, *composite to one another*, *plane*, *solid*, *square*, *cube*, *similar plane* and *solid* numbers, and a *perfect* number, definitions of terms applicable in the numerical theory of proportion, namely *a part* (= a submultiple or aliquot part), *parts* (= a proper fraction), *multiply*, and finally the definition of (four) proportional numbers, which states that 'numbers are proportional when the first is the same multiple, the same part, or the same parts, of the second that the third is of the fourth', i.e. numbers a, b, c, d are proportional if, when $a = \frac{m}{n} b$, $c = \frac{m}{n} d$, where m, n are any integers (although the definition does not in terms cover the case where $m > n$).

The propositions of Book VII fall into four main groups. 1–3 give the method of finding the greatest common measure of two or three unequal numbers in essentially the same form in which it appears in our text-books, Proposition 1 giving the test for two numbers being prime to one another, namely that no remainder measures the preceding quotient till 1 is reached. The second group, 4–19, sets out the numerical theory of proportion. 4–10 are preliminary, dealing with numbers which are 'a part' or 'parts' of other numbers, and numbers which are the same 'part' or 'parts' of other numbers, just as the preliminary propositions of Book V deal with multiples and equimultiples. 11–14 are transformations of proportions corresponding to similar transformations (*separando*, alternately, &c.) in Book V. The following are the results, expressed with the aid of letters which here represent integral numbers exclusively.

If $a:b = c:d \;\; (a > c,\; b > d)$, then

$$(a-c):(b-d) = a:b. \qquad (11)$$

If $a:a' = b:b' = c:c' \ldots$, then each of the ratios is equal to

$$(a+b+c+\ldots):(a'+b'+c'+\ldots). \qquad (12)$$

If $a:b = c:d$, then $a:c = b:d$. $\qquad (13)$

If $a:b = d:e$ and $b:c = e:f$, then, *ex aequali*,

$$a:c = d:f. \qquad (14)$$

If $1:m = a:ma$ (expressed by saying that the third number measures the fourth the same number of times that the unit measures the second), then alternately

$$1:a = m:ma. \qquad (15)$$

The last result is used to prove that $ab = ba$; in other words, that the order of multiplication is indifferent (16), and this is followed by the propositions that $b:c = ab:ac$ (17) and that $a:b = ac:bc$ (18), which are again used to prove the important proposition (19) that, if $a:b = c:d$, then $ad = bc$, a theorem which corresponds to VI. 16 for straight lines.

Zeuthen observes that, while it was necessary to use the numerical definition of proportion to carry the numerical theory up to this point, Proposition 19 establishes the necessary point of contact between the two theories, since it is now shown that the definition of proportion in V, Def. 5, has, when applied to numbers, the same import as that in VII, Def. 20, and we can henceforth without hesitation borrow any of the propositions established in Book V.[1]

Propositions 20, 21 about 'the least numbers of those which have the same ratio with them' prove that, if m, n are such numbers and a, b any other numbers in the same ratio, m

[1] Zeuthen, 'Sur la constitution des livres arithmétiques des Éléments d'Euclide' (*Oversigt over det kgl. Danske Videnskabernes Selskabs Forhandlinger*, 1910, pp. 412, 413).

measures a the same number of times that n measures b, and that numbers prime to one another are the least of those which have the same ratio with them. These propositions lead up to Propositions 22–32 about numbers prime to one another, prime numbers, and composite numbers. This group includes fundamental theorems such as the following. If two numbers be prime to any number, their product will be prime to the same (24). If two numbers be prime to one another, so will their squares, their cubes, and so on generally (27). If two numbers be prime to one another, their sum will be prime to each of them; and, if the sum be prime to either, the original numbers will be prime to one another (28). Any prime number is prime to any number which it does not measure (29). If two numbers are multiplied, and any prime number measures the product, it will measure one of the original numbers (30). Any composite number is measured by some prime number (31). Any number either is prime or is measured by some prime number (32).

Propositions 33 to the end (39) are directed to the problem of finding the least common multiple of two or three numbers; 33 is preliminary, using the G. C. M. for the purpose of solving the problem, 'Given as many numbers as we please, to find the least of those which have the same ratio with them.'

It seems clear that in Book VII Euclid was following earlier models, while no doubt making improvements in the exposition. This is, as we have seen (pp. 215–16), partly confirmed by the fact that in the proof by Archytas of the proposition that 'no number can be a mean between two consecutive numbers' propositions are presupposed corresponding to VII. 20, 22, 33.

Book VIII deals largely with series of numbers 'in continued proportion', i.e. in geometrical progression (Propositions 1–3, 6–7, 13). If the series in G. P. be

$$a^n,\ a^{n-1}b,\ a^{n-2}b^2, \ldots\ a^2b^{n-2},\ ab^{n-1},\ b^n,$$

Propositions 1–3 deal with the case where the terms are the smallest that are in the ratio $a:b$, in which case a^n, b^n are prime to one another. 6–7 prove that, if a^n does not measure $a^{n-1}b$, no term measures any other, but if a^n measures b^n, it measures $a^{n-1}b$. Connected with these are Propositions 14–17 proving that, according as a^2 does or does not measure b^2, a does or does not measure b and vice versa; and similarly, according as a^3 does or does not measure b^3, a does or does not measure b and vice versa. 13 proves that, if a, b, c ... are in G. P., so are a^2, b^2, c^2 ... and a^3, b^3, c^3 ... respectively.

Proposition 4 is the problem, Given as many ratios as we please, $a:b$, $c:d$ ... to find a series p, q, r, ... in the least possible terms such that $p:q = a:b$, $q:r = c:d$, ... This is done by finding the L. C. M., first of b, c, and then of other pairs of numbers as required. The proposition gives the means of compounding two or more ratios between numbers in the same way that ratios between pairs of straight lines are compounded in VI. 23; the corresponding proposition to VI. 23 then follows (5), namely, that plane numbers have to one another the ratio compounded of the ratios of their sides.

Propositions 8–10 deal with the interpolation of geometric means between numbers. If $a:b = e:f$, and there are n geometric means between a and b, there are n geometric means between e and f also (8). If $a^n, a^{n-1}b \dots ab^{n-1}, b^n$ is a G. P. of $n+1$ terms, so that there are $(n-1)$ means between a^n, b^n, there are the same number of geometric means between 1 and a^n and between 1 and b^n respectively (9); and conversely, if $1, a, a^2 \dots a^n$ and $1, b, b^2 \dots b^n$ are terms in G. P., there are the same number $(n-1)$ of means between a^n, b^n (10). In particular, there is one mean proportional number between square numbers (11) and between similar plane numbers (18), and conversely, if there is one mean between two numbers, the numbers are similar plane numbers (20); there are two means between cube numbers (12) and between similar solid numbers (19), and conversely, if there are two means between two numbers, the numbers are similar solid numbers (21). So far as squares and cubes are concerned, these propositions are stated by Plato in the *Timaeus*, and Nicomachus, doubtless for this reason, calls them 'Platonic'. Connected with them are the propositions that similar plane numbers have the same ratio as a square has to a square (26), and similar solid numbers have the same ratio as a cube has to a cube (27). A few other subsidiary propositions need no particular mention.

Book IX begins with seven simple propositions such as that

the product of two similar plane numbers is a square (1) and, if the product of two numbers is a square number, the numbers are similar plane numbers (2); if a cube multiplies itself or another cube, the product is a cube (3, 4); if $a^3 B$ is a cube, B is a cube (5); if A^2 is a cube, A is a cube (6). Then follow six propositions (8–13) about a series of terms in geometrical progression beginning with 1. If $1, a, b, c \dots k$ are n terms in geometrical progression, then (9), if a is a square (or a cube), all the other terms $b, c, \dots k$ are squares (or cubes); if a is not a square, then the only squares in the series are the term after a, i. e. b, and all alternate terms after b; if a is not a cube, the only cubes in the series are the fourth term (c), the seventh, tenth, &c., terms, being terms separated by two throughout; the seventh, thirteenth, &c., terms (leaving out five in each case) will be both square and cube (8, 10). These propositions are followed by the interesting theorem that, if $1, a_1, a_2 \dots a_n \dots$ are terms in geometrical progression, and if a_r, a_n are any two terms where $r < n$, a_r measures a_n, and $a_n = a_r . a_{n-r}$ (11 and Por.); this is, of course, equivalent to the formula $a^{m+n} = a^m . a^n$. Next it is proved that, if the last term k in a series $1, a, b, c \dots k$ in geometrical progression is measured by any primes, a is measured by the same (12); and, if a is prime, k will not be measured by any numbers except those which have a place in the series (13). Proposition 14 is the equivalent of the important theorem that *a number can only be resolved into prime factors in one way.* Propositions follow to the effect that, if a, b be prime to one another, there can be no integral third proportional to them (16) and, if $a, b, c \dots k$ be in G. P. and a, k are prime to one another, then there is no integral fourth proportional to a, b, k (17). The conditions for the possibility of an integral third proportional to two numbers and of an integral fourth proportional to three are then investigated (18, 19). Proposition 20 is the important proposition that *the number of prime numbers is infinite*, and the proof is the same as that usually given in our algebraical text-books. After a number of easy propositions about odd, even, 'even-times-odd', 'even-times-even' numbers respectively (Propositions 21–34), we have two important propositions which conclude the Book. Proposition 35 gives the summation of a G. P. of n terms, and a very elegant solution it is. Suppose that $a_1, a_2, a_3, \ldots a_{n+1}$ are $n+1$ terms in G. P.; Euclid proceeds thus:

We have $$\frac{a_{n+1}}{a_n} = \frac{a_n}{a_{n-1}} = \ldots = \frac{a_2}{a_1},$$

and, *separando*, $$\frac{a_{n+1}-a_n}{a_n} = \frac{a_n - a_{n-1}}{a_{n-1}} = \ldots = \frac{a_2 - a_1}{a_1}.$$

Adding antecedents and consequents, we have (VII. 12)

$$\frac{a_{n+1}-a_1}{a_n + a_{n-1} + \ldots + a_1} = \frac{a_2 - a_1}{a_1},$$

which gives $a_n + a_{n-1} + \ldots + a_1$ or S_n.

The last proposition (36) gives the criterion for *perfect numbers*, namely that, if the sum of any number of terms of the series 1, 2, $2^2 \ldots 2^n$ is prime, the product of the said sum and of the last term, viz. $(1+2+2^2+\ldots+2^n)\,2^n$, is a perfect number, i.e. is equal to the sum of all its factors.

It should be added, as regards all the arithmetical Books, that all numbers are represented in the diagrams as simple straight lines, whether they are linear, plane, solid, or any other kinds of numbers; thus a product of two or more factors is represented as a new straight line, not as a rectangle or a solid.

Book X is perhaps the most remarkable, as it is the most perfect in form, of all the Books of the *Elements*. It deals with irrationals, that is to say, irrational straight lines in relation to any particular straight line assumed as rational, and it investigates every possible variety of straight lines which can be represented by $\sqrt{(\sqrt{a} \pm \sqrt{b})}$, where a, b are two commensurable lines. The theory was, of course, not invented by Euclid himself. On the contrary, we know that not only the fundamental proposition X. 9 (in which it is proved that squares which have not to one another the ratio of a square number to a square number have their sides incommensurable in length, and conversely), but also a large part of the further development of the subject, was due to Theaetetus. Our authorities for this are a scholium to X. 9 and a passage from Pappus's commentary on Book X preserved in the Arabic (see pp. 154–5, 209–10, above). The passage of Pappus goes on to speak of the share of Euclid in the investigation:

'As for Euclid, he set himself to give rigorous rules, which he established, relative to commensurability and incommensurability in general; he made precise the definitions and the distinctions between rational and irrational magnitudes, he set out a great number of orders of irrational magnitudes, and finally he made clear their whole extent.'

As usual, Euclid begins with definitions. 'Commensurable' magnitudes can be measured by one and the same measure; 'incommensurable' magnitudes cannot have any common measure (1). Straight lines are 'commensurable in square' when the squares on them can be measured by the same area, but 'incommensurable in square' when the squares on them have no common measure (2). Given an assigned straight line, which we agree to call 'rational', any straight line which is commensurable with it either in length or in square only is also called rational; but any straight line which is incommensurable with it (i.e. not commensurable with it either in length or in square) is 'irrational' (3). The square on the assigned straight line is 'rational', and any area commensurable with it is 'rational', but any area incommensurable with it is 'irrational', as also is the side of the square equal to that area (4). As regards straight lines, then, Euclid here takes a wider view of 'rational' than we have met before. If a straight line ρ is assumed as rational, not only is $\frac{m}{n}\rho$ also 'rational' where m, n are integers and m/n in its lowest terms is not square, but any straight line is rational which is either commensurable in length or commensurable *in square only* with ρ; that is, $\sqrt{\frac{m}{n}}\cdot\rho$ is rational according to Euclid. In the case of squares, ρ^2 is of course rational, and so is $\frac{m}{n}\rho^2$; but $\sqrt{\frac{m}{n}}\cdot\rho^2$ is not rational, and of course the side of the latter square $\sqrt[4]{\frac{m}{n}}\cdot\rho$ is irrational, as are all straight lines commensurable neither in length nor in square with ρ, e.g. $\sqrt{a} \pm \sqrt{b}$ or $(\sqrt{k} \pm \sqrt{\lambda})\cdot\rho$.

The Book begins with the famous proposition, on which the 'method of exhaustion' as used in Book XII depends, to the effect that, if from any magnitude there be subtracted more than its half (or its half simply), from the remainder more than its half (or its half), and so on continually, there will at length remain a magnitude less than any assigned magnitude of the same kind. Proposition 2 uses the process for finding the G. C. M. of two magnitudes as a test of their commensurability or incommensurability: they are incommensurable if the process never comes to an end, i.e. if no remainder ever measures the preceding divisor; and Propositions 3, 4 apply to commensurable magnitudes the method of finding the G. C. M. of two or three *numbers* as employed in VII. 2, 3. Propositions 5 to 8 show that two magnitudes are commensurable or incommensurable according as they have or have not to one another the ratio of one number to another, and lead up to the fundamental proposition (9) of Theaetetus already quoted, namely that the sides of squares are commensurable or incommensurable in length according as the squares have or have not to one another the ratio of a square number to a square number, and conversely. Propositions 11–16 are easy inferences as to the commensurability or incommensurability of magnitudes from the known relations of others connected with them; e.g. Proposition 14 proves that, if $a : b = c : d$, then, according as $\sqrt{(a^2-b^2)}$ is commensurable or incommensurable with a, $\sqrt{(c^2-d^2)}$ is commensurable or incommensurable with c. Following on this, Propositions 17, 18 prove that the roots of the quadratic equation $ax-x^2 = b^2/4$ are commensurable or incommensurable with a according as $\sqrt{(a^2-b^2)}$ is commensurable or incommensurable with a. Propositions 19–21 deal with rational and irrational *rectangles*, the former being contained by straight lines commensurable in length, whereas rectangles contained by straight lines commensurable in square only are irrational. The side of a square equal to a rectangle of the latter kind is called *medial*; this is the first in Euclid's classification of irrationals. As the sides of the rectangle may be expressed as ρ, $\rho\sqrt{k}$, where ρ is a rational straight line, the *medial* is $k^{\frac{1}{4}}\rho$. Propositions 23–8 relate to medial straight lines and rectangles; two medial straight lines may be either commensurable in length or commensurable in square only:

thus $k^{\frac{1}{4}}\rho$ and $\lambda k^{\frac{1}{4}}\rho$ are commensurable in length, while $k^{\frac{1}{4}}\rho$ and $\sqrt{\lambda} \,.\, k^{\frac{1}{4}}\rho$ are commensurable in square only: the rectangles formed by such pairs are in general *medial*, as $\lambda k^{\frac{1}{2}}\rho^2$ and $\sqrt{\lambda} \,.\, k^{\frac{1}{2}}\rho^2$; but if $\sqrt{\lambda} = k'\sqrt{k}$ in the second case, the rectangle $(k'k\rho^2)$ is rational (Propositions 24, 25). Proposition 26 proves that the difference between two medial areas cannot be rational; as any two medial areas can be expressed in the form $\sqrt{k} \,.\, \rho^2$, $\sqrt{\lambda} \,.\, \rho^2$, this is equivalent to proving, as we do in algebra, that $(\sqrt{k} - \sqrt{\lambda})$ cannot be equal to k'. Finally, Propositions 27, 28 find medial straight lines commensurable in square only (1) which contain a rational rectangle, viz. $k^{\frac{1}{4}}\rho$, $k^{\frac{3}{4}}\rho$, and (2) which contain a medial rectangle, viz. $k^{\frac{1}{4}}\rho, \lambda^{\frac{1}{2}}\rho/k^{\frac{1}{4}}$. It should be observed that, as ρ may take either of the forms a or $\sqrt{A}$, a medial straight line may take the alternative forms $\sqrt{(a\sqrt{B})}$ or $\sqrt[4]{(AB)}$, and the pairs of medial straight lines just mentioned may take respectively the forms

$$(1) \quad \sqrt{(a\sqrt{B})}, \quad \sqrt{\left(\frac{B\sqrt{B}}{a}\right)} \quad \text{or} \quad \sqrt[4]{(AB)}, \quad \sqrt{\left(B\frac{\sqrt{B}}{\sqrt{A}}\right)}$$

$$\text{and } (2) \quad \sqrt{(a\sqrt{B})}, \quad \sqrt{\left(\frac{aC}{\sqrt{B}}\right)} \quad \text{or} \quad \sqrt[4]{(AB)}, \quad \sqrt{\left(\frac{C\sqrt{A}}{\sqrt{B}}\right)}$$

I shall henceforth omit reference to these obvious alternative forms. Next follow two lemmas the object of which is to find (1) two square numbers the sum of which is a square, Euclid's solution being

$$mnp^2 \,.\, mnq^2 + \left(\frac{mnp^2 - mnq^2}{2}\right)^2 = \left(\frac{mnp^2 + mnq^2}{2}\right)^2,$$

where mnp^2, mnq^2 are either both odd or both even, and (2) two square numbers the sum of which is not square, Euclid's solution being

$$mp^2 \,.\, mq^2, \quad \left(\frac{mp^2 - mq^2}{2} - 1\right)^2.$$

Propositions 29–35 are problems the object of which is to find (*a*) two rational straight lines commensurable in square only, (*b*) two medial straight lines commensurable in square only, (*c*) two straight lines incommensurable in square, such that the difference or sum of their squares and the rectangle contained by them respectively have certain characteristics. The solutions are

(*a*) x, y rational and commensurable in square only.

Prop. 29 : ρ, $\rho\sqrt{(1-k^2)}$ [$\sqrt{(x^2-y^2)}$ commensurable with x].

„ 30 : ρ, $\rho/\sqrt{(1+k^2)}$ [$\sqrt{(x^2-y^2)}$ incommensurable with x].

(*b*) x, y medial and commensurable in square only.

Prop. 31 : $\rho(1-k^2)^{\frac{1}{4}}$, $\rho(1-k^2)^{\frac{3}{4}}$ [xy rational, $\sqrt{(x^2-y^2)}$ commensurable with x];

$\rho/(1+k^2)^{\frac{1}{4}}$, $\rho/(1+k^2)^{\frac{3}{4}}$ [xy rational, $\sqrt{(x^2-y^2)}$ incommensurable with x].

„ 32 : $\rho\lambda^{\frac{1}{4}}$, $\rho\lambda^{\frac{1}{4}}\sqrt{(1-k^2)}$ [xy medial, $\sqrt{(x^2-y^2)}$ commensurable with x];

$\rho\lambda^{\frac{1}{4}}$, $\rho\lambda^{\frac{1}{4}}/\sqrt{(1+k^2)}$ [xy medial, $\sqrt{(x^2-y^2)}$ incommensurable with x].

(*c*) x, y incommensurable in square.

Prop. 33 : $\dfrac{\rho}{\sqrt{2}}\sqrt{\left(1+\dfrac{k}{\sqrt{1+k^2}}\right)}$, $\dfrac{\rho}{\sqrt{2}}\sqrt{\left(1-\dfrac{k}{\sqrt{1+k^2}}\right)}$

[(x^2+y^2) rational, xy medial].

„ 34 : $\dfrac{\rho}{\sqrt{\{2(1+k^2)\}}}\cdot\sqrt{\{\sqrt{(1+k^2)}+k\}}$,

$\dfrac{\rho}{\sqrt{\{2(1+k^2)\}}}\cdot\sqrt{\{\sqrt{(1+k^2)}-k\}}$

[x^2+y^2 medial, xy rational].

„ 35 : $\dfrac{\rho\lambda^{\frac{1}{4}}}{\sqrt{2}}\sqrt{\left\{1+\dfrac{k}{\sqrt{(1+k^2)}}\right\}}$, $\dfrac{\rho\lambda^{\frac{1}{4}}}{\sqrt{2}}\sqrt{\left\{1-\dfrac{k}{\sqrt{(1+k^2)}}\right\}}$

[x^2+y^2 and xy both medial and incommensurable with one another].

With Proposition 36 begins Euclid's exposition of the several compound irrationals, twelve in number Those which only differ in the sign separating the two component parts can be taken together. The twelve compound irrationals, with their names, are as follows:

(A_1) Binomial, $\rho + \sqrt{k}\,.\,\rho$ (Prop. 36)
(A_2) Apotome, $\rho \sim \sqrt{k}\,.\,\rho$ (Prop. 73)

(B_1) First bimedial
(B_2) First apotome of a medial
$$k^{\frac{1}{4}}\rho \underset{\sim}{+} k^{\frac{3}{4}}\rho \quad \text{(Props. 37, 74)}$$

(C_1) Second bimedial
(C_2) Second apotome of a medial
$$k^{\frac{1}{4}}\rho \underset{\sim}{+} \frac{\lambda^{\frac{1}{2}}\rho}{k^{\frac{1}{4}}} \quad \text{(Props. 38, 75)}$$

(D_1) Major
(D_2) Minor
$$\frac{\rho}{\sqrt{2}}\sqrt{\left(1+\frac{k}{\sqrt{1+k^2}}\right)} \pm \frac{\rho}{\sqrt{2}}\sqrt{\left(1-\frac{k}{\sqrt{1+k^2}}\right)}$$
(Props. 39, 76)

(E_1) Side of a rational plus a medial area
(E_2) That which 'produces' with a rational area a medial whole
$$\frac{\rho}{\sqrt{2(1+k^2)}}\sqrt{(\sqrt{1+k^2}+k)} \pm \frac{\rho}{\sqrt{2(1+k^2)}}\sqrt{(\sqrt{1+k^2}-k)}$$
(Props. 40, 77)

(F_1) Side of the sum of two medial areas
(F_2) That which 'produces' with a medial area a medial whole
$$\frac{\rho\lambda^{\frac{1}{4}}}{\sqrt{2}}\sqrt{\left(1+\frac{k}{\sqrt{1+k^2}}\right)} \pm \frac{\rho\lambda^{\frac{1}{4}}}{\sqrt{2}}\sqrt{\left(1-\frac{k}{\sqrt{1+k^2}}\right)}$$
(Props. 41, 78).

As regards the above twelve compound irrationals, it is to be noted that

A_1, A_2 are the positive roots of the equation
$$x^4 - 2(1+k)\rho^2\,.\,x^2 + (1-k)^2\rho^4 = 0;$$

B_1, B_2 are the positive roots of the equation
$$x^4 - 2\sqrt{k}(1+k)\rho^2\,.\,x^2 + k(1-k)^2\rho^4 = 0;$$

C_1, C_2 are the positive roots of the equation
$$x^4 - 2\frac{k+\lambda}{\sqrt{k}}\rho^2\,.\,x^2 + \frac{(k-\lambda)^2}{k}\rho^4 = 0;$$

D_1, D_2 are the positive roots of the equation

$$x^4 - 2\rho^2 . x^2 + \frac{k^2}{1+k^2}\rho^4 = 0;$$

E_1, E_2 are the positive roots of the equation

$$x^4 - \frac{2}{\surd(1+k^2)}\rho^2 . x^2 + \frac{k^2}{(1+k^2)^2}\rho^4 = 0;$$

F_1, F_2 are the positive roots of the equation

$$x^4 - 2\surd\lambda . x^2\rho^2 + \lambda\frac{k^2}{1+k^2}\rho^4 = 0.$$

Propositions 42–7 prove that each of the above straight lines, made up of the *sum* of two terms, is divisible into its terms in only one way. In particular, Proposition 42 proves the equivalent of the well-known theorem in algebra that,

if $\quad a + \surd b = x + \surd y$, then $a = x$, $b = y$;

and if $\quad \surd a + \surd b = \surd x + \surd y$,

then $\quad a = x$, $b = y$ (or $a = y$, $b = x$).

Propositions 79–84 prove corresponding facts in regard to the corresponding irrationals with the negative sign between the terms: in particular Proposition 79 shows that,

if $\quad a - \surd b = x - \surd y$, then $a = x$, $b = y$;

and if $\quad \surd a - \surd b = \surd x - \surd y$, then $a = x$, $b = y$.

The next sections of the Book deal with binomials and apotomes classified according to the relation of their terms to another given rational straight line. There are six kinds, which are first defined and then constructed, as follows:

(α_1) First binomial, (α_2) First apotome: $k\rho \pm k\rho\surd(1-\lambda^2)$; (Props. 48, 85)

(β_1) Second binomial, (β_2) Second apotome: $\dfrac{k\rho}{\surd(1-\lambda^2)} \pm k\rho$; (Props. 49, 86)

(γ_1) Third binomial, (γ_2) Third apotome: $m\surd k . \rho \pm m\surd k . \rho\surd(1-\lambda^2)$; (Props. 50, 87)

(δ_1) Fourth binomial, (δ_2) Fourth apotome: $k\rho \pm \dfrac{k\rho}{\sqrt{(1+\lambda)}}$; (Props. 51, 88)

(ϵ_1) Fifth binomial, (ϵ_2) Fifth apotome: $k\rho\sqrt{(1+\lambda)} \pm k\rho$; (Props. 52, 89)

(ζ_1) Sixth binomial, (ζ_2) Sixth apotome: $\sqrt{k}\,.\,\rho \pm \sqrt{\lambda}\,.\,\rho$. (Prop. 53, 90)

Here again it is to be observed that these binomials and apotomes are the greater and lesser roots respectively of certain quadratic equations,

α_1, α_2 being the roots of $x^2 - 2k\rho\,.\,x + \lambda^2 k^2 \rho^2 = 0$,

β_1, β_2 " " $x^2 - \dfrac{2k\rho}{\sqrt{(1-\lambda^2)}}\cdot x + \dfrac{\lambda^2}{1-\lambda^2}k^2\rho^2 = 0$,

γ_1, γ_2 " " $x^2 - 2m\sqrt{k}\,.\,\rho x + \lambda^2 m^2 k \rho^2 = 0$,

δ_1, δ_2 " " $x^2 - 2k\rho\,.\,x + \dfrac{\lambda}{1+\lambda}k^2\rho^2 = 0$,

ϵ_1, ϵ_2 " " $x^2 - 2k\rho\sqrt{(1+\lambda)}\,.\,x + \lambda k^2 \rho^2 = 0$,

ζ_1, ζ_2 " " $x^2 - 2\sqrt{k}\,.\,\rho x + (k-\lambda)\rho^2 = 0$.

The next sets of propositions (54–65 and 91–102) prove the connexion between the first set of irrationals $(A_1, A_2 \ldots F_1, F_2)$ and the second set $(\alpha_1, \alpha_2 \cdots \zeta_1, \zeta_2)$ respectively. It is shown e.g., in Proposition 54, that the side of a square equal to the rectangle contained by ρ and the first binomial α_1 is a binomial of the type A_1, and the same thing is proved in Proposition 91 for the first apotome. In fact

$$\sqrt{\{\rho\,(k\rho \pm k\rho\sqrt{1-\lambda^2})\}} = \rho\sqrt{\{\tfrac{1}{2}k(1+\lambda)\}} \pm \rho\sqrt{\{\tfrac{1}{2}k(1-\lambda)\}}.$$

Similarly $\sqrt{(\rho\beta_1)}$, $\sqrt{(\rho\beta_2)}$ are irrationals of the type B_1, B_2 respectively, and so on.

Conversely, the square on A_1 or A_2, if applied as a rectangle to a rational straight line (σ, say), has for its breadth a binomial or apotome of the types α_1, α_2 respectively (60, 97).

In fact $\quad (\rho \pm \sqrt{k}\,.\,\rho)^2/\sigma = \dfrac{\rho^2}{\sigma}\{(1+k) \pm 2\sqrt{k}\}$,

and $B_1{}^2$, $B_2{}^2$ are similarly related to irrationals of the type β_1, β_2, and so on.

Propositions 66–70 and Propositions 103–7 prove that straight lines commensurable in length with A_1, $A_2 \dots F_1$, F_2 respectively are irrationals of the same type and order.

Propositions 71, 72, 108–10 show that the irrationals A_1, $A_2 \dots F_1$, F_2 arise severally as the sides of squares equal to the sum or difference of a rational and a medial area, or the sum or difference of two medial areas incommensurable with one another. Thus $k\rho^2 \pm \sqrt{\lambda} . \rho^2$ is the sum or difference of a rational and a medial area, $\sqrt{k} . \rho^2 \pm \sqrt{\lambda} . \rho^2$ is the sum or difference of two medial areas incommensurable with one another provided that $\sqrt{k}$ and $\sqrt{\lambda}$ are incommensurable, and the propositions prove that

$$\sqrt{(k\rho^2 \pm \sqrt{\lambda} . \rho^2)} \text{ and } \sqrt{(\sqrt{k} . \rho^2 \pm \sqrt{\lambda} . \rho^2)}$$

take one or other of the forms A_1, $A_2 \dots F_1$, F_2 according to the different possible relations between k, λ and the sign separating the two terms, but no other forms.

Finally, it is proved at the end of Proposition 72, in Proposition 111 and the explanation following it that the thirteen irrational straight lines, the medial and the twelve other irrationals A_1, $A_2 \dots F_1$, F_2, are all different from one another. E.g. (Proposition 111) a binomial straight line cannot also be an apotome; in other words, $\sqrt{x} + \sqrt{y}$ cannot be equal to $\sqrt{x'} - \sqrt{y'}$, and $x + \sqrt{y}$ cannot be equal to $x' - \sqrt{y'}$. We prove the latter proposition by squaring, and Euclid's procedure corresponds exactly to this. Propositions 112–14 prove that, if a rectangle equal to the square on a rational straight line be applied to a binomial, the other side containing it is an apotome of the same order, with terms commensurable with those of the binomial and in the same ratio, and vice versa; also that a binomial and apotome of the same order and with terms commensurable respectively contain a rational rectangle. Here we have the equivalent of rationalizing the denominators of the fractions $\dfrac{c^2}{\sqrt{A} \pm \sqrt{B}}$ or $\dfrac{c^2}{a \pm \sqrt{B}}$ by multiplying the numerator and denominator by $\sqrt{A} \mp \sqrt{B}$ or $a \mp \sqrt{B}$ respectively. Euclid in fact proves that

$$\sigma^2 / (\rho + \sqrt{k} . \rho) = \lambda\rho - \sqrt{k} . \lambda\rho \ (k < 1),$$

and his method enables us to see that $\lambda = \sigma^2 / (\rho^2 - k\rho^2)$. Proposition 115 proves that from a medial straight line an infinite number of other irrational straight lines arise each of which is different from the preceding. $k^{\frac{1}{4}}\rho$ being medial, we take another rational straight line σ and find the mean proportional $\sqrt{(k^{\frac{1}{4}}\rho\sigma)}$; this is a new irrational. Take the mean between this and σ', and so on.

I have described the contents of Book X at length because it is probably not well known to mathematicians, while it is geometrically very remarkable and very finished. As regards its object Zeuthen has a remark which, I think, must come very near the truth. 'Since such roots of equations of the second degree as are incommensurable with the given magnitudes cannot be expressed by means of the latter and of numbers, it is conceivable that the Greeks, in exact investigations, introduced no approximate values, but worked on with the magnitudes they had found, which were represented by straight lines obtained by the construction corresponding to the solution of the equation. That is exactly the same thing which happens when we do not evaluate roots but content ourselves with expressing them by radical signs and other algebraical symbols. But, inasmuch as one straight line looks like another, the Greeks did not get the same clear view of what they denoted (i.e. by simple inspection) as our system of symbols assures to us. For this reason then it was necessary to undertake a classification of the irrational magnitudes which had been arrived at by successive solutions of equations of the second degree.' That is, Book X formed a repository of results to which could be referred problems depending on the solution of certain types of equations, quadratic and biquadratic but reducible to quadratics, namely the equations

$$x^2 \pm 2\mu x \cdot \rho \pm \nu \cdot \rho^2 = 0,$$

and

$$x^4 \pm 2\mu x^2 \cdot \rho^2 \pm \nu \cdot \rho^4 = 0,$$

where ρ is a rational straight line and μ, ν are coefficients. According to the values of μ, ν in relation to one another and their character (μ, but not ν, may contain a surd such as $\sqrt{m}$ or $\sqrt{(m/n)}$) the two positive roots of the first equations are the binomial and apotome respectively of some one of the orders 'first', 'second', ... 'sixth', while the two positive roots of the latter equation are of some one of the other forms of irrationals (A_1, A_2), (B_1, B_2) ... (F_1, F_2).

Euclid himself, in Book XIII, makes considerable use of the second part of Book X dealing with *apotomes*; he regards a straight line as sufficiently defined in character if he can say that it is, e.g., an *apotome* (XIII. 17), a *first apotome* (XIII. 6), a *minor* straight line (XIII. 11). So does Pappus.[1]

Our description of Books XI–XIII can be shorter. They deal with geometry in three dimensions. The definitions, belonging to all three Books, come at the beginning of Book XI. They include those of a straight line, or a plane, at right angles to a plane, the inclination of a plane to a plane (dihedral angle), parallel planes,. equal and similar solid figures, solid angle, pyramid, prism, sphere, cone, cylinder and parts of them, cube, octahedron, icosahedron and dodecahedron. Only the definition of the sphere needs special mention. Whereas it had previously been defined as the figure which has all points of its surface equidistant from its centre, Euclid, with an eye to his use of it in Book XIII to 'comprehend' the regular solids in a sphere, defines it as the figure comprehended by the revolution of a semicircle about its diameter.

The propositions of Book XI are in their order fairly parallel to those of Books I and VI on plane geometry. First we have propositions that a straight line is wholly in a plane if a portion of it is in the plane (1), and that two intersecting straight lines, and a triangle, are in one plane (2). Two intersecting planes cut in a straight line (3). Straight lines perpendicular to planes are next dealt with (4–6, 8, 11–14), then parallel straight lines not all in the same plane (9, 10, 15), parallel planes (14, 16), planes at right angles to one another (18, 19), solid angles contained by three angles (20, 22, 23, 26) or by more angles (21). The rest of the Book deals mainly with parallelepipedal solids. It is only necessary to mention the more important propositions. Parallelepipedal solids on the same base or equal bases and between the same parallel planes (i.e. having the same height) are equal (29–31). Parallelepipedal solids of the same height are to one another as their bases (32). Similar parallelepipedal solids are in the triplicate ratio of corresponding sides (33). In equal parallelepipedal solids the bases are reciprocally proportional to their heights and conversely (34). If four straight lines be propor-

[1] Cf. Pappus, iv, pp. 178, 182.

tional, so are parallelepipedal solids similar and similarly described upon them, and conversely (37). A few other propositions are only inserted because they are required as lemmas in later books, e.g. that, if a cube is bisected by two planes each of which is parallel to a pair of opposite faces, the common section of the two planes and the diameter of the cube bisect one another (38).

The main feature of Book XII is the application of the *method of exhaustion*, which is used to prove successively that circles are to one another as the squares on their diameters (Propositions 1, 2), that pyramids of the same height and with triangular bases are to one another as the bases (3–5), that any cone is, in content, one third part of the cylinder which has the same base with it and equal height (10), that cones and cylinders of the same height are to one another as their bases (11), that similar cones and cylinders are to one another in the triplicate ratio of the diameters of their bases (12), and finally that spheres are to one another in the triplicate ratio of their respective diameters (16–18). Propositions 1, 3–4 and 16–17 are of course preliminary to the main propositions 2, 5 and 18 respectively. Proposition 5 is extended to pyramids with polygonal bases in Proposition 6. Proposition 7 proves that any prism with triangular bases is divided into three pyramids with triangular bases and equal in content, whence any pyramid with triangular base (and therefore also any pyramid with polygonal base) is equal to one third part of the prism having the same base and equal height. The rest of the Book consists of propositions about pyramids, cones, and cylinders similar to those in Book XI about parallelepipeds and in Book VI about parallelograms: similar pyramids with triangular bases, and therefore also similar pyramids with polygonal bases, are in the triplicate ratio of corresponding sides (8); in equal pyramids, cones and cylinders the bases are reciprocally proportional to the heights, and conversely (9, 15).

The method of exhaustion, as applied in Euclid, rests upon X. 1 as lemma, and no doubt it will be desirable to insert here an example of its use. An interesting case is that relating to the pyramid. Pyramids with triangular bases and of the same height, says Euclid, are to one another as their bases (Prop. 5).

It is first proved (Proposition 3) that, given any pyramid, as $ABCD$, on the base BCD, if we bisect the six edges at the

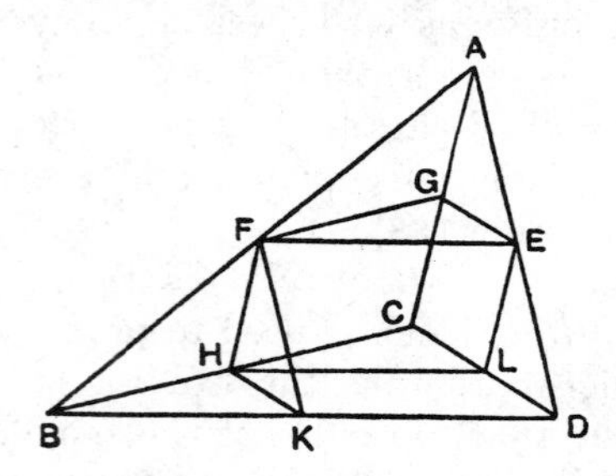

points E, F, G, H, K, L, and draw the straight lines shown in the figure, we divide the pyramid $ABCD$ into two equal prisms and two equal pyramids $AFGE$, $FBHK$ similar to the original pyramid (the equality of the prisms is proved in XI. 39), and that the sum of the two prisms is greater than half the original pyramid. Proposition 4 proves that, if each of two given pyramids of the same height be so divided, and if the small pyramids in each are similarly divided, then the smaller pyramids left over from that division are similarly divided, and so on to any extent, the sums of all the pairs of prisms in the two given pyramids respectively will be to one another as the respective bases. Let the two pyramids and their volumes be denoted by P, P' respectively, and their bases by B, B' respectively. Then, if $B : B'$ is not equal to $P : P'$, it must be equal to $P : W$, where W is some volume either less or greater than P'.

I. Suppose $W < P'$.

By X. 1 we can divide P' and the successive pyramids in it into prisms and pyramids until the sum of the small pyramids left over in it is less that $P' - W$, so that

$$P' > (\text{prisms in } P') > W.$$

Suppose this done, and P divided similarly.

Then (XII. 4)

$$(\text{sum of prisms in } P) : (\text{sum of prisms in } P') = B : B'$$
$$= P : W, \text{ by hypothesis.}$$

But $P >$ (sum of prisms in P):

therefore $\qquad W >$ (sum of prisms in P').

But W is also less than the sum of the prisms in P': which is impossible.

Therefore W is *not* less than P'.

II. Suppose $W > P'$.

We have, inversely,

$$B' : B = W : P$$
$$= P' : V, \text{ where } V \text{ is some solid less than } P.$$

But this can be proved impossible, exactly as in Part I.

Therefore W is neither greater nor less than P', so that

$$B : B' = P : P'.$$

We shall see, when we come to Archimedes, that he extended this method of exhaustion. Instead of merely taking the one approximation, from underneath as it were, by constructing successive figures *within* the figure to be measured and so exhausting it, he combines with this an approximation from *outside*. He takes sets both of inscribed and circumscribed figures, approaching from both sides the figure to be measured, and, as it were, *compresses* them into one, so that they coincide as nearly as we please with one another and with the curvilinear figure itself. The two parts of the proof are accordingly separate in Archimedes, and the second is not merely a reduction to the first.

The object of Book XIII is to construct, and to 'comprehend in a sphere', each of the five regular solids, the pyramid (Prop. 13), the octahedron (Prop. 14), the cube (Prop. 15), the icosahedron (Prop. 16) and the dodecahedron (Prop. 17); 'comprehending in a sphere' means the construction of the circumscribing sphere, which involves the determination of the relation of a 'side' (i.e. edge) of the solid to the radius of the sphere; in the case of the first three solids the relation is actually determined, while in the case of the icosahedron the side of the figure is shown to be the irrational straight line called 'minor', and in the case of the dodecahedron an 'apotome'. The propositions at the beginning of the Book are preliminary. Propositions 1–6 are theorems about straight lines cut in extreme and mean ratio, Propositions 7, 8 relate to pentagons, and Proposition 8 proves that, if, in a regular pentagon, two diagonals (straight lines joining angular points next but one to each other) are drawn intersecting at a point, each of them is divided at the point in extreme and mean ratio, the greater segment being equal to the side of the pentagon. Propositions 9 and 10 relate to the sides of a pentagon, a decagon and a hexagon all inscribed in the same circle, and are preliminary to proving (in Prop. 11) that the side of the inscribed pentagon is, in relation to the diameter of the circle, regarded as rational, the irrational straight line called 'minor'. If p, d, h be the sides of the regular pentagon, decagon, and hexagon inscribed in the same circle, Proposition 9 proves that $h+d$ is cut in extreme and mean ratio, h being the greater segment; this is equivalent to saying that $(r+d)\,d=r^2$, where r is the radius of the circle, or, in other words, that $d = \frac{1}{2}r(\sqrt{5}-1)$. Proposition 10 proves that $p^2 = h^2+d^2$ or r^2+d^2, whence we obtain $p = \frac{1}{2}r\sqrt{(10-2\sqrt{5})}$. Expressed as a 'minor' irrational straight line, which Proposition 11 shows it to be, $p = \frac{1}{2}r\sqrt{(5+2\sqrt{5})}-\frac{1}{2}r\sqrt{(5-2\sqrt{5})}$.

The constructions for the several solids, which have to be inscribed in a given sphere, may be briefly indicated, thus:

1. The regular pyramid or *tetrahedron*.

Given D, the diameter of the sphere which is to circumscribe the tetrahedron, Euclid draws a circle with radius r such that $r^2 = \frac{1}{3}D \,.\, \frac{2}{3}D$, or $r = \frac{1}{3}\sqrt{2} \,.\, D$, inscribes an equilateral triangle in the circle, and then erects from the centre of it a straight line perpendicular to its plane and of length $\frac{2}{3}D$. The lines joining the extremity of the perpendicular to the angular points of the equilateral triangle determine the tetrahedron. Each of the upstanding edges (x, say) is such that $x^2 = r^2+\frac{4}{9}D^2 = 3r^2$, and it has been proved (in XIII. 12) that the square on the side of the triangle inscribed in the circle is also $3r^2$. Therefore the edge a of the tetrahedron $=\sqrt{3} \,.\, r = \frac{1}{3}\sqrt{6} \,.\, D$.

2. The *octahedron*.

If D be the diameter of the circumscribing sphere, a square is inscribed in a circle of diameter D, and from its centre straight lines are drawn in both directions perpendicular to its plane and of length equal to the radius of the circle or half the diagonal of the square. Each of the edges which stand up from the square $=\sqrt{2} \,.\, \frac{1}{2}D$, which is equal to the side of the square. Each of the edges a of the octahedron is therefore equal to $\sqrt{2} \,.\, \frac{1}{2}D$.

3. The *cube*.

D being the diameter of the circumscribing sphere, draw a square with side a such that $a^2 = D \,.\, \frac{1}{3}D$, and describe a cube on this square as base. The edge $a = \frac{1}{3}\sqrt{3} \,.\, D$.

4. The *icosahedron*.

Given D, the diameter of the sphere, construct a circle with radius r such that $r^2 = D \,.\, \frac{1}{5}D$. Inscribe in it a regular decagon. Draw from its angular points straight lines perpendicular to the plane of the circle and equal in length to its radius r; this determines the angular points of a regular decagon inscribed in an equal parallel circle. By joining alternate angular points of one of the decagons, describe a regular pentagon in the circle circumscribing it, and then do the same in the other circle but so that the angular points are not opposite those of the other pentagon. Join the angular points of one pentagon to the nearest angular points of the other; this gives ten triangles. Then, if p be the side of each pentagon, d the side of each decagon, the upstanding sides of the triangles ($= x$, say) are given by $x^2 = d^2 + r^2 = p^2$ (Prop. 10); therefore the ten triangles are equilateral. We have lastly to find the common vertices of the five equilateral triangles standing on the pentagons and completing the icosahedron. If C, C' be the centres of the parallel circles, CC' is produced in both directions to X, Z respectively so that $CX = C'Z = d$ (the side of the decagon). Then again the upstanding edges connecting to X, Z the angular points of the two pentagons respectively ($=x$, say) are given by

$$x^2 = r^2 + d^2 = p^2.$$

Hence each of the edges

$$a = p = \tfrac{1}{2}r\sqrt{(10-2\sqrt{5})} = \frac{D}{2\sqrt{5}}\sqrt{(10-2\sqrt{5})}$$

$$= \tfrac{1}{10}D\sqrt{\{10(5-\sqrt{5})\}}.$$

It is finally shown that the sphere described on XZ as diameter circumscribes the icosahedron, and

$$XZ = r + 2\,d = r + r\,(\sqrt{5}-1) = r \,.\, \sqrt{5} = D.$$

5. The *dodecahedron.*

We start with the cube inscribed in the given sphere with diameter D. We then draw pentagons which have the edges of the cube as diagonals in the manner shown in the figure. If H, N, M, O be the middle points of the sides of the face BF, and H, G, L, K the middle points of the sides of the face BD, join NO, GK which are then parallel to BC, and draw MH, HL bisecting them at right angles at P, Q.

Divide PN, PO, QH in extreme and mean ratio at R, S, T, and let PR, PS, QT be the greater segments. Draw RU, PX, SV at right angles to the plane BF, and TW at right angles to

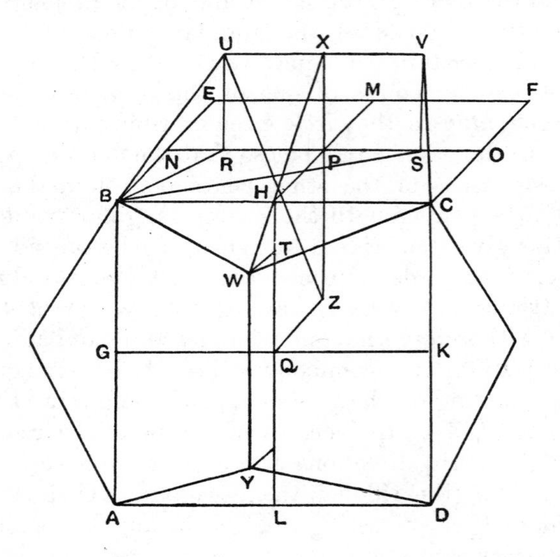

the plane BD, such that each of these perpendiculars $= PR$ or PS. Join UV, VC, CW, WB, BU. These determine one of the pentagonal faces, and the others are drawn similarly.

It is then proved that each of the pentagons, as $UVCWB$, is (1) equilateral, (2) in the same plane, (3) equiangular.

As regards the sides we see, e. g., that

$$BU^2 = BR^2 + RU^2 = BN^2 + NR^2 + RP^2$$

$$= PN^2 + NR^2 + RP^2 = 4\,RP^2 \text{ (by means of XIII. 4)} = UV^2,$$

and so on.

Lastly, it is proved that the same sphere of diameter D which circumscribes the cube also circumscribes the dodecahedron. For example, if Z is the centre of the sphere,

$$ZU^2 = ZX^2 + XU^2 = NS^2 + PS^2 = 3\,PN^2, \quad \text{(XIII. 4)}$$

while

$$ZB^2 = 3\,ZP^2 = 3\,PN^2.$$

If a be the edge of the dodecahedron, c the edge of the cube,

$$a = 2RP = 2 \cdot \frac{\sqrt{5}-1}{4} c$$

$$= \frac{2\sqrt{3}}{3} \cdot \frac{\sqrt{5}-1}{4} D$$

$$= \tfrac{1}{6} D (\sqrt{15} - \sqrt{3}).$$

Book XIII ends with Proposition 18, which arranges the edges of the five regular solids inscribed in one and the same sphere in order of magnitude, while an addendum proves that no other regular solid figures except the five exist.

The so-called Books XIV, XV.

This is no doubt the place to speak of the continuations of Book XIII which used to be known as Books XIV, XV of the *Elements*, though they are not by Euclid. The former is the work of Hypsicles, who probably lived in the second half of the second century B.C., and who is otherwise known as the reputed author of an astronomical tract Ἀναφορικός (*De ascensionibus*) still extant (the earliest extant Greek book in which the division of the circle into 360 degrees appears), besides other works, which have not survived, on the harmony of the spheres and on polygonal numbers. The preface to 'Book XIV' is interesting historically. It appears from it that Apollonius wrote a tract on the comparison of the dodecahedron and icosahedron inscribed in one and the same sphere, i.e. on the ratio between them, and that there were two editions of this work, the first of which was in some way incorrect, while the second gave a correct proof of the proposition that, as the surface of the dodecahedron is to the surface of the icosahedron, so is the solid content of the dodecahedron to that of the icosahedron, 'because the perpendicular from the centre of the sphere to the pentagon of the dodecahedron and to the triangle of the icosahedron is the same'. Hypsicles says also that Aristaeus, in a work entitled *Comparison of the five figures*, proved that 'the same circle circumscribes both the pentagon of the dodecahedron and the triangle of the icosahedron inscribed in the same sphere'; whether this Aristaeus is the same as the Aristaeus of the *Solid Loci*, the elder contemporary of Euclid, we do not know. The proposition of Aristaeus is proved by Hypsicles as Proposition 2 of his book. The following is a summary of the results obtained by Hypsicles. In a lemma at the end he proves that, if two straight lines be cut in extreme and mean ratio, the segments of both are in one and the same ratio; the ratio is in fact $2:(\sqrt{5}-1)$. If then *any* straight line AB be divided at C in extreme and mean ratio, AC being the greater segment, Hypsicles proves that, if we have a cube, a dodecahedron and an icosahedron all inscribed in the same sphere, then:

(Prop. 7) (side of cube) : (side of icosahedron)

$$= \sqrt{(AB^2+AC^2)} : \sqrt{(AB^2+BC^2)};$$

(Prop. 6) (surface of dodecahedron) : (surface of icosahedron)

= (side of cube) : (side of icosahedron);

(Prop. 8) (content of dodecahedron) : (content of icosahedron)

= (surface of dodecahedron) : (surface of icosahedron);

and consequently

(content of dodecahedron) : (content of icosahedron)

$$= \sqrt{(AB^2+AC^2)} : \sqrt{(AB^2+BC^2)}.$$

The second of the two supplementary Books ('Book XV') is also concerned with the regular solids, but is much inferior to the first. The exposition leaves much to be desired, being in some places obscure, in others actually inaccurate. The Book is in three parts unequal in length. The first[1] shows how to inscribe certain of the regular solids in certain others,

[1] Heiberg's Euclid, vol. v, pp. 40–8.

(*a*) a tetrahedron in a cube, (*b*) an octahedron in a tetrahedron, (*c*) an octahedron in a cube, (*d*) a cube in an octahedron, (*e*) a dodecahedron in an icosahedron. The second portion[1] explains how to calculate the number of edges and the number of solid angles in the five solids respectively. The third portion[2] shows how to determine the dihedral angles between the faces meeting in any edge of any one of the solids. The method is to construct an isosceles triangle with vertical angle equal to the said angle; from the middle point of any edge two perpendiculars are drawn to it, one in each of the two faces intersecting in that edge; these perpendiculars (forming the dihedral angle) are used to determine the two equal sides of an isosceles triangle, and the base of the triangle is easily found from the known properties of the particular solid. The rules for drawing the respective isosceles triangles are first given all together in general terms; and the special interest of the passage consists in the fact that the rules are attributed to 'Isidorus our great teacher'. This Isidorus is doubtless Isidorus of Miletus, the architect of the church of Saint Sophia at Constantinople (about A.D. 532). Hence the third portion of the Book at all events was written by a pupil of Isidorus in the sixth century.

The *Data*.

Coming now to the other works of Euclid, we will begin with those which have actually survived. Most closely connected with the *Elements* as dealing with plane geometry, the subject-matter of Books I–VI, is the *Data*, which is accessible in the Heiberg-Menge edition of the Greek text, and also in the translation annexed by Simson to his edition of the *Elements* (although this translation is based on an inferior text). The book was regarded as important enough to be included in the *Treasury of Analysis* (τόπος ἀναλυόμενος) as known to Pappus, and Pappus gives a description of it; the description shows that there were differences between Pappus's text and ours, for, though Propositions 1–62 correspond to the description, as also do Propositions 87–94 relating to circles at the end of the book, the intervening propositions do not

[1] Heiberg's Euclid, vol. v, pp. 48–50. [2] *Ib.*, pp. 50–66.

exactly agree, the differences, however, affecting the distribution and numbering of the propositions rather than their substance. The book begins with definitions of the senses in which things are said to be *given*. Things such as areas, straight lines, angles and ratios are said to be 'given in *magnitude* when we can make others equal to them' (Defs. 1–2). Rectilineal figures are 'given *in species*' when their angles are severally given as well as the ratios of the sides to one another (Def. 3). Points, lines and angles are 'given *in position*' 'when they always occupy the same place': a not very illuminating definition (4). A circle is given *in position and in magnitude* when the centre is given *in position* and the radius *in magnitude* (6); and so on. The object of the proposition called a Datum is to prove that, if in a given figure certain parts or relations are given, other parts or relations are also given, in one or other of these senses.

It is clear that a systematic collection of *Data* such as Euclid's would very much facilitate and shorten the procedure in *analysis*; this no doubt accounts for its inclusion in the *Treasury of Analysis*. It is to be observed that this form of proposition does not actually determine the thing or relation which is shown to be given, but merely proves that it can be determined when once the facts stated in the hypothesis are known; if the proposition stated that a certain thing *is* so and so, e.g. that a certain straight line in the figure is of a certain length, it would be a theorem; if it directed us to *find* the thing instead of proving that it is 'given', it would be a problem; hence many propositions of the form of the *Data* could alternatively be stated in the form of theorems or problems.

We should naturally expect much of the subject-matter of the *Elements* to appear again in the *Data* under the different aspect proper to that book; and this proves to be the case. We have already mentioned the connexion of Eucl. II. 5, 6 with the solution of the mixed quadratic equations $ax \pm x^2 = b^2$. The solution of these equations is equivalent to the solution of the simultaneous equations

$$\left.\begin{aligned} y \pm x &= a \\ xy &= b^2 \end{aligned}\right\},$$

and Euclid shows how to solve these equations in Propositions

84, 85 of the *Data*, which state that 'If two straight lines contain a given area in a given angle, and if the difference (sum) of them be given, then shall each of them be given.' The proofs depend directly upon those of Propositions 58, 59, 'If a given area be applied to a given straight line, falling short (exceeding) by a figure given in species, the breadths of the deficiency (excess) are given.' All the 'areas' are parallelograms.

We will give the proof of Proposition 59 (the case of 'excess'). Let the given area AB be applied to AC, exceeding by the figure CB given in species. I say that each of the sides HC, CE is given.

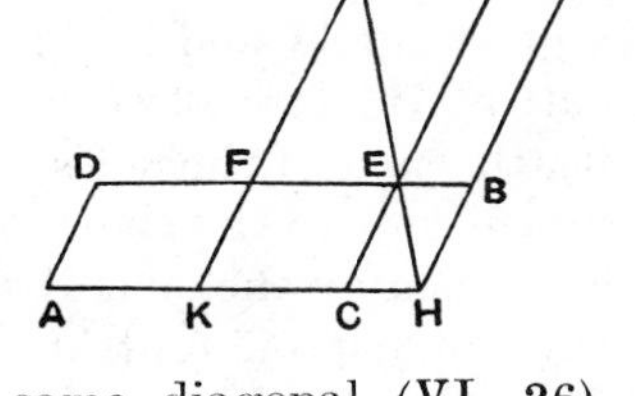

Bisect DE in F, and construct on EF the figure FG similar and similarly situated to CB (VI. 18). Therefore FG, CB are about the same diagonal (VI. 26). Complete the figure.

Then FG, being similar to CB, is given in species, and, since FE is given, FG is given in magnitude (Prop. 52).

But AB is given; therefore $AB+FG$, that is to say, KL, is given in magnitude. But it is also given in species, being similar to CB; therefore the sides of KL are given (Prop. 55).

Therefore KH is given, and, since $KC = EF$ is also given, the difference CH is given. And CH has a given ratio to HB; therefore HB is also given (Prop. 2).

Eucl. III. 35, 36 about the 'power' of a point with reference to a circle have their equivalent in *Data* 91, 92 to the effect that, given a circle and a point in the same plane, the rectangle contained by the intercepts between this point and the points in which respectively the circumference is cut by any straight line passing through the point and meeting the circle is also given.

A few more enunciations may be quoted. Proposition 8 (compound ratio): Magnitudes which have given ratios to the same magnitude have a given ratio to one another also. Propositions 45, 46 (similar triangles): If a triangle have one angle given, and the ratio of the sum of the sides containing that angle, or another angle, to the third side (in each case) be given, the triangle is given in species. Proposition 52: If a (rectilineal) figure given in species be described on a straight line given in magnitude, the figure is given in magnitude. Proposition 66: If a triangle have one angle given, the rectangle contained by the sides including the angle has to the (area of the) triangle a given ratio. Proposition 80: If a triangle have one angle given, and if the rectangle contained by the sides including the given angle have to the square on the third side a given ratio, the triangle is given in species.

Proposition 93 is interesting: If in a circle given in magnitude a straight line be drawn cutting off a segment containing a given angle, and if this angle be bisected (by a straight line cutting the base of the segment and the circumference beyond it), the sum of the sides including the given angle will have a given ratio to the chord bisecting the angle, and the rectangle contained by the sum of the said sides and the portion of the bisector cut off (outside the segment) towards the circumference will also be given.

Euclid's proof is as follows. In the circle ABC let the chord BC cut off a segment containing a given angle BAC, and let the angle be bisected by AE meeting BC in D.

Join BE. Then, since the circle is given in magnitude, and

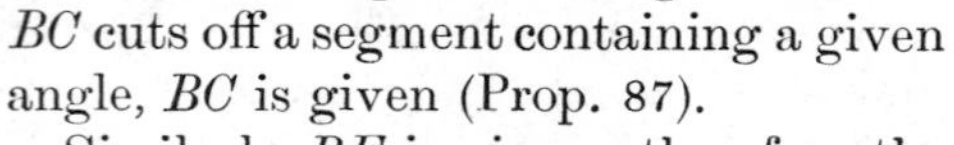

BC cuts off a segment containing a given angle, BC is given (Prop. 87).

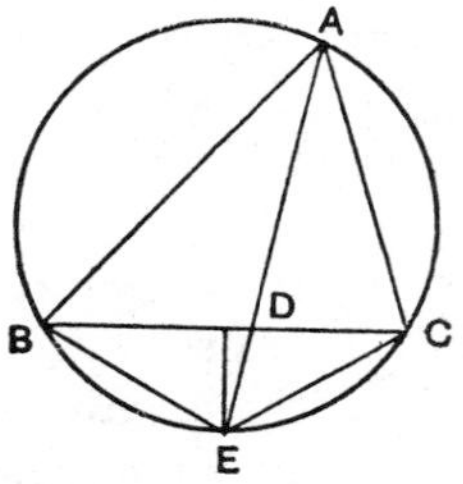

Similarly BE is given; therefore the ratio $BC : BE$ is given. (It is easy to see that the ratio $BC : BE$ is equal to $2 \cos \frac{1}{2} A$.)

Now, since the angle BAC is bisected,

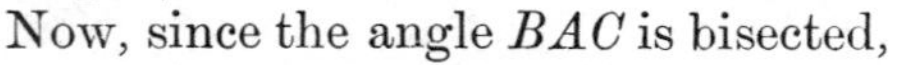

$$BA : AC = BD : DC.$$

It follows that $(BA + AC) : (BD + DC) = AC : DC$.

But the triangles ABE, ADC are similar;

therefore $\quad AE : BE = AC : DC$

$$= (BA + AC) : BC, \text{ from above.}$$

Therefore $(BA + AC) : AE = BC : BE$, which is a given ratio.

Again, since the triangles ADC, BDE are similar,

$$BE:ED = AC:CD = (BA+AC):BC.$$

Therefore $(BA+AC).ED = BC.BE$, which is given.

On divisions (of figures).

The only other work of Euclid in pure geometry which has survived (but not in Greek) is the book *On divisions (of figures)*, *περὶ διαιρέσεων βιβλίον*. It is mentioned by Proclus, who gives some hints as to its content[1]; he speaks of the business of the author being divisions of figures, circles or rectilineal figures, and remarks that the parts may be like in definition or notion, or unlike; thus to divide a triangle into triangles is to divide it into like figures, whereas to divide it into a triangle and a quadrilateral is to divide it into unlike figures. These hints enable us to check to some extent the genuineness of the books dealing with divisions of figures which have come down through the Arabic. It was John Dee who first brought to light a treatise *De divisionibus* by one Muhammad Bagdadinus (died 1141) and handed over a copy of it (in Latin) to Commandinus in 1563; it was published by the latter in Dee's name and his own in 1570. Dee appears not to have translated the book from the Arabic himself, but to have made a copy for Commandinus from a manuscript of a Latin translation which he himself possessed at one time but which was apparently stolen and probably destroyed some twenty years after the copy was made. The copy does not seem to have been made from the Cotton MS. which passed to the British Museum after it had been almost destroyed by a fire in 1731.[2] The Latin translation may have been that made by Gherard of Cremona (1114–87), since in the list of his numerous translations a 'liber divisionum' occurs. But the Arabic original cannot have been a direct translation from Euclid, and probably was not even a direct adaptation of it, since it contains mistakes and unmathematical expressions; moreover, as it does not contain the propositions about the

[1] Proclus on Eucl. I, p. 144. 22–6.

[2] The question is fully discussed by R. C. Archibald, *Euclid's Book on Divisions of Figures with a restoration based on Woepcke's text and on the Practica Geometriae of Leonardo Pisano* (Cambridge 1915).

division of a circle alluded to by Proclus, it can scarcely have contained more than a fragment of Euclid's original work. But Woepcke found in a manuscript at Paris a treatise in Arabic on the division of figures, which he translated and published in 1851. It is expressly attributed to Euclid in the manuscript and corresponds to the indications of the content given by Proclus. Here we find divisions of different rectilinear figures into figures of the same kind, e.g. of triangles into triangles or trapezia into trapezia, and also divisions into 'unlike' figures, e.g. that of a triangle by a straight line parallel to the base. The missing propositions about the division of a circle are also here: 'to divide into two equal parts a given figure bounded by an arc of a circle and two straight lines including a given angle' (28), and 'to draw in a given circle two parallel straight lines cutting off a certain fraction from the circle' (29). Unfortunately the proofs are given of only four propositions out of 36, namely Propositions 19, 20, 28, 29, the Arabic translator having found the rest too easy and omitted them. But the genuineness of the treatise edited by Woepcke is attested by the facts that the four proofs which remain are elegant and depend on propositions in the *Elements*, and that there is a lemma with a true Greek ring, 'to apply to a straight line a rectangle equal to the rectangle contained by AB, AC and deficient by a square' (18). Moreover, the treatise is no fragment, but ends with the words, 'end of the treatise', and is (but for the missing proofs) a well-ordered and compact whole. Hence we may safely conclude that Woepcke's tract represents not only Euclid's work but the whole of it. The portion of the *Practica geometriae* of Leonardo of Pisa which deals with the division of figures seems to be a restoration and extension of Euclid's work; Leonardo must presumably have come across a version of it from the Arabic.

The type of problem which Euclid's treatise was designed to solve may be stated in general terms as that of dividing a given figure by one or more straight lines into parts having prescribed ratios to one another or to other given areas. The figures divided are the triangle, the parallelogram, the trapezium, the quadrilateral, a figure bounded by an arc of a circle and two straight lines, and the circle. The figures are divided into two equal parts, or two parts in a given ratio; or again, a given fraction of the figure is to be cut off, or the figure is to be divided into several parts in given ratios. The dividing straight lines may be transversals drawn through a point situated at a vertex of the figure, or a point on any side, on one of two parallel sides, in the interior of the figure, outside the figure, and so on; or again, they may be merely parallel lines, or lines parallel to a base. The treatise also includes auxiliary propositions, (1) 'to apply to a given straight line a rectangle equal to a given area and deficient by a square', the proposition already mentioned, which is equivalent to the algebraical solution of the equation $ax-x^2 = b^2$ and depends on Eucl. II. 5 (cf. p. 152 above); (2) propositions in proportion involving unequal instead of equal ratios:

If $a \,.\, d >$ or $< b \,.\, c$, then $a:b>$ or $< c:d$ respectively.

If $a:b > c:d$, then $(a\mp b):b > (c\mp d):d$.

If $a:b < c:d$, then $(a-b):b < (c-d):d$.

By way of illustration I will set out shortly three propositions from the Woepcke text.

(1) Propositions 19, 20 (slightly generalized): To cut off a certain fraction (m/n) from a given triangle by a straight

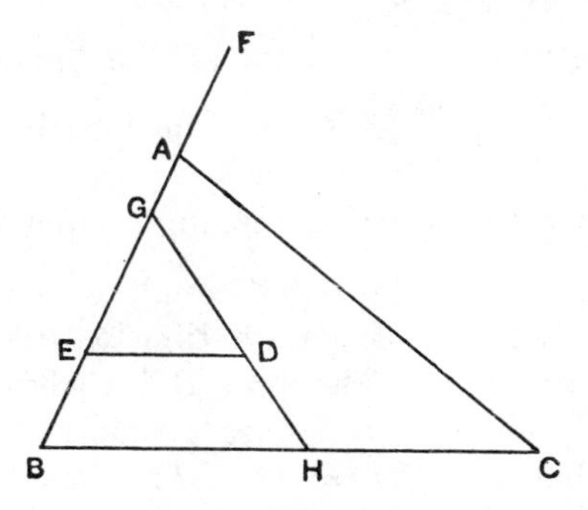

line drawn through a given point within the triangle (Euclid gives two cases corresponding to $m/n = \frac{1}{2}$ and $m/n = \frac{1}{3}$).

The construction will be best understood if we work out the analysis of the problem (not given by Euclid).

Suppose that ABC is the given triangle, D the given

internal point; and suppose the problem solved, i.e. GH drawn through D in such a way that $\Delta\, GBH = \frac{m}{n} \cdot \Delta\, ABC$.

Therefore $GB \,.\, BH = \frac{m}{n} \,.\, AB \,.\, BC$. (This is assumed by Euclid.)

Now suppose that the unknown quantity is $GB = x$, say.
Draw DE parallel to BC; then DE, EB are given.

Now $BH : DE = GB : GE = x : (x - BE)$,

or $$BH = \frac{x \,.\, DE}{x - BE};$$

therefore $$GB \,.\, BH = x^2 \,.\, \frac{DE}{x - BE}.$$

And, by hypothesis, $GB \,.\, BH = \frac{m}{n} \,.\, AB \,.\, BC$;

therefore $$x^2 = \frac{m}{n} \cdot \frac{AB \,.\, BC}{DE}(x - BE),$$

or, if $k = \frac{m}{n} \cdot \frac{AB \,.\, BC}{DE}$, we have to solve the equation

$$x^2 = k\,(x - BE),$$

or $$kx - x^2 = k \,.\, BE.$$

This is exactly what Euclid does; he first finds F on BA such that $BF \,.\, DE = \frac{m}{n} \cdot AB \,.\, BC$ (the length of BF is determined by applying to DE a rectangle equal to $\frac{m}{n} \,.\, AB \,.\, BC$, Eucl. I. 45), that is, he finds BF equal to k. Then he gives the geometrical solution of the equation $kx - x^2 = k \,.\, BE$ in the form 'apply to the straight line BF a rectangle equal to $BF \,.\, BE$ and deficient by a square'; that is to say, he determines G so that $BG \,.\, GF = BF \,.\, BE$. We have then only to join GD and produce it to H; and GH cuts off the required triangle.

(The problem is subject to a διορισμός which Euclid does not give, but which is easily supplied.)

(2) Proposition 28: To divide into two equal parts a given

figure bounded by an arc of a circle and by two straight lines which form a given angle.

Let $ABEC$ be the given figure, D the middle point of BC, and DE perpendicular to BC. Join AD.

Then the broken line ADE clearly divides the figure into two equal parts. Join AE, and draw DF parallel to it meeting BA in F. Join FE.

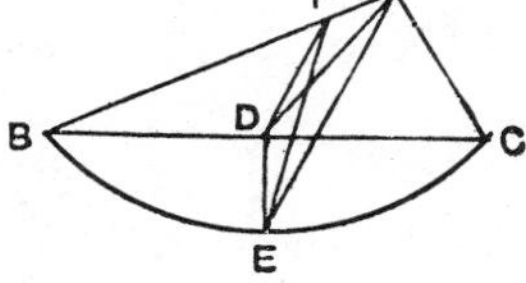

The triangles AFE, ADE are then equal, being in the same parallels. Add to each the area AEC.

Therefore the area $AFEC$ is equal to the area $ADEC$, and therefore to half the area of the given figure.

(3) Proposition 29: To draw in a given circle two parallel chords cutting off a certain fraction (m/n) of the circle.

(The fraction m/n must be such that we can, by plane methods, draw a chord cutting off m/n of the circumference of the circle; Euclid takes the case where $m/n = \frac{1}{3}$.)

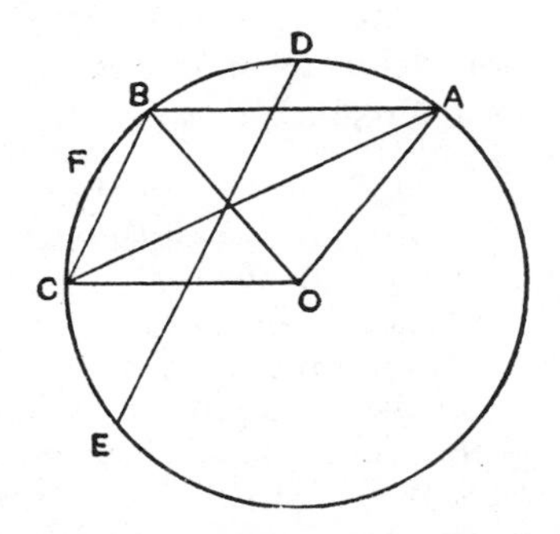

Suppose that the arc ADB is m/n of the circumference of the circle. Join A, B to the centre O. Draw OC parallel to AB and join AC, BC. From D, the middle point of the arc AB, draw the chord DE parallel to BC. Then shall BC, DE cut off m/n of the area of the circle.

Since AB, OC are parallel,

$$\triangle AOB = \triangle ACB.$$

Add to each the segment ADB;

therefore

$$(\text{sector } ADBO) = \text{figure bounded by } AC,\ CB \text{ and arc } ADB$$

$$= (\text{segmt. } ABC) - (\text{segmt. } BFC).$$

Since BC, DE are parallel, (arc DB) = (arc CE);

therefore

(arc ABC) = (arc DCE), and (segmt. ABC) = (segmt. DCE);

therefore (sector $ADBO$), or $\frac{m}{n}$ (circle ABC)

= (segmt. DCE) − (segmt. BFC).

That is BC, DE cut off an area equal to $\frac{m}{n}$ (circle ABC).

Lost geometrical works.

(α) The *Pseudaria*.

The other purely geometrical works of Euclid are lost so far as is known at present. One of these again belongs to the domain of elementary geometry. This is the *Pseudaria*, or 'Book of Fallacies', as it is called by Proclus, which is clearly the same work as the 'Pseudographemata' of Euclid mentioned by a commentator on Aristotle in terms which agree with Proclus's description.[1] Proclus says of Euclid that,

'Inasmuch as many things, while appearing to rest on truth and to follow from scientific principles, really tend to lead one astray from the principles and deceive the more superficial minds, he has handed down methods for the discriminative understanding of these things as well, by the use of which methods we shall be able to give beginners in this study practice in the discovery of paralogisms, and to avoid being ourselves misled. The treatise by which he puts this machinery in our hands he entitled (the book) of Pseudaria, enumerating in order their various kinds, exercising our intelligence in each case by theorems of all sorts, setting the true side by side with the false, and combining the refutation of error with practical illustration. This book then is by way of cathartic and exercise, while the Elements contain the irrefragable and complete guide to the actual scientific investigation of the subjects of geometry.'[2]

The connexion of the book with the *Elements* and the reference to its usefulness for beginners show that it did not go beyond the limits of elementary geometry.

[1] Michael Ephesius, *Comm. on Arist. Soph. El.*, fol. 25v, p. 76. 23 Wallies.

[2] Proclus on Eucl. I, p. 70. 1–18. Cf. a scholium to Plato's *Theaetetus* 191 B, which says that the fallacies did not arise through any importation of sense-perception into the domain of non-sensibles.

We now come to the lost works belonging to higher geometry. The most important was evidently

(β) The *Porisms*.

Our only source of information about the nature and contents of the *Porisms* is Pappus. In his general preface about the books composing the *Treasury of Analysis* Pappus writes as follows[1] (I put in square brackets the words bracketed by Hultsch).

'After the Tangencies (of Apollonius) come, in three Books, the Porisms of Euclid, a collection [in the view of many] most ingeniously devised for the analysis of the more weighty problems, [and] although nature presents an unlimited number of such porisms, [they have added nothing to what was originally written by Euclid, except that some before my time have shown their want of taste by adding to a few (of the propositions) second proofs, each (proposition) admitting of a definite number of demonstrations, as we have shown, and Euclid having given one for each, namely that which is the most lucid. These porisms embody a theory subtle, natural, necessary, and of considerable generality, which is fascinating to those who can see and produce results].

'Now all the varieties of porisms belong, neither to theorems nor problems, but to a species occupying a sort of intermediate position [so that their enunciations can be formed like those of either theorems or problems], the result being that, of the great number of geometers, some regarded them as of the class of theorems, and others of problems, looking only to the form of the proposition. But that the ancients knew better the difference between these three things is clear from the definitions. For they said that a theorem is that which is proposed with a view to the demonstration of the very thing proposed, a problem that which is thrown out with a view to the construction of the very thing proposed, and a porism that which is proposed with a view to the producing of the very thing proposed. [But this definition of the porism was changed by the more recent writers who could not produce everything, but used these elements and proved only the fact that that which is sought really exists, but did not produce it, and were accordingly confuted by the definition and the whole doctrine. They based their definition on an incidental characteristic, thus: A porism is that which falls short of a locus-theorem in

[1] Pappus, vii, pp. 648–60.

respect of its hypothesis. Of this kind of porisms loci are a species, and they abound in the Treasury of Analysis; but this species has been collected, named, and handed down separately from the porisms, because it is more widely diffused than the other species] . . . But it has further become characteristic of porisms that, owing to their complication, the enunciations are put in a contracted form, much being by usage left to be understood; so that many geometers understand them only in a partial way and are ignorant of the more essential features of their content.

'[Now to comprehend a number of propositions in one enunciation is by no means easy in these porisms, because Euclid himself has not in fact given many of each species, but chosen, for examples, one or a few out of a great multitude. But at the beginning of the first book he has given some propositions, to the number of ten, of one species, namely that more fruitful species consisting of loci.] Consequently, finding that these admitted of being comprehended in our enunciation, we have set it out thus:

> If, in a system of four straight lines which cut one another two and two, three points on one straight line be given, while the rest except one lie on different straight lines given in position, the remaining point also will lie on a straight line given in position.

'This has only been enunciated of four straight lines, of which not more than two pass through the same point, but it is not known (to most people) that it is true of any assigned number of straight lines if enunciated thus:

> If any number of straight lines cut one another, not more than two (passing) through the same point, and all the points (of intersection situated) on one of them be given, and if each of those which are on another (of them) lie on a straight line given in position—

or still more generally thus:

> if any number of straight lines cut one another, not more than two (passing) through the same point, and all the points (of intersection situated) on one of them be given, while of the other points of intersection in multitude equal to a triangular number a number corresponding to the side of this triangular number lie respectively on straight lines given in position, provided that of these latter points no three are at the angular points of a triangle (sc. having for sides three of the given straight

lines)—each of the remaining points will lie on a straight line given in position.[1]

'It is probable that the writer of the Elements was not unaware of this, but that he only set out the principle; and he seems, in the case of all the porisms, to have laid down the principles and the seed only [of many important things], the kinds of which should be distinguished according to the differences, not of their hypotheses, but of the results and the things sought. [All the hypotheses are different from one another because they are entirely special, but each of the results and things sought, being one and the same, follow from many different hypotheses.]

'We must then in the first book distinguish the following kinds of things sought:

'At the beginning of the book is this proposition:

> I. *If from two given points straight lines be drawn meeting on a straight line given in position, and one cut off from a straight line given in position* (*a segment measured*) *to a given point on it, the other will also cut off from another* (*straight line a segment*) *having to the first a given ratio.*

'Following on this (we have to prove)

> II. that such and such a point lies on a straight line given in position;

> III. that the ratio of such and such a pair of straight lines is given';

&c. &c. (up to XXIX).

'The three books of the porisms contain 38 lemmas; of the theorems themselves there are 171.'

Pappus further gives lemmas to the *Porisms*.[2]

With Pappus's account of Porisms must be compared the passages of Proclus on the same subject. Proclus distinguishes

[1] Loria (*Le scienze esatte nell'antica Grecia*, pp. 256-7) gives the meaning of this as follows, pointing out that Simson first discovered it: 'If a complete n-lateral be deformed so that its sides respectively turn about n points on a straight line, and $(n-1)$ of its $\frac{1}{2}n(n-1)$ vertices move on as many straight lines, the other $\frac{1}{2}(n-1)(n-2)$ of its vertices likewise move on as many straight lines: but it is necessary that it should be impossible to form with the $(n-1)$ vertices any triangle having for sides the sides of the polygon.'

[2] Pappus, vii, pp. 866-918; Euclid, ed. Heiberg-Menge, vol. viii, pp. 243-74.

the two senses of the word πόρισμα. The first is that of a *corollary,* where something appears as an incidental result of a proposition, obtained without trouble or special seeking, a sort of bonus which the investigation has presented us with.[1] The other sense is that of Euclid's *Porisms.* In this sense

'*porism* is the name given to things which are sought, but need some finding and are neither pure bringing into existence nor simple theoretic argument. For (to prove) that the angles at the base of isosceles triangles are equal is matter of theoretic argument, and it is with reference to things existing that such knowledge is (obtained). But to bisect an angle, to construct a triangle, to cut off, or to place—all these things demand the making of something; and to find the centre of a given circle, or to find the greatest common measure of two given commensurable magnitudes, or the like, is in some sort intermediate between theorems and problems. For in these cases there is no bringing into existence of the things sought, but finding of them; nor is the procedure purely theoretic. For it is necessary to bring what is sought into view and exhibit it to the eye. Such are the porisms which Euclid wrote and arranged in three books of Porisms.'[2]

Proclus's definition thus agrees well enough with the first, the 'older', definition of Pappus. A porism occupies a place between a theorem and a problem; it deals with something already existing, as a theorem does, but has to *find* it (e.g. the centre of a circle), and, as a certain operation is therefore necessary, it partakes to that extent of the nature of a problem, which requires us to construct or produce something not previously existing. Thus, besides III. 1 and X. 3, 4 of the *Elements* mentioned by Proclus, the following propositions are real porisms: III. 25, VI. 11–13, VII. 33, 34, 36, 39, VIII. 2, 4, X. 10, XIII. 18. Similarly, in Archimedes's *On the Sphere and Cylinder,* I. 2–6 might be called porisms.

The enunciation given by Pappus as comprehending ten of Euclid's propositions may not reproduce the *form* of Euclid's enunciations; but, comparing the result to be proved, that certain points lie on straight lines given in position, with the *class* indicated by II above, where the question is of such and such a point lying on a straight line given in position, and

[1] Proclus on Eucl. I, pp. 212. 14; 301. 22.

[2] *Ib.*, p. 301. 25 sq.

with other classes, e.g. (V) that such and such a line is given in position, (VI) that such and such a line verges to a given point, (XXVII) that there exists a given point such that straight lines drawn from it to such and such (circles) will contain a triangle given in species, we may conclude that a usual form of a porism was 'to prove that it is possible to find a point with such and such a property' or 'a straight line on which lie all the points satisfying given conditions', and so on.

The above exhausts all the positive information which we have about the nature of a porism and the contents of Euclid's *Porisms*. It is obscure and leaves great scope for speculation and controversy; naturally, therefore, the problem of restoring the *Porisms* has had a great fascination for distinguished mathematicians ever since the revival of learning. But it has proved beyond them all. Some contributions to a solution have, it is true, been made, mainly by Simson and Chasles. The first claim to have restored the *Porisms* seems to be that of Albert Girard (about 1590–1633), who spoke (1626) of an early publication of his results, which, however, never saw the light. The great Fermat (1601–65) gave his idea of a 'porism', illustrating it by five examples which are very interesting in themselves[1]; but he did not succeed in connecting them with the description of Euclid's *Porisms* by Pappus, and, though he expressed a hope of being able to produce a complete restoration of the latter, his hope was not realized. It was left for Robert Simson (1687–1768) to make the first decisive step towards the solution of the problem.[2] He succeeded in explaining the meaning of the actual porisms enunciated in such general terms by Pappus. In his tract on Porisms he proves the first porism given by Pappus in its ten different cases, which, according to Pappus, Euclid distinguished (these propositions are of the class connected with *loci*); after this he gives a number of other propositions from Pappus, some auxiliary propositions, and some 29 'porisms', some of which are meant to illustrate the classes I, VI, XV, XXVII–XXIX distinguished by Pappus. Simson was able to evolve a definition of a porism which is perhaps more easily understood in Chasles's translation: 'Le porisme est une proposition dans

[1] *Œuvres de Fermat*, ed. Tannery and Henry, I, p. 76–84.
[2] Roberti Simson *Opera quaedam reliqua*, 1776, pp. 315–594.

laquelle on demande de démontrer qu'une chose ou plusieurs choses sont *données*, qui, ainsi que l'une quelconque d'une infinité d'autres choses non données, mais dont chacune est avec des choses données dans une même relation, ont une propriété commune, décrite dans la proposition.' We need not follow Simson's English or Scottish successors, Lawson (1777), Playfair (1794), W. Wallace (1798), Lord Brougham (1798), in their further speculations, nor the controversies between the Frenchmen, A. J. H. Vincent and P. Breton (de Champ), nor the latter's claim to priority as against Chasles; the work of Chasles himself (*Les trois livres des Porismes d'Euclide rétablis* . . . Paris, 1860) alone needs to be mentioned. Chasles adopted the definition of a porism given by Simson, but showed how it could be expressed in a different form. 'Porisms are incomplete theorems which express certain relations existing between things variable in accordance with a common law, relations which are indicated in the enunciation of the porism, but which need to be completed by determining the magnitude or position of certain things which are the consequences of the hypotheses and which would be determined in the enunciation of a theorem properly so called or a complete theorem.' Chasles succeeded in elucidating the connexion between a porism and a locus as described by Pappus, though he gave an inexact translation of the actual words of Pappus: '*Ce qui constitue le porisme est ce qui manque à l'hypothèse d'un théorème local* (en d'autres termes, le porisme est inférieur, par l'hypothèse, au théorème local; c'est à dire que quand quelques parties d'une proposition locale n'ont pas dans l'énoncé la détermination qui leur est propre, cette proposition cesse d'être regardée comme un théorème et devient un porisme)'; here the words italicized are not quite what Pappus said, viz. that 'a porism is that which falls short of a locus-theorem in respect of its hypothesis', but the explanation in brackets is correct enough if we substitute 'in respect of' for 'par' ('by'). The work of Chasles is historically important because it was in the course of his researches on this subject that he was led to the idea of anharmonic ratios; and he was probably right in thinking that the *Porisms* were propositions belonging to the modern theory of transversals and to projective geometry. But, as a restoration of Euclid's work, Chasles's Porisms cannot be regarded as satisfactory. One consideration alone is, to my mind, conclusive on this point. Chasles made 'porisms' out of Pappus's various *lemmas* to Euclid's porisms and comparatively easy deductions from those lemmas. Now we have experience of Pappus's lemmas to books which still survive, e.g. the *Conics* of Apollonius; and, to judge by these instances, his lemmas stood in a most ancillary relation to the propositions to which they relate, and do not in the least compare with them in difficulty and importance. Hence it is all but impossible to believe that the lemmas to the porisms were themselves porisms such as were Euclid's own porisms; on the contrary, the analogy of Pappus's other sets of lemmas makes it all but necessary to regard the lemmas in question as merely supplying proofs of simple propositions assumed by Euclid without proof in the course of the demonstration of the actual porisms. This being so, it appears that the problem of the complete restoration of Euclid's three Books still awaits a solution, or rather that it will never be solved unless in the event of discovery of fresh documents.

At the same time the lemmas of Pappus to the *Porisms* are by no means insignificant propositions in themselves, and, if the usual relation of lemmas to substantive propositions holds, it follows that the *Porisms* was a distinctly advanced work, perhaps the most important that Euclid ever wrote; its loss is therefore much to be deplored. Zeuthen has an interesting remark à propos of the proposition which Pappus quotes as the first proposition of Book I, 'If from two given points straight lines be drawn meeting on a straight line given in position, and one of them cut off from a straight line given in position (a segment measured) towards a given point on it, the other will also cut off from another (straight line a segment) bearing to the first a given ratio.' This proposition is also true if there be substituted for the first given straight line a conic regarded as the 'locus with respect to four lines', and the proposition so extended can be used for completing Apollonius's exposition of that locus. Zeuthen suggests, on this ground, that the *Porisms* were in part by-products of the theory of conics and in part auxiliary means for the study of conics, and that Euclid called them by the same name as that applied to corollaries because they were corollaries with respect to conics.[1] This, however, is a pure conjecture.

(γ) The *Conics*.

Pappus says of this lost work: 'The four books of Euclid's Conics were completed by Apollonius, who added four more and gave us eight books of Conics.'[2] It is probable that Euclid's work was already lost by Pappus's time, for he goes on to speak of 'Aristaeus who wrote the *still extant* five books of Solid Loci *συνεχῆ τοῖς κωνικοῖς*, connected with, or supplementary to, the conics'.[3] This latter work seems to have been a treatise on conics regarded as loci; for 'solid loci' was a term appropriated to conics, as distinct from 'plane loci', which were straight lines and circles. In another passage Pappus (or an interpolator) speaks of the 'conics' of Aristaeus the 'elder',[4] evidently referring to the same book. Euclid no doubt wrote on the general theory of conics, as Apollonius did, but only covered the ground of Apollonius's first three books, since Apollonius says that no one before him had touched the subject of Book IV (which, however, is not important). As in the case of the *Elements*, Euclid would naturally collect and rearrange, in a systematic exposition, all that had been discovered up to date in the theory of conics. That Euclid's treatise covered most of the essentials up to the last part of Apollonius's Book III seems clear from the fact that Apollonius only claims originality for some propositions connected with the 'three- and four-line locus', observing that Euclid had not completely worked out the synthesis of the said locus, which, indeed, was not possible without the propositions referred to. Pappus (or an interpolator)[5] excuses Euclid on the ground that he made no claim to go beyond the discoveries of Aristaeus, but only wrote so much about the locus as was possible with the aid of Aristaeus's conics. We may conclude that Aristaeus's book preceded Euclid's, and that it was, at least in point of originality, more important. When Archimedes refers to propositions in conics as having been proved

[1] Zeuthen, *Die Lehre von den Kegelschnitten im Altertum*, 1886, pp. 168, 173-4.

[2] Pappus, vii, p. 672. 18.

[3] Cf. Pappus, vii, p. 636. 23.

[4] *Ib.* vii, p. 672. 12.

[5] *Ib.* vii, pp. 676. 25-678. 6.

in the 'elements of conics', he clearly refers to these two treatises, and the other propositions to which he refers as well known and not needing proof were doubtless taken from the same sources. Euclid still used the old names for the conic sections (sections of a right-angled, acute-angled, and obtuse-angled cone respectively), but he was aware that an ellipse could be obtained by cutting (through) a cone in any manner by a plane not parallel to the base, and also by cutting a cylinder; this is clear from a sentence in his *Phaenomena* to the effect that, 'If a cone or a cylinder be cut by a plane not parallel to the base, this section is a section of an acute-angled cone, which is like a shield (θυρεός).'

(δ) The *Surface-Loci* (τόποι πρὸς ἐπιφανείᾳ).

Like the *Data* and the *Porisms*, this treatise in two Books is mentioned by Pappus as belonging to the *Treasury of Analysis*. What is meant by surface-loci, literally 'loci on a surface' is not entirely clear, but we are able to form a conjecture on the subject by means of remarks in Proclus and Pappus. The former says (1) that a locus is 'a position of a line or of a surface which has (throughout it) one and the same property',[1] and (2) that 'of locus-theorems some are constructed on lines and others on surfaces'[2]; the effect of these statements together seems to be that 'loci on lines' are loci which *are* lines, and 'loci on surfaces' loci which *are* surfaces. On the other hand, the possibility does not seem to be excluded that loci on surfaces may be loci *traced* on surfaces; for Pappus says in one place that the equivalent of the *quadratrix* can be got geometrically 'by means of loci on surfaces as follows'[3] and then proceeds to use a spiral described on a cylinder (the cylindrical helix), and it is consistent with this that in another passage[4] (bracketed, however, by Hultsch) 'linear' loci are said to be exhibited (δείκνυνται) or realized from loci on surfaces, for the quadratrix is a 'linear' locus, i.e. a locus of an order higher than a plane locus (a straight line or circle) and a 'solid' locus (a conic). However this may be, Euclid's *Surface-Loci* probably included

[1] Proclus on Eucl. I, p. 394. 17.
[2] *Ib.*, p. 394. 19.
[3] Pappus, iv, p. 258. 20–25.
[4] *Ib.* vii. 662. 9.

such loci as were cones, cylinders and spheres. The two lemmas given by Pappus lend some colour to this view. The first of these[1] and the figure attached to it are unsatisfactory as they stand, but Tannery indicated a possible restoration.[2] If this is right, it suggests that one of the loci contained all the points on the elliptical parallel sections of a cylinder, and was therefore an oblique circular cylinder. Other assumptions with regard to the conditions to which the lines in the figure may be subject would suggest that other loci dealt with were cones regarded as containing all points on particular parallel elliptical sections of the cones. In the second lemma Pappus states and gives a complete proof of the focus-and-directrix property of a conic, viz. that *the locus of a point the distance of which from a given point is in a given ratio to its distance from a fixed straight line is a conic section, which is an ellipse, a parabola or a hyperbola according as the given ratio is less than, equal to, or greater than unity.*[3] Two conjectures are possible as to the application of this theorem in Euclid's *Surface-Loci.* (*a*) It may have been used to prove that the locus of a point the distance of which from a given straight line is in a given ratio to its distance from a given plane is a certain cone. Or (*b*) it may have been used to prove that the locus of a point the distance of which from a given point is in a given ratio to its distance from a given plane is the surface formed by the revolution of a conic about its major or conjugate axis.[4]

We come now to Euclid's works under the head of

Applied mathematics.

(*a*) The *Phaenomena*.

The book on *sphaeric* intended for use in astronomy and entitled *Phaenomena* has already been noticed (pp. 349, 351–2). It is extant in Greek and was included in Gregory's edition of Euclid. The text of Gregory, however, represents the later of two recensions which differ considerably (especially in Propositions 9 to 16). The best manuscript of this later recension (b) is the famous Vat. gr. 204 of the tenth century,

[1] Pappus, vii, p. 1004. 17; Euclid, ed. Heiberg-Menge, vol. viii, p. 274.
[2] Tannery in *Bulletin des sciences mathématiques*, 2e série, VI, p. 149.
[3] Pappus, vii, pp. 1004. 23–1014; Euclid, vol. viii, pp. 275–81.
[4] For further details, see *The Works of Archimedes*, pp. lxii–lxv.

while the best manuscript of the older and better version (**a**) is the Viennese MS. Vind. gr. XXXI. 13 of the twelfth century. A new text edited by Menge and taking account of both recensions is now available in the last volume of the Heiberg-Menge edition of Euclid.[1]

(β) *Optics* and *Catoptrica*.

The *Optics*, a treatise included by Pappus in the collection of works known as the Little Astronomy, survives in two forms. One is the recension of Theon translated by Zambertus in 1505; the Greek text was first edited by Johannes Pena (de la Pène) in 1557, and this form of the treatise was alone included in the editions up to Gregory's. But Heiberg discovered the earlier form in two manuscripts, one at Vienna (Vind. gr. XXXI. 13) and one at Florence (Laurent. XXVIII. 3), and both recensions are contained in vol. vii of the Heiberg-Menge text of Euclid (Teubner, 1895). There is no reason to doubt that the earlier recension is Euclid's own work; the style is much more like that of the *Elements*, and the proofs of the propositions are more complete and clear. The later recension is further differentiated by a preface of some length, which is said by a scholiast to be taken from the commentary or elucidation by Theon. It would appear that the text of this recension is Theon's, and that the preface was a reproduction by a pupil of what was explained by Theon in lectures. It cannot have been written much, if anything, later than Theon's time, for it is quoted by Nemesius about A.D. 400. Only the earlier and genuine version need concern us here. It is a kind of elementary treatise on perspective, and it may have been intended to forearm students of astronomy against paradoxical theories such as those of the Epicureans, who maintained that the heavenly bodies *are* of the size that they *look*. It begins in the orthodox fashion with Definitions, the first of which embodies the same idea of the process of vision as we find in Plato, namely that it is due to rays proceeding from our eyes and impinging upon the object, instead of the other way about: 'the straight lines (rays) which issue from the eye traverse the distances (or dimensions) of great

[1] *Euclidis Phaenomena et scripta Musica* edidit Henricus Menge. *Fragmenta* collegit et disposuit J. L. Heiberg, Teubner, 1916.

magnitudes'; Def. 2: 'The figure contained by the visual rays is a cone which has its vertex in the eye, and its base at the extremities of the objects seen'; Def. 3: 'And those things are seen on which the visual rays impinge, while those are not seen on which they do not'; Def. 4: 'Things seen under a greater angle appear greater, and those under a lesser angle less, while things seen under equal angles appear equal'; Def. 7: 'Things seen under more angles appear more distinctly.' Euclid assumed that the visual rays are not 'continuous', i.e. not absolutely close together, but are separated by a certain distance, and hence he concluded, in Proposition 1, that we can never really see the whole of any object, though we seem to do so. Apart, however, from such inferences as these from false hypotheses, there is much in the treatise that is sound. Euclid has the essential truth that the rays are straight; and it makes no difference geometrically whether they proceed from the eye or the object. Then, after propositions explaining the differences in the apparent size of an object according to its position relatively to the eye, he proves that the apparent sizes of two equal and parallel objects are not proportional to their distances from the eye (Prop. 8); in this proposition he proves the equivalent of the fact that, if α, β are two angles and $\alpha < \beta < \frac{1}{2}\pi$, then

$$\frac{\tan\alpha}{\tan\beta} < \frac{\alpha}{\beta},$$

the equivalent of which, as well as of the corresponding formula with sines, is assumed without proof by Aristarchus a little later. From Proposition 6 can easily be deduced the fundamental proposition in perspective that parallel lines (regarded as equidistant throughout) appear to meet. There are four simple propositions in heights and distances, e.g. to find the height of an object (1) when the sun is shining (Prop. 18), (2) when it is not (Prop. 19): similar triangles are, of course, used and the horizontal mirror appears in the second case in the orthodox manner, with the assumption that the angles of incidence and reflection of a ray are equal, 'as is explained in the Catoptrica (or theory of mirrors)'. Propositions 23–7 prove that, if an eye sees a sphere, it sees less than half of the sphere, and the contour of what is seen appears to be a circle; if the eye approaches nearer to the sphere the portion seen becomes less, though it appears greater; if we see the sphere with two eyes, we see a hemisphere, or more than a hemisphere, or less than a hemisphere according as the distance between the eyes is equal to, greater than, or less than the diameter of the sphere; these propositions are comparable with Aristarchus's Proposition 2 stating that, if a sphere be illuminated by a larger sphere, the illuminated portion of the former will be greater than a hemisphere. Similar propositions with regard to the cylinder and cone follow (Props. 28–33). Next Euclid considers the conditions for the apparent equality of different diameters of a circle as seen from an eye occupying various positions outside the plane of the circle (Props. 34–7); he shows that all diameters will appear equal, or the circle will really look like a circle, if the line joining the eye to the centre is perpendicular to the plane of the circle, *or*, not being perpendicular to that plane, is equal to the length of the radius, but this will not otherwise be the case (35), so that (36) a chariot wheel will sometimes appear circular, sometimes awry, according to the position of the eye. Propositions 37 and 38 prove, the one that there is a locus such that, if the eye remains at one point of it, while a straight line moves so that its extremities always lie on it, the line will always *appear* of the same length in whatever position it is placed (not being one in which either of the extremities coincides with, or the extremities are on opposite sides of, the point at which the eye is placed), the locus being, of course, a circle in which the straight line is placed as a chord, when it necessarily subtends the same angle at the circumference or at the centre, and therefore at the eye, if placed at a point of the circumference or at the centre; the other proves the same thing for the case where the line is fixed with its extremities on the locus, while the eye moves upon it. The same idea underlies several other propositions, e.g. Proposition 45, which proves that a common point can be found from which unequal magnitudes will appear equal. The unequal magnitudes are straight lines BC, CD so placed that BCD is a straight line. A segment greater than a semicircle is described on BC, and a similar segment on CD. The segments will then intersect at F, and the angles subtended by BC and CD at F are equal. The rest of the treatise is of the same character, and it need not be further described.

The *Catoptrica* published by Heiberg in the same volume is not by Euclid, but is a compilation made at a much later date, possibly by Theon of Alexandria, from ancient works on the subject and mainly no doubt from those of Archimedes and Heron. Theon[1] himself quotes a *Catoptrica* by Archimedes, and Olympiodorus[2] quotes Archimedes as having proved the fact which appears as an axiom in the *Catoptrica* now in question, namely that, if an object be placed just out of sight at the bottom of a vessel, it will become visible over the edge when water is poured in. It is not even certain that Euclid wrote *Catoptrica* at all, since, if the treatise was Theon's, Proclus may have assigned it to Euclid through inadvertence.

(γ) *Music.*

Proclus attributes to Euclid a work on the *Elements of Music* (*αἱ κατὰ μουσικὴν στοιχειώσεις*)[3]; so does Marinus.[4] As a matter of fact, two musical treatises attributed to Euclid are still extant, the *Sectio Canonis* (Κατατομὴ κανόνος) and the *Introductio harmonica* (Εἰσαγωγὴ ἁρμονική). The latter, however, is certainly not by Euclid, but by Cleonides, a pupil of Aristoxenus. The question remains, in what relation does the *Sectio Canonis* stand to the 'Elements' mentioned by Proclus and Marinus? The *Sectio* gives the Pythagorean theory of music, but is altogether too partial and slight to deserve the title 'Elements of Music'. Jan, the editor of the *Musici Graeci*, thought that the *Sectio* was a sort of summary account extracted from the 'Elements' by Euclid himself, which hardly seems likely; he maintained that it is the genuine work of Euclid on the grounds (1) that the style and diction and the form of the propositions agree well with what we find in Euclid's *Elements*, and (2) that Porphyry in his commentary on Ptolemy's *Harmonica* thrice quotes Euclid as the author of a *Sectio Canonis*.[5] The latest editor, Menge,

[1] Theon, *Comm. on Ptolemy's Syntaxis*, i, p. 10.
[2] *Comment. on Arist. Meteorolog.* ii, p. 94, Ideler, p. 211. 18 Busse.
[3] Proclus on Eucl. I, p. 69. 3.
[4] Marinus, *Comm. on the Data* (Euclid, vol. vi, p. 254. 19).
[5] See Wallis, *Opera mathematica*, vol. iii, 1699, pp. 267, 269, 272.

points out that the extract given by Porphyry shows some differences from our text and contains some things quite unworthy of Euclid; hence he is inclined to think that the work as we have it is not actually by Euclid, but was extracted by some other author of less ability from the genuine 'Elements of Music' by Euclid.

(δ) Works on mechanics attributed to Euclid.

The Arabian list of Euclid's works further includes among those held to be genuine 'the book of the Heavy and Light'. This is apparently the tract *De levi et ponderoso* included by Hervagius in the Basel Latin translation of 1537 and by Gregory in his edition. That it comes from the Greek is made clear by the lettering of the figures; and this is confirmed by the fact that another, very slightly different, version exists at Dresden (Cod. Dresdensis Db. 86), which is evidently a version of an Arabic translation from the Greek, since the lettering of the figures follows the order characteristic of such Arabic translations, *a, b, g, d, e, z, h, t*. The tract consists of nine definitions or axioms and five propositions. Among the definitions are these: Bodies are equal, different, or greater in size according as they occupy equal, different, or greater spaces (1–3). Bodies are equal in *power* or in *virtue* which move over equal distances in the same medium of air or water in equal times (4), while the *power* or *virtue* is greater if the motion takes less time, and less if it takes more (6). Bodies are *of the same kind* if, being equal in size, they are also equal in *power* when the medium is the same; they are different in kind when, being equal in size, they are not equal in *power* or *virtue* (7, 8). Of bodies different in kind, that has more *power* which is more dense (*solidius*) (9). With these hypotheses, the author attempts to prove (Props. 1, 3, 5) that, of bodies which traverse unequal spaces in equal times, that which traverses the greater space has the greater *power* and that, of bodies of the same kind, the *power* is proportional to the size, and conversely, if the *power* is proportional to the size, the bodies are of the same kind. We recognize in the *potentia* or *virtus* the same thing as the δύναμις and ἰσχύς of Aristotle.[1] The

[1] Aristotle, *Physics*, Z. 5.

property assigned by the author to bodies *of the same kind* is quite different from what we attribute to bodies of the same specific gravity; he purports to prove that bodies of the same kind have *power* proportional to their size, and the effect of this, combined with the definitions, is that they move at speeds proportional to their volumes. Thus the tract is the most precise statement that we possess of the principle of Aristotle's dynamics, a principle which persisted until Benedetti (1530–90) and Galilei (1564–1642) proved its falsity.

There are yet other fragments on mechanics associated with the name of Euclid. One is a tract translated by Woepcke from the Arabic in 1851 under the title 'Le livre d'Euclide sur la balance', a work which, although spoiled by some commentator, seems to go back to a Greek original and to have been an attempt to establish a theory of the lever, not from a general principle of dynamics like that of Aristotle, but from a few simple axioms such as the experience of daily life might suggest. The original work may have been earlier than Archimedes and may have been written by a contemporary of Euclid. A third fragment, unearthed by Duhem from manuscripts in the Bibliothèque Nationale in Paris, contains four propositions purporting to be 'liber Euclidis de ponderibus secundum terminorum circumferentiam'. The first of the propositions, connecting the law of the lever with the size of the circles described by its ends, recalls the similar demonstration in the Aristotelian *Mechanica*; the others attempt to give a theory of the balance, taking account of the weight of the lever itself, and assuming that a portion of it (regarded as cylindrical) may be supposed to be detached and replaced by an equal weight suspended from its middle point. The three fragments supplement each other in a curious way, and it is a question whether they belonged to one treatise or were due to different authors. In any case there seems to be no independent evidence that Euclid was the author of any of the fragments, or that he wrote on mechanics at all.[1]

[1] For further details about these mechanical fragments see P. Duhem, *Les origines de la statique*, 1905, esp. vol. i, pp. 61–97.

BECOME

like Jesus

Regaining Clarity on Transformation in the Local Church

Mindy Caliguire

BECOME like Jesus: Regaining Clarity on Transformation in the Local Church

Request for information should be addressed to:
Willow Creek Association, 67 E. Algonquin Road, South Barrington, IL 60010

Cover & Interior Design by: Rebecca Gallagher, Fresh Folio Design

Printed in the United States of America.

12 13 14 15 16/1 2 3 4 5 6 7 8

If this work is dedicated to anyone, it's to you—to leaders who care about transformation. This book is dedicated to you, to your efforts, to who you are, and to who you might

by God's grace

BECOME.

Acknowledgements

This project simply would not have materialized if not for the vision, effort, and disciplined focus of many associates at the Willow Creek Association (WCA), and I'd like to specifically acknowledge a few.

- Our President, Jim Mellado, and COO, Craig Terrill's passionate leadership towards Matthew 28 in the Body of Christ has inspired and propelled me.
- Kim Anderson's grace, creativity, and talent for research, editing and writing made this possible, and incredibly fun along the way.
- As the leader of publishing for WCA, Nancy Raney's vision for this project and her focused guidance during the effort made this possible.

Thanks also to our massive team of intercessors, who have been praying daily for this effort to effectively serve church leaders.

Lastly, I want to acknowledge the many church leaders around the world who have shared their stories with me and allowed me to "see" into their world. The future of the Body of Christ is in good hands.

Mentors, friends, "desert guides," and my family near-and-far should be acknowledged for doing life with me, and for holding out hope for who I might yet become.

*All these efforts have contributed to whatever value you might find in these pages. Whatever is confusing, incomplete, understated, overstated, or in error—that's mine.

Table of Contents

Foreword

My parents moved more than 40 times in their 55 years of marriage. I was along for the ride for 23 of those years experiencing 18 countries while under their care. I thought it was normal that families moved every other year! It was such an adventure and I loved it. But, it never ceased to amaze me that wherever we went, how quickly our family connected to a local church that provided immediate relationship, connection and purpose.

Oddly enough, I remember the various churches by their floors from dirt to clay brick, tile, carpet and stained concrete. Some were start-up churches and others were multiple generations old and long on tradition. I remember being a part of churches with a couple dozen in attendance to hundreds, to thousands, and even tens of thousands. In between moves, my family was church. Often times we would have services with just the six of us having communion, singing worship songs and opening the Bible for teaching.

I can truly say that most if not all of my defining life decisions were made in the context of my experience with a faith community. I grew up with a love and respect for the beauty and potential of local churches... all kinds of churches. I saw the church working the way it was designed to work with the poor being served, the lost being found, and wounds being healed. I also had a front row seat to the church when it wasn't working right... when the pastor left his wife and ran off with another woman taking with them money from the church. I saw the struggle of building programs never completed, church splits and human politics becoming the practice of the day. I could not have had a more contrasting experience of both extremes!

Local churches do have the amazing potential to be the miraculous Body-of-Christ in this world, but they don't drift into that reality. For the last twenty years, we at the Willow Creek Association (WCA) have been serving and studying thousands of churches around the world from almost 100 Christian denominations. We can say with great certainty that the difference we see between effective and ineffective churches is the presence of a core of gifted Christ-centered leaders creating amazing environments that move people to meet Jesus and over time reflect his character. That's why those of us at the WCA have given ourselves fully to implement a very simply but effective strategy to develop these kinds of leaders and churches. This is how Bill Hybels put it:

> *"When God transforms the life of just one leader, that leader can transform a church. When one church is transformed, you can transform a community. And when enough churches are thriving, you can effect a region, country and eventually the entire world with the positive, life-changing power of Jesus Christ and the redeeming and restoring work of his people."*

Our vision is that one day it would be normal for every church to prevail. For that to happen, a movement of *first responders* needs to rise up to follow the fresh winds of the Spirit in establishing new Biblical patterns of church life that blaze new trails for others to follow. These are the kinds of pioneering leaders and churches we are called to inspire and equip at the WCA. ***BECOME like Jesus*** is a strategy book that describes the path toward leadership and church transformation that we will align the ministry of the WCA to serve. If you and your church are called to be one these *first responders* because you're serious about living out the Matthew 28 mandate to *go and make disciples of all nations,* then we're serious about serving you!

Jim Mellado

President, Willow Creek Association

Introduction

When you hear the word "country," what comes to mind? Do you picture a pasture of cows grazing in a field? Or do you sing along to the latest Rascal Flatts' song? Perhaps you envision an actual mass of land that you'd like to someday visit. Funny how the same word offers different realities.

What if I said the word "church"? Is your first thought around a service, like "What time is church on Sunday?"

Or ...

Is church a building? Is it the building over on Main Street?

Or ...

Is church comprised of a staff, pastor and a particular organizational chart? Do people notice and ask, "Hey, look at what that church is doing!"

Or ...

Do you think of the local church as *ecclesia*, the called out ones, a worshipping group of people, meaningfully connected in a local setting, with God at their center? Those who are bringing the presence and activity of Jesus to their community?

By virtue of habit, most often we tend to think of the first three meanings of "church": "I drive to church." "I attend a church." "I work for (or at) a church." Likewise, we often evaluate: "Church was good today!" "I like that church." "That's an innovative, or traditional, or a leadership-oriented church." But those are not the real the church.

How can we tell? If we canceled every service, the building burned to the ground, and the entire staff quit, we'd still have church, the ecclesia. People would still gather to care for one another, pray and connect with each other. The ecclesia would continue to move toward God together, and be salt and light to the community around them (Matthew 6).

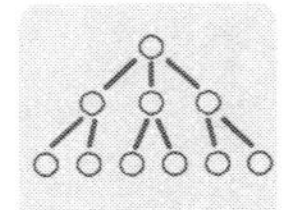
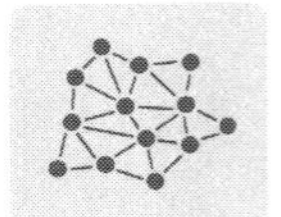

When we wonder about that definition of church, and ask "How is our ecclesia doing," the answer is more difficult to discern. Yet it remains an important question, one that leaders will be held accountable for by God.

So how can we, as leaders, help the local church be more effective in causing that fourth definition of church to flourish and grow? How do we help people discover and deepen life in that community with God at its center? As part of this vibrant community, what can we expect to be happening in the lives of those involved?

Answers to those questions will be impacted by our view of the human soul.

Member of the Ecclesia. A Closer Look at the Soul...

Who are these members of the ecclesia? They are the community of the redeemed. Their souls are "saved," for sure. What does that mean, exactly? What does that mean for transformation?

When we think about our soul, we should consider it in the context of our whole personhood.

Philosopher and theologian Dallas Willard explains:

> *What is running your life at any given moment is your soul. Not external circumstances, or your thoughts, or your intentions, or even your feelings, but your soul. The soul is that aspect of your whole being that* correlates, integrates, *and* enlivens *everything going on in the various dimensions of the self. It is the life-center of the human being.*[1]

Sometimes Christians tend to think of the soul only with reference to whether or not it is "saved" or "born again" or "heaven-bound." And while the eternal destiny of a soul is of utmost urgency, something else is quite true about souls, and vitally important in the here and now: Souls are living.

The Old Testament Hebrew word for soul is *nephesh*. The Scriptures use this word when God formed Adam out of dust in Genesis 2:7. God breathed into Adam,and Adam came to life, a "living being"—a *nephesh*.

In the New Testament, the Greek word for soul, *psuche*, is also translated as life. One's soul and one's life are woven together.

As believers in Christ, our souls have been brought to new life—regeneration has begun (Ephesians 2). The integrated soul is driving everything about that life. And so, like all living things, our souls can flourish or they can be diminished. When we pay attention to the health of our own soul and discover (experience) ways for it to grow. We gradually and naturally become the kind of people who can live as Jesus would in our context. Transformation happens.

This alone explains the redemptive power and potential of the ecclesia.

What Might We Become? A New Lens

We need to start our exploration of this topic with a broad view of the church as the ecclesia, made up of redeemed living souls. Perhaps in the past we've presented an accurate, but incomplete picture of church and what—exactly—we're inviting people into, as followers, or disciples of Jesus. We've often minimized the spiritual dynamics at work in a redeemed soul. Perhaps these are clues to some frustration felt in recent years. What could happen if that began to shift? What might become of us?

My prayer for this book is that we might view "church" and "soul" and the relationship between them in a new way. We want our efforts as leaders to serve God's purposes in this world, to indeed bear fruit: as we seek the lost, search after the strays, bind up the injured, heal the wounded, strengthen the weak, and provide spiritual nourishment (Ezekiel 34).

We want to become, as individuals and as a community, those who are capable of participating in God's work here on earth.

I love imagining: What might we become?

CHAPTER 1: The Hope of Becoming

Promises of transformation surround us. Television ads market the latest anti-wrinkle cream with guaranteed results in just two days. If you want a beach-ready body in no time, you'll find the newest ab machine on late-night infomercials. We can cover the grey, lose a few inches and simplify our lives—often in three quick steps. If only transformation was that easy. We long for improvement. We long to shed the emptiness and the ordinariness of our lives for beauty, fulfillment and meaning.

The hope of becoming something new and improved fuels entire industries: executive coaching, management consulting, health clubs, and "aesthetic surgery" clinics.

As Christ-followers, our hope grows even more profound. We hope to become something new and improved where it matters most—on the inside. We want to fulfill our potential and become all that God created us to be. We feel both the desire and the responsibility to reflect the character of Christ in our world.

And if we have been called by God to some realm of influence in the church, we also bear some responsibility for this internal development in other's lives.

The good news is: It's possible. Maybe not in three easy steps, or for $250 per month, but true lasting transformation is a very real hope. God promises, and is already at work to accomplish, just such a transformation within His followers. We can be confident of this: *"he who began a good work in you will carry it on to completion until the day of Christ Jesus"* (Philippians 1:6).

But what path should we take as we attempt to traverse the gap between who we are right now and the person we might become? As church leaders, how can we best help those we lead to become more like Christ?

Spiritual Formation, Discipleship and Transformation: What Do They All Mean?

Many terms have been associated with this *process of becoming* in the soul of one who is "in Christ." You've likely heard these: sanctification, transformation, discipleship, and even spiritual formation. While they may have come to mean slightly different things within various ministries or church traditions, I find it helpful to think of them as essentially similar. They all speak to the process by which a human soul, following the way of Jesus, is being re-shaped at the core of their being, in a direction that reflects the character of God.

What would that look like? Being shaped that way on the interior? Bill Hybels, Senior Pastor at Willow Creek Community Church, gives great language to this when he speaks of the "Law of Spiritual Formation." It is during which that we are formed in the image of Christ, so that we would begin to naturally *"love the things that God loves, disdain the things that God would disdain, have compassion on that which God has compassion, and so on..."*[1]

Dallas Willard defines it this way: *"Spiritual formation for the Christian basically refers to the Spirit-driven process of forming the inner world of the human self in such a way that it becomes like the inner being of Christ himself."*[2]

The diagram below is a simple, but hopefully not simplistic, way of depicting this process of becoming.

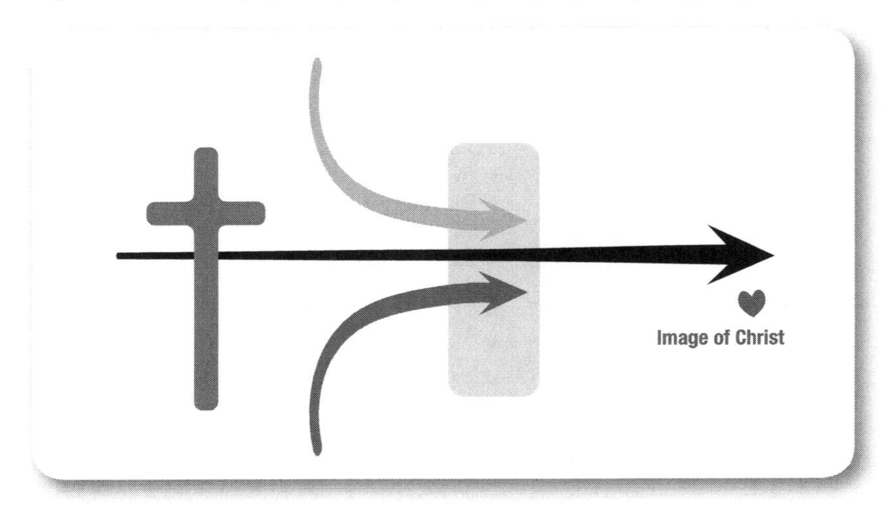

- The main horizontal line simply symbolizes a human soul changing over time. We don't sign up for that process; it's a fact of being human! Think of it this way: Human beings are highly malleable. We are *always* becoming a certain kind of person.
- The cross lines represent the beginning of a soul's relationship with God through the person of Jesus. When grace invades a human soul, everything changes! The former darkness turns to light, the soul has gone from death to life (Ephesians 2).

Prior to entering into a relationship with God through the redemptive work of Christ, the soul's transformation is random. It is not necessarily focused on a specific goal or outcome. Once grace has infused this soul, a few new things can be said of the process of transformation in the life of someone who is "in Christ" (Ephesians 1).

- The arrowhead at the right end of the horizontal line reflects the direction in which someone "in Christ" is becoming. The soul is being formed into the image of Christ (Galatians 4:19). It is a process. Over time, that seed of grace transforms us.
- The curved arrow swooping downward from above shows the indwelling activity of the Holy Spirit. He is exerting energy in this process of transformation (Philippians 1:6).
- The curved arrow swooping upward shows the energy the individual exerts towards that same goal of transformation (Philippians 2:12). There is significant effort we expend as God's character is pressed through our DNA—not to work for our salvation, but to work out our salvation. But while God's activity is constant, our awareness of and responsiveness to God's activity fluctuates, thus limiting our openness to new growth.
- The large vertical rectangle brings together this ongoing activity of God and our own efforts. It represents the role of spiritual practices in the life of someone who is "in Christ." We become aware and responsive to the work of God in our life as we regularly connect with Him. That's why spiritual practices are so important as a way to keep our soul open and yielded.

Broadly defined, a spiritual practice is anything we are intentional about doing, to become aware of and responsive to this ongoing work of God in our lives. (We'll talk more about spiritual practices in Chapter 8.)

So whether you call it spiritual formation, or transformation, or discipleship, or even sanctification, this process is at work in those who are redeemed (Galatians 4:19). And it matters that they traverse this journey, and that they do so in the context of a vibrant community. Together, they are the ecclesia.

This entire journey is represented in Jesus' mandate to His followers. His final words of sending—of commissioning—His followers were:

> *"Therefore go and make disciples of all nations, baptizing them in the name of the Father and of the Son and of the Holy Spirit, and teaching them to obey everything I have commanded you. And surely I am with you always, to the very end of the age" (Matthew 28: 19 – 20).*

While many leaders might think of this passage as Jesus' mandate for evangelism, it expresses His expectation of the whole journey. Not merely converts, but disciples. Conversion is the first step in becoming a disciple.

Perhaps Willard best describes what we want to build and offer to those we lead: "*Fortunate or blessed are those who are able to find or are* ***given a path of life*** *that will form their spirit and inner world in a way that is truly strong and good and directed Godward*" (emphasis mine).[3]

A Path Big Enough to Include the Whole Journey

I'm from the Northeastern part of the U.S., a place where roads and highways are generally named for what they are: the Mass Turnpike, the New York State Thruway, the Long Island Expressway.

But now I live near Chicago. And while streets here tend to be more organized and better marked than in New England (no cow-paths-turned-superhighways or evasive signage just to confuse visitors), there is one notable exception: the Elgin-O'Hare Expressway. You see, I've discovered since moving here that the Elgin-O'Hare Expressway goes neither to Elgin nor O'Hare.

I've looked into it. The vision for the highway was clear. Government leaders and city planners had a grand vision to join the western suburbs of Chicago as far west as Elgin with the O'Hare International Airport. Somehow that goal got lost along the way. It really should be called the Hanover Park-Itasca Expressway. That doesn't exactly roll off the tongue.

> "Conversion is the first step in becoming a disciple."

Obviously, people still make the journey between O'Hare and Elgin. Thousands of drivers every day actually use the completed portion to get where they're going.

Recently, as I entered the ramp to the ill-named expressway, I felt a pang of conviction. Almost every ministry I've ever been involved with or helped launch began with a clear and well-articulated vision.

In every case, on one end we wanted to reach the lost and at the other end of the spectrum, we wanted to nurture believers into mature disciples. We wanted to obey Jesus' command in Matthew 28. But I've come to realize that while the roads we've built do serve *most* people on *part* of that journey, oftentimes we never really reach either end.

Just like that part of the Illinois highway system, people obviously still move from being deeply irreligious to becoming fully devoted followers. The question becomes, are our pathways helping?

The Ecclesia at Its Best

A dramatic shift has happened in just one generation to reclaim the church's vital role in reaching the lost. No longer ignored, or left to the realm of para-church organizations, the church has, by and large, rightly focused considerable efforts on evangelism—with tremendous impact!

Efforts to be more relevant, to communicate more clearly, and to intentionally reach beyond the comfort zone of the "holy huddle" have fueled much of the true success and effectiveness of modern mega-churches and missionally-minded efforts. In many places we've built a highway starting right at the O'Hare International Airport! Constant new efforts and creative ways should be explored, but by and large *we get it*.

But more and more church leaders around the world are wondering if their effectiveness in making disciples, not merely converts, has not been as strong as it could be and should be. The on-ramps, signage, and pavement for the rest of the journey have been only partially completed, or even left undone.

Leading for Transformation

For the past few years, I've been honored to share in strategic conversations at the Willow Creek Association about Jesus' mandate in Matthew 28:19-20. We've been asking, *how might we help local churches become even more effective in carrying out **ALL** of Matthew 28*?

In the past, it would have been nearly impossible to assess how effectively a church was supporting this entire spiritual journey in the lives of its people.

Cally Parkinson, the church consultant and co-author of REVEAL, offers this insight:

> *Since the birth of the Christian church two thousand years ago, we have judged our ability to help people grow by asking, in essence, one question: How many?*
>
> *How many decisions for Christ? How many baptisms? How many attend our weekend services, small groups and Bible classes? How many volunteers? How many tithe?*
>
> *But "how many"—by itself—doesn't completely address what the church is called to do. That question is a good start, but it measures only what we can see.*
>
> *When it comes to spiritual growth, we need to be able to measure the unseen. We need a glimpse of people's attitudes, thoughts and feelings. We need words that reveal their hearts for God and for others.*

sidebar continued on next page

sidebar continued

We need a fresh view. We need another question—one that helps us understand the spiritual journey so we can help and encourage people in their pursuit of Christ.

The question is simple: It's the central question between God and us. And God was the first one to ask it (in Genesis 3:9): Where are you?

The health of your church is not about the numbers. It's about the movement of your people toward Christ, toward deep love of God and genuine love for others.[5]

Converts might be "counted," but how could we know if an entire group of people was more like Christ, and how could we know if our church's efforts had helped in that regard? We relied on conjecture, anecdotal stories, or conventional measures of "ministry effectiveness."

The ground-breaking research published in REVEAL (Cally Parkinson and Greg Hawkins) has finally given us a new lens to "see" into this elusive world of the ecclesia.

Through research with over 1,000 churches and 250,000 individuals, it has been found that everyone is on a spiritual journey to some degree and God is the leader of each of those journeys. Spiritual growth *"is not linear or predictable. It is a complex process as unique as each individual, and it progresses at a pace determined by each person's circumstances and the activity of the Holy Spirit."*[4] The REVEAL research, which we will explore more fully in Chapters 2 and 8, offered new insights into the spiritual journey, and new data to help church leaders see how effectively they are helping their congregation traverse this journey.

I'm not suggesting we develop a mechanical conveyor belt of programs at churches that move people automatically from pre-baptism towards maturity. That would be entirely incongruous with the mystery and ongoing process of an authentic spiritual life.

We must ask, is there more we can do to increase our effectiveness? How might the leadership of a local church lead in such a way that it truly extends a path of life comprehensive enough to encompass those far from God and those who are fully surrendered "in Christ"?

This is an important question because leaders have been entrusted with a unique role in shaping the future of the local church. They have to lead.

Whose Job Is It?

". . . . if [your spiritual gift] is to lead, do it diligently" (Romans 12:8).

Bill Hybels often talks about this verse at the annual Global Leadership Summit held at Willow Creek Community Church:

> *Romans 12:8 tells those of us who have the gift of leadership that we had better sit up and take notice, we had better "lead with diligence." Why? Because the Church, the bride of Christ, upon which the eternal destiny of the world depends, will flourish or falter largely on the basis of how we lead. My ultimate concern is not leadership. For me, the bottom line is the Acts 2 church. But I am absolutely convinced that the church will never reach her full redemptive potential until men and women with the leadership gift step up and lead.*[6]

If you are reading this, chances are you are actively involved in some part of the process. Perhaps you're even a senior leader, assuming significant accountability for the whole. Whatever your role, it matters that you lead. It matters that you lead with diligence. It matters that you lead with a clear eye to what is really going on as human souls receive the invitation to new life in Christ, and then traverse an entire journey of transformation in the context of your ecclesia.

What new pathways might need to be created and what will be your strategy for helping your congregation move along this path? Who will cast the vision for the entire journey? How will your community reach out to the lost, to the strays? Who will teach the "found" the critical truths and practices necessary for maturity in Christ? What ought to be true of the inner life of leaders and teams in order to create environments conducive to growth?

Before we explore actual strategies and a framework for taking next steps, let's explore more specifically what current research (primarily in the United States) tells us about the spiritual vitality of the church. As we move "from here to there," we need to really understand where we *are*, both in the Christian community at large as well as in local congregations.

So let's take a look at a few indicators of the condition of the church.

CHAPTER 2: A Life-Changing Message

For a few summers during college, I worked as a lifeguard at a Christian camp in the scenic Adirondack Mountains of New York. Each morning, staff and guests together joined in chapel services to receive life-changing teaching from an exceptional visiting faculty of seminary professors and pastors.

Topics included character studies on the life of Moses, exegetical teaching on Galatians, Fruit of the Spirit, and the universal call to evangelism and missions. Day after day, week after week, one powerful life-changing message after another. Did I mention they were life-changing?

In the early weeks of that first summer, I sat enraptured and took copious notes. Until I got completely depressed. Despite all those great life-changing messages, I was still pretty much the same person. My actual life hadn't changed much at all.

Except now I knew things. I *knew* of Moses' leadership and faith, I *understood* Paul's challenge to the Galatians, I *grasped* what the Fruit of the Spirit *ought* to look like in my life, and became *keenly aware* of God's heart for those who were lost.

But on the inside I was pretty much the same, if not worse, since I saw my immaturity with new clarity. As I looked around for cues of how to handle this lack of transformation in my own life, I came pretty quickly to a conclusion: it didn't matter.

No one seemed to care about actual transformation at all. Of course, no one would say that out loud, but the main point seemed to be *learning*. So I kept taking those copious notes, listened really well, and studied the various passages as suggested. I noticed that conversations after chapel, if they mentioned the life-changing message at all, centered on the ideas themselves, how well they were presented, and whether we agreed or not.

A similar dynamic can unknowingly settle in to the ministry of a local church. When we substitute *transfer of information* for *transformation* the impact is predictable. We can be surrounded by life-changing messages, but remained relatively unchanged.

What do we need to understand in order to move past the transfer of information and embrace what is catalytic for genuine transformation? Three current bodies of research speak to transformation from three different perspectives. These include the The Barna Group, The Fuller Youth Institute, and the Willow Creek Association's REVEAL research.

Data from the General Population of Christians in the U.S.

Researcher George Barna and his team at the Barna Group surveyed 15,000 people to find out whether or not people's *lives* were transformed by their faith in and relationship with God. His findings suggest a clear pathway of spiritual development, but he notes that many who profess Christianity have not actually taken significant steps along that journey.

They may take part in recommended church activities, but in his words, *"you can be the Church Lady and yet be no further along the journey than [noted atheist] Richard Dawkins. Simply attending church activities and classes does not guarantee or necessarily enhance one's transformational experience. Wholeness requires more than simply showing up and gorging on the religious-activities menu."* [1]

Hopefully the thought that church programs do not necessarily inspire life-change will not come as a surprise, but what should catch our attention is the scope he suggests: some 20 percent of self-identified Christians have not progressed substantially in their transformation journey. Despite belief in a life-changing message, lives remain largely unchanged.

Data from Christian High School Graduates

Similarly, the Fuller Youth Institute recently completed extensive research on the staying power of high school students' faith commitment, once they start college. Does faith stick?

Their findings, detailed in *Sticky Faith* by Dr. Kara Powell and Dr. Chap Clark, reveal that 20 percent of college students who leave the faith *planned* to do so while they were in high school. The remaining 80 percent *intended to stick with their faith but did not.* [2]

For the few youth group graduates who do stay in deep connection with God and community, one common thread (there are a few), is that those students had formed a 'sticky web of relationships.'

> "When we substitute *transfer of information* for *transformation* the impact is predictable."

Perhaps, Powell and Clark suggest, if churches could more intentionally foster these kinds of relationships, students would more naturally transition their experience of faith into their future. Sadly, that kind of intentional relationship development often does not occur. In fact, we traditionally do just the opposite.

Says Powell: *"In an effort to offer relevant and developmentally appropriate teaching and fellowship for children and teenagers, we have segregated—and I use that verb intentionally but not lightly—kids from the rest of the church.... It's no wonder students have a hard time finding a church. Those who have been sitting at the youth ministry 'kids' table' don't know church. They know youth group, not church."* [3] Though they've absorbed many life-changing messages, their participation in church will not necessarily "stick."

Again, we're back to our opening question: What do we mean by the word "church"? What exactly have these students left? What exactly inspires them to stay connected? They leave programs and buildings. They remain connected to genuine disciples of Jesus. Can you really blame them?

Data from and for Churches

In 2007, the Willow Creek Association published the findings from research conducted on an unprecedented scale. It has now extended to 1500 churches, representing some 400,000 individual surveys taken, and in five countries. The groundbreaking "REVEAL" data focused entirely on patterns that emerged when congregants were asked about their spiritual lives in the context of their church.[4]

The REVEAL research provides a very unique lens of where the people in a congregation see themselves with reference to their depth of surrender to Jesus. The report generated for *a* local church, based on *its* people's data, provides a completely new perspective on *that* church's effectiveness in helping its people grow.

So, what did REVEAL reveal? Some learnings relate to the spiritual journey itself, and some relate to the local church. Much has been written regarding all of the findings. For our purposes, let's highlight a few high-level findings.

Let's start with REVEAL's findings about the spiritual journey:

1. *There is a journey.* As never before, REVEAL data established the presence of a broad-spectrum journey—not simply the prior categories of "lost" and "found," but clear stages of growth and development, even among the "found."

- The diagram above identifies the four broad segments of people on a spiritual continuum who are Exploring Christ; Growing in Christ; Close to Christ and Christ-Centered.

2. *It matters that God's people make this journey to Christ-centeredness.* Why? We cannot just skip this point and assume that, of course, people should grow. *Why does this really matter?* Isn't it enough that they're saved?

   REVEAL research confirmed that only those who truly surrender their personal agendas for life—at the deepest levels—can lead out as sacrificial, joy-filled, courageous, free, and vibrant agents of grace and change. They are world-changing people of significant evangelistic and societal influence.

Much more can be said of the fascinating findings uncovered through REVEAL, but for now those two points are our focus for the individual journey. There is a journey, and it matters that we take it.

Beyond the individual's journey, some of what REVEAL *revealed* relates to a church's journey. After a congregation takes REVEAL, the leadership team may discover (for the first time using actual data) that they're not as effective in helping various stages of the journey as they had assumed or hoped. This awareness often provokes a season of reflection and discernment among leaders, as together they consider their responsibility, direction, and initiatives for the future.

Some REVEAL churches discovered that their efforts towards discipleship were in fact supporting the spiritual growth of its congregants in *exceptional* ways. In fact, now the WCA could actually identify those leadership teams and discover what it was about *them* and about *their ministry* that contributed to such a strong impact for transformation.

Those churches became the subject of even further research, leading to the "best practices" outlined in the 2011 award-winning book, *MOVE*. We will discuss those best practices in greater detail in Chapter 8.

But every church leadership team I've talked to, no matter their current level of effectiveness, wants to keep growing. They want to steward their resources even more wisely, in a more focused direction, to accomplish what God has called them to do. They are not satisfied to simply transfer the information of a life-changing message. They want themselves and their people to be transformed.

Why Has Transformation Been So Elusive?

Let's unpack a few potential reasons why genuine transformation in the church has been so elusive, despite being surrounded by a life-changing message.

I'd like to highlight three contributing factors which offer an explanation as to why we wrestle with the topic of real transformation:

- **Partial communication** of the Gospel
- **False assumptions** about genuine spiritual growth
- **Lack of relationships** in the ecclesia

Partial Communication of the Gospel

Let's start with the basics. What is the invitation we extend to those considering a relationship with God?

In recent years, much has been written and discussed on how the gospel that Jesus preached is different than the gospel many churches preach today. It's a difficult and at times, controversial issue. Everyone who loves and serves Jesus Christ cares deeply about the essence and careful communication of His message, the Good News. This is a topic for thoughtful consideration and, when needed, respectful disagreement. So I move forward gently.

Personally, I first learned how to "share the Gospel" in a para-church ministry deeply committed to evangelism and discipleship. Many of my friends came to faith through this ministry. Tremendous and lasting life-change resulted.

I sat dumbfounded, therefore, many years later as I heard John Ortberg teach a mid-week series at Willow Creek Community Church. The series looked at the Gospel according to Jesus. In the very first message, by simply reading Jesus' own words from the Scripture, John showed us what should have been obvious, but wasn't.

When Jesus announced His good news, or explained the "reason He came," or described the invitation He extended, it had relatively little to do with offering forgiveness for sins in a conversion experience. There was no reference to the idea of "saving faith." Rather, the majority of Jesus' words had to do with announcing the availability of "The Kingdom." This was not the definition of "sharing the Gospel" I had been accustomed to.

More recently, New Testament scholar and professor Scot McKnight extends this idea further in his book, *The King Jesus Gospel.* He describes how he believes Christians have built a "salvation culture" but not a "gospel culture":

> *Our emphasis on the call to personal faith has created a "salvation culture," a culture that focuses on and measures people on the basis of whether they can witness to an experience of personal salvation. Our salvation culture tends toward asking one double-barreled question: "Who is in and who is out?" Or more personally, "Are you in or out?"* [5]

In introducing Scot's new book, Anglican Bishop and leading New Testament scholar, N.T. Wright supports this understanding with a terrific metaphor:

> *For many people, "the gospel" has shrunk right down to a statement about Jesus' death and its meaning, and a prayer with which people accept it. That matters, the way the rotor blades of a helicopter matter. You won't get off the ground without them. But rotor blades alone don't make a helicopter. And a microcosmic theory of atonement and faith don't, by themselves, make up "the gospel."* [6]

If we hope for individuals to experience transformation, we may need to first "renegotiate the contract," as Willard has said. This may not be what congregants thought they signed up for, if they responded to an invitation to salvation only. They are holding rotor blades, but not learning to fly.

As Gabe Lyons has observed, *"Jesus's atonement was not only meant to be a simple ticket to heaven—it carried consequence for how Christians live their lives on earth today."* [7] Turns out, this is not widely known. Rightly understood, atonement and transformation are inseparable parts of a whole, a more complete, life-changing message.

Confusion about Genuine Spiritual Growth

Since salvation and our here-and-now lives have not been connected, we struggle to understand how, or sometimes even why, we would need to grow once we are "saved." Except that we tend to get stuck. Drs. Henry Cloud and John Townsend have helped millions of people get unstuck, even those who have spent their lives surrounded by –and even teaching—those life-changing messages.

I took careful notice when Cloud and Townsend compiled some of their findings in a book entitled, *False Assumptions: Twelve Christian Beliefs that will Drive You Crazy.* As experts in authentic spiritual growth, they addressed toxic false assumptions they noticed among Christians. By extension, these assumptions showed themselves in communities of Christians. These specific beliefs would not only drive you crazy, but would also severely limit one's potential for growth.

Each "belief" is addressed in an entire chapter, but simply reviewing this table of contents will help you recognize these very commonly held beliefs. Do you identify with any of these? Did you at one time believe these? Do you still? Remember, these are *false* assumptions about spiritual growth.

Twelve Christian Beliefs That Will Drive You Crazy: Relief from False Assumptions:

1. It's selfish to have my needs met.
2. If I'm spiritual enough, I will have no pain or sinfulness.
3. If I change my behavior, I will grow spiritually and emotionally.
4. I just need to give it to the Lord.
5. One day, I'll be finished with recovery.
6. Leave the past behind.
7. If I have God, I don't need people.
8. "Shoulds" are good.
9. Guilt and shame are good for me.
10. If I make right choices, I will grow spiritually.
11. Just doing the right thing is more important than why I do it.
12. If I know the truth, I will grow.[8]

Every single one of these false assumptions has been present in many of the ministries I've been a part of and led!

Consider just the final assumption: *If I know the truth, I will grow.* Perhaps this, more than all the others, has held captive the imagination of many. To be sure, truth and information are vital to the transformation process, but they are *not* the *only* elements in that process.

Lack of Community in the Ecclesia

A third contributing force to lack of transformation in the lives of our congregants emerges in part from the culture around us. Most adults lack deep relationships.

Do you know your neighbors? Can you name the families on your street? Or sitting next to you in church services? For more and more of us these days, it's not unusual to live in a neighborhood and not know the people next-door, never mind down the block. We may not know those we sit near in church.

You may have heard the phrase "refrigerator rights" coined several years ago by Dr. Will Miller and Dr. Glenn Sparks. It refers to people you're so comfortable with that you can go into each other's refrigerators, make a sandwich and grab a drink with no questions asked.

Miller and Sparks explain how the increasing lack of "refrigerator rights" with our friends and neighbors affects us:

> *"... the core emotional problem of modern life is this: a pervasive personal detachment and aloofness from other people. And this characterizes life for too many of us. We no longer live in physical or emotional closeness to the people who helped shape us, including our family of origin, friends, neighbors, and the acquaintances of our childhood. And we have failed to replace our social network with new people. It's not just about moving away. It's about being away, being apart, being isolated and too alone. It is about the loss of 'refrigerator rights' with others.*[9]

As many of us know all too well, this deepening cultural isolation all too often exists within the church. What was meant to be—and often has been—a vibrant, alive community with God at its center, has become highly individualistic, even isolated. As a result, Christians only rarely experience the kinds of relationships that stimulate spiritual growth.

The cultural trend towards isolation coupled with the spiritual dynamics at work with shame, guilt, manipulation, striving, and pretense that often exist in Christian relationships keep many deeply isolated while appearing so "together."

In case you doubt this point, consider Barna's research. It discovered that *"only one out of every five self-identified Christians believes that spiritual maturity requires a vital connection to a community of faith."*[10]

This statistic alarms me more than any other. It alerts me to the deep need in our day to cast a new vision for soul-level transformation in the context of community. This is an enormous topic, and we will address it more fully in Chapter 8. For now, suffice it to say that the potential for true spiritual growth as God intended in the body of Christ is severely limited because we are so deeply disconnected.

Moving on

And yet, there is hope. In individuals and in communities, throughout church history and in our day the transforming power of the Gospel is being proclaimed, believed, followed, and obeyed. But I believe there is more we can do to learn, to lead, to follow, and to help others follow. How can we as leaders lead in such a way to unlock this transforming power?

CHAPTER 3: A Powerful Key

They asked if I'd drive one of two cars, the "black car," back from the airport after transporting a group.

"Glad to!" I eagerly replied. But when I walked outside, I immediately noticed this was not just a black car. This was a Porsche. At the time, I drove a black Ford Taurus. That's the kind of car you call a "black car." Not a Porsche. Whatever color it might be, you call it a Porsche.

If I had known, I might not have agreed to help, since Porsches can be slightly intimidating to Taurus drivers. I had only even *been* in a Porsche when riding shotgun to sales calls with my manager at a Fortune 100 company.

But I had agreed, and there were no other drivers left to return the car. So I was handed the key.

I dropped down low into the leather bucket seat, surrounded by a dizzying number of gauges and controls on the dashboard, most of which made no sense to me whatsoever. What I did remember from my years in sales was that the ignition on a Porsche was in an unusual place. I just couldn't remember where.

Drive shaft? Nope. Under the steering wheel? Nope. On the dashboard? Nothing. I couldn't find the ignition anywhere.

My anxiety rose, as I knew the other driver had already left and was waiting for me at the end of the parking lot. I saw him turn his car around and come back to see why I wasn't moving. After all, this was a Porsche. Not made for sitting still.

Finally, with a huge sigh of relief and satisfaction, I noticed the key hole under my right arm on the console between the seats! But then there was a new problem: the key fit, but it wouldn't turn. I knew that some cars have special buttons you have to press as you turn the key, or turn the wheel a certain way. I tried anything and everything as my guide finally got out of his car and walked over to see what was taking me so long.

Me: "I figured out where the ignition is, but I'm not sure how to start it!?!"

Him: "Yeah. That's the glove box."

Without looking back toward me, he pointed out the ignition on the dash to the *left* of the steering wheel and walked back to his car. I wanted to die.

Looking for the Key

I wonder how many of us in ministry have looked for one "key" or another over the years, trying to harness this very powerful engine we know is supposed to do amazing things, even more amazing than a Porsche.

For a while, we thought the key would be seeker services and many churches focused a majority of their efforts in creating dynamic services to attract the lost. And the lost were drawn to the church and lives were transformed by the power of God.

Others thought the key might be in small groups. If everyone could be in a well-led small group, people would grow and become fully-devoted followers of Christ. And many people did join small groups and grew in their relationship with God.

For some, missional communities were the missing key in their churches or contemporary worship or new discipleship methods. All of these ideas bore fruit and helped people grow in their love for God and others.

We thought these keys were the hope to unlock the vast potential we believed possible in the church. Potential that seemed to be obvious, but somehow lacking in our actual experience.

In recent years, many have wondered if the key might be available to us in the realm of spiritual formation, which in itself, can mean many things to many people. Some think of it as contemplative styles of worship while for others it means the recovery of ancient writings and "rules". Still others think of spiritual formation as adopting spiritual practices that aid one's personal life with God.

Personally, I'm a big fan of all of these. I do believe the broad field of spiritual formation has important things to offer the local church in our day.

But like me in the Porsche, we really don't know where to go with it. Which button should we push? What do we do next? How does the possession of this key really help us get to where we're going?

Time to Shift

In exploring how the realities of spiritual formation might inform or shape the way we do ministry, I'd like to suggest some important areas of change that would and should occur. You could think of them as significant shifts that need to be made.

These shifts would not necessarily replace any of the "keys" we've mentioned earlier, because these shifts are deeper than the level of various church programs and structures. They represent foundational values that undergird behaviors, relationships, and strategies and stem directly from a broad understanding of spiritual formation.

The chart below lists four necessary shifts the church needs to make if we desire to more effectively guide those we serve into genuine, soul-level transformation.

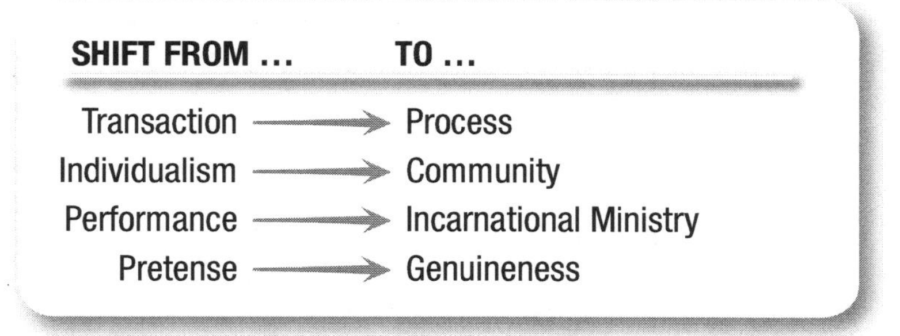

As we explore these four shifts, I want to ask your forgiveness in advance. I won't be exploring these areas fully in this short chapter.

Most notably, I will not necessarily suggest *why* these patterns exist, or at this point *how* these might change. Those topics are both important and deserve full treatment; my purpose here is simply to *name* these existing patterns, suggest how they severely inhibit transformation, and offer an alternative.

Transaction to Process

Inviting people into a life-saving relationship with Christ is one of the greatest privileges we have as believers. As ministry leaders, it's also our privilege to help those new to faith discover ways to grow spiritually.

As described in Chapter 2, there is a partial communication of the Gospel when we focus on inviting people into a salvation experience, but fail to invite them into a transformational journey as a disciple of Jesus. A ministry based on extending an invitation to salvation (defined as forgiveness only) tends to be very transaction-oriented. I repent, accept Jesus, am forgiven and receive eternal life. I pray a prayer, walk the aisle, cross the line.

In becoming more process-oriented, we cast vision and extend the invitation to the *rest* of the spiritual journey. It is an integrated whole process. Referring to the REVEAL framework described earlier, it matters that we look at all the stages and determine catalysts to help each of these four segments of people take spiritual next steps. REVEAL co-author Cally Parkinson has studied literally hundreds of survey results from churches who've taken REVEAL and explains why this matters for church leaders:

> *Identifying these segments gives pastors and church leaders a good framework to figure out where people are in terms of spiritual maturity. But just knowing where they are is not enough. Our ultimate goal is to understand how people actually grow to become more like Jesus. We want a deep understanding of the most significant catalysts of spiritual growth across the continuum so we can really help people move from one segment to the next.*[1]

As we move from transaction to process, we can expect to see people at all stages—those far from God and those in deep relationship with God—continually inspired to take their next step, whatever that may be.

Individualism to Community

We've already established the deep tendency towards isolation in our culture, as well as in the church. But before we lean too heavily on things like friendship and community and refrigerator rights, we must look at this from another important angle: recovering our identity as God's people.

"Community is the one indispensible condition for human flourishing."

The transactional invitation to a *personal* plan of salvation results in losing any sense of *who we are* in community, or why that matters. You know it well as pastors and leaders. We struggle to motivate Christ followers to build strong relationships. The transactional invitation to a personal plan of salvation is essentially a personal thing. Nothing about the message of Jesus' forgiveness compels us into deep, transformational community, and it frequently doesn't emanate from an understanding of our new identity in Christ.

One thing I've often wished for is a version of the New Testament that would allow English readers to see how often the pronoun "you" is used to refer to a group of people. In our western way of thinking, we hyper-individualize biblical truths never originally intended exclusively for singular or individual application.

The Hebrew way of understanding, the culture into which Jesus lived and taught, and, thus, the writers of the New Testament, would never have imagined the individualistic interpretation of these verses which is so common in Western culture. We are simply not used to thinking in the plural sense, and we are not used to thinking of our identity as the people of God in a given community, not just individual people who sit in the same building and worship at the same time.

Even when we do acknowledge each other, the focus on community tends to remain consumeristic in its mindset. We speak of benefits to belonging in a church. We work hard to persuade people to pursue deeper relationships, usually by helping them see how it will make their lives better. It remains about having a better life. Optional.

And yet these deep relationships are absolutely vital to the process of spiritual formation and the activity of God in a particular community. A shift in this direction would involve communicating this greater sense of *who we are*. It would also involve a greater sense of *how we are*, as individuals within a community.

In *Everybody's Normal Until You Get to Know Them*, John Ortberg writes, *"Community is what you were created for. It is God's desire for your life. It is the one indispensible condition for human flourishing."* [2]

What if, from the very introduction to the Gospel, embracing the Christian faith necessarily involved a new network of relationships? What if participation in the church isn't optional after salvation? We are only scratching the surface here, and will address this more in Chapter 8. Any serious reading of the New Testament will further this point. The Gospel isn't simply a system of beliefs. It is an invitation to a whole new way of life, a transforming way of life, including an entirely new way of relating.

These relationships are both a vision of how we are to be with each other and also God's mechanism by which we *become* that vision (Ephesians 4).

When my husband and I were planting a church in Boston, one of our favorite visionary authors was Elton Trueblood. See if this doesn't get your heart pumping as well.

> *"Once the church was a brave and revolutionary fellowship, changing the course of history by the introduction of discordant ideas; today it is a place where people go and sit on comfortable benches, waiting patiently until time to go home to their Sunday dinners.*
>
> *The idea of a redemptive fellowship, so amazingly central to Christianity, involves an entire philosophy of civilization. How is civilization changed? It is changed, early Christianity answers, by the creation of fellowships which eventually become infectious in the entire cultural order."* [3]

We didn't give our lives for bench-warming Christianity, and neither did you. It's about the community, even though much of our language and practice reflects far too great a focus on the individual. If the church is to become the revolutionary fellowship it can and should be, this needs to shift.

Performance to Incarnational Ministry

Most of us see, or have experienced, the high-performance culture found in many churches today. Teaching pastors feel the pressure to "hit it out of the park" each week. Worship leaders, youth directors, children's leaders, feel the need to entertain and "take it up a notch," presumably so that attendees will come back the next week.

At its core, it begins to feel, and even look, like effectiveness is all about our performance. If we do well, people will become Christians, or worship more authentically, or join our initiatives, or donate generously. And the reverse is true. If we don't do well, they won't. It appears the work of the kingdom is solely in our hands.

I'm overstating for effect here, but I want to make this point clearly. If values consistent with spiritual formation are to govern our leadership, then we must root out the performance mentality. Outcomes must remain solidly in God's hands.

And we must stay clear on what energies within us produce the work. Incarnational ministry recognizes and depends on the activity of God's indwelling presence within us as the agent of change. The Apostle Paul instructed the church in Corinth on this dynamic, by explaining how Spiritual Gifting would work: *"Now to each one the manifestation of the Spirit is given for the common good"* (1 Corinthians 12:7). It's the manifestation of God's Spirit within us that produces "results." We work hard, of course. But in some mystical way, we swing the bat… but God knocks it out of the park.

Obviously, it matters that we prepare and put forth our best effort, but from the onset we acknowledge God as the one who brings fruit. Not us. We are serving an audience of One.

How has this performance mentality harmed the cause of transformation? In my own life, and the lives of many leaders I've worked with, greater focus on what we do for God has all but squashed the life of God in us. What if the most important thing we have to offer a person, or a congregation, or a decision, or a conflict is the life of God in us? That's a vision for incarnational ministry.

In bringing together the worlds of the doing and the being, I often think there needs to be a holy collision between the realm of the jet-fuel drinkers (those high-performance "ministry machines") and the candle-lighters (those "resting in God" spiritual formation types).

What could such a collision yield? I'd like to think it would yield leaders with inexplicably high levels of enormous impact. They would experience even higher capacities than they even realized, lead with a power not their own, lead from a place of deep soul health, and lead out of an overflow of spiritual vitality. It's an entirely different way of life and way of leadership, and is necessary in our day. We'll discuss this more in Chapter 6, and in truth this is where we must begin.

But let's be clear, the performance mentality at best robs our sanity, and at worst actually blocks the power of the Holy Spirit. Incarnational leadership extends a far different invitation.

Pretense to Genuineness

You've seen them in your church. You know, the people who show up week after week and always seem to have it together: the perfect career, the perfect hair, the perfect ministry team, the perfect kids, the perfect marriage. When you ask, "How are you?" they answer, "Great!" Life seems to be smooth sailing year-round. Church leaders can play this charade, too.

But for any real transformation to take place, honesty in relationships is required. We cannot grow or allow the healing and re-shaping power of God to work in our lives, if we will not face the truth about who we are. We also must bring that truth into community. As the saying in twelve-step recovery groups goes, "we're only as sick as our secrets."

Unfortunately, the typical Christian "community" is often marked more by pretense than by genuine, honest interaction. In intimate, safe, life-giving relationships the seed of grace that has been planted deep inside us finally has the fertile soil in which to grow.

If we desire to create communities where transformation occurs naturally, rather than as an occasional phenomenon, we will need to shift our strong inclination towards pretense to humility and appropriate self-disclosure.

People who experience transformation are people who have learned to be genuine.

Before We Move on . . .

We've already covered a lot of ground. I hope at least some of these ideas have caused you to reflect on how these important shifts are currently expressed in your church. If you're anything like me, you're ready to dive into strategy and figure out ways to align resources and people to begin taking action to even further stimulate change.

But first, let's make sure we're placing this powerful key into the right keyhole, rather than pushing harder and harder in the wrong place and wonder why it doesn't "work."

Let's press the pause button in order to slow down and process what God might be stirring up. Let's be reminded of where God is in all this. What is His heart towards the spiritual growth of His people? And towards you as a leader?

CHAPTER 4: God's Obsession

If you want to cause a bit of controversy, ask a room full of pastors whether they should be learning principles for ministry from business leaders. Or political leaders. Or academic leaders.

Chances are, the room will polarize pretty quickly, and both sides will feel strongly that their view is right. As you well know, the often-ugly debate rages on blogs and on Christian radio, in private conversations and in board rooms.

I'm not sure it's really the best question. Let's consider another instead: Is our leadership truly accomplishing what God wants accomplished, regardless of how we learned?

What does God see the role of leadership as being? When the mantle is given, or removed, what is in God's mind as the objective? I'm learning that this raises a lesser-spoken-of role: that of a shepherd. The spiritual gift of "shepherding" in most spiritual gift assessments is generally described as this:

> *Shepherding—The gift of shepherding is manifested in persons who look out for the spiritual welfare of others. Although pastors, like shepherds, do care for members of the church, this gift is not limited to a pastor or staff member (Ephesians 4:11).* [1]

And a bit of Greek to set our thinking on the right path:

> *"The Greek word "poimen" means pastor. In Paul's spiritual gifts listing in Ephesians 4:11, this term is translated "pastor." Although the word "poimen" is translated pastor only one time in Scripture it is used sixteen additional times. The remaining sixteen are all translated "shepherd." Therefore, we are actually discussing the GIFT of shepherding, not the POSITION of pastor. Though a good pastor must have the gift of shepherding, everyone who has the gift of shepherding is not called to be pastor. The gift can be used in many positions in a church."*[2]

In my twenties, I participated in leadership meetings where we, as emerging leaders, were challenged to determine whether we were called to be "shepherds" which in this context meant: *lead only a few people* or if we were called to be "ranchers," which meant *lead a lot of people*, including the shepherd. I remember being slightly confused at the time, and unsure about the metaphor. I'm still not sure about it.

One *poimen* cannot provide all spiritual care and oversight on his or her own. Additional volunteers and even staff join in this effort to help a church community grow and remain strong. But at some level, don't all leaders in ministries retain the role of shepherd? Is that too small for him or her? Too mundane?

It doesn't appear to be too small for God. Apparently, He doesn't relegate that to the "little people" with small leadership capacities. In fact, if God doesn't like how His people are being led, it appears God will step in and do it directly. Shepherding can't remain in the hands of those who jeopardize the health of the flock.

We've been considering how congregations might become places where true transformation into the likeness of Jesus is routinely experienced. In this chapter, I want to pause our high-voltage conversation. Let's re-orient our minds and hearts back to God's desire for the flock. And for ourselves, as members of that flock.

God's Challenge to the Shepherds

Ezekiel 34 is one scripture passage that is difficult to teach in full. Or at least I find it difficult. I love the inverse "job description" God provides for those who lead a congregation. But I'll admit, the judgment and even anger I sense in God's "voice" in verses one through ten is hard to read, and hard to read out loud. Look up the entire chapter and try it for yourself sometime. It's hard to speak these words.

They're harsh. They're convicting to human leadership systems that seek their own welfare rather than the health of the flock. While we never set out to have it be "about us," sometimes egos and unhealthy ambitions and deep insecurities end up ruling our interior worlds. This is often to the detriment of the flock. It usually takes a lot of deep reflection and soul-work as a leader to notice and yield these areas to God for our own healing and transformation. In my own life, I can vouch, it is not just a one-time thing. My motives can, and far too often do, change for the worse in an instant.

At the same time, I absolutely love this passage. I'm compelled by it. It provides a powerful, clear view into what God expects to be happening in a flock under the care of good, rather than self-serving, shepherds. God is obsessed about it.

It also makes clear that God will hold leaders accountable to this vision of shepherding. God "fires" all the current shepherds, and tells us exactly what is on the "to do" list to rectify the situation.

Take a moment to read Ezekiel 34:11-16:

> [11] *"'For this is what the Sovereign LORD says: I myself will search for my sheep and look after them.* [12] *As a shepherd looks after his scattered flock when he is with them, so will I look after my sheep. I will rescue them from all the places where they were scattered on a day of clouds and darkness.* [13] *I will bring them out from the nations and gather them from the countries, and I will bring them into their own land. I will pasture them on the mountains of Israel, in the ravines and in all the settlements in the land.* [14] *I will tend them in a good pasture, and the mountain heights of Israel will be their grazing land. There they will lie down in good grazing land, and there they will feed in a rich pasture on the mountains of Israel.* [15] *I myself will tend my sheep and have them lie down, declares the Sovereign LORD.* [16] ***I will search for the lost and bring back the strays. I will bind up the injured and strengthen the weak, but the sleek and the strong I will destroy. I will shepherd the flock with justice.****'"*

Did you notice how the flock will be cared for? God will:

- Provide nourishment for all
- Search for the lost
- Bring back the strays
- Bind up the injured
- Strengthen the weak
- Guarantee justice

How does this picture of a shepherd's role align with your desire to help people move toward maturity in Christ? Does this feel a bit overwhelming or does it make your heart skip a beat with passion for the lost, the strays, or the injured, knowing this is God's heart as well?

Joining the Good Shepherd

God is doing this good work, but He invites us to join Him in caring for the flock.

In 1 Peter 5:2-4, we find this charge for shepherds:

> [2] *Be shepherds of God's flock that is under your care, watching over them—not because you must, but because you are willing, as God wants you to be; not pursuing dishonest gain, but eager to serve;* [3] *not lording it over those entrusted to you, but being examples to the flock.*

God cares quite a bit that a flock is cared for well. God also cares about the *way* leaders (shepherds) exert their leadership. Toward what end, and in what way, are we being obedient to this charge in 1 Peter?

A Lonely Road

While the call to shepherd the flock moves us forward, it is true that leadership can be lonely, thankless work. Do you wonder if you are making any difference at all and if anyone really notices, much less cares, about your investments in God's people? Shepherding God's flock is hard work.

But I want to remind you that for every ...

> Lost person you've found ...
> Stray you've sought after...
> Struggling marriage you've counseled ...
> Disillusioned teen you've redirected ...
> Forgotten person you've noticed ...
> Recovering addict you've brought to healing ...
> Desperate prayer you've prayed and
> Hardened heart you've helped soften ...
> Church you've pulled out of a gutter ...

God has seen it all. All. Everything. Every late night phone call, every prayer whispered over grieving parents, every time you stepped into a thorny conflict hoping against hope that love would prevail, God has seen it all. Every time you cast vision for what could be. Every time you examined your soul and brought forth fresh teaching that inspired growth in others. God has seen it all.

Your acts of service have not gone unnoticed. Not only that, God receives these acts of love as demonstrations of your affection for Him.

Hebrews 6:10 offers these words of encouragement to you:

> *God is not unjust; he will not forget your work and the love you have shown him as you have helped his people and continue to help them.*

Getting Back on Course

As we return again to our conversation about effective strategies, let's keep in mind the main point. God is obsessed with the kind of spiritual care that results in health and vitality of His flock. We are simply temporary stewards under the leadership of the Good Shepherd. Which is very good news to us, since we ourselves are sheep in need of care.

If you've gotten off course in any way, perhaps now would be a good time to set this book down for awhile and pick up a journal instead. Take time to listen. Listen carefully. What does God have to say to you, personally? What might God, as your shepherd, want you to know?

Just as Jesus told Peter three times in John 21 to feed His lambs and take care of His sheep, you've also been called to do just that. No matter how far you've strayed.

Pick up John 21 and hear Jesus' directives afresh. May this next season of ministry find you and your flock flourishing and resting in the care and provision of our Good Shepherd, who has laid down His own life for us, the sheep.

Isaiah 40:11

He tends his flock like a shepherd:
He gathers the lambs in his arms
and carries them close to his heart;
he gently leads those that have young.

John 10:11

"I am the good shepherd. The good shepherd lays down his life for the sheep."

1 Peter 2:25

For you were like sheep going astray, but now you have returned to the Shepherd and Overseer of your souls.

Revelation 7:17

"For the Lamb at the center of the throne
will be their shepherd;
he will lead them to springs of living water.
And God will wipe away every tear from their eyes."

CHAPTER 5: An Organizing Framework for Change

Bill Hybels regularly reminds church leaders that *"the local church is the hope of the world, because Christ is at the center. And her future rests in the hands of her leaders."* [1]

We have taken up that challenge to guard and protect the future. We increase focus on leading so that the entire spectrum of spiritual development will naturally occur—from those exploring the claims of Christ, to those who are Christ-centered. This is Jesus' *full* Matthew 28 mandate.

So what will be our strategy to accomplish this great dream, God's dream, for transformation in the local church?

Richard Rumelt is recognized as one of the world's most influential thinkers on strategy and management. In his book, *Good Strategy/Bad Strategy*, Rumelt explains that: *"Good strategy is coherent action backed up by an argument, an effective mixture of thought and action with a basic underlying structure I call the kernel."* [2] In his view, the kernel of *good* strategy contains three necessary elements:

- A *diagnosis* that defines or explains the nature of the challenge.
- A *guiding policy* for dealing with the challenge.
- A set of *coherent actions* that are designed to carry out the guiding policy.[3]

Rumelt illustrates this point with many examples, one being the battle between David and Goliath. Here, we have an unlikely match-up, with a clear (presumed) winner and loser. A quick set-up for failure.

The surprise in the story comes because, despite the odds against him, David wins. He has all the necessary elements of the kernel of good strategy. He has a clear diagnosis (bully enemy is threatening God's people), a guiding policy (aim at the unprotected forehead), and coherent actions (forget the heavy armor, move quickly, secure lightweight ammo, and throw a hard-hitting stone from a safe distance). As my teenage sons would say, that's an EPIC WIN.

In previous chapters, we've discussed the diagnosis. Now, in the remaining chapters, let's enter the realm of a guiding policy, or a point of view, that stands to actually make a difference. The elements of our guiding policy or framework include:

- **Leader's soul** Many dimensions of leadership development matter, but here we will pay careful attention to the spiritual dynamics at work in the leader's soul, and how this impacts effectiveness in transformation.
- **Culture** The culture of a ministry and leadership team can make or break the success of efforts to provide environments for authentic transformation. Before you build the system, consider the team.
- **Strategy** Finally, the realm of HOW. How will we cast vision for the journey? How will we leverage our gatherings for opportunities to encounter God? How will we help our flock get into transformation-oriented relationships? How will we help them take next steps, whatever they may be?

From there, we can discuss "coherent actions," though these will be highly specific to each church leadership team and setting.

Let's begin with the most personal, and often the most neglected element: the leader's soul.

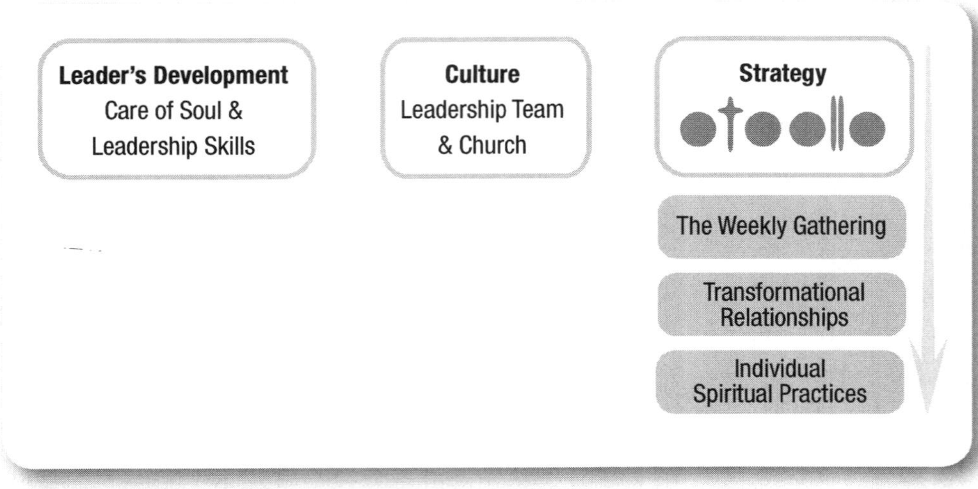

Leader's Development
Care of Soul &
Leadership Skills

Culture
Leadership Team
& Church

Strategy

The Weekly Gathering

Transformational Relationships

Individual Spiritual Practices

CHAPTER 6: The Leader's Soul

Sometimes my friends and I wonder why I care so much about people's souls. About leader's souls. Then I remember. I care because I've been brought back to life. I'm an accidental messenger, of an unintended message. Allow me to explain.

My husband and I moved to Massachusetts in August of 1991 to start a new church. We possessed many things a church-planting team could count in their favor to achieve great results. We had Ivy League educations, marketplace success and seminary training, internship experience at one of the most visionary churches of that time, and a network of family and friends who shared our vision enough to support the work with their time, talents, and treasure.

The first Spring after we moved, I had the opportunity to watch a portion of the Boston Marathon right around the corner from our apartment in Newton.

If you're not a runner, you may not know much about the Boston Marathon. It's considered one of the most difficult marathons in the world. Only elite runners are even eligible to participate. A qualifying time from another marathon is needed *just to enter* the 26.2 mile race.

The topography of the race course sets this challenging marathon apart. At mile 16, runners start a five-mile series of up and down inclines. The last and longest of these hills is called "Heartbreak Hill."

Along with thousands of other onlookers we watched and waited near mile 20—at the start of Heartbreak Hill—for the runners to pass by. Wild cheers erupted as the frontrunners appeared, still sprinting even after all the miles they'd just run.

One of the runners caught my eye. He was a top runner from Kenya and in the lead pack. I noticed him because he started to wobble back and forth, sideways across the road, struggling to stay upright. This high-performing, world-class athlete stumbled, faltered over the curbside and collapsed on someone's lawn—right in front of us.

Many from the crowd gathered around him as we gasped in horror. This exhausted runner twitched violently, his eyes rolling back into their sockets. I thought he was dying.

An emergency team ran to him, wrapped him in a foil survival blanket and whisked him to the hospital. Later we learned that he fully recovered. But Heartbreak Hill had gotten the best of him in that race.

My Own Heartbreak Hill

Little did I know on that April day how the image of a runner on the sidelines would play out in my own life just a few years later. I, too, was headed for my own Heartbreak Hill.

I had been running a different kind of race, but at a pace that was not healthy. Like that elite runner, I pushed beyond my reserves and lost sustainability. Despite intelligence, a fierce devotion to God and His purposes, and a strong grasp of scripture and theology, I too wound up twitching on the sidelines of my own life. I was unable to care for myself or others, and unsure whether I would ever "run" again.

For several months I was completely out of the race and faced numerous hospital stays, ongoing tests and scans. I struggled with unexplained but severe vertigo. Because no one knew what was wrong, we did not have any assurance I would regain my ability to function.

In hindsight, I realize I actively ignored every symptom of soul neglect. My physical health had been compromised for some time, but I just pushed through. Relationally, I had been in "pretending mode" for far too long. Emotionally, I was numb, barely going through the motions of ministry, motherhood and family life. My mind was highly distracted, and yet I was utterly unable and unwilling to stop my drivenness to work harder to build the church. I felt responsible for *everything*.

I lacked joy, or hope, or even true perspective. I felt competitive, "untethered," powerless, and terrorized by the "what if's." I was a good soldier, doing what good soldiers do. Soldiering on. I had completely lost myself in the midst of my own life.

As I've spent significant energy reflecting back on this time, I discovered a great irony. Despite being "saved," things were not, in fact, well with my soul. I thought that running at a fast pace and pushing beyond limits was spiritual obedience, even virtuous sacrifice. I was attempting to bear fruit. It was all for the sake of the cause.

Didn't God want me to give an all-out effort for Him?

Finding a New Way to Run

No one enters ministry with the intention to trash the well-being of his or her soul. Yet we often find ourselves in a spiritual vitality death-spiral, facing ever-increasing ministry demands while yielding ever diminishing returns. But we march dutifully onward. We assume that our spiritual state, a neglected soul, is somehow part of the "deal" in a life devoted to ministry.

Tragically, we often neglect the care of our own souls in our attempts to care for the souls of others. We might even think this is effective leadership, because we see no other way to meet the mounting demands.

As I gradually began to heal, I discovered a rather feisty resolve emerge from within. I refused to live, or lead, this way any longer. I had to find a new way to run the race. I had to abandon all the intentionality and strategy around ministry and instead abandon myself to God. He could do through me whatever He wanted. I was done caring. I know that may sound harsh, or irresponsible, but that's how it happened. God mercifully led me on a healing path. I've been amazed and blessed by the opportunities He's given me to serve His people once more, and even to share what I've learned with others who face similar challenges.

As I talk with many leaders—often forerunners in their fields—I find many nearing exhaustion or wavering like that Kenyan runner. Maybe you can relate. You know what it feels like to face impossible demands. Perhaps you are experiencing that early fatigue and can feel the lactic acid building up. You're running on fumes and dangerously low on physical and spiritual reserves.

Maybe you've even begun to waver and stumble, and you're scared because you only know how to keep running.

Please pay attention to this! Learn to recognize these symptoms for what they are: legitimate cries of a soul thirsty for re-connection with God. Knowledge of Scripture and character strength alone are not sufficient fuel to run the race you've been assigned. You and I need a vital, real-time connection with God in order to run. But that requires some effort, and usually, a lot of faith.

> "Once we clearly acknowledge the soul, we can learn to hear its cries."
>
> Dallas Willard
> *Renovation of the Heart*[1]

An Epidemic?

I wish my story were an anomaly. Instead, I find these same tendencies and patterns in many leaders who talk with me. Perhaps this is why I feel so strongly about a necessary shift to incarnational ministry, away from performance. *There has to be a better way.* Consider these recent statistics compiled from Barna, Focus on the Family and Fuller Seminary that paint a pretty stark picture:

- Over 1,500 pastors leave the ministry each month due to moral failure, spiritual burnout, or contention in their churches (18,000 per year).
- Fifty percent of pastors' marriages will end in divorce.
- Fifty percent of pastors are so discouraged that they would leave the ministry if they could, but have no other way of making a living.
- Eighty percent of seminary and Bible school graduates who enter the ministry will leave the ministry within the first five years.
- Seventy percent of pastors constantly fight depression.
- Almost forty percent polled said they have had an extramarital affair since beginning their ministry.
- Seventy percent said the only time they spend studying the Word is when they are preparing their sermons.[2]

Matthew 16:26 says, *"And how will you benefit if you gain the whole world, but forfeit your soul? Or what can you give in exchange for your soul?"*

Who is Jesus talking to in this passage? His disciples. Those close-in. Rather than simply an appeal to those far from God, this is a sober warning for disciples and especially leaders.

Sometimes I change the words slightly, "And how could you possibly benefit if you built a great ministry but forfeited the well-being of your soul? What could you give in exchange for your soul?" It's a rhetorical question posed by Jesus. The answer is NOTHING. Nothing is worth that toxic exchange. That is not God's heart for you, or for those you're leading.

Some of the deepest change needed in the church centers around the spiritual vitality of leaders. Your ultimate source of life is God. Not knowledge about God, but God. Connection with God.

That's why Jesus talked so much about the metaphor of a vine.

'Apart from Me You Can Do Nothing'

"I am the vine; you are the branches. If you remain in me and I in you, you will bear much fruit; apart from me you can do nothing." John 15:5

Have you ever had God whisper something important during those mid-night hours when you can't sleep? During one of those sleepless nights when I was still immobilized by vertigo, God gently reminded me of John 15:5. "Mindy, what part of nothing didn't you understand?"

I've come to realize there was quite a bit about nothing that I didn't understand. I had no theology for the life of the redeemed soul. For the needs of a soul that is "in Christ."

In fact, the idea of "apart from me" was confusing. I knew there was "nothing that could separate me" from the love of God (Romans 8). I knew there was nowhere I could go from God's presence

(Psalm 139). I knew he would never leave me nor forsake me (Hebrews 13:5). What could it possibly mean for Jesus to warn, "apart from me you can do nothing"? There was a lot about nothing I didn't understand.

Several years ago, a strong storm hit my neighborhood and badly damaged a large tree in our backyard. The tree had two central branches and the storm blew one of them entirely off. It filled the backyard. Anyone looking at the branch on the ground knew it was doomed. It was completely, irrevocably, disconnected from its source of life.

But nobody told the leaves. Nothing on that disconnected branch even *wilted* for days—it stayed quite green and lush. Several weeks passed before the inevitable signs of disconnection could be seen.

God kept drawing my attention back to this fallen branch. It visually represented the parts of "nothing" I didn't understand. Like that tree, when I disconnect from God, there's generally enough "grace in the system" so that my life doesn't fail and fall apart right away.

But unlike the tree in my yard, my soul actually has the capacity to re-connect to my source of life. How do we re-connect? Look back to the diagram on page 14. Our ability to reconnect with the source of life is often found in spiritual practices. We'll talk more in depth about spiritual practices in Chapter 8.

Because Jesus created the human soul, he knows we have this capacity to connect and then disconnect. He doesn't shame us for this tendency to disconnect. Rather, he consistently invites us to connect: "return to me" (Acts 3:19), "come to me" (Matthew 11:28), "I'm standing at the door knocking—dine with me" (Revelation 3:20), "come away with me to a quiet place" (Mark 6:31).

STAYING CONNECTED TO THE VINE

"In a vineyard or a vine bearing much fruit, the owner is glorified, as it tells of his skill and care. In the disciple who bears much fruit, the Father is glorified." [3]

"O my Father, watch over me, and keep me, and let nothing ever for a moment hinder the freshness that comes from a full abiding in the Vine. Let the very thought of a withered branch fill me with holy fear and watchfulness." [4]

The metaphor stands. The human soul is healthy to the extent it experiences a strong connection to and receptivity to that Vine. Under those conditions, the soul is most alive and fruitful. A rather simple spiritual concept, but often so hard to live out.

As an aside, connection is not necessarily "having a daily morning quiet time." Unfortunately, if we're in "transfer of information" mode, it is possible to spend an entire 45 minutes of "quiet time" and never actually connect with God. Daily solitude, Bible reading, or quiet time needs to open my soul to God and connect with Him, in order for my soul to be truly nourished.

It is also important to note that Jesus' example demonstrates how a deepening connection with God pulls us into a deeper engagement with the world, not less. In Jesus' patterns of withdrawal and engagement we see a great picture emerging. The inward journey into the heart of God naturally leads us outward as we take on the heart of God. One friend described it this way: Radical self-care leads to radical self-sacrifice. It reveals another paradox of the Kingdom.

As we become more and more like the heart of God, we gain clarity on God's purposes in our world. Our souls grow in their willingness and strength to become actively involved.

We are ready to run the race, but on a completely different source of strength.

Living and Leading in the Paradox

Leaders are faced with the same paradox that confronts all of God's people. Our source of power is accessed through our dependence, not strength. You can't go to graduate school to learn this. It is learned by living. We relinquish power and, paradoxically, discover power.

Since regaining the health and life of my physical body, Galatians 2:20 has reminded me of what is most true. *"I have been crucified with Christ and I no longer live, but Christ lives in me. The life I now live in the body, I live by faith in the Son of God, who loved me and gave himself for me."*

The "life I now live in the body" (and I am so thankful to have it back!), I want to live by faith. I want to lead by faith. I want to have it be true that I no longer live, but Christ lives in me.

I'm guessing that's what you want as well.

What if you could leave the responsibility of fruit-bearing to God? What if you could focus instead on the quality of the connection through which that fruit will be borne?

What would that look like for you? Would that be a relief? Does it raise fears of losing your effectiveness?

This is what the LORD says:
"Stand at the crossroads and look;
ask for the ancient paths,
ask where the good way is,
and walk in it,
and you will find rest for your souls.
Jeremiah 6:16

Maybe you just need to reconnect. Even now as you read. If your soul resembles the statistics listed on page 48 maybe it's time for you to experience your own deep refusal. Refuse to keep living and leading in a way that is sucking the life right out of you. I don't think that's what anyone, and especially not God, wants for you.

You really can lead from a healthy soul.

First, listen well to the cries of your soul. Then, give yourself permission to do whatever it takes—rearrange whatever is needed—to discover what brings your soul into deep connection with God. Build that into your way of life. Depend on it like you'd depend on an oxygen tank as a scuba diver.

You, as strong and intelligent and godly as you are, remain fundamentally dependent. You can either resist that dependence, or build a life around it.

As for me, I choose life. And I hope many more leaders will as well.

The world has enough examples of a life without God. Don't, for the sake of the church, provide another.

Next Steps

Take at least 30 minutes in a quiet place, maybe with a journal and a Bible. Specifically ask God to help you sense the current state of your soul. In my experience, God will reveal what you need to hear. Gently, but clearly. Write about what you're sensing.

During that time, also pause to note the primary ways God connects with you personally. You could consider various spiritual pathways, or specific spiritual practices, or even recall a very meaningful time in your journey when your sense of God was quite strong. Under what circumstances has your sense of God being with you, and for you, been strong?

Finally, talk with someone else about how your soul is really doing. Whether with a friend, or a group of ministry peers, being known in relationship is one of the most important ways to move out of isolation and into community that connects you with others and God.

CHAPTER 7: Culture that Nurtures Transformation

A few years ago our family made the completely spontaneous decision to become a "dog family." I didn't grow up in a dog family, so no one was more surprised than me when I found myself campaigning for the cute puppy I discovered in the adopt-a-pet area at the pet store one fateful Saturday. I went to get a hamster.

Now, a few years and another dog later, I have learned some things about dogs. Dogs have remarkable abilities! Some dogs are sled dogs, some are bird-dogs, some are search and rescue dogs, some are even seeing-eye dogs. I, on the other hand, have eat-the-furniture dogs. Eat the furniture, eat the walls, eat the shoes, eat my journal, eat my Bible. But we love them anyway.

I've also learned a lot about how the *culture* of a family can be impacted by the presence of furry, smelly, shedding, licking, jumping, biting, not-very-bright, but irrepressible, irresistible, unconditional love. These dogs bring a presence into our home that generally makes us better. They just do.

For those of you who are not (yet) in a dog family, I've discovered there is a warmth and welcome that awaits whenever we arrive. Any of us. Any of our friends. There is joy upon arrival, no matter how late we are, how much we screwed up at school that day, no matter how mean we were to our brothers, or how we forgot to finish folding the laundry.

There is often reason to laugh at an upturned belly, a noxious smell, a furrowed brow, or a desperate plea for bad pizza. With our dogs around, not only do we have more fun; I actually think we *are* more fun. We are more gracious, more willing to look beyond stuff and see each other, more loved and wanted in spite of our faults, and generally that makes us better people. We work harder. We play better. We forgive better. We enjoy life and God and each other more.

Culture Matters

The culture of a group of people is a tricky thing to describe, but we all know what it feels like. The culture of a family can be welcoming and gracious, or it can be perfectionistic, stingy and harsh. The culture of a church can be the same way. And so can the culture of a team.

A family, a church and a team can all "look good on paper," but step into the kitchen, the staff meeting, or the sanctuary, and you can quickly tell what's actually going on. No matter how good it looks on paper, culture will often predict more about that group's impact for good than anything else.

A visionary friend recently sent me this quote regarding culture:

The one unique role of leadership is to manage culture. When an organization's culture is dysfunctional or maladaptive, it is the responsibility of the leader to break that culture and replace it with a new one.
-Edgar Schien, Professor, MIT's Sloan School

Culture is difficult to ascertain and even more difficult to change, but it needs to be a leader's obsession. It may seem quite counter-intuitive, but if we learn to focus on the *how (culture)*, the *what (strategy)* has a much better chance of being accomplished. Both are important for Holy Spirit-directed, Holy Spirit-empowered leadership.

The Undead

Do you want to know what catapulted my 13-year-old son to new levels of popularity in middle school? He toted around, *The Zombie Handbook*. I'm frankly a bit embarrassed at the amount of time our family has spent at meals discussing the finer points of zombie etiquette and defense mechanisms.

Zombies. The undead. They are forever depicted in my mind by Captain Barbosa from "Pirates of the Caribbean." He tells Elizabeth Swan that she better get used to ghost stories because she's in one! As he speaks, he shockingly reveals the true nature of his crew. They look perfectly alive and "normal." But in direct moonlight she could see the truth. Inside they are dead and nothing but bones. Everything else is an illusion because they are under a curse. This is zombie reality.

I wonder if a similar revelation doesn't eventually occur to many of us who serve the local church. When you catch us in the right light, you can see us for what we are—dry bones. The show goes on, the work gets done, we look good on paper, but our ability to feel pain or feel pleasure is gone. We are numb. Dead on the inside. We could write the zombie handbook.

What are some clues that you're on a team of the undead and some symptoms to pay attention to?

- Drivenness and exhaustion
- Shrill "vision casting" and striving
- Folks who are slow to listen, quick to speak and quick to become angry
- Prevalence of an apocalyptic fear that drives everything
- The behind-closed-doors presence of what one business writer calls the "gallows humor of the doomed."[1]

Elements of Culture

When we speak of the culture of a church, we can explore two areas: the entire church, or the ministry team(s) that lead the church. While they're inter-related, the leadership team culture more often dictates the church culture, not the reverse.

We'll focus our attention in this chapter on important elements of a ministry leadership team culture. In the end, that leadership team's culture will impact the whole ministry, like yeast in bread dough.

Three dimensions of culture help us see this more clearly. They are: the *organizational* dimension, the *relational* dimension, and the *spiritual* dimension. They are deeply interrelated, but will be addressed separately.

Organizational Culture

Organizational culture speaks to invisible work realities that deeply impact the effectiveness of a team. These realities include things like job performance, accountability, role clarity, and decision-making processes.

Even healthy organizations can get sideways on these important organizational characteristics of culture, particularly following significant change. The near-constant change in church leadership poses a unique challenge!

And in case you think the constant re-shuffling of teams and labor is a new thing, you'll love this quote from Michael Hyatt's leadership blog:

> *We trained hard, but it seemed that every time we were beginning to form up in teams we would be reorganized. I was to learn later in life that we tend to meet any new situation by reorganizing. And a wonderful method it can be for creating the illusion of progress whilst producing confusion, inefficiency and demoralization."* Gaius Petronius, AD 66 [2]

In my work at the Willow Creek Association over the last few years, we've been addressing many elements of culture. After experiencing tremendous change, we uncovered several points of pain in our culture. We have been working to make solid improvements. We used an online survey to benchmark our current culture. We drew wisdom for our process of culture change from books that spoke to the topic.

In doing so, we learned about key areas of our *organizational* culture that required improvement. These included internal communication, role clarity, sustainable management structures, and ongoing development for staff.

Development plans, job descriptions, and process for decision-making may not seem terribly connected to our grand vision for transformation in the Body of Christ. But organizational mission cannot be achieved without healthy organizational culture. And since the local church's mission concerns Matthew 28, it follows that if our organizations do not have the attributes of healthy organizational culture, our ability to fulfill that mission is greatly compromised.

If we want to lead for transformation, we need to aspire to the healthiest levels of organization possible.

Relational Culture

Interpersonal qualities like authenticity, openness, and love are required for authentic transformation. Therefore, we run into brick walls of resistance when church leadership teams remain marked by high walls of mistrust, perfectionism, self-protection, and pretense.

A congregation can't "go there" in pursuit of transformation if the leaders can't, or won't. Consider the experience of Peter Scazzero, founder and senior pastor of New Life Fellowship Church in Queens, New York, and author of *The Emotionally Healthy Church*. He describes his former leadership style as task-driven. As a result, the people around him felt used rather than loved.

> *It became difficult to distinguish between loving people for who they are versus using them for how they could join with the mission. Did I need these people to be converted to Christ in order to build the church or my program? Or could I simply delight in them as created beings made in the image of God? I was so deeply involved in getting Christ's work done that the line became impossible to distinguish.*[3]

God intervened to help Peter discover that the essence of his own of spiritual maturity is not in his gifts, power or accomplishments. It is in his capacity to truly love. That changed everything for him and for his team.

People, even those working at a ministry, are not always easy to love! How should we think about the relationships that undergird our work together? Do they really matter? If so, why? And what might that mean for our dream of transformation in a congregation?

Dr. Henry Cloud does not define dysfunction in relationships by the presence of conflict, brokenness, or failure. That's to be expected. Dysfunction exists when efforts to improve that conflict, brokenness, or failure do not help, but make matters worse. When those attempted hard conversations don't solve the problem, that's dysfunction.

A healthy relational culture is not shocked by problems. It constructively solves them in a way that honors God, people, and the work to which those people are called.

While much more can be said about relational culture, let's bear in mind the directives given to the young New Testament church. The vast majority had to do with a *way of relating* that would mark these revolutionary new communities.

If that way of relating is to mark an entire community, it must first mark its leaders.

Spiritual Culture

Deep connection with God brings vitality to the soul and strengthens the heart of a leader. So also a strong spiritual culture impacts the spiritual vitality and effectiveness of an organization.

International Justice Mission (IJM) is a great example of a Christian ministry that models an integrated, vibrant spiritual culture. IJM has been called by God to an urgent, difficult, and at times dangerous work. It intervenes to free oppressed people from slavery and abuse around the world. Gary Haugen, IJM's president, works hard to cultivate an organizational spiritual rhythm to sustain their work in the face of such an audacious goal.

I wonder how many churches similarly recognize that their mission requires this level of spiritual sustenance. If we could see and hear the threats to our work and the amazing potential that exists when our ministries step boldly into their callings, would we perhaps behave differently? It does take eyes of faith.

Haugen and his team bring considerable human strength and expertise to their work. And yet, they intentionally make space to commune with God as the primary source of their strength, not only as individuals but together. They represent a fantastic example of a high-performance team (in all the right senses) fully and explicitly depending on incarnational, spiritual dynamics. Gary shared his perspective on spiritual culture:

> *I always ask new staff, "Where does the power to do this work come from?" It comes from God. It's not my job to rid the world of injustice. There's a God of justice and it's His job, but I need to ask, "Lord, what do you want me to do as a part in that?"*
>
> *We have a daily rhythm as a staff team. At 8:30 each morning, we have stillness. We sit still and prepare spiritually for the day. I believe it's the bare minimum to get through the day. What is really going to advance the Kingdom work? Answering 15 emails or being still for 30 minutes? We also stop work at 11 a.m. every day and gather to pray for 30 minutes. We call these two customs remembrance and reorientation.*
>
> *This all came about when I did a spiritual audit. I reflected over the past years and asked God, "What did we do together? What was of You and what*

was not of You? I realized that prayerless striving was not of God and there were times I didn't invite God into our work. That's when I introduced some changes and we started the daily prayer time.

I've seen deeper spiritual health among my staff. I can't force the 8:30 stillness, but I've invited my staff to participate and given it some structure. We offer small groups that meet on Tuesdays during stillness and once a quarter, we hold a prayer retreat that meets offsite for a whole day of prayer and directed reflection. Once a year, we gather all our prayer partners for a long weekend of prayer in Washington, D.C.

I, along with the staff, have been able to do ministry sustainably with joy. How could we possibly do this work without constant refreshment through God? [4]

How daunting is the task you've been asked to lead? What seen or unseen threats exist? What or who stands to be set free? What are you doing to cultivate a collective dependence on God for your work together?

How would your organization fare if you bravely faced a "spiritual audit"? Chances are, *as a team* you need wisdom, resources, clarity, and strength. *As a team*, you need mountains to be moved and vast chasms to be bridged. *As a team*, you need spiritual strength for the spiritual work before you.

There's only one source for that kind of power and strength.

Leading for Collective Soul Health

The organizational culture must be strong for mission achievement. This is true of any organization. In addition to that, relationships ought to reflect the way of the Kingdom, if we hope the dynamics of the Spirit to be at work. And, finally, our experience of shared dependency on God can and should permeate our work. It is not all up to us. We move, together, in God's strength. That's the spiritual dimension of culture.

Some steps you can take to strengthen your culture would be to **benchmark** your current culture, **define** the culture you want to create, and then **align** behaviors, policies, systems, and structure to that new desired culture. It's a huge task!

My prayer for leaders everywhere is that healthy culture becomes the new normal. We are no longer writing *The Zombie Handbook*, but writing a collective story of cultural health. It can be done.

A friend of mine, Lance Witt, leads a ministry for church leaders called Replenish. He previously served as the executive pastor and teaching pastor at Saddleback Church and has served in pastoral leadership of the local church for 25 years. He has talked with many leaders over the years. Lance says:

You can't diagnose the health of a ministry by its size or rate of growth. Nor can you diagnose it by what happens on the platform on Sunday mornings. If you want to talk about an organization's true spiritual health, you have to look at the health of the team that leads it.[5]

It might also help to buy your team a dog.

Next Steps

The time has come for you to develop your own plan for shifting the culture of your organization. This exercise will help you create a workable "blueprint" for how you plan to proceed with reshaping the culture you lead.

The assignment will focus on the three areas of culture we studied in Chapter 7:

1. Organizational culture
2. Relational culture
3. Spiritual culture

Within each of these areas, develop a plan for change with four parts:

A. **Define reality.** Clearly identify the strengths and weaknesses of your culture. What is the natural "drift" of the culture? The key is to be painfully honest and sober minded in your assessment.

B. **Identify key leaders.** Who are the two or three critical people that must be involved in order for any cultural change in this area to be successful? What gifts, skills or influence makes each person a vital partner in this work?

C. **What to stop.** Discern what behaviors, programs, structures, attitudes, or even people must be "pruned" in order to foster change and growth. As each is identified, indicate whether removing it is a 30, 90 or 180 day goal, and then organize the plan into a logical sequence.

D. **What to start.** You've no doubt dreamed about the culture you'd like to see in your organization. Develop a plan for what new things must be put in place for a cultural transformation to succeed. Once again as each element is outlined, indicate whether implementing it will take 30, 90 or 180 days.

CHAPTER 8: A Strategy for Transformation

Dallas Willard has said, *"A simple goal for the leaders of a particular group would be to bring all those in attendance to understand clearly what it means to be a disciple of Jesus and to be solidly committed to discipleship in their whole life."* [1]

It is simple to state that goal. But, we all know the process to achieve it is complex. So far, we've discussed two very important and yet invisible ways we can impact the hope of real transformation in our church. These are the leader's soul, including all aspects of development, and the culture of the ministry team.

Now we arrive at this very important question. What do we actually do, as shepherds of a particular community, to pursue the goal Dallas Willard suggests above? I imagine leaders saying, "Okay, now that my soul's in a healthy place and our culture is strong, what are we actually supposed to *do*?!?"

It's a completely fair and appropriate question. The answer does not need to be vague, obscure, or even particularly mystical. We all recognize the Holy Spirit as the one who truly causes growth. It remains, however, the responsibility of shepherds to tend a flock towards health, growth, and life. It's work. Work we can do.

In this chapter we will explore some perspectives on strategy. I hope they will be helpful to you as you consider your current and future plans to help your congregation recover and live into the vision to become like Jesus.

Let's first look at what the Willow Creek Association discovered about churches whose REVEAL data indicated they are exceptionally effective in helping their people along a spiritual journey. Then, we'll discuss some key environments that are especially important in the process of transformation for an individual.

The Top Five Percent of REVEAL Churches

REVEAL research identifies best practices in the top 5 percent of REVEAL churches. Let's take a closer look. These best practices are explained in great detail in the newest REVEAL book, *MOVE*, by Cally Parkinson and Greg Hawkins.

Four of the five best practices discovered relate directly to the "strategy" we are discussing in this chapter. The other best practice relates to our discussion on the leader's soul in Chapter 6. Let's start there.

Best Practice #1: Lead from Christ-Centered Core

Best practice churches were all led by what Cally and Greg refer to simply as: Christ-centered leadership. As individuals and as a team they are surrendered to following Christ's call on their lives and ministry.

> *"...we identified one overarching leadership principle that emerged in our interaction with the senior pastors of these top-5 percent churches," Hawkins says. "These churches are led by individuals* consumed *with making disciples. Absolutely consumed. Making disciples of Christ was unquestionably their most important aspiration and the deepest desire of their hearts. And that characteristic fueled all four of the practices..."*[2]

We will return to Greg's powerful challenge to church leaders in *MOVE* at the end of this chapter. In light of this overarching leadership principle and best practice of Christ-centered leadership. I don't think we can hear this challenge enough.

But for now let's explore the additional four best practices that can give us insight as we explore the work of strategy.

Best Practice #2: Get People Moving

Do you remember our opening illustration of the Elgin-O'Hare Expressway? Churches that "get people moving" have very clear on-ramps into their church and into the process of becoming a disciple. They provide clear next steps for all stages of the spiritual journey. They envision and resource the congregation for full-spectrum spiritual growth.

Importantly, those involved see themselves as being ON a spiritual journey, and that the journey requires their involvement. I've had pastors ask me, "How do I help people know they're on a spiritual journey? Every week, I stand and teach in front of a sea of passionless faces. I don't think they're interested in that idea at all": Their hopelessness is deep and palpable.

Passive participation isn't really an option at a church that "gets people moving." From the very beginning of someone's involvement, the vision of a life-long journey is made clear. Congregants take ownership for their spiritual growth. They come to understand the role the church plays in supporting their growth.

To outline a reasonable preferred path that feels authentic and rally resources to communicate that path requires careful thought and planning. If this path is not currently defined or well communicated, this important work is one of the first areas to direct your strategic efforts.

Your people need to learn—as a part of your community—that they're on a journey. They need to know you can help them take next steps in that community that are not mysterious or weird. These steps are clear and within reach.

Best Practice #3: Embed the Bible

One striking similarity of all high-scoring REVEAL churches was their creativity and resolve to embed the Bible in *everything*. Whether through devotionals created for sermon series, or daily emails of Scripture, or Scripture sung or spoken or read or taught in weekly worship gatherings, the Word of God pervaded these best practice churches.

Engagement with Scripture is the most powerful catalyst for spiritual growth at all stages of the spiritual journey. REVEAL data suggests this embedding of Scripture especially helps growing Christians. Similarly, a key driver to spiritual growth for the growing segment is the adoption of personal spiritual practices, with reflection on Scripture leading the pack.

Best Practice #4: Create Ownership

Do your people attend church, or do they see themselves as the church?

Best practice churches have successfully conveyed the essential identity of their congregation as the ecclesia. They do not attend the church, they *are* the church. This makes perfect sense in the life of someone who is becoming a disciple in their whole life.

They gather for an hour or so on Sunday and give generously to the work of the ministry. But as they live in neighborhoods, work in offices, attend local schools, and work out in health clubs, they see themselves as the church. This marks a key development in their understanding of their identity and calling as disciples of Jesus and as part of a local congregation.

Best Practice #5: Pastor the Local Community

This profound level of identity in "ownership" is naturally expressed in local communities. God's love and ways are expressed when the local church pastors the community.

Only those who have truly died to their own agenda can live sacrificially as agents of a Kingdom agenda in their communities. These are the Revolutionary Communities Elton Trueblood spoke of on page 33. Indeed, they notice and serve the world, starting with the world at their doorstep.

These churches aren't interested in building their own empire. Such a statement wouldn't even make sense to them. They are focused on building up and blessing the community around them. They have become like Jesus, living as Jesus would if He were there in their place.

Consider your strategy for aiding the process of transformation in your church. Consider how you might take next steps along the lines of these best practices. They are obviously, and intentionally, not prescriptive at all. This requires deep reflection, discernment, prayer, observation, and creativity from your leadership team. But the data tells a clear story. These practices are *always* present in churches that are highly effective in helping a congregation experience real transformation.

Power-Filled Environments

Now that we have identified best practices, the next step is to consider specific initiatives along the lines of these best practices. Three environments are especially important to leverage in your initiatives. They each open up peoples' hearts to the Holy Spirit to bring about transformation. The three environments are:

- The Weekly Gathering
- Transformational Relationships
- Individual Spiritual Practices

Let's begin with The Weekly Gathering.

The Weekly Gathering

When we come together as the local church for times of corporate teaching, worship, prayer, service, creative elements, participation in the sacraments, and more, our hearts are softened to the activity of God in our lives. We invite God's Holy Spirit to move, and He does! Powerfully.

Bill Hybels recently said this about the importance of the Gathering:

> *A different kind of thing happens when you experience the presence of God in community in a church gathering. When Jesus said, "When two or three gather in my Name…," something gets reinforced. You stand strong among dozens or hundreds or thousands—it doesn't matter the size—and you get reinforced by each other's convictions. You hear other people sing and give witness to the power and promises of God. When we gather, something happens that we can't experience when we're ungathered.*[3]

Most of us can remember an experience like this when God did a new thing in us. Grace invaded for the very first time, as we finally understood and actually experienced the love of God. Forgiveness crept into the dark recesses of our resentments and they miraculously disappeared. That tight grasp we held on possessions and prestige, as if our very life depended on it, was released. And we discovered real life.

Whatever the topic, we left that gathering as people changed forever. The weekly gathering also changes who we are as a community. Who we are becoming *together.* When a church recovers the vision to become like Jesus, the teaching and experiences in the weekly gathering are crucial. They will become a main *way* to lead through those important shifts described in Chapter 2.

Next Steps

Think about your weekly gathering. What is working and what might need a few adjustments? Here are several questions to process with your team:

- All stages: To what extent does your typical weekend experience provide something to challenge next steps for each segment on the spiritual continuum? (Spiritual continuum illustrated on page 24.) Or when the entire body gathers, do you primarily serve just one or two stages in the journey?
- A clear invitation: How closely does your "invitation to the gospel" resemble the transaction-only message? How might you intentionally expand that invitation to include the broader process of becoming a disciple?
- Targeted teaching: If someone "sat under" the teaching of your church for a year, what would they learn about the process of transformation? What would you want them to learn?
- Modeling connection: What would take your people beyond merely learning about connecting? What could you do to help your people actually *connect* with God in worship, prayer, silence, reflection on Scripture, or journaling?

The gathering is one of the most powerful environments in which God transforms human souls.

Transformational Relationships

The REVEAL data validated another vitality important environment. It is relationships, and specifically spiritually-oriented relationships. This ought not surprise us as we study Scripture and consider our own journey with God.

When I ask groups, "*What or who had the greatest impact in shaping who you are as a Christ-follower today?*" most people give names, not events. Relationships, not authors. By God's design, the human soul is made for deep connection. God's power is profound when released through others as they help us "become" like Jesus.

"The gathering is one of the most powerful environments in which God transforms human souls."

This begins when we are far from God, and remains throughout our lives. Alpha is an international ministry designed to give people the opportunity to explore the meaning of life. I believe one of the reasons for Alpha's tremendous impact in helping many millions of people experience growth is that they wisely crafted a pathway supported by relationships (www.alpha.org). Our fundamental need for community remains the same as we mature. However, the kinds of relationships, and the way those relationships help us, change over time. There is a certain way in which the living God shapes us through the eyeballs and ears, the laughter and tears, of someone who loves us, knows our story, and increasingly wants for us what God wants for us.

Larry Crabb writes about the power of relationships in his book, *Connecting*:

> *This kind of relating depends entirely on deep fellowship with Christ and then spills over on to other people with the power to change their lives, not always on our timetable or in the ways we expect but as the sovereign Spirit moves.*[4]

In my observation, many devoted Christ-followers remain stunted and stagnant in their growth because they are fundamentally isolated. For whatever reason, and there are many, they do not move into the kinds of relationships that can actually help them become who God intends for them to become. That is why the shift from isolation to community (Chapter 2) is one of the most important changes facing the church.

Church leadership teams who want to support spiritual growth must be strategic about helping people at each stage of the spiritual journey. They need to help them understand their need for these kinds of relationships, and then have clear next steps to develop these highly-transformational connections.

From REVEAL, we learned that spiritual relationships play an *increasing* role as people advance across the spiritual continuum. *"Spiritual community shows up as a growth catalyst in all three movements, though the specific form of that community shifts as people grow in faith from more casual friendships to mentor relationships, relationships that typically involve greater accountability and intimacy."*[5]

Next Steps

What is the relational culture like in your church? Is community seen as vitally important for transformation? Here are a few questions to think about:

- Vision: How and when is vision given to explain the importance of relationships in becoming like Jesus? Does the teaching from the pulpit match what the typical congregant will experience if they take a risk and get involved?
- On-Ramps: What environments and systems for relational connections are available? If someone eagerly desired to take a next step, how hard would they have to work to take it?

- Training: How will leaders be equipped to provide support and guidance to others at varying stages of the spiritual journey? Different facilitation and leadership skills are needed to teach, offer spiritual direction, listen effectively, etc.
- Environment: How might you create space and resources for relationships to form, grow, and deepen?

Recently, I worked with one large church in the southeast part of the United States whose leaders decided to creatively change their membership on-ramp. They did this to reverse declining enrollment and retention in the membership class. What whiz-bang strategy did they employ? New technology? New highly-produced compelling videos? No. They got people talking.

Simply by decreasing the amount of teaching, placing participants at round tables instead of lecture-style seats, and allowing time for spontaneous relationships to emerge as a result of table conversations, they have seen a dramatic reversal. They're exploding with new ideas of how to launch and support these new, almost unintended small groups. Best of all, transformation stories emerge on a weekly basis.

They would tell you this success wasn't about the quality of those discussion questions. Rather, believing in the power of connecting, they very deliberately and strategically crafted an experience to leverage the powerful way God works through people.

That's just one example, one *kind* of relationship a church has creatively made possible. What are other kinds of relationships a church might support, in order to help facilitate transformation for their people? This is a huge topic. It would expand to include the role of seeker groups, family groups, serving groups, small groups, missional groups, prayer groups, Bible study groups, mentors, spiritual friends, accountability partners, spiritual directors, spiritual direction groups, and beyond!

For now, the question is, *How could you be even more strategic about leveraging relationships?* It will be worth the time and energy you spend to re-evaluate prior ways and create new ways. The right kinds of relationships, at the right stage of the journey, stand as one of the main environments in which the Holy Spirit moves to cause growth.

Individual Spiritual Practices

We know the power of God moves in the gathered community. Most of us have experienced the deep soul-shaping that occurs in God-focused relationships. But the final environment to explore is the most personal. There are certain ways in which my soul—anyone's soul—needs to be met and shaped by God that can only happen in the most intimate of connections: just me and God. This is the realm of personal spiritual practices.

One of the great contributions of the REVEAL research, beyond exposure of the journey itself, was the finding that personal spiritual practices are catalytic to spiritual growth at all stages of that journey. They are especially catalytic for the second transition, as those who are "growing" in Christ take increasing ownership of their spiritual life and become "close" to Christ.

> "The Disciplines allow us to place ourselves before God so that he can transform us."

Why are personal spiritual practices so powerful? A spiritual practice can be thought of as *anything I do to intentionally carve out time and space to become aware of and responsive to the ongoing work of God in my life.*

Take a look again at the spiritual formation diagram on page 14. The big rectangle represents those spiritual practices. They do not change us or change our standing in the eyes of God. They bring us to a place of connection with God and openness to His changing.

"God has given us the Disciplines of the spiritual life as a means of receiving his grace," says Richard Foster in his classic book, *Celebration of Discipline. "The Disciplines allow us to place ourselves before God so that he can transform us."*[6]

What might this mean for church strategy? As it turns out, quite a bit. Since the publication of REVEAL, many churches have once again taken up the conversation of core spiritual practices. These churches have renewed perspective on their vital importance in the spiritual journey.

It is vitally important for a church to strategically help their entire congregation understand spiritual practices. Individuals at all four stages of the spiritual continuum need to understand the role of spiritual practices, how to do them, and ultimately how to incorporate them into their way of life.

One step churches often take as a starting point is to identify a limited number of spiritual practices they would like to embed into their cultural. The "best practice" of embed the Scripture speaks to the power of this as an initiative. Rather than overwhelm people with too many spiritual practices, it's helpful to focus.

I have found the following core individual spiritual practices to be helpful as a starting point for focus:

- Prayer
- Use of Scripture
- Silence and solitude
- Self-examination
- Simplicity (way of life)

Prayer

Prayer is the language of our relationship with God. It takes the focus off our self-sufficiency tendencies and deepens our dependency on God.

When I teach classes on ways of prayer, I am often surprised to learn how adults have learned so little about prayer by attending a church. If they did not grow up in the church, and sometimes even if they did, how will they learn to pray, even in the most basic sense? What will be your strategy to help them in this? For those who learned important rubrics such as ACTS, what will take them deeper into other ways of prayer?

Use of Scripture

We know that any encounter with God's Word places us in direct contact with God himself! Scripture is living and active. But how will those beginning a journey learn? How might we increase biblical literacy in a congregation? How can we keep reflection on Scripture fresh in the hearts of those who are far enough along to have encountered a "dark night of the soul" experience?

These questions speak to our strategy. *"Reflection on Scripture is, by far, the most influential personal spiritual practice for every segment. We know that reading and reflecting on the Scriptures is critical to spiritual growth, and the REVEAL results confirm that there may be nothing more important we can do with our time and effort than encouraging and equipping our people in this practice."*[7]

Silence and Solitude

Solitude is extended time alone with God. We intentionally withdraw from the many other forces that would otherwise distract us and simply "be" with God. Sometimes we deepen this experience by choosing not to speak, even in the presence of others.

Catholic priest and author Henri Nouwen helps us understand the connection between solitude and our spiritual journey. He writes, *"Solitude is the furnace of transformation. Without solitude we remain victims of our society and continue to be entangled in the illusions of the false self."*[8]

In recent years, my own church has intentionally built times for silence right in our main services. As a congregant, I find that even just a few minutes of stillness settles my spirit and opens me up exactly as Nouwen describes. The tangled illusions of my false self quiet down a bit, and I can actually open myself to God–right then and there.

What might your plan be for helping people experience God in solitude and silence, whether at the gathering, with a group, or on their own?

Self-examination

A friend of mine who knows first-hand about the power of twelve-step recovery groups asked me, "*Have you ever wondered why the people who visit the basement of the church get well, but the people upstairs never do?*"

Ouch. Yes, I have wondered. And I believe there are many reasons, but here's just one: Those twelve steps *build* the process of self-examination (inventorying) and confession *directly* into their path. Not surprisingly, those who walk that path find life.

We all need those times of introspection leading to confession! Self-examination is an honest look inside guided by the Holy Spirit. Psalm 139:23-24 is the heart-cry of someone correctly moving into a time of Spirit-directed soul searching :

> [23] *Search me, God, and know my heart;*
>
> *test me and know my anxious thoughts.*
>
> [24] *See if there is any offensive way in me,*
>
> *and lead me in the way everlasting.*

If you chose this as a spiritual practice to embed within your ministry, how would you go about doing it? How would you model this practice, as part of your way of life, so your church can understand and follow your example? How would you provide teaching and opportunities for self-examination?

Simplicity

Biblical simplicity is all about singularity of focus. It is about complete surrender to running the race God has marked out for you (Hebrews 12). It is knowing what to say "yes" to and what to say "no" to. This, of course, requires a grace-filled environment for people to discover the race God has called them to and then make decisions that align with that. It is that "leap of faith" taken by those who have truly died to self and live for Christ. This discipline of Simplicity also undergirds the behaviors seen in churches that create ownership and pastor the community. This comes as their identity and focus become more aligned with God's identity and focus.

What will your strategy be for embracing Simplicity in a way that your people can grasp and follow? How will you provide guidance and grace for people to discern their focus and calling? How might you support people in their practice of Simplicity?

Next Steps

Please don't assume that people in your congregation know how to pray or that they regularly read the Bible or practice solitude. As a leader, it's important to model spiritual practices in your own life and teach about them throughout the year. Think about these questions:

- Focus. Assuming reflection on Scripture, what additional spiritual practices do you want to focus on to embed in the near future?
- Train. How are leaders trained to teach others about the value of spiritual practices? In your setting, what are the most effective environments in which to teach core spiritual practices?
- Embed. What creative means can you use to allow first Scripture and then any other core spiritual practices into the life of your community? How about in your own personal life? Remember, it all starts with you.

'The One Main Thing'

When we invite people into the journey of spiritual transformation, we have a tremendous responsibility to help and lead people all along that journey.

Let's allow Greg Hawkins to remind us of the one main thing we must do:

> *The first step to building a great, spiritually vital church is for you—and the leaders around you—to follow Christ with your whole heart every day of your lives. To die to your own agendas and follow Christ, one day at a time. To declare that your relationship with Christ is the most important relationship in your life. To pursue intimacy with Christ with your entire mind, body, heart, soul, and strength. To allow nothing, absolutely nothing, to stop you from this one main thing.*
>
> *You can do this. And for the sake of your congregation and those you lead, you must do this. You can try other paths, find a new strategy, perhaps, or hire some really talented staff members. But in the end, if your church is not led by people completely devoted to Jesus—people who prioritize their relationship with him above anything else—it will not work. It will not produce life. It will not change the world. And we all know how badly this world needs changing."*[9]

You might want to stop right now and offer yourself, once again, to God.

It all starts there.

Years ago as a college student this prayer was given to me. I call it my "Daily Prayer of Surrender," and highly recommend it to you as well:

Jesus,

i am a humble, lowly servant.

take me... all of me.

add anything, take anything away.

at any cost. with any price.

make me yours completely... wholly.

may I not be remembered for

the way i wear my hair,

or the shape of my face,

or the people i know,

or the crowds i've addressed.

may i be known for loving You...

for carrying a dream...

for building bridges

to the hurt and broken and lost in the world.

make me what You would be

if you lived in person where i do.

may everything accomplished through my simple life

bring honor and glory to You.

take my human flaws and failures

and use them to remind those who know me

that only You are God and i will always just be ________.

amen

CHAPTER 9: Regaining Clarity on Transformation in the Local Church

One of my greatest joys is the privilege of working with ministry leaders and teams along the lines of the themes in this book. I'm often asked to work with a church staff team to discover specific ways to care for their souls given the demands of leadership.

You could call what I do *training*. At least that's what the immigration officers in Canada recently decided to call it. And it had unprecedented consequences. They refused my entry into Vancouver and sent me home!

I had been detained all day in a crowded immigration office while a group of ministry leaders awaited my arrival. I have been to Canada quite frequently for this kind of event. Consequently, I was shocked and stunned when the officers handed me their decision and a set of boarding passes back to O'Hare. I couldn't get into Canada.

In a nutshell, a high-octane immigration officer determined that the work I was doing didn't qualify as clergy work. Instead, my work might actually add value to the people of their country. It might help them to become something different or better. I was rejected based on the laws governing *that* kind of work.

But did you catch that? If I could have persuaded them that I was clergy, simply coming to fill in for another clergy member in Canada, that would be permissible. "Because that wouldn't actually make a difference," she said. That comment was the most stunning and painful thing I heard that day.

I couldn't get it out of my head. To this day it still haunts me.

It actually reminded me of another striking comment that came from a close, very secular friend. In a Facebook comment lamenting the way presidential candidates are dragging their "religion" into public discourse, this friend wrote:

> *I can't think of any other job interview process where a candidate's religious beliefs are relevant (aside of working for a religious organization). The presidency of the U.S. is no different. And yet I am hearing U.S. presidential candidates (and their detractors as well as their supporters) trying to out do one another as to issues of religious belief. This will not create economic opportunity, feed the hungry, improve our infrastructure, educate anyone, or contribute to peace or security. Stick to the issues. There are plenty of them.*

Did you notice that? *This will not create economic opportunity, feed the hungry, improve our infrastructure, educate anyone, or contribute to peace or security.*

Does the Church Matter?

These two recent experiences have made me reflect even further on the ideas we've discussed in this book. Is the local church making a difference in our world? Does fulfilling *all* of Jesus' mandate in Matthew 28 make a difference?

While we might be sad, given the research we've seen, we certainly can't be surprised by these secular assumptions. Until the church rises up and is known for something other than just "filler," and utterly inconsequential beliefs, the secular world will never have reason to change its perceptions of the church, and by extension, Jesus.

As my flight back home pulled away from the gate, I found myself praying for that immigration officer. It hadn't occurred to me all day long. I was just praying to be let in.

Oh Lord, let the change begin with me.

Recovering the Vision of the Ecclesia

Jesus intends for the church to be so much more. In the words of N.T. Wright:

> *It's a place of welcome and laughter, of healing and hope, of friends and family and justice and new life. It's where the homeless drop in for a bowl of soup and the elderly stop by for a chat. It's where one group is working to help drug addicts and another is campaigning for global justice. It's where you'll find people learning to pray, coming to faith, struggling with temptations, finding new purpose, and getting in touch with a new power to carry that purpose out. It's where people bring their own small faith and discover, in getting together with others to worship the one true God, that the whole becomes greater than the sum of its parts.*[1]

As we regain clarity on transformation in the local church, Philippians 2:15 offers God's vision of how His communities might be seen in this dark world:

"...so that you may become blameless and pure, children of God without fault in a warped and crooked generation, in which you shine like stars in the universe."

Begin with Perspective

It's a privileged calling to help further the activity of God in your church and to "shine like stars." It's hard work as you well know.

It's one of the reasons the Willow Creek Association organized an annual conference called *The Church Transformation Intensive* to help churches serious about authentic spiritual growth. Churches sign up as a team. Over three days they process together how their church can develop and strengthen an integrated plan for church-wide spiritual transformation.

I've enjoyed being one of the facilitators along with a terrific group of church leaders, authors and thought leaders in the spiritual formation and church leadership realm.

One of the teaching faculty, Pete Richardson of the Paterson Center, introduces a six-phase strategic planning process called the Paterson Process™. This six-phase process releases a team to lead and manage a church as a system comprised of multiple programs, ministries and functions, rather than dis-integrated parts. Pete said: *"A good plan created in isolation almost guarantees a lack of ownership among those who are expected to implement it."* [2] Working as a team is so important.

Pete explains the critical first step in the process, gaining *perspective* on the issues being discussed.

Matthew 18:3
And he said: "Truly I tell you, unless you change and **become** like little children, you will never enter the kingdom of heaven."

Matthew 20:26
"Not so with you. Instead, whoever wants to **become** great among you must be your servant..."

John 1:12
Yet to all who did receive him, to those who believed in his name, he gave the right to **become** children of God—

John 3:30
He must **become** greater; I must **become** less.

John 4:14
"...but whoever drinks the water I give them will never thirst. Indeed, the water I give them will **become** in them a spring of water welling up to eternal life."

John 12:36a
"Believe in the light while you have the light, so that you may **become** children of light."

Ephesians 2:22
And in him you too are being built together to **become** a dwelling in which God lives by his Spirit.

Ephesians 4:15
Instead, speaking the truth in love, we will grow to **become** in every respect the mature body of him who is the head, that is, Christ.

Titus 3:7
...so that, having been justified by his grace, we might **become** heirs having the hope of eternal life.

It must come *before* any strategic planning. Leaders cannot move forward until they have enough clarity and perspective to develop a wise path forward. Perspective answers the question, Where are we now? He describes perspective this way:

> *"Perspective is seeing things without distortion, correctly reading the signals of what is unfolding. Perspective is the result of finding truth and even sensing future realities before they have happened. It is also a matter of squarely facing existing facts."* [3]

Honestly facing the facts and the trajectories of current initiatives can be painful and trying. Facing current culture can be difficult for many reasons, and even facing the true well being of our own souls can cause us to flinch.

It's important that church teams gain new perspective on all dimensions of how spiritual transformation is really happening—or not happening—in the lives of their congregants. Following their participation in the Church Transformation Intensive, one team from Michigan gained this insight:

> *We came thinking we needed a point person to drive the transformation initiative in our church. We left realizing that everything we do is about spiritual transformation, and that every staff member has to be on board. We have lots of work to do to develop a comprehensive effective plan, but this conference gave us the vision and will do to it.*

Perspective is all about seeing. Recovery of vision is one of the most important things we can gain.

Recovery of Vision

In the darkest moments of my bout with vertigo and subsequent soul implosion, I feared how my story might end. Life on the sidelines isn't a picture of widespread impact in the local church, much less the world. I began to wonder if I wouldn't also lose my sanity if the vision, balance, and focus weren't restored. I couldn't walk, drive, read, watch TV, and only eat if I was blindfolded. Even then, the nausea always won. I couldn't take care of myself, much less anyone else hour after hour, day after day. I could only lie in bed, eyes closed, doing nothing.

Prayer felt awkward and difficult. I knew all too well that God doesn't physically heal every calamity His children suffer. Why should God heal me, when many others don't get well? There are no human answers to those questions.

On one sleepless night, I decided it would be important to at least ask. To admit, I really do want to get well. To have my life back again. Somehow, I felt prompted to ask in this way, *"God, I'm just going on record: I really do want to be well. I understand you don't heal everyone. But if you do choose to heal me, here's my request, please make it matter that I went through this. Please make it matter—somehow."*

If I have anything to offer God's people in this season of my life, it's a direct answer to that prayer. My personal journey to healing became the way I was called to serve. I suppose even this book flows from God's answer to that prayer—and I am so humbled, and grateful. I really "get" how dependant I am.

Over time, God graciously and lovingly restored me to full health and set me on a new path. I came off the sidelines, ready to run a new kind of race in a way that breathed life into my soul.

How about you? What kind of story is God writing with your life? What's your vision for who you might become? And how about your church? Are you regaining clarity for what your people will become?

God can heal your vision, and even *make it matter* that you've been on whatever journey has brought you this far.

Be Thou My Vision

Deuteronomy 30:20 reminds me of the truth I've learned about God in that season, and every season since. This may seen innocuous, but consider carefully God's word to us—especially the bolded phrase.

> *and that you may love the LORD your God, listen to his voice, and hold fast to him.* ***For the LORD is your life****, and he will give you many years in the land he swore to give to your fathers, Abraham, Isaac and Jacob.*

God is my vision. God is my life. All other visions, and there are many good ones, are secondary to that vision. Perhaps that's why Paul's prayer below for the Ephesians is directed first to the inner life of those in the church with acknowledgment of the outward fruit that is borne in a community.

Listen well for the connection, and keep in mind that every time he says "you," the pronoun is plural. He's praying for them. And yes, this is also my prayer for you [plural]:

> *[16] I pray that out of his glorious riches he may strengthen you with power through his Spirit in your inner being, [17] so that Christ may dwell in your hearts through faith. And I pray that you, being rooted and established in love, [18] may have power, together with all the Lord's holy people, to grasp how wide and long and high and deep is the love of Christ, [19] and to know this love that surpasses knowledge—that you may be filled to the measure of all the fullness of God.*
>
> *[20] Now to him who is able to do immeasurably more than all we ask or imagine, according to his power that is at work within us, [21] to him be glory in the church and in Christ Jesus throughout all generations, for ever and ever! Amen.*
>
> *Ephesians 3:16-21.*

Notes

Introduction

1. Dallas Willard, *Renovation of the Heart* (Colorado Springs, CO: NavPress, 2002), 199.

Chapter 1

1. Bill Hybels, "Law of Spiritual Formation" message at Willow Creek Community Church.
2. Willard, *Renovation of the Heart*, 22.
3. Ibid, 20.
4. Cally Parkinson and Greg Hawkins, *Move* (Grand Rapids, MI: Zondervan), 26.
5. Parkinson, "REVEAL: Where Are You?" article, Willow Creek Association, 2010, 1.
6. Bill Hybels, *Courageous Leadership* (Grand Rapids, MI: Zondervan, 2002), 27.

Chapter 2

1. George Barna, "Research on How God Transforms Lives Reveals a 10-Stop Journey," Barna Group, March 17, 2011. This article highlights findings from Barna's book, *Maximum Faith: Live Like Jesus.*
2. Dr. Kara Powell and Dr. Chap Clark, *Sticky Faith* (Grand Rapids, MI: Zondervan, 2011), 16.
3. Powell, 95 and 100.
4. For more information on REVEAL and the REVEAL Spiritual Life Survey, visit www.revealnow.com.
5. Scot McKnight, *The King Jesus Gospel* (Grand Rapids, MI: Zondervan, 2011), 30.
6. N.T. Wright as quoted in *The King Jesus Gospel,* 13.
7. Gabe Lyons, *The Next Christians* (New York: Doubleday, 2010), 201.
8. Dr. Henry Cloud and Dr. John Townsend, *12 "Christian" Beliefs That Can Drive You Crazy: Relief From False Assumptions* (Grand Rapids, MI: Zondervan, 1994), table of contents.
9. Dr. Will Miller and Dr. Glenn Sparks, *Refrigerator Rights* (Barrington, IL: Willow Creek Association, 2008), 12-13.
10. Barna, "Self-Described Christians Dominate America but Wrestle with Four Aspects of Spiritual Depth," Barna Group, September 13, 2011.

Chapter 3

1. Parkinson, "REVEAL: Where Are You?" article, Willow Creek Association, 2010, 2
2. John Ortberg, *Everybody's Normal Till You Get to Know Them* (Grand Rapids, MI: Zondervan, 2003), 32.
3. Elton Trueblood, *Alternative to Futility* (New York: Harper & Brothers, 1948), 31.

Chapter 4

1. Definition of the spiritual gift of shepherding as defined in *Network: The Right People, in the Right Places, for the Right Reasons, at the Right Time* by Bruce Bugbee and Don Cousins.
2. http://www.churchgrowth.org/giftslist.php

Chapter 5
1. Bill Hybels has quoted this phrase hundreds of times to church leaders around the world. It's a message he believes with all his heart.
2. Richard Rumelt, *Good Strategy/Bad Strategy* (New York: Crown Business, 2011), 77.
3. Ibid, 77.

Chapter 6
1. Willard, *Renovation of the Heart*, 209,
2. © 2007 (research from 1989 to 2006) R. J. Krejcir Ph.D. *Francis A. Schaeffer Institute of Church Leadership Development www.churchleadership.org*
3. Andrew Murray, *The True Vine* (Chicago: Moody Publishers, 2007), 87.
4. Ibid, 77.

Chapter 7
1. Rosabeth Moss Cantor, *Confidence: How Winning Streaks and Losing Streaks Begin and End* (New York: Crown Business, 2004), 97.
2. Michael Hyatt's leadership blog post on June 1, 2009 called "On Reorganization": http://michaelhyatt.com/on-reorganization.html
3. Peter Scazzero, *The Emotionally Healthy Church* (Grand Rapids, MI: Zondervan, 2003), 185.
4. I interviewed Gary Haugen at Willow Creek on April 26, 2010 for a Willow Creek Association videocast.
5. Lance Witt, *Replenish* (Grand Rapids, MI: Baker Books, 2011), 187.

Chapter 8
1. Willard, *Renovation of the Heart*, 244.
2. Parkinson and Hawkins, *Move*, 25.
3. Bill Hybels said this at a weekend service at Willow Creek Community Church.
4. Larry Crabb, *Connecting* (Nashville: W Publishing Group, 1997), 5.
5. Parkinson and Hawkins, *Move*, 120.
6. Richard Foster, *Celebration of Discipline: The Path to Spiritual Growth* (San Francisco: Harper & Row, 1978), 7.
7. Parkinson and Hawkins, *Move*, 117 and 119.
8. Henri Nowen, *The Way of the Heart* (New York: Ballantine Books, 1981), 13.
9. Parkinson and Hawkins, *Move*, 254.

Chapter 9
1. N.T. Wright, *Simply Christians: Why Christianity Makes Sense* (New York: HarperOne, 2006), 123.
2. Pete Richardson is a Master Facilitator at the Paterson Center. His training has helped the WCA in countless ways and equipped the staff with an effective strategic planning process. For more information on the Paterson Process, visit www.patersoncenter.com.
3. Richardson, Paterson Center.

Maximizing the Transformative Power of the Local Church

Since 1992, Willow Creek Association (WCA) has supported the life-changing work of local churches—churches that are fully committed to the cause of engaging people far from God on a journey toward Christ-centered living.

We believe churches and leaders like these are at the very center of God's plan to redeem and restore this broken, hurting world for Christ. Carrying Christ's message of hope and salvation, they also fuel a cascading movement of transformation as they bring grace, healing, and life change into their communities, regions, and the world.

Each year, more than 160,000 Christian leaders in 20,000 churches attend WCA training events or use WCA resources. More than 8,000 of these churches, representing 90 different denominations and networks, are also connected to WCA and to one another through WCA membership.